Climate Change Effect
on Crop Productivity

Climate Change Effect on Crop Productivity

Edited by
Rakesh S. Sengar and Kalpana Sengar

CRC Press
Taylor & Francis Group
Boca Raton London New York

CRC Press is an imprint of the
Taylor & Francis Group, an **informa** business

CRC Press
Taylor & Francis Group
6000 Broken Sound Parkway NW, Suite 300
Boca Raton, FL 33487-2742

First issued in paperback 2017

ISBN-13: 978-1-4822-2920-2 (hbk)
ISBN-13: 978-1-138-74822-4 (pbk)

Library of Congress Cataloging-in-Publication Data

Climate change effect on crop productivity / editors: Rakesh S. Sengar, Kalpana Sengar.
    pages cm
  Includes bibliographical references and index.
    ISBN 978-1-4822-2920-2 (hardcover : alk. paper) 1. Crops and climate. 2. Crop yields.
  I. Sengar, Rakesh S. II. Sengar, Kalpana.

S600.7.C54C654 2015
630.2'515--dc23
                                                2014027320

Visit the Taylor & Francis Web site at
http://www.taylorandfrancis.com

and the CRC Press Web site at
http://www.crcpress.com

Contents

Foreword

 I am happy to know Dr. R.S. Sengar, associate professor and officer-in-charge, Dr. Kalpana Sengar, women scientist fellow (DST), Department of Agriculture and Biotechnology, Sardar Vallabhbhai Patel University of Agriculture and Technology, Meerut and their colleagues who have taken the initiative to bring out a book to enhance the understanding of the phenomenon of climate change and crop productivity.

Climate change is impacting all aspects of our day-to-day existence and more profoundly the crop production system, which is very critical for the food security of any nation. The main objective of this book is to provide the recent developments on this subject and to increase the understanding of its readers in a very holistic manner. I am confident that this book will serve as a useful reference material for researchers, students and consultants engaged in this field, particularly those concerned with the science of biotechnology, climate change and plant productivity. It will certainly assist faculty members in improving teaching the more effectively. It will also provide a better insight and greater to all the students pursuing this subject.

I congratulate Dr. Sengar and his young and enthusiastic faculty colleagues for their sincere efforts, diligence and hard work in bringing out this all-important book.

A.K. Singh

Preface

The impact of climate change at the global level is severe, especially in the developing world, and the social and economic implications in China, India, Brazil, and the poor countries of the tropical belt in Africa and Latin America. This book about the stakes for world agriculture makes a major contribution on this score. This analysis has significant implications for all concerned about global warming effect on crop productivity due to climate change and long-term agriculture technique development. This study confirms the asymmetry between potentially severe agricultural damages such as the effect on crop yield due to variation in temperature.

Agriculture sustainability has become the basic principle of modern agriculture and it implies the successful management of agricultural resources to satisfy human needs of food, fiber, fruits, forest and fuel without endangering the environment. This has achieved relatively higher importance in the last few decades and has become an inevitable part of the syllabus of any agricultural course in universities and institutes. This book is written in the context of the 'global climate change and effect on crop yield and agriculture productivity', and all the chapters are contributed by experts scientists, professors and researchers in their respective field. This book is intended to provide relevant information and opportunities for productive engagement and discussion among government negotiators, experts, stakeholders, and others involved and interested in climate change and agriculture. For this report, the institute convened a team of international, independent expert authors, and facilitated the

work of the author's team as well as a series of informal dialogues with a broad range of country negotiators, nongovern-
mental organisations and agricultural experts.

We are highly thankful and obliged to Dr. H.S. Gaur, vice
chancellor of SVPUA&T, Meerut, for his persistent encouragement and valuable suggestion for the successful completion of
this manuscript. We have received generous help from many
senior scientists and fellow teachers for the preparation of this
book under the guidance of Professor Shivendra Vikram Sahi,
head, Department of Biotechnology, Ogden College of Science
and Engineering, Western Kentucky University, Bowling Green,
Kentucky, USA. We will remain ever-indebted to our respected
teacher Professor V.P. Singh, head and dean, Department of Plant
Science, MJP Rohilkhand University, Bareilly, Dr. Anil Gupta,
professor and head, Department of Molecular Biology and
Biotechnology, College of Basic and Tech, Pantnagar, Udham
Singh Nagar; Professor Vinay Kumar Sharma, Department
of Bioscience and Biotechnology, Banasthali, Rajasthan;
Professor R.P. Singh, Department of Biotechnology, Indian
Institute of Technology (IIT) Roorkee; Professor P.K. Gupta,
Chaudhary Charan Singh (CCS) University, Meerut; Professor
N.S. Sikhawat, Department of Biotechnology, Jai Narayan
Vyas (JNV) University, Jodhpur; Dr. N.K. Singh and Dr. T.R.
Sharma, principal scientist, biotechnology, National Research
Centre on Plant Biotechnology, New Delhi; Professor Akhilesh
Tyagi, director, National Institute of Plant Genome Research
(NIPGR), New Delhi, Chandigarh; Professor B.D. Singh,
Banaras Hindu University (BHU), Varanasi; Professor R.L.
Singh, Dr. Ram Manohar Lohia Avadh (RMLA) University,
Faizabad; Dr. Sundeep Kumar Sharma and Dr. Rakesh Singh,
National Bureau of Plant Genetic Resources (NBPGR), New
Delhi; Dr. Dharmendra Singh, Indian Agriculture Research
Institute (IARI), New Delhi and Dr. A.K. Sharma, Ramie
Research Station (ICAR), Sorbhog, Barpeta, Assam, for their
extraordinary help in shaping this book. We are highly thank-
ful to our colleagues and faculty members of the College of
Biotechnology and Department of Agriculture Biotechnology,
SVPUA&T, Meerut for extending valuable comments and for
helping directly and indirectly. It would be rather impossible to
list all those who have provided encouragement and help in the
preparation of this book. I am extremely thankful to all of them.

The first author will ever remain grateful to his reverend
parents (Dr. Sanwal Singh Sengar and Smt. Kamla Sengar)
who kindly inspired him for this contribution, his wife Sarita,
and kids (Divyanshu and Kartikey) who gave him persistent

encouragement by their smiling faces to enable him to write the manuscript. As a first author, I express my due gratitude to my younger brother (Dr. Rajesh Singh Sengar), his better half (Smt. Kalpana Sengar) and kids (Saumya and Amranshu) for their affection, continuous help, and cooperation during the preparation of the manuscript. We are very grateful to Dr. Gagandeep Singh, Jennifer Stair, Kate Gallo, Arlene Kopeloff, Florence Kizza, Karolina, Syed Mohamad Shajahan and Taylor & Francis for their kind cooperation and support from idea to bringing out this publication in a presentable form.

I would like to express my sincere gratitude towards my mentor Late Professor H.S. Srivastav, who has always been a source of strength and moral support in all my endeavours, especially in the field of life sciences. I hope this book will satisfy the needs of the majority of academicians, scholars and students. Benediction of many dignitaries from ICAR Institute, South Asian University (SAU) and IIT has led me to the completion of this herculean task well in time. In the future, I expect their same cooperation for quality improvements and wish that I serve the society at large in a better way. I am indebted to everyone who is directly or indirectly involved in the successful completion of this book.

Dr. Rakesh Singh Sengar

About the Book

This book reports on the results of experiments to assess the effects of global climate change on crop productivity. It covers issues such as CO_2, ozone on plants, productivity fertilisation effect, UV (ultraviolet) radiation, temperature and stress on crop growth. Agriculture is a complex sector involving different driving parameters (environmental, economic and social). It is now well recognised that crop production is very sensitive to climate change with different effects according to the region. This book underlines such concerns about the current status of our environment and agriculture.

This book analyses the global consequences to crop yields, production and risk of hunger of linked socio-economic and climate scenarios. The potential impacts of climate change are estimated for use to evaluate consequent changes in global cereal production, cereal prices and the number of people at risk from hunger. The crop yield results elucidate the complex regional patterns of projected climate variables, CO_2 effects and agricultural systems that contribute to aggregations of global crop production.

This book contains 19 chapters and discusses the impact of changing climate on agriculture, environment stress physiology, adaptation mechanism, climate change data of recent years, impact of global warming and climate change on different crops such as sugarcane, wheat, rice and medicinal plants. The concluding chapter gives an idea of the overall global picture in terms of the effect already discussed in response of crops to climate change during abiotic stress. This book also attempts to underline various strategies for reducing agriculture's vulnerability to climate change and for adaptation to the ongoing climate change.

This book will be useful for agriculturists, environmentalists, climate change specialists, policy makers and research scholars engaged in research on climate agriculture-related issues.

Editors

Dr. Rakesh Singh Sengar, the editor of the book, is the associate professor, Department of Agriculture Biotechnology, College of Agriculture in Sardar Vallabhbhai Patel University of Agriculture and Technology, Meerut. Dr. Sengar has 19 years of teaching, research and extension experience to his credit. He remained associated with teaching for a period of about 10 years at Govind Ballabh Pant University of Agriculture and Technology, Pantnagar, Udham Singh Nagar (Uttarakhand). He has supervised five PhD students and has published more than 48 research papers in India and foreign journals of repute. Dr. Sengar has contributed 105 abstracts/papers to different conferences/symposiums/workshops organised at national and international forums. He has published 780 popular articles in Hindi and English languages in important journals and magazines. He has also published four books for the scientific community and farmers. Dr. Sengar has delivered more than 200 talks at All India Radio and television at Delhi, Lucknow, Rampur and Bareilly centres. Dr. Sengar has keenly worked as one of the active members of the advisory committee of Doordarshan, New Delhi. Dr. Sengar is a life member of several professional societies. He is a member of the editorial boards and review committees of few journals. Dr. Sengar is on the panel of examiners of several colleges/institutions and universities. He is one of the members of RDC of Uttarakhand Technical University, Dehradun (Uttarakhand). He has been the principal and co-investigator of a few projects financed by various government departments. Dr. Sengar is the recipient of the 'Best writer' award from Vishwa Agro Marketing and Communication, Kota, Rajasthan, 'Kunwar Saxena Bahadur SRDA' award from the society for Recent Development in Agriculture, Meerut, 'Man of the Year' and 'Research Board of Advisors' awards from American Biographical Institute, Inc., USA, 'Aryabhat' 2010 award from Vigyan Bharti, New Delhi and Fellow of Society of Plant Research, Bareilly. Dr. Sengar has also received a gold medal in 2011 from Hitech Horticulture Society, Meerut and Dr. J.C. Edward Medal 2012 from Bioved Research Society, Allahabad and 'Scientist of the Year—2013' from the Academy of Environmental Biology, Lucknow.

Kalpana Sengar, the co-editor of this book, is the youngest recipient of the Women Scientist Fellowship award given by the Department of Science and Technology, Government of India. She has seven years research experience to her credit in biotechnology. She completed her MSc in biotechnology at the CCS University and her PhD from MJP Roheilkhand University and is working as a young scientist at Sardar Vallabhbhai Patel University of Agriculture and Technology. She has published more than 20 papers in reputed national and international journals and has presented papers in several national and international conferences. She has received the best report diploma award by the Bioinfobank library, Poland in 2009, the Flemish Institute for Biotechnology (VIB), a PhD scholarship for attending the international symposium in Belgium in 2010 and a scholarship to attend the 17th annual conference of the International Sustainable Development Research Society (ISDRS) hosted by the Earth Institute, Columbia University. She has research interest in agriculture, human nutrition, biotechnology and plant tissue culture and so on.

Contributors

Sandeep Arora
Department of Molecular Biology and Genetic Engineering
Govind Ballabh Pant University of Agriculture and
    Technology
Uttarakhand, India

B. Kalyana Babu
Molecular Biology and Genetic Engineering
Govind Ballabh Pant University of Agriculture and
    Technology
Uttarakhand, India

Sangeeta Baruah
Ramie Research Station
Central Research Institute for Jute and Allied Fibres
Indian Council of Agricultural Research
Assam, India

H.S. Bhadoria
Department of Agriculture Engineering
R.V.S. Krishi Vishwa Vidyalaya
Madhya Pradesh, India

Arpan Bhowmik
Indian Agricultural Statistics Research Institute
New Delhi, India

Sumit Chakravarty
Department of Forestry
Uttar Banga Krishi Viswavidyalaya
West Bengal, India

Amit Chauhan
Central Institute of Medicinal and Aromatic Plants
Research Centre
Uttarakhand, India

Netra Chhetri
School of Geographical Sciences and Urban Planning
Arizona State University
Tempe, Arizona

Pooja Choudhary
NRC on Plant Biotechnology, IARI Campus
New Delhi, India

Manish Das
Horticulture Division
Indian Council of Agricultural Research
New Delhi, India

S.K. Guru
Department of Plant Physiology
Govind Ballabh Pant University of Agriculture and
    Technology
Uttarakhand, India

Seema Jaggi
Indian Agricultural Statistics Research Institute
New Delhi, India

Pradeep K. Jain
NRC on Plant Biotechnology, IARI Campus
New Delhi, India

Amit Kumar
College of Agriculture Biotechnology
Sardar Vallabhbhai Patel University of Agriculture and
    Technology
Uttar Pradesh, India

Arvind Kumar
Department of Agricultural Meteorology
Narendra Deva University of Agriculture and Technology
Uttar Pradesh, India

Vivek Kumar
Amity Institute of Microbial technology
Amity University
Uttar Pradesh, India

R.S. Kureel
Department of Agriculture Biotechnology
Naredra Dev University of Agriculture and Technology
Uttar Pradesh, India

Kanhaiya Lal
Sugarcane Research Institute
Uttar Pradesh, India

Samarendra Mahapatra
Department of Agribusiness Management
Orissa University of Agriculture and Technology
Orissa, India

Anusaya Mallick
Department of Environmental Science
University of Kalyani
West Bengal, India

Rajeev Nayan
Department of Molecular Biology and Genetic Engineering
Govind Ballabh Pant University of Agriculture and
    Technology
Uttarakhand, India

Rajendra Chandra Padalia
Central Institute of Medicinal and Aromatic Plants
Research Centre
Uttarakhand, India

Anjali Pande
Department of Molecular Biology and Genetic Engineering
Govind Ballabh Pant University of Agriculture and
    Technology
Uttarakhand, India

Anju Puri
Barring Union Christian College
Punjab, India

Vivekanand Pratap Rao
Tissue Culture Lab
Sardar Vallabhbhai Patel University of Agriculture and
    Technology
Uttar Pradesh, India

Madhu Rawat
Department of Molecular Biology and Genetic Engineering
Govind Ballabh Pant University of Agriculture and
    Technology
Uttarakhand, India

Bård Romstad
CICERO
Oslo, Norway

Shivendra Vikram Sahi
Applied Research and Technology Programme
Western Kentucky University
Bowling Green, Kentucky

A.C. Samal
Department of Environmental Science
University of Kalyani
West Bengal, India

S.C. Santra
Department of Environmental Science
University of Kalyani
West Bengal, India

M.K. Sarma
Ramie Research Station
Central Research Institute for Jute and Allied Fibres
Indian Council of Agricultural Research
Assam, India

Kalpana Sengar
Department of Agriculture Biotechnology
Sardar Vallabhbhai Patel University of Agriculture and
    Technology
Uttar Pradesh, India

Rakesh Singh Sengar
Department of Agriculture Biotechnology
Sardar Vallabhbhai Patel University of Agriculture and
    Technology
Uttar Pradesh, India

A.K. Sharma
Ramie Research Station
Central Research Institute for Jute and Allied Fibres
Indian Council of Agricultural Research
Assam, India

Dinesh K. Sharma
Centre of Environment Science and Climate Resilient
    Agriculture (CESCRA)
IARI
New Delhi, India

Gopal Shukla
Department of Forestry
Uttar Banga Krishi Viswavidyalaya
West Bengal, India

Ashu Singh
Department of Agriculture Biotechnology
Sardar Vallabhbhai Patel University of Agriculture and
    Technology
Uttar Pradesh, India

D.P. Singh
National Agricultural Research and Extension Institute
East Coast Demerara, Guyana

Uma Maheswar Singh
Molecular Biology and Genetic Engineering
Govind Ballabh Pant University of Agriculture and
    Technology
Uttarakhand, India

Rakesh Srivastava
Department of Molecular Genetics
Ohio State University
Columbus, Ohio

Rashmi Srivastava
National Bureau of Fish Genetic Resources
Uttar Pradesh, India

Gunjan Tiwari
Genetics and Plant Breeding
Govind Ballabh Pant University of Agriculture and
    Technology
Uttarakhand, India

Asbjørn Torvanger
CICERO
Oslo, Norway

Padmakar Tripathi
Department of Agricultural Meteorology
Narendra Deva University of Agriculture and Technology
Uttar Pradesh, India

Michelle Twena
CICERO
Oslo, Norway

Eldho Varghese
Indian Agricultural Statistics Research Institute
New Delhi, India

Ram Swaroop Verma
Central Institute of Medicinal and Aromatic Plants
Research Centre
Uttarakhand, India

CHAPTER ONE

Climate change vis-à-vis agriculture

Indian and global view— implications, abatement, adaptation and trade-off

Sumit Chakravarty, Anju Puri and
Gopal Shukla

Contents

Abstract

Climate change is looming large towards humanity in the coming decades. Agriculture also produces significant effects on climate change as a possible contributor of greenhouse gasses to the atmosphere and as an industry that is highly sensitive to climatic changes. Climate is significant in the distribution, productivity and security of food. There should be a realisation that climate is both a resource to be managed wisely and a hazard to be dealt with. Thus, a portfolio of assets to prepare for climate change is needed. This chapter discusses in global detail, with special reference to India, the contribution of agriculture towards climate change, its implication, abatement, trade-off, adaptation and adjustment with barriers and policy recommendations towards achieving climate-smart agriculture.

1.1 Introduction

The major environmental problem of our time is the threat of global climate change, which is due to anthropogenic modification of the atmosphere (Anonymous, 2001a–e; 2007a–c, 2008a; Chakravarty and Mallick, 2003). Fossil fuel consumption coupled with deforestation has increased the concentration of CO_2 in the atmosphere by some 25% (Wittwer, 1990; Chakravarty et al., 2012). The increasing greenhouse gases (GHGs) resulted in global warming by 0.74°C over the past 100 years and 11 of the 12 warmest years were recorded between 1995 and 2006 (Anonymous, 2007b). The IPCC projections on temperature predict an increase of 1.8–4.0°C by the end of this century (Anonymous, 2007b). The issues of climate change due to this warming have led to a serious concern of agricultural productivity worldwide, because agriculture is both a possible contributor of GHGs to the atmosphere (Waggoner, 1992; Duxbury et al., 1993; Jackson and Geyer, 1993; Bakken et al., 1994; Jallow, 1995; Krapfenbauer and Wriessnig, 1995; Zeddies, 1995; Tinker et al., 1996; Boyle and Lavkulich, 1997; Fearnside, 1997) and an industry that is highly sensitive to climatic variates (Rogers and Dahlman, 1993; Hofreither and Sinabell, 1996).

Global warming and climate change are often interchangeably used and understood, but these terms are not identical. Climate change includes both warming and cooling conditions, while global warming pertains only to climatic changes related to increase in temperatures (Grover, 2004). The climatic system is a complex interactive system consisting of the atmosphere, land surface, snow and ice, oceans and other bodies of water and living things. The atmospheric component of the climatic system most obviously characterises climate. It is often defined as 'average weather'. Climate is usually described in terms of the mean and variability of temperature, precipitation and wind over a period of time ranging from months to millions of years (Anonymous, 2007a).

First of all, all over the world, there has been a slow but steady rise in temperature over the last few decades. Moreover, alongside this warming, the globe has also been subject to a general decline in rainfall since the first half of the nineteenth century (Nicholson, 1994, 2001). While one may be inclined to think only in terms of more dramatic weather events such as floods, droughts, storms and hurricanes, adversely affecting agricultural production, it is important to note that even small changes in climate could feasibly have substantial effects, particularly if countries do not have the necessary technology and/or endowments to deal with these. Indeed, agronomic models of climate sensitivity suggest that climate changes in most developing countries are likely to be harmful and can make agricultural areas less productive (Reilly et al., 1994; Rosenzweig and Parry, 1994).

In India, an increasing trend in temperature and no significant changes in rainfall (with some regional variations) has been reported during the last 100 years (Mooley and Parthasarathy, 1984; Hingane et al., 1985; Thapliyal and Kulshrestha, 1991; Rupa Kumar et al., 1992, 1994, 2003; Kripalani et al., 1996; Pant et al., 1999; Singh et al., 2001; Stephenson et al., 2001; May, 2002; Singh and Sontakke, 2002; Mall et al., 2006, 2007) and projected during the last part of the twenty-first century in the range of 0.7–5.8°C (Bhaskaran et al., 1995; Lal et al., 1995, 2001; Lonergan, 1998; Anonymous, 2001e; Rupa Kumar and Ashrit, 2001; Rupa Kumar, 2002; Rupa Kumar et al., 2003). Analyses done for India generally show temperature, heat waves, droughts and floods and sea levels increasing, while glaciers decrease, which is similar to that reported by IPCC (Ninan and Bedamatta, 2012). The magnitude of the change varies in some cases. However, some regional patterns were noted. The areas along the west coast, North Andhra Pradesh and North-West India, reported an increase in monsoon rainfall. Some places across East Madhya Pradesh and adjoining areas, North-East India and parts of Gujarat and Kerala (–6% to –8% of normal over 100 years), recorded a decreasing trend. Surface air temperature for the period 1901–2000 indicates a significant warming of 0.4°C over 100 years. The spatial distribution of changes in temperature indicated a significant warming trend along the west coast, Central India, interior Peninsula and North-East India. However, a cooling trend was observed in the North-West and some parts of Southern India. There is evidence that the glaciers in the Himalayas are receding at a rapid pace. It is projected that by the end of the twenty-first century, rainfall will increase by 15–31% and the mean annual temperature will increase by 3–6°C. The warming will be more pronounced over land areas with the maximum increase in northern India. The warming is also projected to be relatively greater in the winter and post-monsoon seasons (Ninan and Bedamatta, 2012).

1.2 Agriculture as a contributor

At the same time, agriculture has not only been shown to produce significant effects on climate change primarily through
the production and release of GHGs, such as carbon dioxide,
methane and nitrous oxide, but also by altering the Earth's land
cover that can change its ability to absorb or reflect heat and
light, thus contributing to radiative forcing. Land use change
such as deforestation and desertification, together with the use
of fossil fuels are the major anthropogenic sources of carbon
dioxide. Besides the problems associated with land use through
deforestation for example, can translate into increased erosion. Agriculture itself is the major contributor to increasing
methane and nitrous oxide concentrations in the Earth's atmosphere. Agricultural practices themselves have often added to
the water shortage problem as in Africa or other arid/semi-
arid areas more than anywhere else due to the differences in
property rights. More precisely, because farmers are often not
the owners of the land they work on, the preservation of natural resources is generally viewed as a secondary objective. In
addition, pressures represented by increasing populations and
changing technology add to the problem of land deterioration
related to agricultural practices (Drechsel et al., 2001). Another
illustration of environment-damaging agricultural practices
is the intense use of fertilisers in low-quality lands. As yields
increase, so will water consumption, thus creating a vicious
circle (Gommes and Petrassi, 1996).

Global trends   Agriculture releases into the atmosphere significant amounts
of GHGs, that is, CO_2, CH_4 and N_2O (Cole et al., 1997;
Anonymous, 2001a,b; Paustian et al., 2004). CO_2 is released
from microbial decay or burning plant litter and soil organic
matter (Janzen, 2004; Smith, 2004a–c); CH_4 from fermenta-
tive digestion by ruminants, stored manures, paddy cultivation
or decomposition of organic materials in anaerobic conditions
(Mosier et al., 1998) and N_2O from microbial transformation
of nitrogen in soils and manures especially under wet conditions where available nitrogen exceeds plant requirements
(Smith and Conen, 2004; Oenema et al., 2005). Direct agricultural emissions were 10–12% of the total anthropogenic
GHG emissions in 2005, that is, 5.1–6.2 Pg CO_2-eq. (Smith
et al., 2007a). CH_4 contributes 3.3 GtCO_2-eq. year^{-1} and N_2O
2.8 GtCO_2-eq. year^{-1}. Of the global anthropogenic emissions in
2005, agriculture accounted for about 60% of N_2O and about
50% of CH_4 (Denman et al., 2007). Globally, agricultural

CH_4 and N_2O emissions have increased by nearly 17% from 1990 to 2005, an average annual emission increase of about 60 $MtCO_2$-eq. year^{-1} (Anonymous, 2006a,b). Three sources—biomass burning (N_2O and CH_4), enteric fermentation (CH_4) and soil N_2O emissions—together explained 88% of the increase. Livestock (cattle and sheep) account for about one-third of global anthropogenic emission of CH_4 (Murray et al., 1976; Kennedy and Milligan, 1978; Crutzen, 1995; Anonymous, 2006a). Agricultural lands generate very large CO_2 fluxes both to and from the atmosphere (Anonymous, 2001a) but the net flux is small (Smith et al., 2007a), which amounts to 40 $MtCO_2$-eq. in 2000, less than 1% of global anthropogenic CO_2 emissions (Anonymous, 2006b). GHG emissions from deforestation mainly in tropical countries contributed an additional 5.9 Pg CO_2-eq. per year (with an uncertainty range of ±2.9 Pg CO_2-eq.), thus equalling or exceeding emissions from all other agricultural sources combined.

Agricultural N_2O emissions will increase 35–60% till 2030 due to increasing use of nitrogenous fertiliser and animal manure production (Mosier and Kroeze, 2000; Anonymous, 2003, 2006a). If the demands for food increase and the diet shifts as projected, then annual emissions of GHGs from agriculture may escalate further (Smith et al., 2007a). If CH_4 emissions increase proportionately with increasing livestock, then it is projected that CH_4 emission will increase by 60% till 2030 (Anonymous, 2003) while both enteric fermentation and manure management will increase CH_4 emission by 21% from 2005 to 2020 (Anonymous, 2006a). Further, although global rice production areas will increase to 4.5% by 2030, substantial CH_4 emission is not expected, which may be due to less rice grown in continuous flooding under future water-scarce conditions or due to rice cultivars emitting less CH_4 (Wang et al., 1997). But a sustained increase in the area of irrigated rice between 2005 and 2020, a 16% increase in CH_4 emission is projected (Anonymous, 2006a). The baseline 2020 emissions for non-CO_2 GHGs is 7250 $MtCO_2$-eq. Non-CO_2 GHG emissions in agriculture are projected to increase by about 13% from 2000 to 2010 and by 13% from 2010 to 2020 (Anonymous, 2006b). Unfortunately, for non-CO_2 GHG emission estimates, there is no baseline for 2030. Assuming a similar rate of increase from 2000 to 2020, the 2030 global agricultural non-CO_2 GHG emissions were projected to increase 13% during 2000–2010 and 2010–2020, while 10–15% increase were projected for 2020–2030, that is, from 8000 to 8400 with a mean of 8300 $MtCO_2$-eq. by 2030 (Anonymous, 2006a). Moreover, the future evolution of CO_2 emissions from agriculture is uncertain (Smith et al., 2007a). Fortunately, stable/declining deforestation (Anonymous, 2003) and increased adoption of conservation tillage practices (Anonymous, 2001c) will decrease CO_2 emission (Smith et al., 2007a).

Regional trends The magnitude of emissions and relative importance of the different sources vary widely among 10 world regions: developing countries of South Asia, developing countries of East Asia, sub-Saharan Africa, Latin America and the Caribbean, Middle East and North Africa, Caucasus and Central Asia, Western Europe (EU 15, Norway and Switzerland), Central and Eastern Europe, OECD Pacific (Australia, New Zealand, Japan and Korea) and OECD North America, that is, Canada, the United States and Mexico (Anonymous, 2006a). Non-Annex I countries comprising five regions contributed 74% of total agricultural emissions. N_2O emissions from soils primarily due to N fertilisers and manures were the main GHG source from seven regions, while CH_4 from enteric fermentation was the main GHG source in the other three regions (Latin America and Caribbean, the countries of Eastern Europe, the Caucasus and Central Asia and OECD Pacific). This was due to 24% and 36% of global sheep and cattle population in these three regions (Anonymous, 2003).

Rice production emitted 97% and biomass burning emitted 92% of the total world CH_4 emissions in developing countries, while South and East Asia dominated the emissions from rice production with 82% and emissions from biomass burning dominated with 74% in sub-Saharan Africa, Latin America and the Caribbean. Developed regions with 52% of total emissions from only manure management were higher than the developing regions with 48% of total emissions (Anonymous, 2006a). However, CO_2 emissions and removal from agricultural lands in these 10 regions are uncertain as some countries reported net emissions while some reported net removals, but countries from Eastern Europe, the Caucasus and Central Asia had an annual emission of 26 $MtCO_2$ year^{-1} in 2000 (Anonymous, 2006b).

The Middle East, North Africa and sub-Saharan Africa were the highest emitters of GHGs with a combined 95% increase in the period 1990–2020 (Anonymous, 2006a). The per capita food production is either declining or at levels lesser than adequate in sub-Saharan Africa (Scholes and Biggs, 2004) due to low and declining soil fertility along with inadequate fertiliser inputs (Sanchez, 2002; Smith et al., 2007a). The rising wealth of urban populations in this region (South and Central Africa, including Angola, Zambia, Democratic Republic of

Congo, Mozambique and Tanzania) will increase the demand for livestock products, intensifying and expanding agriculture to still largely unexploited areas, thereby resulting in higher GHG emissions (Smith et al., 2007a). In East Asia, with a 4 and 12 times increase of milk and meat production, respectively, from 1961 to 2004 (Anonymous, 2006c) and its projected continued increase in consumptions, the GHG emissions are expected to increase 86% and 153%, respectively, from enteric fermentation and manure management, during 1990 to 2020 (Anonymous, 2006a). In a pursuit to ensure food security for its teeming population, South Asia will be using more and more nitrogenous fertiliser and manure, thereby increasing its GHG emission (Anonymous, 2006a).

Deforestation of cropland and grassland in the Latin America and Caribbean resulted in increased emissions of GHG, mainly CO_2 and N_2O (Anonymous, 2006c). N_2O emissions have significantly reduced in the countries of Central and Eastern Europe, the Caucasus and Central Asia for their decreased use of nitrogenous fertilisers since 1990. However, driven by favourable economic conditions in these countries, the use of nitrogenous fertilisers may shoot up again, which will result in 32% increase of N_2O emissions from soils by 2020 (Anonymous, 2006c). Non-CO_2 GHG emissions increased consistently from the agricultural sector between 1990 and 2020 in OECD North America and OECD Pacific with an 18% and 21% increase, respectively. These emissions were from manure management and N_2O from soils. In Oceania, emission increased due to exponential increases of nitrogenous fertiliser use, while in North America it increased due to management of manure from cattle, poultry and swine production along with manure application to soils. Fortunately, CO_2 emission from land conversion has been reduced in both these regions having active vegetation policies restricting further clearing (Anonymous, 2006a). The only region in the globe with decreased projection of GHG emissions from agriculture till 2020 is Western Europe due to its adoption of climate-specific and environmental policies along with economic constraints on agriculture (Anonymous, 2006a).

1.3 Global agricultural land use change and implication

Out of the total agricultural land globally in 2002 (5023 Mha), pasture has 3488 Mha or 69% dominated, followed by cropland with 1405 Mha or 28% (Anonymous, 2006c). During the past four decades, driven by increasing population pressure, agricultural land increased by 500 Mha added from other land uses, especially in the developing world with the conversion of 6 Mha forests and 7 Mha other land annually, with a projection of an additional 500 Mha up to 2020 mostly in Latin America and sub-Saharan Africa (Fedoroff and Cohen, 1999; Rosegrant et al., 2001; Huang et al., 2002; Trewavas, 2002; Green et al., 2005). Technological progress has enabled remarkable improvements in land productivity and increasing per capita food availability, although from a per capita declining land availability. The share of animal diets has consistently increased, particularly poultry in the developing countries of South and South-East Asia and sub-Saharan Africa due to economic progress and changing lifestyles (Rosegrant et al., 2001; Roy et al., 2002; Smith et al., 2007a). Annual GHG emissions from agriculture are expected to rise in the coming decades due to escalating demands of food and shifts in diet. The main trend in the agricultural sector with the implications for GHG emissions or removals can be outlined as follows:

- Driven by the declining rate of land productivity, the use of marginal lands (increase in the risk of soil erosion/degradation), more irrigation, fertiliser and energy (for moving and manufacturing fertiliser) will increase CO_2 emission (Schlesinger, 1999; Mosier, 2001; Lal 2004a,b; Van Oost et al., 2004).

- Globally, in 1999, the area under zero-tillage was 50 Mha, which was 3.5% of the total arable land and is increasingly adopted (Anonymous, 2001c). However, such practices are frequently combined with periodical tillage, making the assessments of GHG balance highly uncertain (Smith et al., 2007a). Furthermore, the use of agricultural products such as bio-plastics, bio-fuels and biomass as substitutes for fossil fuel-based products is an encouraging trend that has a potential to reduce GHG emissions.

- Growing demand of animal products may further accelerate land use change (from forest to grassland) and larger herds of livestock with higher manure production and management thus will increase GHG emission (Smith et al., 2007a).

- In a more open-market economy-driven world with changes in policies (e.g. subsidies), encouraging international trade of agricultural products caused by regional increase of production and demands will increase CO_2 emission further due to greater use of energy for transportation (Smith et al., 2007a).

1.4 Climate change impacts

Climate change will have a profound impact on human and ecosystems during the coming decades through variations in global average temperature and rainfall (Anonymous 2001a, d,e). The temperature of the temperate and polar regions will increase (Wittwer, 1990), decreasing the snowing period (Seino, 1995); thereby increasing the length and intensity of growing periods (Wittwer, 1980; Decker et al., 1985) and growing degree units (Rosenzweig, 1985). The consequences include melting glaciers, sinking of oceans, more precipitation, more and more extreme and unpredicted weather events, shifting seasons, increasing incidences and resurgence of pests, weeds and diseases (Chakravarty and Mallick, 2003; Goulder and Pizer, 2006; Ninan and Bedamatta, 2012). Tropical countries are likely to be affected more than the countries in the temperate regions (Anonymous, 2007a,b). Climate change poses unprecedented challenges to human society and ecosystems in the twenty-first century, particularly in the developing nation in the tropics (Parry, 1990; Parry et al., 1992, 2004, 2005; McCarthy et al., 2001). The accelerating pace of climate change combined with global population and income growth threatens food security (Nelson et al., 2009). It will also affect livelihoods and human well-being (Ninan and Bedamatta, 2012). Populations in the developing world which are already vulnerable and food insecure are likely to be more seriously affected.

The impact of climate change will persist. This will affect the basic elements of life around the world such as access to water, food production, healthcare and the environment (Ninan and Bedamatta, 2012). Millions of people could suffer from hunger, water shortage and coastal flooding as the world gets warmer. The overall costs and risks of climate change are expected to be equivalent to losing at least 5% of global GDP each year, if we do not act now. If a wider range of risks is taken into account, the estimated damage could rise to 20% of GDP or more (Stern, 2006, 2007). There are certain regions, sectors, ecosystems and social groups which will be affected the most by climate change and the consequences of economic globalisation. Managing the impact of climate change, therefore, poses a challenge to governments and societies (Ninan and Bedamatta, 2012).

In 2005, nearly half of the economically active population in developing countries (2.5 billion) relied on agriculture for its livelihood (Nelson et al., 2009). Today three-fourths of the world's poor population live in rural areas (Anonymous, 2008b). Agriculture and allied sectors are highly sensitive and vulnerable to climate change (Adams et al., 1998) as these changes will have an impact on agriculture by affecting crops, soil, livestock, fisheries and pests, directly and indirectly (Anonymous, 2007b; Ninan and Bedamatta, 2012). Global warming due to the greenhouse effect is expected to affect the hydrological cycle namely, precipitation, evapotranspiration and soil moisture, which will pose new challenges for agriculture. The Food Policy Report 2009 suggested that agriculture and human well-being will be negatively affected by climate change (Nelson et al., 2009) and summarises the following impact:

- In developing countries, climate change will cause yield declines for the most important crops. South Asia will be particularly hard hit.
- Climate change will have varying effects on irrigated yields across regions, but irrigated yields for all crops in South Asia will experience large declines.
- Climate change will result in additional price increases for the most important agricultural crops such as rice, wheat, maize and soya beans. Higher feed prices will result in higher meat prices. As a result, climate change will reduce the growth in meat consumption slightly and cause a more substantial fall in cereals consumption.
- Calorie availability in 2050 will not only be lower than in the no-climate-change scenario. It will actually decline relative to 2000 levels throughout the developing world.
- By 2050, the decline in calorie availability will increase child malnutrition by 20%, relative to a world with no climate change. Climate change will eliminate much of the improvement in child malnourishment levels that would occur with no climate change.
- Thus, aggressive agricultural productivity investments of US$ 7.1–7.3 billion are needed to raise calorie consumption enough to offset the negative impacts of climate change on the health and well-being of children.

The brunt of environmental changes on India is expected to be very high due to greater dependence on agriculture, limited natural resources, alarming increase in human and livestock population, changing patterns in land use and socio-economic factors that pose a great threat in meeting the growing food, fibre, fuel and fodder requirements (Ninan and Bedamatta, 2012). Droughts, floods, tropical cyclones, heavy precipitation events, hot extremes and heat waves are known to impact agricultural production and farmer's livelihood negatively as all agricultural commodities even today are sensitive to such variability. Increasing glacier melt in the Himalayas will change the availability of irrigation especially in the Indo-Gangetic plains affecting food production. Further warming is likely to lead to a loss of 1.6 million tonnes of milk production in India by 2020 (Ninan and Bedamatta, 2012). Total farm-level net-revenue loss of 8.4–25% is projected for the country in an event of 2°C temperature rise along with a 7% precipitation increase, which will amount to a loss of *₹81–195 billion (Kavi Kumar and Parikh, 1998, 2001a; Sanghi et al., 1998; Kavi Kumar, 2009).

Impacts on agriculture

Globally   Climate and climatic resources change can affect agriculture of both developing and developed countries in a variety of ways (Downing, 1996; Watson et al., 1996; Cline, 2008). Climate change and agriculture are interrelated processes, both of which takes place on a global scale (Anonymous, 2007c). Climate change is projected to have significant impacts on conditions affecting crop and livestock production, including temperature, carbon dioxide, glacial run-off, precipitation hydrologic balances, input supplies, other components of agricultural systems and the interaction of these elements (Adams et al., 1998; Webster, 2008; Gornall et al., 2010). For example, crop and livestock yields are directly affected by changes in climatic factors such as temperature and precipitation and the frequency and severity of extreme events such as droughts, floods and wind storms/tropical cyclones. Beyond a certain range of temperatures, warming tends to reduce yields because crops speed through their development, producing less grain in the process. It was estimated that warming since 1981 has resulted in annual combined yield losses of 40 million tonnes or US $5 billion (Lobell and Field, 2007).

Higher temperatures also interfere with the ability of plants to get and use moisture. Evaporation from the soil accelerates when temperatures rise and plants increase transpiration. These conditions determine the carrying capacity of the biosphere to produce enough food for the human population and domesticated animals. Despite technological advances such as improved varieties, genetically modified organisms and irrigation systems, weather is still a key factor in agricultural productivity, as well as soil properties and natural communities (Curry et al., 1990; Curtis and Wang, 1998). The effect of climate on agriculture is related to variabilities in local climates rather than in global climate patterns (Kaufmann and Snell, 1997; Freckleton et al., 1999; Gadgil et al., 1999; Tan and Shibasaki, 2006). The international aspect of trade and security in terms of food implies the need to also consider the effects of climate change on a global scale. The poorest countries would be hardest hit with reductions in crop yields mostly in tropical and subtropical regions due to decreased water availability and new or changed insect pest incidence (Anonymous, 2001a,b; Cline, 2007, 2008). Marine life and the fishing industry will also be severely affected in some places. Climate change induced by increasing GHGs is likely to affect crops differently from region to region. A decrease in potential yields is likely to be caused by the shortening growth period, decreases in water availability and poor vernalisation. Climatic change would affect agriculture in several ways as

- Productivity, in terms of quantity and quality of crops
- Agricultural practices through changes of water use (irrigation) and agricultural inputs such as herbicides, insecticides and fertilisers
- Environmental effects relating to frequency and intensity of soil drainage (leading to nitrogen leaching), soil erosion, reduction of crop diversity
- Rural space through the loss and gain of cultivated lands, land speculation, land renunciation and hydraulic amenities
- Adaptation, that is, organisms may become more or less competitive, as well as humans' urgency to develop more competitive organisms, such as flood-resistant or salt-resistant varieties of rice.

The possible changes to climate and atmosphere in the coming decades may influence GHG emissions from agriculture and the effectiveness of practices adopted to minimise those (Smith et al., 2007a). The concentration of CO_2 is projected to double within the next century. This will influence the plant growth rates, plant litter composition, drought tolerance and nitrogen demands (Torbert et al., 2000; Norby et al., 2001; Jensen and Christensen, 2004; Henry et al., 2005; Van Groenigen et al., 2005; Long et al., 2006). Increasing temperatures may not only increase crop production in colder regions due to a longer growing season (Smith et al., 2005a,b) but also could accelerate decomposition of soil organic matter, releasing stored soil carbon into the atmosphere (Fang et al., 2005; Knorr et al., 2005; Smith et al., 2005b). Moreover, changes in precipitation patterns could change the adaptability of crops or cropping systems selected to reduce GHG emissions (Smith et al., 2007a).

Agriculture will have a two-sided effect: an increased CO_2 climate change, first, directly by the fertilising effect creating a higher level of ambient CO_2 (both positively and negatively) in the atmosphere and, second, indirectly by the effect of change in climate on crop, livestock, insect pests, diseases, weeds, soils and water supplies (Easterling et al., 1989; Parry et al., 1989, 1990). These impacts classified as both direct (positive and negative) and indirect are listed in Tables 1.1 through 1.3.

The effects of climate change on agriculture vary by region and by crop (Adams et al., 1998). Higher growing season temperatures can significantly impact agricultural productivity, farm incomes and food security (Battisti and Naylor, 2009). In mid and high latitudes, the suitability and productivity of

Table 1.1   Positive impacts on agriculture

S. no.	Evidence of climate change	Impact on agricultural production
1	Longer frost-free periods	Use of higher-yielding genetics
2	Lower daily maximum temperature in summer	Reduced plant stress
3	More freeze/thaw cycles in winter	Increased soil tilt and water infiltration
4	More summer precipitation	Reduced plant stress
5	More soil moisture	Reduced plant stress
6	Higher dew point temperatures	Reduced moisture stress
7	Higher intensity of solar output	Increased degree days
8	More diffuse light (increased cloudiness)	Reduced plant stress
9	Higher water-use efficiency	Higher yields
10	Warmer spring soil temperatures	Use of higher-yielding genetics
11	Reduced risk of late spring or early fall frosts	Use of higher-yielding genetics
12	Increased atmospheric CO_2 levels	Increased photosynthesis and yields

Table 1.2 Negative impacts on agriculture

S. no.	Evidence of climate change	Impact on agricultural production
1	More spring precipitation causes water logging of soils	Delay planting, reduced yields, compaction, change to lower-yielding genetics
2	Higher humidity promotes disease and fungus	Yield loss, increased remediation costs
3	Higher night-time temperatures in summer	Plant stress and yield loss
4	More intense rain events at the beginning of crop cycle	Re-planting and field maintenance costs; loss of soil productivity and soil carbon
5	More droughts	Yield loss; stress on livestock; increase in irrigation costs; increased costs to bring feed and water to livestock
6	More floods	Re-planting costs, loss of soil productivity and soil carbon; damage to infrastructure and logistics
7	More over-wintering of pests due to warmer winter low temperature	Yield loss, increased remediation costs
8	More vigorous weed growth due to temperature, precipitation and CO_2 changes	Yield loss, increased remediation costs
9	Summer time heat stress on livestock	Productivity loss, increase in miscarriages, may restrict cows on pasture
10	Temperature changes increase disease among pollinators	Losses to cropping (forage, fruits, vegetables) systems
11	Increased taxes or regulations on energy-dependent inputs to agriculture (e.g. nitrogen fertiliser)	Profitability impacts on producers; loss of small-scale farm supply dealers
12	New diseases or re-emergence of diseases that had been eradicated or under control	Enlarged spread pattern, diffusion range and amplification of animal diseases

crops are projected to increase and extend northwards, especially for cereals and cool season seed crops (Maracchi et al., 2005; Tuck et al., 2006). Crops prevalent in southern Europe such as maize, sunflower and soya beans could also become viable further north and at higher altitudes (Hildén et al., 2005; Audsley et al., 2006). Here, yields could increase by as much as 30% by the 2050s, depending on the crop (Alexandrov et al., 2002; Ewert et al., 2005; Richter and Semenov, 2005; Audsley et al., 2006). Large gains in potential agricultural land was projected for the Russian Federation in the coming century (64% increase over 245 million hectares by the 2080s) due to its

Table 1.3 Indirect impacts on agriculture

S. no.	Situational change	Impact on agriculture
1	Regulation involving greenhouse gas emissions	Potential increased costs to meet new regulations; opportunities to participate in new carbon markets and increase profits
2	Litigation from damages due to extreme events or management of carbon markets	Legal costs may increase
3	New weed and pest species migration	Control strategies will have to be developed; increased pest management costs as well as crop losses
4	Vigorous weed growth results in increased herbicide use	Increase in resistance or reduction in time to development of resistance; regulatory compliance costs or litigation over off-site damages from pesticides
5	Possibility of increased inter-annual variability of weather patterns	Increased risk in crop rotation, genetic selection and marketing decisions
6	Increased global demand for food production due to climate and demographic changes	New markets; increase in intensification of production; increase in absentee ownership
7	Increased period for forage production	Decreased need for large forage storage across winter for livestock operations

longer planting windows and generally more favourable growing conditions under warming (Fischer et al., 2005). However, technological developments could outweigh these effects, resulting in combined wheat yield increases of 37–101% by the 2050s (Ewert et al., 2005).

A record crop yield loss of 36% occurred in Italy for corn grown in the Po valley where extremely high temperatures prevailed (Ciais et al., 2005). It is estimated that such summer temperatures in Europe are now 50% more likely to occur as a result of anthropogenic climate change (Stott et al., 2004). In areas where temperatures are already close to the physiological maxima for crops such as seasonally arid and tropical regions, higher temperatures may be more immediately detrimental, increasing the heat stress on crops and water loss by evaporation (Gornall et al., 2010). A 2°C local warming in the mid latitudes could increase wheat production by nearly 10%, whereas at low latitudes the same amount of warming may decrease yields by nearly the same amount. Different crops show different sensitivities to warming. It is important to note the large uncertainties in crop yield changes for a given level of warming.

Water is vital to plant growth, so varying precipitation patterns forcing a northward advance of monsoon rainfall further into Africa and Asia, increasing the occurrence of total rainfall, will have a significant impact on agriculture (Parry et al., 1988, 1989; Wittwer, 1990). This rainfall will also be more intense in its occurrence and therefore will propagate flooding and erosion. Food production can also be impacted by too much water (Gornall et al., 2010). Heavy rainfall events leading to flooding can wipe out entire crops over wide areas and excess water can also lead to other impacts, including soil water logging, anaerobicity and reduced plant growth. Indirect impacts include delayed farming operations. Agricultural machinery may not be adapted to wet soil conditions. The proportion of total rain falling in heavy rainfall events appears to be increasing and this trend is expected to continue as the climate continues to warm. A doubling of CO_2 is projected to lead to an increase in intense rainfall over much of Europe. In the higher end projections, rainfall intensity increases by over 25% in many areas important for agriculture. As over 80% of total agriculture is rain-fed, projections of future precipitation changes often influence the magnitude and direction of climate impacts on crop production (Olesen and Bindi, 2002; Tubiello et al., 2002). The impact of global warming on regional precipitation is difficult to predict owing to strong dependencies on changes in atmospheric circulation, although there is growing confidence in projections of a general increase in high-latitude precipitation, especially in winter and an overall decrease in many parts of the tropics and sub-tropics (Anonymous, 2007b).

Precipitation is not the only influence on water availability. Increasing evaporative demands owing to rising temperatures and longer growing seasons could increase crop irrigation requirements globally by between 5% and 20% or possibly more by the 2070s or 2080s, but with large regional variations, increasing in the Middle East and North Africa and South-East Asia (Doll, 2002; Abou-Hadid et al., 2003; Arnell et al., 2004; Fischer et al., 2006) and decreasing in China (Tao et al., 2003). The temperature increase due to elevated CO_2 will also induce higher rates of evapotranspiration causing reduction in soil moisture (Schlesinger and Mitchell, 1985; Kellogg and Zhao, 1988; Zhao and Kellogg, 1988; Parry et al., 1990). The areas which may suffer due to reduced soil moisture between December and February are southern and western Africa, South-East Asia, the Arabian peninsula, eastern Australia and southern North America, while between June and August are West Africa, western Europe, China, Soviet Central Asia, South-West United

States, Mexico, Central America, eastern Brazil and north-eastern and western Australia (Parry et al., 1990).

Some major rivers such as the Indus and Ganges are fed by mountain glaciers with approximately one-sixth of the world's population currently living in glacier-fed river basins (Stern, 2007). Populations are projected to rise significantly in major glacier-fed river basins such as the Indo-Gangetic plain. These river basins are irrigated agricultural land comprising less than one-fifth of all cropped area, but produce between 40% and 45% of the world's food (Doll and Siebert, 2002). The majority of observed glaciers around the globe are shrinking (Zemp et al., 2008) due to changes in atmospheric moisture, particularly in the tropics (Bates et al., 2008). Melting glaciers will initially increase river-flow, although the seasonality of flow will be enhanced bringing with it an increased flood risk (Juen et al., 2007). In the long term, glacial retreat is expected to be enhanced further, leading to an eventual decline in run-off, although the greater time scale of this decline is uncertain. As such, changes in remote precipitation and the magnitude and seasonality of glacial melt waters could, therefore, potentially impact food production for many people.

Water for irrigation is largely often extracted from rivers such as the Nile and the Ganges, which depend upon distant climatic conditions (Gornall et al., 2010). Agriculture along the Nile in Egypt and in the Indo-Gangetic plains in India depends on rainfall from the upper reaches of the Nile and the Ganges in the Ethiopian Highlands and the Himalayas, respectively. These areas are mostly between mid and high latitudes, where predictions for warming are the greatest. Warming in winter means that less precipitation falls as snow and that which accumulates melts earlier in the year. The changing patterns of snow cover fundamentally alter how such systems store and release water. Changes in the amount of precipitation affect the volume of run-off, particularly near the end of the winter at the onset of snow melt. Temperature changes mostly affect the timing of run-off with earlier peak flow in the spring. Although additional river-flow can be considered beneficial to agriculture, this is only true if there is an ability to store run-off during times of excess to use later in the growing season.

Thus, climate changes remote from production areas is also critical. In rivers such as the Nile, climate change will increase flow throughout the year that will benefit agriculture, but in the Ganges, run-off increases in peak flow during monsoon season while in the dry season river-flow is very low. Without sufficient storage of peak season flow, water scarcity will affect agricultural productivity despite overall increases in annual water availability. Moreover, increases at peak flow will cause damage to croplands through flooding (Gornall et al., 2010). Globally, only a few rivers currently have adequate storage to cope with large shifts in seasonal run-off (Barnett et al., 2005). Where storage capacities are not sufficient, much of the winter run-off will immediately be lost to the oceans. The water from these glaciers feeding large rivers such as the Indus, Ganges and Brahmaputra is likely to be contributing a significant proportion of seasonal river-flow, although the exact magnitude is unknown. Currently, nearly 500 million people are reliant on these rivers for domestic and agricultural water resources. Climate change may mean the Indus and Ganges become increasingly seasonal rivers ceasing to flow during the dry season (Kehrwald et al., 2008). Combined with a rising population, this means that water scarcity in the region would be expected to increase in the future.

Under rising atmospheric CO_2 and climate change, the potential impacts of devastating pathogens and insect pests may change. This will change the crop–pest relationship because climate (mainly temperature) plays a dominant role in the distribution and development of pests in the following ways: increases in the rate of development and number of generations produced per year, extension of the geographical range beyond the present margin of distribution, earlier establishment of pest populations in the growing season and increases in the risk of migrant invasion and exotic species (Parry et al., 1990; Wittwer, 1990; Rosenzweig and Hillel, 1993). A major factor in global warming could be greater survival through over-wintering and persistence of plant diseases and insects. With higher atmospheric concentration of CO_2, plants will grow faster and accumulate more carbohydrates and nitrogen (Bhattacharya and Geyer, 1993), changing the feeding habit of insects (Fajer et al., 1989), which will lead to higher pest density and intense damage.

An increase in the over-wintering range and population density of *Heliothis zea* in the US grain belt will increase the damage to soya beans (Anonymous, 1989). Pests, such as aphids (Newman, 2004) and weevil larvae (Staley and Johnson, 2008), respond positively to higher CO_2. Increased temperatures also reduced the over-wintering mortality of aphids, enabling earlier and potentially more widespread dispersion (Zhou et al., 1995). The sub-Saharan Africa migration patterns of locusts are influenced by rainfall patterns (Cheke and Tratalos, 2007). Warming or drought may change the resistance of crops to specific diseases or through increased pathogenicity of organisms by mutation induced by environmental stress (Gregory et al., 2009). The severity of disease in oilseed rape could increase within its existing range and can also spread northward over the next 10–20 years (Evans et al., 2008). Changes in climate variability may also be significant affecting the predictability and amplitude of outbreaks (Gornall et al., 2010).

Of all crop pests, weeds are the most damaging, but they will cease to be so in the event of CO_2 increases in the atmosphere (Chakravarty and Mallick, 2003). This is because the condition that will generally favour crop production over weed growth. Out of the world's 18 most noxious weeds, 14 are C_4 plants, and out of the 20 most important food crops, 16 are C_3 plants (Wittwer, 1990). As discussed earlier, C_3 plants will respond more vigorously in elevated atmospheric CO_2 dominating the C_4 weeds (Chakravarty and Mallick, 2003). Climate change will shift the agricultural potential to new regions, but may also shift with the introduction of new pest species. Barley now growing as a highly marginal crop in Iceland may become cultivable throughout lowland Iceland due to longer growing seasons, but losses to pest and diseases will rise up to 15% from today's minimal level (Bergthorsson et al., 1988).

Sea level will rise owing to thermal expansion of the existing mass of ocean water and water flowing in after melting of land ice due to the warming climate inundating coastal land (Titus, 1990; Gornall et al., 2010). The potential sea-level rise due to melting and discharge of West Antarctic, East Antarctic and Greenland ice sheets will be approximately 2 m by 2100 (Pfeffer et al., 2008; Rohling et al., 2008). The crop production will be vulnerable, where large sea-level rise occurs in conjunction with low-lying coastal agriculture, such as in major river deltas, which are valuable agricultural land owing to the fertility of fluvial soils. Sea-level rise threatens to inundate agricultural lands and salinise groundwater in the coming decades in the United States (Park et al., 1988) and 20% of farmland in Bangladesh and Egypt (Broadhus et al., 1986). Although inundation may not pose a major threat to US agriculture, it would be a major threat for countries such as Bangladesh and Egypt whose nationwide productivity mainly depends on cultivated river deltas (Chakravarty and Mallick, 2003).

Moreover, the main culprit of climate change, the carbon emissions, can also help agriculture by enhancing biomass production due to the net increase in photosynthetic gain (especially in C_3 crops such as wheat, rice and soya beans) because of the reduction in photorespiration, increase water-use efficiencies and reduce transpiration by closing stomata through increased stomatal resistance (Rogers et al., 1980, 1981, 1983, 1984; Kimball and Idso, 1983; Acock, 1990; Goudriaan and Unsworth, 1990; Parry et al., 1990; Rosenberg et al., 1990; Stockle et al., 1992a,b; Field et al., 1995; Grant et al., 1999; Norby et al., 2001; Tubiello and Ewert, 2002; De Costa et al., 2003; Widodo et al., 2003; Ewert, 2004; Parry et al., 2004; Ainsworth and Long, 2005). The present levels of CO_2 are suboptimal for photosynthesis, but other major factors influencing growth such as light, water, temperature and nutrients (Osmond et al., 1980; Downton et al., 1981). Plants will respond differently according to their biochemical pathways for photosynthesis. C_3 plants will respond vigorously while C_4 plants will not respond to elevated CO_2 concentration (Alien, 1979; Kimball, 1983a,b; Morison and Gifford, 1983; Acock and Alien, 1986; Cure, 1986; Alien et al., 1987; Parry et al., 1990).

Increasing atmospheric CO_2 concentrations to 550 ppm could increase photosynthesis in C_3 crops by nearly 40% (Long et al., 2004) and these plants have the capacity to carry out photosynthesis up to 1000 ppm CO_2, that is, the CO_2 compensation point is higher than the C_4 plants (Kimball, 1985, 1986; Wittwer, 1985, 1986). There will be no physiological benefits with rising CO_2 concentrations in C_4 crops such as maize, millet, sorghum and sugarcane as CO_2 is concentrated to 3–6 times the atmospheric concentrations (von Caemmerer and Furbank, 2003). These crops at such conditions will receive the required CO_2 in less time, opening the stomata for a shorter duration, making them more water-use efficient. This may marginally increase their yields (Long et al., 2004). Crop yield increase is lower than the photosynthetic response; increase in atmospheric CO_2 to 550 ppm would on average increase C_3 crop yields by 10–20% and C_4 crop yields by 0–10% (Gifford, 2004; Long et al., 2004; Ainsworth and Long, 2005). Plants of C_3 photosynthetic pathway may benefit in dry matter production from an increase in atmospheric concentration of CO_2 through enhancement of leaf expansion, an increase in photosynthetic rate per unit leaf area and an increase in water-use efficiency (Wittwer, 1990). However, different responses of photosynthesis and RuBisco will be encountered among C_3 plant species as a result of future increases in CO_2 and temperature (Vu et al., 1997).

The other beneficial effects are usually increases in leaf area and thickness, stem height, branching, seed and fruit number and weight, C:N ratio, organ size with higher root-to-shoot ratios, harvest index or yields of marketable products notably with earlier maturity, particularly in potatoes (Arteca et al., 1979), sweet potatoes (Bhattacharya et al., 1985) and increases in biological nitrogen fixation in soya beans (Hardy and Havelka, 1975; Phillips et al., 1976; Finn and Brun, 1982; Lamborg et al., 1983). Elevated CO_2 concentration may increase carbohydrates, but may reduce chlorophyll, proteins, amino acids, carotene and mineral nutrients (Bhattacharya and Geyer, 1993). Despite the potential positive effects on yield quantities, elevated CO_2 may be detrimental to the yield quality of certain crops, as in wheat through protein content reduction (Sinclair et al., 2000).

Elevated CO_2, besides affecting the crop, also affects the environment, which in turn may have either beneficial or damaging effects on agricultural production (Lemon, 1983; Morison, 1987; Peiris et al., 1996; Rosenzweig and Hillel, 1998). The projection of global-scale yields can be determined through the strength of CO_2 fertilisation (Parry et al., 2004; Nelson et al., 2009). North America and Europe will benefit from climate change with strong CO_2 fertilisation but Africa and India will lose 5% even with strong CO_2 fertilisation. These projected losses will be up to 30% if the CO_2 fertilisation effects are not considered. The crop response to elevated CO_2 may be actually lower than previously thought with consequences for crop modelling and projections of food supply (Long et al., 2004, 2009). This is because of many limiting factors such as pests and weeds, nutrients, competition for resources, soil water and air quality, which are neither well understood at large scale nor well implemented in leading models. The science, however, is far from certain on the benefits of carbon fertilisation (Cline, 2008).

However, the nature of the biophysical effects and human responses to agriculture are complex and uncertain (Adams et al., 1998). There are large uncertainties to uncover; particularly, because there is lack of information on many specific local regions and include the uncertainties on the magnitude of climate change, the effects of technological changes on productivity, water and fertiliser application strategies, changes in pest and disease occurrence, global food demands and the numerous possibilities of adaptation (Cannon, 2003; Engvild, 2003; Fuhrer, 2003). Most agronomists believe that agricultural production will be mostly affected by the severity and pace of climate change, not so much by gradual trends in climate. If change is gradual, there may be enough time for biota adjustment. Rapid climate change, however, would harm agriculture in many countries, especially those that are already suffering from rather poor soil and climate conditions because there is less time for optimum natural selection and adaption (Mendelsohn and Schlesinger, 1999; Chakravarty and Mallick, 2003). The overall effect of climate change on agriculture will depend on the balance of these effects. An assessment of the effects of global climate changes on agriculture might help to properly anticipate and adapt farming to maximise agricultural production.

On India   Indian agriculture is more vulnerable to climate change as it is economically associated, where poverty is strongly related with the agricultural performance of the country (Ninan and Bedamatta, 2012). Future crop production losses of 10–40% are associated with an increase in temperature from 2080 to 2100 (Parry et al., 1992; Aggarwal and Kalra, 1994; Dinar et al., 1998; Kavi Kumar and Parikh, 2001a,b; Anonymous, 2007a; Kavi Kumar, 2009). The projected impact of climate change on agriculture varies across regions because India has an immense climatic/geographic diversity (Kavi Kumar, 2007; Ninan and Bedamatta, 2012). In the arid regions, even small increases in temperature will decline agricultural production, but the same rate of increase in the Himalayas will increase agricultural production (Anonymous, 2009a). Studies conducted with the Ricardian approach projected an increase of 2.0–3.5°C with less rainfall will result in a 3–26% loss of net agricultural revenue (Sanghi et al., 1998; Kavi Kumar and Parikh, 1998, 2001a,b; Kavi Kumar, 2007, 2009). Increasing climatic variability associated with global warming, nevertheless, will result in considerable seasonal/annual fluctuations in food production (Mall et al., 2006).

Rice production in many parts of India projected between 2010 and 2070 would increase by 26% in an optimistic and 9–30% in a pessimistic scenario (Aggarwal and Mall, 2002; Kalra et al., 2007). In Kerala, rice production will decrease by 6% with a 1.5°C temperature rise (Saseendran et al., 2000). Rice and wheat production in north-western India will not be affected by the doubling of CO_2 and temperature increases of 2–3°C, but will decrease by 20% with water shortages (Lal et al., 1998). Every 1°C temperature rise during the growing period will result in loss of 4–5 million tonnes in future wheat production (Kalra et al., 2007). Haryana, Punjab, western Uttar Pradesh and coastal Tamil Nadu will have negative effects on wheat production, but eastern districts of West Bengal and parts of Bihar will have benefits from projected climate change (Kavi Kumar and Parikh, 2001a). Soya bean production in Madhya Pradesh will reduce by 4% with a 10% decline in daily rainfall, doubling CO_2 and a 3°C temperature rise (Lal et al., 1999). The production of maize and jowar also will decline with an increase in temperature (Kalra et al., 2007). The effects of temperature and CO_2 on other crops are listed in Table 1.4. Losses for other crops are still uncertain, but they are expected to be relatively smaller, especially for kharif crops (Ninan and Bedamatta, 2012). There is an increasing effect of climate change on Indian agriculture in spite of the possible advances made through technology and the country's overall development (Kavi Kumar, 2009). This warrants further systematic studies to understand the impact of future climate change on India as the agricultural sector is extremely sensitive to climate variability.

1.5 Adjustment

It is certain that in an event of any climate change, agriculture will adjust to meet these changes, which is likely with a spatial shift of crop potential. Areas under today's climatic conditions suited for a crop or combination of crops may no longer be suitable after a climatic shift, or otherwise, an area today not suitable for growing particular crop(s), may be suitable tomorrow (Chakravarty and Mallick, 2003). Maize growing successfully in South England at present may shift northward with a rise in temperature, wheat will shift eastward in the United States (Decker et al., 1985), north and southward in India and maize northward in the United States (Newman, 1980). Similar northward shifts are projected for sunflower in the UK (Parry et al., 1989), citrus, olives and vines in southern Europe (Imeson et al., 1987), while a southward shift of land use is projected in the Southern Hemisphere (Salinger, 1988). This might also expand successful commercial production of mangoes, papayas, litchis, bananas, pineapples and other fruits from sub-tropical and tropical to temperate areas (Chakravarty and Mallick, 2003).

Changes in climate would influence agriculture by changing the length of the growing season, crop yield, agricultural potential and shifting the geographical area (Hogg, 1992). Many crops can adjust to possible climate change. However, the magnitude of the projected climate will shift northward as change will vary from location to location and the influence will be a function of the change in climate to the existing condition (Chakravarty and Mallick, 2003). So at mid latitudes, agricultural potential would decrease toward poles due to smaller

Table 1.4 Climate change impacts on Indian crop production

Crop	Region	Temperature	CO_2 level	Production impact	Reference
Rice	Uttarakhand	Increased	Doubled	Positive	Achanta (1993)
	India	High	Increased	Increase in production under the global circulation model (GCM) scenario	Mohandass et al. (1995)
	Kerala	Up to 5°C rise	425 ppm	Increase in production due to fertilisation effect of CO_2 up to 2°C and also enhances water-use efficiency but up to 5°C temperature, there is a continuous decline in rice yield	Saseendran et al. (2000)
	Central, South and North-West	0.7°C and 1°C increase	—	Central and South India production will increase but in the North-West, production will decrease significantly under irrigated condition due to decrease in monsoon rainfall. Reduction in crop duration all over due to temperature increase	Rathore et al. (2001)
	Northern, southern, western and eastern	1–4°C rise	450 ppm	Eastern and western less affected, northern moderate and southern severely affected. With CO_2 of 450 ppm and 1.9–2°C increase, production increased in all regions	Aggarwal and Mall (2002)
	India	Atmospheric brown clouds (ABC)		Increase	Auffhammer et al. (2006)
Rice, wheat	Northern coastal regions	0.5–2°C rise		Decrease in rice and wheat yields and would reduce wheat crop duration by 7 days	Sinha and Swaminathan (1991)
	North-West India	2–3°C rise	Increased	Doubling CO_2 increases rice and wheat yield but at 3°C for wheat and 2°C for rice nullified the positive effects of increased CO_2	Lal et al. (1998)

continued

Table 1.4 (continued) Climate change impacts on indian crop production

Crop	Region	Temperature	CO_2 level	Production impact	Reference
Wheat, rice, maize, groundnut	Punjab	1°C, 2°C and 3°C increase	Increased	Yield reduction in all the crops. Increased CO_2 increases crop production	Hundal and Kaur (1996)
Wheat, winter maize	Bihar	Increased	Increased	Yield increase in winter maize but decrease in wheat	Abdul Haris et al. (2013)
Wheat	India	1°C increase	—	Yield reduction	Saini and Nanda (1986)
	Northern India	0–2°C rise	425 ppm	At 1°C yield increased but reduced at 2°C with reduced evapotranspiration	Aggarwal and Sinha (1993)
	India	Increased	—	Yield decreased	Gangadhar Rao and Sinha (1994)
	North, Central India at high and lower latitude; tropical and sub-tropical	0–2°C rise	425 ppm	At 1°C yield increased but reduced at 2°C temperature. Sub-tropical reduction by 1.5–5.8 % but in tropical decrease was 17–18 %. Slightly increased at high latitudes but significant decrease at lower latitude. 10–15 % yield reduction in Central India but non-significant in north	Aggarwal and Kalra (1994)
Sorghum	Hyderabad, Akola and Sholapur	Three scenarios of climate change	Increased	Yield reduction at Hyderabad and Akola but at Sholapur marginally increased	Gangadhar Rao et al. (1995)
	India	1–2°C rise	50–700 ppm	7–12% reduction in yield. 50 ppm CO_2 increase, yield increased by 0.5% but was nullified with a rise of 0.08°C and 700 ppm by a rise of 0.9°C	Chatterjee (1998)

Maize	India	Up to 4°C rise	350–700 ppm	Decreased yield in both irrigated and rainfed conditions. With 350 ppm CO_2 and up to 4°C yield decreased by 30% but at 700 ppm CO_2 yield increased by 9% over present day condition	Sahoo (1999)
Brassica	North	Varied	Varied	Production is likely to increase and extend its range at relatively drier regions	Uprety et al. (1996)
Chickpea, pigeon pea	India	Up to 2°C rise	350–700 ppm	Chickpea yield not increased with reduction in total crop duration but increase in CO_2 increased yield while yield of pigeon pea decreased at 1°C increase	Mandal (1998)
Soya bean	Central	3°C rise with 10% decline in rainfall	Doubled	50% yield increase in Central India but at 3°C increase no effect of doubling CO_2 as yield reduced which was acute at reduction of daily rainfall decreasing yield by 32%	Lal et al. (1999)

thermal inputs, while the same increase of temperature will have a greater relative effect at higher latitudes than at lower latitudes due to greater temperature increases at higher latitudes (Parry et al., 1990). Considering the examples of the adjustment and yield increase of wheat in India, maize in Iowa, United States and northern Europe, rice in Philippines and Indonesia, soya bean in Brazil, sunflower in the Red Valley of the United States, oil palm in Malaysia and canola (rape) in Canada, the future adjustment of agricultural crop production can also be indexed by an already observed rate of change (Wittwer, 1990). Global warming projection, especially during winter months at high latitudes (Williams and Oakes, 1978; Parry et al., 1988, 1989; Wittwer, 1990), will extend the efficient crop ecological zone indicating a significant northward shift of balance of agricultural resources (Parry et al., 1989).

The extent of this crop ecological zone may not make the introduction of new genetic material necessary as it would advance the thermal limit of cereal cropping in mid-latitude Northern Hemisphere regions by about 150–200 km and raise the altitudinal limit by about 150–200 m in the European Alps, making it similar to the Pyrenees located 300 km south of the Alps (Parry et al., 1989). A rise of temperature in cool temperate and cold regions will lengthen the potential growing season and increase growth rates. This will shorten the required growing period (except where moisture is a limiting factor), as in Finland where yields of barley and oat will increase by 9–18% (Kethunen et al., 1988), in Iceland where the carrying capacity of grasslands for sheep will increase by two and half times (Bergthorsson et al., 1988) and critically low-yield steppe regions will have a twofold increase in yields (Sirotenko et al., 1997). In areas presently with low precipitation, the elevated CO_2 concentration will be beneficial to the crop yields as in China during summer monsoon where an increase in 100 mm rainfall with 1°C temperature rise will increase yields of rice, maize and wheat by 10% (Zhang, 1989) and similarly in Japan (Yoshino et al., 1988). The projected climate change due to an increase in CO_2 concentration will favour a change from the existing, often quick maturing cultivars to be grown for a longer and more intense growing season and late maturing varieties will be more suitable for such conditions. For instance, late growing rice cultivars presently in Central Japan will have a yield increase of 26% and quick maturing varieties now growing in northern Japan will have increase of only 4% (Yoshino et al., 1988). Similarly, a switch to winter-sown cereals (wheat, barley and oats), as in the case of wheat in Saskatchewan and

Central Russia, will give higher yields than spring cereals because of longer growing seasons and reduced damage by high evapotranspiration rates (Pitovranov et al., 1988).

The establishment of new zones of agricultural potential is likely to bring about changes in crop location and crop varieties. These changes will however be influenced by the regional pattern of rainfall or variation in soils and competitiveness of different crops (Parry et al., 1990). For instance, cereal crop production in Europe will not be influenced as significantly as elsewhere. Crop production will suffer most severely in the inherently vulnerable regions of Africa, South America, Middle East, Asia Pacific, South-East and Central Asia where changes in temperature and precipitation will further stress the already limited productive capacity of these regions. Cold and marginal regions of both the Northern and Southern Hemisphere (Canada, Alaska, Iceland, Scandinavia, Russia, New Zealand, Tasmania and others) will benefit from higher temperatures and its associated optimum conditions, such as longer growing seasons, higher growing degree units and more frost-free periods with higher yields (Smit et al., 1989). If the climate change will occur as predicted, the agricultural production is likely to increase in North America, northern Europe, Commonwealth of Independent States, China and South America (Rosenzweig, 1985; Wilks, 1988; Wittwer, 1990). The crop yield of the Soviets and other European countries will boost by 50% while China and India will benefit with enhanced production of soya beans, winter wheat, rice, corn and cotton with northern migration. But there are also areas where productivity of some crops will not change after an increase in CO_2 concentration; for example, wheat production in major areas of the United States would remain the same (Hansen et al., 1981).

The projected climate change will bring about a large number of changes in crop management that will modify the climate change on agriculture. Some regions and crops are critically more vulnerable than others (Chakravarty and Mallick, 2003). Resources for crop production are usually most critical in agriculturally developing countries than in developed countries (Oram, 1985). The climate change scenarios considered by various models would relax the current constraints imposed by a short and cool frost-free season, but without adjustive measures drier conditions and accelerated crop development rates were estimated to offset potential gains stemming from elevated CO_2 concentrations (Brklacich et al., 1996). Under such conditions, higher crop yield would require greater amounts of fertiliser and water (Wittwer, 1990). The yields of major crops in dry and arid tropical and sub-tropical areas will decrease as irrigation water will become limiting because of additional stress on crops already affected by higher temperatures (Beran and Arnell, 1989). A substantial increase in cost and management of irrigation water is likely to occur in these areas. A northern migration of agriculture would increase irrigation and fertiliser in sandy soils, which may create worse groundwater problems (Wittwer and Robb, 1964). Such a situation is most likely in Punjab and surrounding areas (Chakravarty and Mallick, 2003). In areas where the amount or intensity of rainfall will increase, management would be oriented in a way to prevent soil erosion. Moreover, increases in fertiliser use may be required in such areas. Thus, the agricultural productivity impacts in most developing countries of Central and South America, Africa, South-East Asia and the Pacific Islands will be minimal through a combination of agricultural zones and adjustments in agricultural technology and management (Parry et al., 1990; Wittwer, 1990).

1.6 Abatement/mitigation

With continuous population growth and improving incomes, but with no increase in arable land, the primary objective of agriculture is to satisfy production demand of 50–80% above today's level through sustained productivity improvement by 2050 (Anonymous, 2003, 2006d, 2009b; Müller, 2009). This will require optimising agricultural productivity using techniques that minimise unwanted impacts such as GHG emissions, eutrophication and acidification. Agricultural systems are dynamic and managed ecosystems, critical to the human response regarding production and food supply in the era of climate change. This is because producers and consumers are continuously responding to changes in crop and livestock yields, food prices, input prices, resource availability and technological changes. A fundamental question with regard to climate change is whether agriculture can adapt quickly and autonomously or will the response be slow and dependent on structural policies and programmes? Accounting for these adaptations and adjustments is difficult but necessary in order to measure climate change impacts accurately. Failure to account for human adaptations, either in the form of short-term changes in consumption and production practices or long-term technological changes, will overestimate the potential damage from climate change and underestimate its potential benefits (Adams et al., 1998).

In fact the stability and predictability of the climate and the ability of farmers to adapt their practices to changing climatic conditions will ensure the future of food production and global food security (Chakravarty and Mallick, 2003). The overall level of agricultural GHG emissions will continue to rise for the foreseeable future as agricultural production expands to keep pace with growing food, feed, fibre and bioenergy demand. Increasing agricultural efficiency is critical to keep overall emissions as low as possible and to reduce the level of emissions per unit of agricultural output. Efficient and responsible production, distribution and use of water, fertilisers and other inputs are central to achieving these goals. Agricultural systems can adapt to offset the negative effects of climate change, but not without costs for changes in technology involving research and development and farm-level adoption, including possible physical and human capital investments (Anonymous, 1992; Rosenberg, 1992; Easterling et al., 1993; Kaiser et al., 1993; Mendelsohn et al., 1994; Easterling, 1996; Adams et al., 1998).

Mitigation is unlikely to occur without action, and higher emissions are projected in the future if current trends are left unconstrained. Global population will increase by 50% from present, reaching nine billion by 2050 (Lutz et al., 2001; Cohen, 2003). This enormous population pressure will require double production of cereals and other animal-based foods during the coming decades, which will require more use of N fertiliser and livestock increasing N_2O and CH_4 emissions from enteric fermentation unless more efficient fertilisation/management techniques and products can be found (Tilman et al., 2001; Mosier, 2002; Roy et al., 2002; Galloway, 2003; Green et al., 2005). CH_4 and N_2O emissions vary greatly with land use depending on trends towards globalisation or regionalisation and on the emphasis placed on material wealth relative to sustainability and equity (Strengers et al., 2004). Trends in GHG emissions in the agricultural sector depend mainly on the level and rate of socio-economic development, human population growth and diet, application of adequate technologies, climate and non-climate policies and future climate change. Consequently, mitigation potentials in the agricultural sector are uncertain, making a consensus difficult to achieve and hindering policy making. Opportunities for mitigating GHGs in agriculture fall into three broad categories (Smith et al., 2007a,b,c; Niggli et al., 2009) based on the underlying mechanisms.

Reducing emissions

The fluxes of GHGs can be reduced by efficient management of carbon and nitrogen flows in agricultural ecosystems. The practices that deliver added N more efficiently to crops often reduce N_2O emissions (Bouwman, 2001), and managing livestock to make most efficient use of feeds often reduces amounts of CH_4 produced (Clemens and Ahlgrimm, 2001). The approaches that best reduce emissions depend on local conditions and, therefore, vary from region to region.

Enhancing removals

Agricultural ecosystems stock large carbon reserves mostly in the form of soil organic matter (Anonymous 2001a) which are lost more than 50 Pg C (Paustian et al., 1998; Lal, 1999, 2001a,b; 2002, 2003, 2004a–e, 2005; Lal and Bruce, 1999; Lal et al., 2003). This loss can be recovered through improved management, thereby withdrawing atmospheric CO_2. These practices can be adopted locally to increase the photosynthetic input of carbon and/or slow the return of stored carbon to CO_2 through respiration, fire or erosion. This will increase carbon reserves by sequestering carbon or stocking carbon sinks (Lal, 2004a) through agro-forestry or other perennial plantings on agricultural lands (Albrecht and Kandji, 2003). Agricultural and forest lands also remove CH_4 from the atmosphere by oxidation, but this effect is small compared to other GHG fluxes (Smith and Conen, 2004; Tate et al., 2006).

Avoiding/ displacing emissions

The combustion of bioenergy feedstock used as a source of fuel either directly or after conversion releases CO_2 (Schneider and McCarl, 2003; Cannell, 2003). The net benefit of bioenergy sources to the atmosphere is equal to the fossil-derived emissions displaced, which are less than any emissions from producing, transporting and processing. Conserving forest, grassland and other non-agricultural vegetation or discouraging further agricultural management practices into new lands can restrict GHG emissions (Foley et al., 2005).

The net benefit of these practices so adopted will depend on the combined effects on all gases (Robertson and Grace, 2004; Schils et al., 2005; Koga et al., 2006), which may either reduce emissions indefinitely or temporarily (Marland et al., 2001, 2003a; Six et al., 2004). Where a practice affects radiative forcing through other mechanisms such as aerosols or albedo, those impacts also need to be considered (Marland et al., 2003b; Andreae et al., 2005). The broad categories of options mentioned above can be adopted through any one or combination of the management practices discussed below.

Cropland management

Mitigation practices in cropland management include the following practices:

Agronomy: Improved agronomic practices such as using improved crop varieties, extending crop rotations especially with perennial crops (produce more below ground carbon), rotation with legumes, growing 'cover' or 'catch' crops, efficient fertiliser/nutrient, pesticides and other input management and avoiding or reducing fallow which not only increases yields but also increases soil carbon storage through higher residue production (Follett, 2001; Izaurralde et al., 2001; West and Post, 2002; Lal, 2003, 2004a; Barthès et al., 2004; Freibauer et al., 2004; Paustian et al., 2004; Smith, 2004a,b; Alvarez, 2005). However, N benefits (also with legume-derived N) can be offset by emissions of higher soil N_2O and CO_2 from fertiliser manufacture (Schlesinger, 1999; Pérez-Ramírez et al., 2003; Robertson, 2004; Gregorich et al., 2005; Rochette and Janzen, 2005). The catch or cover crops can extract available N unused by the preceding crop, thereby reducing N_2O emissions (Barthès et al., 2004; Freibauer et al., 2004).

Nutrient management: Crops cannot always use applied nitrogen that emits out of the soil as N_2O efficiently (Galloway et al., 2003, 2004; Cassman et al., 2003; McSwiney and Robertson, 2005). Nitrogen-use efficiency can be improved by reducing leaching and volatile losses, applying the precise crop need, using slow/controlled-release forms or nitrification inhibitors (slowing the microbial processes leading to N_2O formation), applying just prior to plant uptake (least susceptible to loss), placing precisely for accessibility to roots and avoiding excess application during immediate plant requirements, which will directly reduce N_2O emissions and indirectly reduce GHG emissions from N fertiliser manufacture (Cole et al., 1997; Schlesinger, 1999; Dalal et al., 2003; Paustian et al., 2004; Robertson, 2004; Monteny et al., 2006).

Tillage/residue management: Minimal or zero tillage generally results in soil carbon gain and reduced CO_2 and N_2O emissions through enhanced decomposition of retained crop residues and erosion due to less disturbance of soil and less energy use (Marland et al., 2001, 2003b; West and Post, 2002; Cassman et al., 2003; Cerri et al., 2004; Smith and Conen, 2004; Alvarez 2005; Gregorich et al., 2005; Helgason et al., 2005; Li et al., 2005; Madari et al., 2005; Ogle et al., 2005; Koga et al., 2006). Residue burning should be avoided to prevent emissions of aerosols and GHGs generated from fire (Cerri et al., 2004).

Water management: Supplementary irrigation provides water to 18% of the world's cropland (Anonymous, 2005a). Improving the efficiency of this irrigation system supplementary through delivery and drainage management along with further extension of irrigated area will boost yield and residue returns, thereby increasing soil carbon and also suppressing N_2O emissions by improving aeration (Follett, 2001; Reay et al., 2003; Lal, 2004a; Monteny et al., 2006). However, the energy used for water delivery or higher moisture and fertiliser N inputs may offset this gain through CO_2 and N_2O emissions, respectively (Schlesinger 1999; Liebig et al., 2005; Mosier et al., 2005).

Rice management: CH_4 emission from cultivated wetland rice soil can be reduced by growing low exuding cultivars, draining once or several times during the growing season, using efficient water management during off-season by keeping the soil dry or avoiding waterlogging and incorporating properly composted organic materials/residues (may be by producing biogas) during the dry period (Wang and Shangguan, 1996; Yagi et al., 1997; Wassmann et al., 2000; Aulakh et al., 2001; Cai et al., 2000, 2003; Xu et al., 2000, 2003; Kang et al., 2002; Yan et al., 2003; Cai and Xu, 2004; Smith and Conen, 2004; Khalil and Shearer, 2006). Frequently, however, draining may be constrained by water supply and may partly offset the reduced CH_4 emission benefit by increasing N_2O emissions (Akiyama et al., 2005). These practices will also increase productivity by enhancing soil organic carbon stocks (Pan et al., 2006).

Agro-forestry: Planting trees and other perennial species in an agro-forestry system also increases soil carbon sequestration (Guo and Gifford, 2002; Paul et al., 2003; Oelbermann et al., 2004; Mutuo et al., 2005), but the effects on N_2O and CH_4 emissions are not well known (Albrecht and Kandji, 2003).

Land cover (use) change: Increasing the land cover or changing land use, similar to the native vegetation over the entire land area ('set-asides') or in localised spots such as grassed waterways, field margins and shelterbelts effectively converts drained croplands back to wetlands, reducing emissions and increasing carbon storage (Follett, 2001; Ogle et al., 2003; Falloon et al., 2004; Freibauer et al., 2004; Lal, 2004b; Paustian et al., 2004). Converting drained croplands back to wetland, however, may stimulate CH_4 emissions because waterlogging creates anaerobic conditions (Paustian et al., 2004).

Grazing land management and pasture improvement

Globally, the area under grazing lands is more than croplands and is usually managed less intensively (Anonymous, 2006c). The practices that reduce emissions and enhance removals of GHG are discussed below.

Grazing intensity: Grazing intensity and timing influence the removal, growth, carbon allocation and flora of grasslands, affecting the amount of carbon accrual in soils (Conant et al.,

2001, 2005; Rice and Owensby, 2001; Conant and Paustian, 2002; Freibauer et al., 2004; Reeder et al., 2004; Liebig et al., 2005). The effects are inconsistent as there are many types of grazing practices involving diversified plant species, soil and climate (Schuman et al., 2001; Derner et al., 2006).

Increasing productivity: Carbon stock of grazing lands can be increased by improving its productivity through alleviating nutrient and moisture deficiencies (Conant et al., 2001; Schnabel et al., 2001). Adding nitrogen and energy use for irrigation stimulates N_2O and CO_2 emissions, which may, however, offset some of the benefits (Schlesinger, 1999; Conant et al., 2005).

Nutrient management: The practices (discussed for crop-land) that improve the plant nutrient uptake can reduce N_2O emissions (Follett et al., 2001; Dalal et al., 2003). Nutrient management on grazing lands is made complicated through deposition of faeces and urine from livestock that, too, are uncontrolled and randomly added (Oenema et al., 2005).

Fire management: Anthropogenic or natural on-site biomass burning either contributes to climate change through GHG emission, production of smoke aerosols (have either warming or cooling effects on the atmosphere), albedo reduction of the land surface for several weeks (causing warming) and disturbed woody versus grass cover proportion, particularly in savan-nahs which occupy about one-eighth of the global land surface (Andreae, 2001; Andreae and Merlet, 2001; Menon et al., 2002; Anderson et al., 2003; Beringer et al., 2003; Jones et al., 2003; Van Wilgen et al., 2004; Andreae et al., 2005; Venkataraman et al., 2005). Therefore, reducing the frequency or intensity of fires through more effective fire suppression, reducing fuel load by vegetation management and burning at a time of year when less CH_4 and N_2O are emitted can restrict these processes along with an increased CO_2 sink into soil and biomass (Scholes and van der Merwe, 1996; Korontzi et al., 2003).

Species introduction: Introducing grass species with higher productivity (legumes) or carbon allocation to deeper roots can increase soil carbon (Fisher et al., 1994; Davidson et al., 1995; Conant et al., 2001; Machado and Freitas, 2004; Soussana et al., 2004), and perhaps also can reduce emissions from fer-tiliser manufacture if biological N_2 fixation displaces applied N fertiliser (Sisti et al., 2004; Diekow et al., 2005).

Management of organic/peaty soils

Organic or peaty after draining can be used for agriculture, but the accelerated aeration decomposition in these soils results in high CO_2 and N_2O fluxes (Kasimir-Klemedtsson et al., 1997). The drainage of such soils should either be avoided in the first place or a higher water table be re-established. If not, emissions can be reduced to some extent by avoiding deep ploughing, discouraging row crops and tubers and maintaining a shallower water table (Freibauer et al., 2004).

Restoration of degraded lands

Degraded agricultural lands can be partially restored and carbon storage can be improved by re-vegetation, nutrient amendments, application of organic substrates (manures, biosolids, composts), minimum/zero tillage, retaining crop residues and water conservation (Batjes, 1999; Bruce et al., 1999; Lal, 2001a,b, 2003, 2004b; Olsson and Ardö, 2002; Paustian et al., 2004; Foley et al., 2005). Where these practices involve higher nitrogen amendments, the benefits of carbon sequestration may be partly offset by higher N_2O emissions (Smith et al., 2007a).

Livestock management

The practices for reducing CH_4 and N_2O emissions from livestock (cattle and sheep) rearing are categorised as improved feeding practices, use of specific agents or dietary additives and long-term management changes and animal breeding (Monteny et al., 2006; Soliva et al., 2006).

Improved feeding practices: Improving pasture quality (in less developed regions to improve animal productivity), replacing forages with concentrates, supplementing certain oils or oilseeds to the diet and optimising protein intake (reduce N excretion) can reduce CH_4 and N_2O emissions, but may increase daily methane emissions per animal (Blaxter and Claperton, 1965; Leng, 1991; Johnson and Johnson, 1995; McCrabb et al., 1998; Machmüller et al., 2000; Phetteplace et al., 2001; Lovett et al., 2003; Beauchemin and McGinn, 2005; Clark et al., 2005; Alcock and Hegarty, 2006; Jordan et al., 2006a–c).

Specific agents and dietary additives: Dietary additives fed to the animals can suppress methanogenesis to reduce CH_4 emissions. These are ionophores (antibiotics—banned in the EU); halogenated compounds (inhibit methanogenic bacteria—can have side effects such as reduced intake); novel plant compounds such as condensed tannins, saponins and essential oils (side effect—reduced digestibility); probiotics (yeast culture); propionate precursors (fumarate or malate—expensive); vaccines (against methanogenic—not yet commercially available) and bovine somatotropin and hormonal growth implants (Wolin et al., 1964; Benz and Johnson, 1982; Rumpler et al., 1986; Johnson et al., 1991; Bauman, 1992; Schmidely, 1993; Van Nevel and Demeyer, 1995, 1996; McCrabb, 2001; Newbold et al., 2002; Lila et al., 2003; Pinares-Patiño et al., 2003; McGinn et al., 2004; Wright et al., 2004; Newbold et al., 2005;

Hess et al., 2006; Kamra et al., 2006; Newbold and Rode, 2006; Patra et al., 2006).

Long-term management changes and animal breeding: Breeding for high-yielding varieties and better management practices for improved efficiency (producing meat—animals reach slaughter weight at a younger age) and decreasing replacement heifers reduces methane emission per unit of animal product (Lovett and O'Mara, 2002; Boadi et al., 2004; Kebreab et al., 2006; Lovett et al., 2006).

Manure management

CH_4 or N_2O emissions from stored manure can be reduced by cooling, use of solid covers, mechanically separating solids from slurry, composting (solidifying), anaerobical digestion to capture CH_4 for renewable energy source or by altering feeding practices (Amon et al., 2001; Clemens and Ahlgrimm, 2001; Gonzalez-Avalos and Ruiz-Suarez, 2001; Monteny et al., 2001, 2006; Külling et al., 2003; Paustian et al., 2004; Chadwick, 2005; Pattey et al., 2005; Amon et al., 2006; Clemens et al., 2006; Hindrichsen et al., 2006; Kreuzer and Hindrichsen, 2006; Xu et al., 2007). However, globally for most animals there is limited opportunity for manure management, treatment or storage as excretion happens in the field, and handling for fuel or fertility amendments occur when it is dry and methane emissions are negligible.

Bioenergy

Facing pollution threats from fossil fuels, forest/agricultural crops and residues are now being increasingly used as green fuel for a viable alternative (Rogner et al., 2000; Cerri et al., 2004; Edmonds, 2004; Hamelinck et al., 2004; Hoogwijk, 2004; Paustian et al., 2004; Richter, 2004; Sheehan et al., 2004; Dias de Oliveira et al., 2005; Eidman, 2005; Hoogwijk et al., 2005; Anonymous, 2006e; Faaij, 2006). Biofuels also release CO_2 but this CO_2 is of recent atmospheric origin, which displaces CO_2 released from fossil carbon. The net benefit to atmospheric CO_2, however, depends on energy used in growing and processing the bioenergy feedstock (Spatari et al., 2005).

Some mitigation measures operate predominantly on one GHG (e.g. dietary management of ruminants to reduce CH_4 emissions) while others have impacts on more than one GHG (e.g. rice management). Moreover, practices may benefit more than one gas while others involve a trade-off between gases (e.g. restoration of organic soils). Consequently, a practice that is highly effective in reducing emissions at one site may be less effective or even counterproductive elsewhere. This means that there is no universally applicable list of mitigation practices and the proposed practices will need to be evaluated for individual agricultural systems according to the specific climatic, edaphic, social settings, and historical land use and management (Smith et al., 2007a). The effectiveness of mitigation strategies also changes with time. Some practices such as those which elicit soil carbon gain have diminishing effectiveness after several decades while others can reduce energy use restricting emissions indefinitely. For instance, there is a strong time dependency of emissions from no-till agriculture, in part because of changing influence of tillage on N_2O emissions (Six et al., 2004). Many of the climate change effects have high levels of uncertainty but demonstrate that the practices chosen to reduce GHG emissions may not have the same effectiveness in coming decades. Consequently, programmes to reduce emissions in the agricultural sector will need to be designed with flexibility for adaptation in response to climate change (Smith et al., 2007a).

1.7 Co-benefits and trade-offs of mitigation options

The merits of an agricultural GHG emission mitigation practice cannot be judged solely on the effectiveness of GHG mitigation. Agro-ecosystems are inherently complex and very few practices yield purely beneficial outcomes, but instead involve some trade-offs above certain levels or intensities of implementation (DeFries et al., 2004; Viner et al., 2006). The co-benefits and trade-offs of a practice may vary from place to place because of differences in climate, soil or the way the practice is adopted (Smith et al., 2007a). Land use changes and agricultural management can either be beneficial or harmful to biodiversity; for instance, loss of biodiversity due to intensification of agriculture or large-scale production of biomass energy crops while perennial crops often used for energy production can favour biodiversity if they displace annual crops or degraded areas (Berndes and Börjesson, 2002; Anonymous, 2006e; Feng et al., 2006; Xiang et al., 2006). Agricultural mitigation practices may influence non-agricultural ecosystems. Increasing the productivity on existing croplands may 'spare' some forest or grasslands (West and Marland, 2003; Balmford et al., 2005; Mooney et al., 2005); however, the net effect of such trade-offs has not yet been fully quantified (Huston and Marland, 2003; Green et al., 2005).

Implementation of agricultural GHG mitigation measures may allow expanded use of fossil fuels and may have some negative effects through emissions of sulphur, mercury and other pollutants (Elbakidze and McCarl, 2007). In producing bioenergy, feedstock can either be crop residue or a densely rooted perennial crop which either will reduce soil quality by depleting soil organic matter or may improve soil quality by replenishing organic matter, respectively (Paustian et al., 2004). A key potential trade-off is between the production of bioenergy crops and food security. Food insecurity is determined more by inequity of access to food than by absolute food production insufficiencies, so the impact of this trade-off depends, among other things, on the economic distributional effects of bioenergy production (Smith et al., 2007a). Efficiently managed bioenergy plantations cannot only reduce nutrient leaching and soil erosion but also increase nutrient recirculation, stock soil carbon, improve soil fertility, remove cadmium or other heavy metals from soils or wastes, aid in the treatment of nutrient-rich wastewater and sludge and provide habitats for biodiversity in the agricultural landscape (Berndes and Börjesson, 2002; Berndes et al., 2004; Börjesson and Berndes, 2006).

The practices maintaining/increasing crop productivity can also ensure food security during the coming decades (Anonymous, 2003, 2005a; Rosegrant and Cline, 2003; Lal, 2004a,b; Follett et al., 2005; Sanchez and Swaminathan, 2005). Carbon conserving practices also sustain or enhance fertility, productivity and resilience of soil resources (Díaz-Zorita et al., 2002; Cerri et al., 2004; Freibauer et al., 2004; Kurkalova et al., 2004; Lal, 2004a; Paustian et al., 2004). Agro-ecosystems are primarily dependent on manufactured fertilisers (Galloway et al., 2003; Galloway, 2003). The practices that improve nitrogen-use efficiency reduces N_2O emission, thus it also reduces GHG emissions from fertiliser manufacture and avoids deleterious effects on water and air quality from N pollutants (Dalal et al., 2003; Paustian et al., 2004; Oenema et al., 2005; Olesen et al., 2006). However, where productivity is improved by increasing inputs, soil acidification or salinisation may occur (Barak et al., 1997; Connor, 2004; Díez et al., 2004). Fresh water is becoming scarce and agricultural practices for mitigation of GHGs can either have negative or positive effects on fresh water conservation and quality (Rockström, 2003; Rosegrant and Cline, 2003). Some practices could intensify water use by reducing stream flow or groundwater reserves (Unkovich, 2003; Dias de Oliveira et al., 2005) while some may affect water quality through enhanced leaching of pesticides and nutrients (Machado and Silva, 2001; Freibauer et al., 2004). Highly productive, evergreen, deep-rooted bioenergy plantations generally have a higher water use than the land cover they replace (Berndes, 2002; Jackson et al., 2005).

1.8 Adaptation

Adaptation could be possible at the farmer, economic agent and the macro levels with short- and long-term approaches for autonomous and policy-driven adaptations (Stern, 2007). The response time for adoption and technological response in Indian agriculture is 5–15 years for the productive life of farm assets, crop rotation cycles and recovery from major disasters (Jodha, 1989). Table 1.5 gives details of these adaptation practices.

Some of the broad categories of responses which could be beneficial regardless of how or whether climate changes as identified by Ninan and Bedamatta (2012) are given below:

- Improved training and general education of populations dependent on agriculture
- Identification of the present vulnerabilities of agricultural systems
- Agricultural research to develop new crop varieties
- Food programmes and other social security programmes to provide insurance against supply changes
- Transportation, distribution and market integration to provide the infrastructure to supply food during crop shortfalls
- Removals of subsidies, which can, by limiting changes in prices, mask the climate change signal in the marketplace

Table 1.5 Adaptation in practice

Climate change	Autonomous adaptation	Policy-driven adaptation
Short-run	Making short-run adjustment—changing crop planting dates Spreading the loss—pooling risk through insurance	Understanding climate risks—researching risks and vulnerability assessment Improving emergency response—early warning systems
Long-run	Investment in climate resilience if future effects relatively well understood and benefits easy to capture fully localized irrigation on farm	Investing to create or modify major infrastructure- reservoir storage, increased drainage capacity, higher sea walls Avoiding impacts—land use planning to restrict development in flood plains or in areas of increasing aridity

Gradually, the farmers and governments may adopt technologies and production techniques that take the climatic changes into account and thus reduce its impact. From the perspective of a developing country like India, a sustainable development agenda will be the prudent way to address the concerns over climate change (Sathaye et al., 2006). Finally, it is important to consider how the agricultural sector in developing countries may have directly responded to climatic changes. The adaptation in African countries is minimal relative to Asian countries due to poor economic policies which have undermined any incentives to appropriate adaptation to climatic change in the agricultural sector (Anonymous, 2001d). The response to losses in agricultural production to climatic changes that could dampen their effects may, at least in the short run, be an adjustment of prices. However, in these countries, since agricultural products are also for export and these countries tend to be price makers on the world commodities market, a loss in production is unlikely to have any effect on prices for most agricultural products for most countries (Reilly et al., 1994; Deaton, 1999).

Agriculture is not well prepared to cope with climate change especially in Southern Africa and Asia (Lobell et al., 2008). This means that our food systems must focus on building resilience as well as the ability to adapt to a warming climate. As these attributes become more appreciated, they also will lead to greater innovation in agriculture and food sectors (Niggli et al., 2009). Intensive agriculture has neglected traditional skills and knowledge. Organic agriculture always has been based on practical farming skills, observation, personal experience and intuition without reliance on modern inputs, which needs to be adopted in today's climate-changed scenario for manipulating complex agro-ecosystems, breeding locally adapted seeds and livestock, producing on-farm fertilisers and inexpensive nature-derived pesticides (Tengö and Belfrage, 2004).

1.9 Agricultural GHG mitigation potential

Farming practices that conserve and improve soil fertility are important for the future of agriculture and food production. Organic agriculture systems are built on a foundation of conserving and improving diversity by using diverse crops, rotations and mixed farm strategies. The diversity of landscapes, farming activities, fields and agro-biodiversity is greatly enhanced in organic agriculture (Niggli et al., 2008), which makes these farms more resilient to unpredictable weather patterns that result from climate change (Bengtsson et al., 2005; Hole et al., 2005). Enhanced biodiversity reduces pest outbreaks (Wyss et al., 1995; Pfiffner et al., 2003; Pfiffner and Luka, 2003; Zehnder et al., 2007). Similarly, diversified agro-ecosystems reduce the severity of plant and animal diseases while improving utilisation of soil nutrients and water (Altieri et al., 2005). Better soil structure, friability, aeration and drainage, lower bulk density, higher organic matter content, soil respiration (related to soil microbial activity), more earthworms and a deeper topsoil layer are all associated with the lower irrigation need (Proctor and Cole, 2002). Under conditions in which water is limited during the growing period, yields of organic farms are equal or significantly higher than those of conventional agriculture common in developing countries (Badgley et al., 2007). Water capture in organic plots was twice as high as in conventional plots during torrential rains significantly reducing the risk of floods (Lotter et al., 2003).

In Switzerland and the United States, organic matter, water percolation through top layer and soil structure stability were higher in organically managed soils than in conventional soils (Mäder et al., 2002; Marriott and Wander, 2006), making organic fields less prone to soil erosion (Reganold et al., 1987; Siegrist et al., 1998) and resulted significantly in higher yields of corn and soya bean in dry years (Lotter et al., 2003; Pimentel et al., 2005). In Tigray Province, one of the most degraded parts of Ethiopia, agricultural productivity was doubled by soil fertility techniques such as compost application and introduction of leguminous plants into the crop sequence instead of using purchased mineral fertilisers (Edwards, 2007). These reports recommend the practice of organic farming to improve soil fertility through green manuring, leguminous intercropping, composting and recycling of livestock manure for reducing GHGs, while also increasing global food productivity.

Eventually, a complete conversion to organic agriculture could decrease global yields by 30–40% in intensively farmed regions under the best geo-climate conditions (Niggli et al., 2009). In the context of subsistence agriculture and in regions with periodic disruptions of water supply brought on by droughts or floods, organic agriculture is competitive to conventional agriculture and often superior with respect to yields (Halberg et al., 2006; Badgley et al., 2007; Sanders, 2007; Anonymous, 2008c). Organic agriculture has a huge potential for climate change mitigation strategies in agricultural production (Pimentel et al., 1995; Niggli et al., 2008, 2009):

- It reduces wind, water and overgrazing erosion of 10 million ha annually, essential for ensuring future food security.
- It rehabilitates poor soils, restores organic matter content and brings such soils back into productivity.
- It is inherently based on lower livestock densities and can compensate for lower yields by a more effective vegetable production. Organic agriculture has a land use ratio of 1:7 for vegetable and animal production.
- The potential productivity of organic farms and organically managed landscapes can be improved considerably by scientific agro-ecological research.
- It conserves agricultural biodiversity, reduces environmental degradation impacts and integrates farmers into high-value food chains.

Numerous attempts particularly on soil carbon sequestration have been made to assess the technical potential for GHG mitigation in agriculture (Anonymous, 1996; Boehm et al., 2004; Caldeira et al., 2004; Ogle et al., 2004, 2005; Smith et al., 2007b,c). Mitigation potentials for CO_2 represent the net change in soil carbon pools reflecting the accumulated difference between carbon inputs to the soil after CO_2 uptake by plants and the release of CO_2 by decomposition in soil. Mitigation potentials for N_2O and CH_4 depend solely on emission reductions. As mitigation practices can affect more than one GHG; it is important to consider the impact of mitigation options on all GHGs (Robertson et al., 2000; Smith et al., 2001; Gregorich et al., 2005).

It was estimated that 400–800 MtC year^{-1} (equivalent to about 1400–2900 MtCO$_2$-eq. year^{-1}) could be sequestered in global agricultural soils. In addition, 300–1300 MtC (equivalent to about 1100–4800 MtCO$_2$-eq. year^{-1}) from fossil fuels could be offset by using 10–15% of agricultural land to grow energy crops in which crop residues will contribute 100–200 MtC (equivalent to about 400–700 MtCO$_2$-eq. year^{-1}) to fossil fuel offsets if recovered and burned. CH_4 emissions from agriculture would be reduced by 15–56% through improved nutrition of ruminants and better management of paddy. Improved management would reduce N_2O emissions by 9–26%. The global 2030 technical potential for mitigation options in agriculture considering no economic and other barriers for all gases was estimated to be 4500–6000 MtCO$_2$-eq. year^{-1} or 89% from soil carbon sequestration, 9% from mitigation of methane and 2% from mitigation of soil N_2O emissions (Caldeira et al., 2004; Smith et al., 2007b).

The contribution of agriculture to global GHG emissions ranges from 5.1 to 6.1 Gt CO_2-eq. The global potential of arable and permanent cropping systems to sequester is 200 kg C ha^{-1} year^{-1} and pasture systems is 100 kg ha^{-1} year^{-1}; the world's carbon sequestration will amount to 2.4 Gt CO_2-eq. year^{-1} (Lal, 2004a; Niggli et al., 2009). This minimum scenario for a conversion to organic farming would mitigate 40% of the world's agriculture GHG emissions (Niggli et al., 2009). The sequestration rate on arable land adopting organic farming with reduced tillage techniques will be 500 kg C ha^{-1} year^{-1}, which will contribute 65% mitigation of the agricultural GHG and, thus, total global organic mitigation would be 4 Gt CO_2-eq. year^{-1}. This indicates that application of sustainable management techniques to build up soil organic matter have the potential to balance a large part of the agricultural emissions although their effect over time may be reduced as soils are built up (Foereid and Høgh-Jensen, 2004). By a conversion to organic farming, another approximately 20% of the agricultural GHG could be reduced by abandoning industrially produced nitrogen fertilisers as is practiced by organic farms. This encouraging figure strongly supports the reality of low GHG agriculture and the possibility of climate neutral farming.

1.10 Agricultural GHG mitigation economic potential

Estimates of agricultural mitigation potential at various assumed carbon prices for N_2O and CH_4 (not for soil carbon sequestration) were worked out (Anonymous, 2001b, 2006a,b; McCarl and Schneider, 2001; Manne and Richels, 2004; DeAngelo et al., 2006; Rose et al., 2007; Smith et al., 2007b). It was estimated that the economic mitigation potential for soil carbon sequestration is 27 US$/t$CO_2$-eq. (Manne and Richels, 2004). The 2030 global economic mitigation potential of 1500–4300 MtCO_2-eq. year^{-1} is at carbon prices of 20–100 US$/t$CO_2$-eq. (Smith et al., 2007b).

1.11 Barriers and opportunities/implementation issues

Changes in climate may add stress to local and regional agricultural economies already dealing with long-term economic changes in agriculture. In addition, there may be barriers to adaptation that limit responses such as the availability of and access to financial resources and technical assistance, as well as the availability of other inputs, such as water and fertiliser. Uncertainty about the timing and rate of climate change also limits adaptation and, if expectations are incorrect, could contribute to the costs associated with transition and disequilibrium. The barriers to adoption of carbon sequestration activities on agricultural lands following Smith et al. (2007a) are discussed below.

Maximum storage

Carbon sequestration in soils or terrestrial biomass is a rapid and cheap available option that needs 15–60 years to reach a maximum capacity for the ecosystem, depending on management practice, management history and the system (West and Post, 2002; Caldeira et al., 2004; Sands and McCarl, 2005).

Reversibility

Mostly, agricultural mitigation options are reversible and a change in management can reverse the gains in carbon sequestration. Reduction in N_2O and CH_4 emissions, avoiding emissions as a result of agricultural energy efficiency gains and the substitution of fossil fuels by bioenergy are non-reversible (Smith et al., 2007a).

Baseline

The GHG net emission reduction is assessed relative to a baseline, but selecting an appropriate baseline is a problem (Smith et al., 2007a).

Uncertainty

Complex biological and ecological processes involved in GHG emissions and carbon storage is complex and less understood (mechanism uncertainty). This makes investors shy away from the agricultural mitigation options. Moreover, agricultural systems exhibit substantial variability between seasons and locations, creating high variability in offset quantities at the farm level (measurement uncertainty), which can be reduced by increasing the geographical extent and duration of the accounting unit (Kim and McCarl, 2005).

Displacement of emissions

Adopting certain agricultural mitigation practices may reduce production within implementing regions. However, this benefit may be offset by increased production outside the project region unconstrained by GHG mitigation objectives reducing the net emission. 'Wall-to-wall' accounting can detect this and crediting correction factors may need to be employed (Murray et al., 2004; Anonymous, 2005b).

Transaction costs

Under an incentive-based system such as a carbon market, the amount of money farmers receive is not the market price but the market price, less brokerage cost. This may be substantial and is an increasing fraction as the amount of carbon involved diminishes, creating a serious entry barrier for smallholders. In developing countries, this could involve many thousands of farmers (Smith et al., 2007a).

Measurement and monitoring costs

Such costs can be either minimal (Mooney et al., 2004) or large (Smith, 2004c). In general, measurement costs per carbon-credit sold decrease as the quantity of carbon sequestered and area sampled increase. Methodological advances in measuring soil carbon may reduce costs and increase the sensitivity of change detection. However, improved methods to account for changes in soil bulk density remain a hindrance to quantification of changes in soil carbon stocks (Izaurralde and Rice, 2006). With the development of remote sensing, new spectral techniques to measure soil carbon and modelling offer opportunities to reduce costs, but will require evaluation (Ogle and Paustian, 2005; Brown et al., 2006; Izaurralde and Rice, 2006; Gehl and Rice, 2007).

Property rights

Property rights, landholdings and the lack of a clear single-party land ownership in certain areas may inhibit implementation of management changes (Smith et al., 2007a).

Other barriers

The other possible barriers to implementation include the availability of capital, the rate of capital stock turnover, the rate of technological development, risk attitudes, need for research and outreach, consistency with traditional practices, pressure for competing uses of agricultural land and water, demand for agricultural products, high costs for certain enabling technologies and ease of compliance (e.g. straw burning is quicker than residue removal and can also control some weeds and diseases, so farmers favour straw burning) (Smith et al., 2007a).

Considering the growing concern of elevated atmospheric GHGs, the complex economics and availability of fossil fuels and the deterioration of the environment and health conditions with a shift away from intense reliance on heavy chemical inputs to an intense biologically-based agriculture and food system is possible today (Niggli et al., 2009). Sustainable and organic agriculture offers multiple opportunities to reduce GHGs and counteract global warming. Organic agriculture reduces energy requirements for production systems by 25–50% compared to conventional chemical-based agriculture. Reducing GHGs through their sequestration in soil has even greater potential to mitigate climate change. Soil improvement is essential for agriculture in developing countries where crop inputs (chemical fertilisers and pesticides) are costly and unavailable. Further, this requires special equipment and knowhow for their proper application which is not widespread.

Productive and ecologically sustainable agriculture is crucial to reduce trade-offs among food security, climate change and ecosystem degradation. Organic agriculture therefore represents a multi-targeted and multi-functional strategy. It offers a proven alternative concept that is being implemented quite successfully by a growing number of farms and food chains. Currently, 1.2 million farmers practice organic agriculture on 32.2 million ha of land (Willer and Kilcher, 2009). Many components of organic agriculture can be implemented within other sustainable farming systems. The system-oriented and participative concept of organic agriculture combined with new sustainable technologies (such as no tillage) offer greatly needed solutions in the face of climate change (Niggli et al., 2009).

1.12 Strategy recommendations

International Food Policy Research Institute 2009 suggested the below-given policy and mitigation programme recommendations (Nelson et al., 2009).

Efficient developmental policies and programmes

Given the current uncertainty about location-specific effects of climate change, good development policies and programmes are the best climate-change adaptation investments. A pro-growth, pro-poor development agenda that supports agricultural sustainability also contributes to food security and climate-change adaptation in the developing world.

Higher investments

Climate change places new and more challenging demands on agricultural productivity, requiring investments for enhancing research on rural and irrigation infrastructure and technology dissemination. The International Food Policy Research Institute recommends at least $7 billion per year additional fund support to finance the research, rural infrastructure and irrigation investments to offset the negative effects of climate change on human well-being.

Reinvigorate research and extension programmes

Partnerships with other national systems and international centres along with investment in laboratory scientists and infrastructure are needed. Strong extension linkages among the stakeholders is essential for transferring technology, facilitating interaction, building capacity among farmers and encouraging farmers to form their own networks.

Efficient data collection, dissemination and analysis

Strengthened global efforts are required to collect and disseminate spatial data on agriculture through remote sensing. Statistical programmes should be increased and encouraged through funding.

Agricultural adaptations in international climate negotiation process

International climate negotiations are forums for governments and civil society organisations to negotiate proposals and put forth practical actions on adaptation in agriculture.

Climate change adaptation and food security are interrelated

Agricultural adaptation practices to manage climate change will also enhance food security, especially through access of resources to the poor, which in turn will help them adapt to climate change.

Encourage and support community-based adaptation strategies

International development agencies and national governments should encourage and support community participation in adaptation planning and execution through technical, financial and capacity building of local communities. This will help rural communities strengthen their capacity to cope with disasters, improve their land management skills and diversify their livelihoods.

1.13 Conclusion

Climate change warming or cooling, if they are to occur as projected are likely to influence plant production and management under both well-watered and drought conditions in developed and developing countries of the world. It seems likely that a higher fertility level, a higher cost of irrigation, soil conservation and increased pesticide input will be required to sustain a higher rate of crop growth at elevated CO_2 concentrations. Productivity can likely be maintained with an increased cost of management by the developed regions of the world. But in the developing regions where present levels of productivity are low due to these reasons, future maintenance of productivity is unlikely. This is because of increasing poverty due to population increase and increasing political and economic instability.

Moreover, farmers in these regions cannot afford to apply these management inputs due to their increasing costs (Chakravarty and Mallick, 2003). The rich can adjust by adapting to suitable management practices but the poor cannot because they cannot afford to. The Earth will suffer irrespective of being rich or poor and the overall impact on humanity will be drastic because more than three-fourths of the world's population live in poorer nations.

Not only the agricultural system but also all the biotic systems will fail to adapt to change in the environment, inviting a total catastrophe in the future. At the same time, we also strongly believe that we can sustain our biosphere by population control, judicious natural resource management, proper land use and efficient waste management; that is by overall sustainable development of human, animal and plant resources and through equitable socio-economic development. We should better understand the concept of 'global village' to avert such a global problem as each and every ecosystem and economy of this globe is inter-connected and inter-dependent (Chakravarty and Mallick, 2003). Climate is significant in the distribution, production and security of food. There should be a realisation that climate is both a resource to be managed wisely and a hazard to be dealt with (Wittwer, 1995). Thus, a portfolio of assets to prepare for climate change is needed. The assets are land, water, energy, physical infrastructure, genetic diversity, research capacity, information systems, human resources, political institutions and the world market. The value of each asset for adapting to climate change and policy steps to increase their flexibility is necessary (Waggoner, 1992).

References

ABDUL HARIS, A. V.; BISWAS, S.; CHHABRA, V.; ELANCHEZHIAN, R. and BHATT, B. P. 2013. Impact of climate change on wheat and winter maize over a sub-humid climatic environment. *Current Science* **104:** 206–214.

ABOU-HADID, A. F.; MOUGOU, R.; MOKSSIT, A. and IGLESIAS, A. 2003. *Assessment of Impacts, Adaptation and Vulnerability to Climate Change in North Africa: Food Production and Water Resources*. AIACC AF90 Semi-Annual Progress Report.

ACHANTA, A. N. 1993. An assessment of the potential impact of global warming on Indian rice production. In: *The Climate Change Agenda: An Indian Perspective*, ed. Achanta, A. N. TERI, New Delhi.

ACOCK, B. 1990. Effects of carbon dioxide on photosynthesis, plant growth and other processes. In: *Impact of Carbon Dioxide, Trace Gases and Climate Change on Global Agriculture*, eds. Kimball, B. A.; Rosenberg, N. I. and Alien, L. H. Jr. Special Publication no. 53. ASA, Madison, WI, USA. pp. 45–60.

ACOCK, B. and ALIEN, L. H. JR. 1986. Crop responses to elevated carbon dioxide concentrations. In: *Direct Effects of Increasing Carbon Dioxide on Vegetation*, eds. Strain, B. R. and Cure, J. D. National Technical Information Service. US Department of Commerce, Springfield VA.

ADAMS, R. M.; HURD, B. H.; LENHART, S. and LEARY, N. 1998. Effects of global climate change on agriculture: An interpretative review. *Climate Research* **11**: 19–30.

AGGARWAL, P. K. and KALRA, N. 1994. *Simulating the Effect of Climatic Factors, Genotype and Management on Productivity of Wheat in India*. Indian Agricultural Research Institute Publication, New Delhi. 156p.

AGGARWAL, P. K. and MALL, R. K. 2002. Climate change and rice yields in diverse agro-environments of India. II. Effect of uncertainties in scenarios and crop models on impact assessment. *Climatic Change* **52**: 331–343.

AGGARWAL, P. K. and SINHA, S. K. 1993. Effect of probable increase in carbon dioxide and temperature on productivity of wheat in India. *Journal of Agricultural Meteorology* **48**: 811–814.

AINSWORTH, E. A. and LONG, S. P. 2005. What have we learned from 15 years of free-air CO_2 enrichment (FACE)? A meta-analytic review of the responses of photosynthesis, canopy. *New Phytology* **165**: 351–371.

AKIYAMA, H.; YAGI, K. and YAN, X. 2005. Direct N_2O emissions from rice paddy fields: Summary of available data. *Global Biogeochemical Cycles* **19**, GB1005, doi: 10.1029/2004GB002378.

ALBRECHT, A. and KANDJI, S. T. 2003. Carbon sequestration in tropical agroforestry systems. *Agriculture, Ecosystems and Environment* **99**: 15–27.

ALCOCK, D. and HEGARTY, R. S. 2006. Effects of pasture improvement on productivity, gross margin and methane emissions of a grazing sheep enterprise. In: *Greenhouse Gases and Animal Agriculture: An Update*, eds. Soliva, C. R.; Takahashi, J. and Kreuzer, M. International Congress Series No. 1293. Elsevier, the Netherlands. pp. 103–106.

ALEXANDROV, V.; EITZINGER, J.; CAJIC, V. and OBERFORSTER, M. 2002. Potential impact of climate change on selected agricultural crops in north-eastern Austria. *Global Change Biology* **8**: 372–389.

ALIEN, L. H. 1979. Potentials for carbon dioxide enrichment. In: *Modification of the Aerial Environment of Crops*, eds. Barfield, B. J. and Gerber, J. F. American Society of Agricultural Engineering, St. Joseph Mich. pp. 500–519.

ALIEN, L. H. JR.; BOOTE, K. J.; JONES, J. W.; JONES, P. H.; VALLE, R. R.; ACOCK, B.; ROGERS, H. H. and

DAHLMAN, R. C. 1987. *Responses of Vegetation to Carbon dioxide*. US Department of Agriculture and US Department of Energy, Washington DC.

ALTIERI, M. A.; PONTI, L. and NICHOLLS, C. 2005. Enhanced pest management through soil health: Toward a belowground habitat management strategy. *Biodynamics* (summer): 33–40.

ALVAREZ, R. 2005. A review of nitrogen fertilizer and conservative tillage effects on soil organic storage. *Soil Use and Management*, **21:** 38–52.

AMON, B.; AMON, T.; BOXBERGER, J. and WAGNER-ALT, C. 2001. Emissions of NH_3, N_2O and CH_4 from dairy cows housed in a farmyard manure tying stall (housing, manure storage, manure spreading). *Nutrient Cycling in Agro-Ecosystems* **60:** 103–113.

AMON, B.; KRYVORUCHKO, V.; AMON, T. and ZECHMEISTER-BOLTENSTERN, S. 2006. Methane, nitrous oxide and ammonia emissions during storage and after application of dairy cattle slurry and influence of slurry treatment. *Agriculture, Ecosystems & Environment* **112:** 153–162.

ANDERSON, T. L.; CHARLSON, R. J.; SCHWARTZ, S. E.; KNUTTI, R.; BOUCHER, O.; RODHE, H. and HEINTZENBERG, J. 2003. Climate forcing by aerosols—A hazy picture. *Science* **300:** 1103–1104.

ANDREAE, M. O. 2001. The dark side of aerosols. *Nature* **409:** 671–672.

ANDREAE, M. O. and MERLET, P. 2001. Emission to trace gases and aerosols from biomass burning. *Global Biogeochemical Cycles* **15:** 955–966.

ANDREAE, M. O.; JONES, C. D. and COX, P. M. 2005. Strong present-day aerosol cooling implies a hot future. *Nature* **435:** 1187.

ANONYMOUS. 1989. *The Potential Effects of Global Climate Change on the United States, Vol. 1, Regional Studies*. Draft Report to Congress, US Environmental Protection Agency. 91p.

ANONYMOUS. 1992. *Preparing U.S. Agriculture for Global Climate Change*. Task Force Report No. 119. Council for Agricultural Science and Technology, Ames, IA.

ANONYMOUS. 1996. *Climate Change 1995: The Science of Climate Change. Contribution of Working Group I to the Second Assessment Report of the Intergovernmental Panel on Climate Change (IPCC)*. Cambridge University Press, Cambridge.

ANONYMOUS. 2001a. *Climate Change 2001: The Scientific Basis. Contribution of Working Group I to the Third Assessment Report of the Intergovernmental Panel on Climate Change*. IPCC, Geneva.

ANONYMOUS. 2001b. *Climate Change 2001: Mitigation: Contribution of Working Group III to the Third Assessment Report of the Intergovernmental Panel on Climate Change*. IPCC, Geneva.

ANONYMOUS. 2001c. *Soil Carbon Sequestration for Improved Land Management*. World Soil Resources Reports No. 96. FAO, Rome 58p.

ANONYMOUS. 2001d. *Climate Change 2001: Impacts, Adaptation and Vulnerability. Contribution of Working Group II to the Third Assessment Report of the Intergovernmental Panel on Climate Change*, Cambridge: Cambridge University Press. http://www.ipcc.ch/ipccreports/tar/wg2/index.htm

ANONYMOUS. 2001e. *Climate Change 2001: Impacts, Adaptation, and Vulnerability.* Cambridge University Press, Cambridge.

ANONYMOUS, 2003. *World Agriculture: Towards 2015/2030.* An FAO Perspective. FAO, Rome. 97p.

ANONYMOUS. 2005a. *Ecosystems and Human Well-Being: Current State and Trends. Findings of the Condition and Trends Working Group.* Millennium Ecosystem Assessment Series, Island Press, Washington DC. 815p.

ANONYMOUS. 2005b. *Greenhouse Gas Mitigation Potential in U.S. Forestry and Agriculture.* United States Environmental Protection Agency, Washington DC. http://epa.gov/sequestration/pdf/ghg_part2.pdf

ANONYMOUS. 2006a. *Global Anthropogenic Non-CO$_2$ Greenhouse Gas Emissions: 1990–2020.* United States Environmental Protection Agency, 430-R-06–003. Washington DC. http://www.epa.gov/nonco2/econ-inv/downloads/GlobalAnthro EmissionsReport.pdf

ANONYMOUS. 2006b. *Global Mitigation of Non-CO$_2$ Greenhouse Gases.* United States Environmental Protection Agency, 430-R-06–005. Washington DC. http://www.epa.gov/nonco2/econ-inv/downloads/GlobalMitigationFullReport.pdf.

ANONYMOUS. 2006c. *FAOSTAT Agricultural Data.* http://faostat.fao.org

ANONYMOUS. 2006d. *Livestock's Long Shadow: Environmental Issues and Options.* FAO, Rome, Italy.

ANONYMOUS. 2006e. *How Much Biomasses can Europe Use Without Harming the Environment?* European Environment Agency Briefing 2/2006. http://reports.eea.europa.eu/briefing_2005_2/en

ANONYMOUS. 2007a. *Climate Change 2007: Climate Impacts, Adaptation and Vulnerability* Working Group II to the Intergovernmental Panel on Climate Change Fourth Assessment Report, DRAFT technical summary 2006. Intergovernmental Panel on Climate Change, Geneva.

ANONYMOUS. 2007b. *Climate Change 2007: The Physical Science Basis: Summary for Policymaker's Working Group I to the Intergovernmental Panel on Climate Change Fourth Assessment Report.* Intergovernmental Panel on Climate Change, Geneva.

ANONYMOUS. 2007c. *Intergovernmental Panel on Climate Change Special Report on Emissions Scenarios.* IPCC, Geneva. http://www.grida.no/climate/ipcc/emission/076.

ANONYMOUS. 2008a. *Climate Change. Ministry of Environment and Forests.* Government of India, New Delhi. http://envfor.nic.in/cc/what.htm.

ANONYMOUS. 2008b. *Agriculture for Development.* World Development Report 2008. The World Bank, Washington DC.

ANONYMOUS, 2008c. *Organic Agriculture and Food Security in Africa*. UNEP-UNCTAD Capacity-building Task Force on Trade, Environment and Development. http://www.unep-unctad. org/cbtf/publications/UNCTAD_DITC_TED_2007_15.pdf.

ANONYMOUS. 2009a. *South Asia: Shared Views on Development and Climate Change*. The World Bank, Washington DC.

ANONYMOUS. 2009b. *Fertilizers, Climate Change and Enhancing Agricultural Productivity Sustainably*. International Fertilizer Industry Association, Paris.

ARNELL, N. W.; LIVERMORE M. J. L.; KOVATS, S.; LEVY, P. E.; NICHOLLS, R.; PARRY, M. L. and GAFFIN, S. R. 2004. Climate and socio-economic scenarios for global-scale climate change impacts assessments: Characterizing the SRES storylines. *Global Environmental Change—Human Policy Dimensions* **14**: 3–20.

ARTECA, R. N.; POOVAIAH, B. W. and SMITH, O. E. 1979. Changes in carbon dioxide fixation, tuberization and growth induced by CO_2 applications to the root zones of potato plants. *Science* **205**: 1279–1280.

AUDSLEY, E.; PEARN, K. R.; SIMOTA, C.; COJOCARU, G.; KOUTSIDOU, E.; ROUNSEVELL, M. D. A.; TRNKA, M. and ALEXANDROV, V. 2006. What can scenario modelling tell us about future European scale agricultural land-use and what not? *Environmental Science Policy* **9**: 148–162.

AUFFHAMMER, M.; RAMANATHAN, V. and VINCENT, J. R. 2006. Integrated model shows that atmospheric brown cloud and GHGs have reduced rice harvests in India. *PNAS* **103**: 52.

AULAKH, M. S.; WASSMANN, R.; BUENO, C. and RENNENBERG, H. 2001. Impact of root exudates of different cultivars and plant development stages of rice (*Oryza sativa* L.) on methane production in a paddy soil. *Plant and Soil* **230**: 77–86.

BADGLEY, C.; MOGHTADER, J.; QUINTERO, E.; ZAKEM, E.; CHAPPELL, M. J.; AVILÉS-VÀZQUEZ, K.; SAMULON, A. and PERFECTO, I. 2007. Organic agriculture and the global food supply. *Renewable Agriculture and Food Systems* **22**: 86–108.

BAKKEN, L.; REFSGAARD, K.; CHRISFENSEN, S.; VATN, A.; MANNETJEL, T. and FRAME, J. 1994. Energy use and emission of greenhouse gases from grassland and agriculture. In: *Grassland and Society*. Proceedings of the 15th General Meeting of the European Grassland Federation, eds. Bakken, L.; Refsgaard, K.; Christensen, S.; Vatn, A. and Mannetjel, T. Agricultural University Norway, 1432 Aas, Norway. pp. 361–376.

BALMFORD, A.; GREEN, R. E. and SCHARLEMANN, J. P. W. 2005. Sparing land for nature: Exploring the potential impact of changes in agricultural yield on the area needed for crop production. *Global Change Biology* **11**: 1594–1605.

BARAK, P.; JOBE, B. O.; KRUEGER, A. R.; PETERSON, L. A. and LAIRD, D. A. 1997. Effects of long-term soil acidification due to nitrogen fertilizer inputs in Wisconsin. *Plant and Soil* **197**: 61–69.

BARNETT T. P.; ADAM J. C. and LETTENMAIER, D. P. 2005. Potential impacts of a warming climate on water availability in snow-dominated regions. *Nature* **438:** 303–309.

BARTHÈS, B.; AZONTONDE, A.; BLANCHART, E.; GIRARDIN, C.; VILLENAVE, C.; LESAINT, S.; OLIVER, R. and FELLER, C. 2004. Effect of a legume cover crop (*Mucuna pruriens* var. *utilis*) on soil carbon in an Ultisol under maize cultivation in southern Benin. *Soil Use and Management* **20:** 231–239.

BATES, B. C.; KUNDZEWICZ Z. W.; WU, S. and PALUTIKOF, J. P. 2008. *Climate Change and Water.* Technical Paper of the Intergovernmental Panel on Climate Change. IPCC Secretariat Geneva, Switzerland.

BATJES, N. H. 1999. *Management Options for Reducing CO₂ Concentrations in the Atmosphere by Increasing Carbon Sequestration in the Soil.* Dutch National Research Programme on Global Air Pollution and Climate Change, Project executed by the International Soil Reference and Information Centre, Wageningen, the Netherlands. 114p.

BATTISTI, D. S. and NAYLOR, R. L. 2009. Historical warnings of future food insecurity with unprecedented seasonal heat. *Science* **323:** 240–244.

BAUMAN, D. E. 1992. Bovine somatotropin: Review of an emerging animal technology. *Journal of Dairy Science* **75:** 3432–3451.

BEAUCHEMIN, K. and McGINN, S. 2005. Methane emissions from feedlot cattle fed barley or corn diets. *Journal of Animal Science* **83:** 653–661.

BENGTSSON, J.; AHNSTRÖM, J. and WEIBULL, A. C. 2005. The effects of organic agriculture on biodiversity and abundance: A meta-analysis. *Journal of Applied Ecology* **42:** 261–269.

BENZ, D. A. and JOHNSON, D. E. 1982. The effect of monensin on energy partitioning by forage fed steers. *Proceedings of the West Section of the American Society of Animal Science* **33:** 60.

BERAN, M. A. and ARNELL, N. W. 1989. *Effect of Climatic Change on Quantitative Aspects of United Kingdom Water Resource.* Department of the Environment and Water Directorate, Institute of Hydrology. Wallingford. 93p.

BERGTHORSSON, P.; BJORNSSON, H.; DYRMUNDSON, B.; HELGADOTTIR, A. and JONMUNDS, J. V. 1988. The effect of climatic variations on agriculture in Iceland. In: *The Impact of Climatic Variations on Agriculture. Vol. I: Assessment in Cool Temperature and Cold Regions*, eds. Parry, M. L.; Carter, T. R. and Konijn, N. T. Kluwer, Dordrecht, the Netherlands. pp. 383–509.

BERINGER, J.; HUTLEY, L. B.; TAPPER, N. J.; COUTTS, A.; KERLEY, A. and O'GRADY, A. P. 2003. Fire impacts on surface heat, moisture and carbon fluxes from a tropical savanna in northern Australia. *International Journal of Wildland Fire* **12:** 333–340.

BERNDES, G. 2002. Bioenergy and water: The implications of large-scale bioenergy production for water use and supply. *Global Environmental Change* **12:** 253–271.

BERNDES, G. and BÖRJESSON, P. 2002. *Multi-Functional Biomass Production Systems.* http://www.elkraft.ntnu.no/eno/konf_pub/ISES2003/full_paper/6%20MISCELLANEOUS/O6%204.pdf

BERNDES, G.; FREDRIKSON, F. and BORJESSON, P. 2004. Cadmium accumulation and Salix-based phytoextraction on arable land in Sweden. *Agriculture, Ecosystems & Environment* **103:** 207–223.

BHASKARAN, B.; MITCHELL, J. F. B.; LAVERY, J. R. and LAL. M. 1995. Climatic response of Indian subcontinent to doubled CO_2 concentrations. *International Journal of Climatology* **15:** 873–892.

BHATTACHARYA, N. C.; BISWAS, P. K.; BHATTACHARYA, S.; SIONIT, N. and STRAIN, R. R. 1985. Growth and yield responses of sweet potato *Ipomea batatas* to atmospheric CO_2 enrichment. *Crop Science* **25:** 975–981.

BHATTACHARYA, N. C. and GEYER, R. A. 1993. *Prospects of Agriculture in a Carbon Dioxide-Enriched Environment.* A global warming forum: Scientific, economic and legal overview. pp. 487–505.

BLAXTER, K. L. and CLAPERTON, J. L. 1965. Prediction of the amount of methane produced by ruminants. *British Journal of Nutrition* **19:** 511–522.

BOADI, D.; BENCHAAR, C.; CHIQUETTE, J. and MASSÉ, D. 2004. Mitigation strategies to reduce enteric methane emissions from dairy cows: Update review. *Canadian Journal of Animal Science* **84:** 319–335.

BOEHM, M.; JUNKINS, B.; DESJARDINS, R.; KULSHRESHTHA, S. and LINDWALL, W. 2004. Sink potential of Canadian agricultural soils. *Climatic Change* **65:** 297–314.

BÖRJESSON, P. and BERNDES, G. 2006. The prospects for willow plantations for waste water treatment in Sweden. *Biomass & Bioenergy* **30:** 428–438.

BOUWMAN, A. 2001. *Global Estimates of Gaseous Emissions from Agricultural Land.* FAO, Rome. 106p.

BOYLE, C. A. and LAVKULICH, L. 1997. Carbon pool and dynamics in the lower framer basin from 1827 to 1990. *Environment and Management* **21:** 443–455.

BRKLACICH, M.; CURRAN, P.; BRUNT, D. and CARTER, T. R. 1996. The application of agricultural land rating and crop models to CO_2 and climate change issues in northern regions: The Mackenzie Basin case study. *Agriculture and Food Science Finlan*d **5:** 351–365.

BROADHUS, J.; MILLMAN, J.; EDWARDS, S. F.; AUBREY, D. G. and GABLE, F. 1986. Sea level rise and damming of rivers: Possible effects of Egypt and Bangladesh. In: *Effects of Changes in Stratospheric Ozone and Global Climate*, ed. Titus, J. Vol. 4. Environmental Protection Agency and UNEP, Washington DC. pp. 165–189.

BROWN, D. J.; SHEPHERD, K. D.; WALSH, M. G.; MAYS, M. D. and REINSCH, T. G. 2006. Global soil characterization with VNIR diffuses reflectance spectroscopy. *Geoderma* **132:** 273–290.

BRUCE, J. P.; FROME, M.; HAITES, E.; JANZEN, H.; LAL, R. and PAUSTIAN, K. 1999. Carbon sequestration in soils. *Journal of Soil and Water Conservation* **54:** 382–389.

CAI, Z. C.; TSURUTA, H.; GAO, M.; XU, H. and WEI, C. F. 2003. Options for mitigating methane emission from a permanently flooded rice field. *Global Change Biology* **9:** 37–45.

CAI, Z. C.; TSURUTA, H. and MINAMI, K. 2000. Methane emissions from rice fields in China: Measurements and influencing factors. *Journal of Geophysical Research* **105 D13:** 17231–17242.

CAI, Z. C. and XU, H. 2004. Options for mitigating CH_4 emissions from rice fields in China. In: *Material Circulation through Agro-Ecosystems in East Asia and Assessment of Its Environmental Impact*, ed. Hayashi, Y. NIAES Series 5. Tsukuba. pp. 45–55.

CALDEIRA, K.; MORGAN, M. G.; BALDOCCHI, D.; BREWER, P. G.; CHEN, C. T. A.; NABUURS, G. J.; NAKICENOVIC, N. and ROBERTSON, G. P. 2004. A portfolio of carbon management options. In: *The Global Carbon Cycle. Integrating Humans, Climate, and the Natural World*, eds. Field, C. B. and Raupach, M. R. SCOPE 62. Island Press, Washington DC. pp. 103–129.

CANNELL, M. G. R. 2003. Carbon sequestration and biomass energy offset: Theoretical, potential and achievable capacities globally, in Europe and the UK. *Biomass and Bioenergy* **24:** 97–116.

CANNON, R. J. C. 2003. The implications of predicted climate change for insect pests in the UK, with emphasis on non-indigenous species. *Global Change Biology* **4:** 785–790.

CASSMAN, K. G.; DOBERMANN, A.; WALTERS, D. T. and YANG, H. 2003. Meeting cereal demand while protecting natural resources and improving environmental quality. *Annual Review of Environment and Resources* **28:** 315–358.

CERRI, C. C.; BERNOUX, M.; CERRI, C. E. P. and FELLER, C. 2004. Carbon cycling and sequestration opportunities in South America: The case of Brazil. *Soil Use and Management*, **20:** 248–254.

CHADWICK, D. R. 2005. Emissions of ammonia, nitrous oxide and methane from cattle manure heaps: Effect of compaction and covering. *Atmospheric Environment* **39:** 787–799.

CHAKRAVARTY, S.; GHOSH, S. K.; SURESH, C. P.; DEY, A. N. and SHUKLA, G. 2012. Deforestation: Causes, effects and control strategies. In: *Global Perspectives on Sustainable Forest Management*, ed. Okia, C. A. Intech Publishers, Croatia. pp. 3–28.

CHAKRAVARTY, S. and MALLICK, K. 2003. Agriculture in a greenhouse world, what really will be? In: *Environmental Challenges of the 21st Century*, ed. Kumar, A. A. P. H. Publishing Corporation, New Delhi. pp. 633–652.

CHATTERJEE, A. 1998. *Simulating the Impact of Increase in Temperature and CO$_2$ on Growth and Yield of Maize and Sorghum.* M.Sc. Thesis. Indian Agricultural Research Institute, New Delhi. Unpublished.

CHEKE, R. A. and TRATALOS, J. A. 2007. Migration, patchiness and population processes illustrated by two migrant pests. *Bioscience* **57**: 145–154.

CIAIS, P.; REICHSTEIN, M.; VIOVY, N.; GRANIER, A.; OGEE, J.; ALLARD, V.; AUBINET, M. et al. 2005. Europe-wide reduction in primary productivity caused by the heat and drought in 2003. *Nature* **437**: 529–533.

CLARK, H.; PINARES, C. and DE KLEIN, C. 2005. Methane and nitrous oxide emissions from grazed grasslands. In: *Grassland. A Global Resource,* ed. McGilloway. D. Wageningen Academic Publishers, Wageningen, the Netherlands. pp. 279–293.

CLEMENS, J. and AHLGRIMM, H. J. 2001. Greenhouse gases from animal husbandry: Mitigation options. *Nutrient Cycling in Agroecosystems* **60**: 287–300.

CLEMENS, J.; TRIMBORN, M.; WEILAND, P. and AMON, B. 2006. Mitigation of greenhouse gas emissions by anaerobic digestion of cattle slurry. *Agriculture, Ecosystems and Environment* **112**: 171–177.

CLINE, W. R. 2007. *Global Warming and Agriculture: Impact Estimates by Country.* Center for Global Development and Peterson Institute for International Economics, Washington.

CLINE, W. R. 2008. Global warming and Agriculture. *Finance and Development.* **3**: 23–27.

COHEN, J. E. 2003. Human population: The next half century. *Science* **302**: 1172–1175.

COLE, C. V.; DUXBURY, J.; FRENEY, J.; HEINEMEYER, O.; MINAMI, K.; MOSIER, A.; PAUSTIAN, K. et al. 1997. Global estimates of potential mitigation of greenhouse gas emissions by agriculture. *Nutrient Cycling in Agroecosystems* **49**: pp. 221–228.

CONANT, R. T. and PAUSTIAN, K. 2002. Potential soil carbon sequestration in overgrazed grassland ecosystems. *Global Biogeochemical Cycles* **16**: 1143.

CONANT, R. T.; PAUSTIAN, K.; DEL GROSSO, S. J. and PARTON, W. J. 2005. Nitrogen pools and fluxes in grassland soils sequestering carbon. *Nutrient Cycling in Agroecosystems* **71**: 239–248.

CONANT, R. T.; PAUSTIAN, K. and ELLIOTT, E. T. 2001. Grassland management and conversion into grassland: Effects on soil carbon. *Ecological Applications* **11**: 343–355.

CONNOR, D. J. 2004. Designing cropping systems for efficient use of limited water in southern Australia. *European Journal of Agronomy* **21**: 419–431.

CRUTZEN, P. J. 1995. The role of methane in atmospheric chemistry and climate. In: *Ruminant Physiology: Digestion, Metabolism, Growth and Reproduction*, eds. Von Engelhardt, W.; Leonhard-

Marek, S.; Breves, G. and Giesecke. D. Ferdinand Enke Verlag, Stuttgart. pp. 291–316.

CURE, J. D. 1986. Carbon dioxide doubling responses: A crop survey. In: *Direct Effect of Increasing CO_2 on Vegetation*, eds. Strain, B. R. and Cure, J. D. National Technical Information Service. US Department of Commerce, Springfield VA.

CURRY, R. B.; PEART, R. M.; JONES, J. W.; BOOTE, K. J. and ALLEN JR., L. H. 1990. Response of crop yield to predicted changes in climate and atmospheric CO_2 using simulation. *Trans. ASAE* **33:** 981–990.

CURTIS, P. S. and WANG, X. 1998. A meta-analysis of elevated CO_2 effects on woody plant mass, form, and physiology. *Oecologia* **113:** 299–313.

DALAL, R. C.; WANG, W.; ROBERTSON, G. P. and PARTON, W. J. 2003. Nitrous oxide emission from Australian agricultural lands and mitigation options: A review. *Australian Journal of Soil Research* **41:** 165–195.

DAVIDSON, E. A.; NEPSTAD, D. C.; KLINK, C. and TRUMBORE, S. E. 1995. Pasture soils as carbon sink. *Nature* **376:** 472–473.

DE COSTA, W. A. J. M.; WEERAHOON, W. M. W.; HERATH, H. M. L. K. and ABEYWARDENA, M. I. 2003. Response of growth and yield of rice to elevated atmospheric carbon dioxide in the sub humid zone of Sri Lanka. *Journal of Agronomy and Crop Science* **189:** 83–95.

DEANGELO, B. J.; DE LA CHESNAYE, F. C.; BEACH, R. H.; SOMMER, A. and MURRAY, B. C. 2006. Methane and nitrous oxide mitigation in agriculture. Multi-Greenhouse Gas Mitigation and Climate Policy. *Energy Journal,* Special Issue **#3**. http://www.iaee.org/en/publications/journal.aspx.

DEATON, A. 1999. Commodity prices and growth in Africa. *Journal of Economics Perspectives.* **13:** 23–40.

DECKER, W.L.; JONES, V. and ACHUTUNI, R. 1985. The impact of CO_2 induced climate change on US agriculture. In: *Characteristics of Information Requirements for Studies of CO_2 Effects: Water Resources, Agriculture, Fisheries, Forests and Human Health,* ed. White, M. R. National Technical Information Service. US Department of Commerce, Springfield, VA. pp. 69–93.

DEFRIES, R. S.; FOLEY, J. A. and ASNER, G. P. 2004. Land-use choices: Balancing human needs and ecosystem function. *Frontiers in Ecology and Environment* **2:** 249–257.

DENMAN, K. L.; BRASSEUR, G.; CHIDTHAISONG, A.; CIAIS, P.; COX, P. M.; DICKINSON, R. E.; HAUGLUSTAINE, D. et al. 2007. Couplings between changes in the climate system and biogeochemistry. In: *Climate Change 2007: The Physical Science Basis. Contribution of Working Group I to the Fourth Assessment Report of the Intergovernmental Panel on Climate Change*, eds. Solomon, S.; Qin, D.; Manning, M.; Chen, Z.; Marquis, M.; Averyt, K. B.; Tignor, M. and Miller, H. L. Cambridge University Press, Cambridge.

DERNER, J. D.; BOUTTON, T. W. and BRISKE, D. D. 2006. Grazing and ecosystem carbon storage in the North American Great Plains. *Plant and Soil* **280:** 77–90.

DIAS DE OLIVEIRA, M. E.; VAUGHAN, B. E. and RYKIEL, JR., E. J. 2005. Ethanol as fuel: Energy, carbon dioxide balances and ecological footprint. *BioScience* **55:** 593–602.

DÍAZ-ZORITA, M.; DUARTE, G. A. and GROVE, J. H. 2002. A review of no till systems and soil management for sustainable crop production in the sub-humid and semi-arid Pampas of Argentina. *Soil and Tillage Research* **65:** 1–18.

DIEKOW, J.; MIELNICZUK, J.; KNICKER, H.; BAYER, C.; DICK, D. P. and KÖGEL-KNABNER, I. 2005. Soil C and N stocks as affected by cropping systems and nitrogen fertilization in a southern Brazil Acrisol managed under no-tillage for 17 years. *Soil and Tillage Research* **81:** 87–95.

DÍEZ, J. A.; HERNAIZ, P.; MUÑOZ, M. J.; DE LA TORRE, A.; VALLEJO, A. 2004. Impact of pig slurry on soil properties, water salinization, nitrate leaching and crop yield in a four-year experiment in Central Spain. *Soil Use and Management* **20:** 444–450.

DINAR, A.; MENDELSOHN, R.; EVENSON, R.; PARIKH, J.; SANGHI, A.; KUMAR, K.; McKINSEY, J. and LONERGAN. S. 1998. *Measuring the Impact of Climate Change on Indian Agriculture.* World Bank Technical Paper No. 402. Washington, DC.

DÖLL, P. 2002. Impact of climate change and variability on irrigation requirements: A global perspective. *Climate Change* **54:** 269–293.

DÖLL, P. and SIEBERT, S. 2002. *Global Modeling of Irrigation Water Requirements.* Water Resource. Research, 38.

DOWNING, T. E. (ED.) 1996. *Climate Change and World Food Security.* NATO ASI Series, Series 1. Global Environmental Change. Springer, Berlin. 662p.

DOWNTON, W. J. S.; BORKMAN, O. and PIKE C. 1981. Consequences of increased atmospheric concentration of carbon dioxide for growth and photosynthesis of higher plants. In: *Carbon Dioxide and Climate: Australian Research*, ed. Pearman, G. I. Australian Academy of Science, Canberra. pp. 143–153.

DRECHSEL, P.; GYIELE, L.; KUNZE, D. and COFIE, O. 2001. Population density, soil nutrient depletion, and economic growth in sub-Saharan Africa. *Ecological Economics* **38:** 251–258.

DUXBURY, J. M.; HARPER, L. A. and MOSIER, A. R. 1993. *Contributions of Agroecosystems to Global Climate Change. Agricultural Ecosystem Effects on Trace Gases and Global Climate Change.* ASA Special Publication no. 55. pp. 1–18.

EASTERLING, W. E. 1996. Adapting North American agriculture to climate change. *Agriculture and Forest Meteorology* **80:** 9–11.

EASTERLING, W. E.; CROSSON, P. R.; ROSENBERG, N. J. McKENNY, M. S.; KATZ, L. A. and LEMON, K. A. 1993. Paper 2. Agricultural impacts of and responses to climate change in the Missouri-Iowa-Nebraska-Kansas (MINK) region. *Climate Change* **24:** 23–61.

EASTERLING, W. E. III.; PARRY, M. L. and CROSSON, P. R. 1989. Adapting future agriculture to changes in climate. In: *Greenhouse Warming: Abatement and Adaptation*, ed. Rosenberg, N. J. Resources for the Future, Washington DC. pp. 91–104.

EDMONDS, J. A. 2004. Climate change and energy technologies. *Mitigation and Adaptation Strategies for Global Change* **9:** 391–416.

EDWARDS, S. 2007. The impact of compost use on crop yields in Tigray, Ethiopia. Institute for Sustainable Development (ISD). In: *Proceedings of the International Conference on Organic Agriculture and Food Security*. FAO, Rome. ftp://ftp.fao.org/ paia/organicag/ofs/02-Edwards.pdf

EIDMAN, V. R. 2005. Agriculture as a producer of energy. In: *Agriculture as a Producer and Consumer of Energy*, eds. Outlaw, J. L.; Collins, K. J. and Duffield, J. A. CABI Publishing, Cambridge. pp. 30–67.

ELBAKIDZE, L. and MCCARL, B. A. 2007. Sequestration offsets versus direct emission reductions: Consideration of environmental co-effects. *Ecological Economics* **60:** 564–571.

ENGVILD, K. C. 2003. A review of the risks of sudden global cooling and its effects on agriculture. *Agriculture and Forest Meteorology* **115:** 127–137.

EVANS, N.; BAIERL, A.; SEMENOV, M. A.; GLADDERS, P. and FITT, B. D. L. 2008. Range and severity of a plant disease increased by global warming. *Journal of Royal Society Interface* **5:** 525–531.

EWERT, F. 2004. Modeling plant responses to elevated CO_2: How important is leaf area index. *Annals of Botany* **93:** 619–627.

EWERT, F.; ROUNSEVELL, M. D. A.; REGINSTER, I.; METZGER, M. J. and LEEMANS, R. 2005. Future scenarios of European agricultural land use. I. Estimating changes in crop productivity. *Agricultural Ecosystem and Environment* **107:** 101–116.

FAAIJ, A. 2006. Modern biomass conversion technologies. *Mitigation and Adaptation Strategies for Global Change* **11:** 335–367.

FAJER, E. D.; BOWERS, M. D. and BAZZAZ, F. A. 1989. The effects of enriched CO_2 atmospheres on plant insect herbivore interactions. *Science* **243:** 1198–1200.

FALLOON, P.; SMITH, P. and POWLSON, D. S. 2004. Carbon sequestration in arable land: The case for field margins. *Soil Use and Management* **20:** 240–247.

FANG, C.; SMITH, P.; MONCRIEFF, J. B. and SMITH, J. U. 2005. Similar response of labile and resistant soil organic matter pools to changes in temperature. *Nature* **433:** 57–59.

FEARNSIDE, P. M. 1997. Greenhouse gases from deforestation in Brazilian Amazon: Net committed emissions. *Climate Change* **35:** 321–360.

FEDOROFF, N. V. and COHEN, J. E. 1999. Plants and population: Is there time? *Proceedings of the National Academy of Sciences, USA* **96:** 5903–5907.

FENG, W.; PAN, G. X.; QIANG, S.; LI, R. H. and WEI, J. G. 2006. Influence of long-term fertilization on soil seed bank diversity of a paddy soil under rice/rape rotation. *Biodiversity Science* **14**: 461–469.

FIELD, C.; JACKSON, R. and MOONEY, H. 1995. Stomatal responses to increased CO_2: Implications from the plant to the global scale. *Plant Cell Environment* **18**: 1214–1255.

FINN, G. A. and BRUN, W. A. 1982. Effect of atmospheric CO_2 enrichment on growth, nonstructural carbohydrate content and root nodule activity in soybean. *Plant Physiology* **69**: 327–331.

FISCHER, G.; SHAH, M.; TUBIELLO, F. N. and VAN VELHUIZEN, H. 2005. Socio-economic and climate change impacts on agriculture: An integrated assessment, 1990–2080. *Philosophical Transactions of Royal Society Biological Sciences* **360**: 2067–2083.

FISCHER, G.; TUBIELLO, F.; VAN VELHUIZEN, H. and WIBERG, D. 2006. Climate change impacts on irrigation water requirements: Effects of mitigation, 1990–2989. *Technological Forecasting and Social Change* **74**: 1083–1107.

FISHER, M. J.; RAO, I. M.; AYARZA, M. A.; LASCANO, C. E.; SANZ, J. I.; THOMAS, R. J. and VERA, 1994. Carbon storage by introduced deep-rooted grasses in the South American savannas. *Nature* **371**: 236–238.

FOEREID, B. and HØGH-JENSEN, H. 2004. Carbon sequestration potential of organic agriculture in northern Europe—A modelling approach. *Nutrient Cycling in Agroecosystems* **68**: 13–24.

FOLEY, J. A.; DEFRIES, R.; ASNER, G.; BARFORD, C.; BONAN, G.; CARPENTER, S. R.; CHAPIN, F. S. et al. 2005. Global consequences of land use. *Science* **309**: 570–574.

FOLLETT, R. F. 2001. Organic carbon pools in grazing land soils. In: *The Potential of U.S. Grazing Lands to Sequester Carbon and Mitigate the Greenhouse Effect*, eds. Follett, R. F.; Kimble, J. M. and Lal, R. Lewis Publishers, Boca Raton, Florida. pp. 65–86.

FOLLETT, R. F.; KIMBLE, J. M. and LAL, R. 2001. The potential of U. S. grazing lands to sequester soil carbon. In: *The Potential of U.S. Grazing Lands to Sequester Carbon and Mitigate the Greenhouse Effect*, Follett, R. F.; Kimble, J. M. and Lal, R. Lewis Publishers, Boca Raton, Florida. pp. 401–430.

FOLLETT, R. F.; SHAFER, S. R.; JAWSON, M. D. and FRANZLUEBBERS, A. J. 2005. Research and implementation needs to mitigate greenhouse gas emissions from agriculture in the USA. *Soil and Tillage Research* **83**: 159–166.

FRECKLETON, R. P.; WATKINSON, A. R.; WEBB, D. J. and THOMAS, T. H. 1999. Yield of sugarbeet in relation to weather and nutrients. *Agriculture and Forest Meteorology* **93**: 39–51.

FREIBAUER, A.; ROUNSEVELL, M.; SMITH, P. and VERHAGEN, A. 2004. Carbon sequestration in the agricultural soils of Europe. *Geoderma* **122**: 1–23.

FUHRER, J. 2003. Agroecosystem responses to combinations of elevated CO_2, ozone, and global climate change. *Agricultural Ecosystem Environment* **97**: 1–20.

GADGIL, S.; RAO, P. R. S. and SRIDHAR, S. 1999. Modeling impact of climatic variability on rainfed groundnut. *Current Science* **76**: 557–569.

GALLOWAY, J. N. 2003. The global nitrogen cycle. *Treatise on Geochemistry* **8**: 557–583.

GALLOWAY, J. N.; ABER, J. D.; ERISMAN, J. W.; SEITZINGER, S. P.; HOWARTH, R. W.; COWLING, E. B. and COSBY, B. J. 2003. The nitrogen cascade. *Bioscience* **53**: 341–356.

GALLOWAY, J. N.; DENTENER, F. J.; CAPONE, D. G.; BOYER, E. W.; HOWARTH, R. W.; SEITZINGER, S. P.; ASNER, G. P. et al. 2004. Nitrogen cycles: Past, present, and future. *Biogeochemistry* **70**: 153–226.

GANGADHAR RAO, D.; KATYAL, J. C.; SINHA, S. K. and SRINIVAS, K. 1995. Impacts of Climate Change on Sorghum Productivity in India: Simulation Study. In: *Climate Change and Agriculture: Analysis of Potential International Impacts.* ASA Special Publ. No. 59. pp. 325–337.

GANGADHAR RAO, D. and SINHA, S. K. 1994. Impact of climate change on simulated wheat production in India. In: *Implications of Climate Change for International Agriculture: Crop Modelling Study*, eds. Rosenzweig, C. and Iglesias, I. USEPA, Washington DC. pp. 1–17.

GEHL, R. J. and RICE, C. W. 2007. Emerging technologies for *in situ* measurement of soil carbon. *Climatic Change* **80**: 43–54.

GIFFORD, R. M. 2004. The CO_2 fertilizing effect—Does it occur in the real world? *New Phytology* **163**: 221–225.

GOMMES, R. and PETRASSI, F. 1996. Rainfall variability and drought in sub-Saharan Africa since 1960. *FAO Agrometeorology Series 9*, Food and Agriculture Organization of the United Nations.

GONZALEZ-AVALOS, E. and RUIZ-SUAREZ, L. G. 2001. Methane emission factors from cattle in Mexico. *Bioresource Technology* **80**: 63–71.

GORNALL, J.; BETTS, R.; BURKE, E.; CLARK, R.; CAMP, J.; WILLETT, K. and WILTSHIRE, A. 2010. Implications of climate change for agricultural productivity in the early twenty-first century. *Philosophical Transactions of the Royal Society of London Biological Sciences* **365**: 2973–2989.

GOUDRIAAN, J. and UNSWORTH, M. U. 1990. Implications of increasing CO_2 and climate change for agricultural productivity and water resources. In: *Impact of Carbon Dioxide; Trace Gases and Climate Change on Global Agriculture.* Special Publication no. 53. ASA, Madison. pp. 111–130.

GOULDER, L. H. and PIZER, W. A. 2006. *The Economics of Climate Change.* Discussion Paper 06, Resources for the Future (RFF). Washington, DC.

GRANT, R. F.; WALL, G. W.; KIMBALL, B. A. and FRUMAU, K. F. A. 1999. Crop water relations under different CO_2 and irrigation: Testing of ecosystem with the free air CO_2 enrichment

(FACE) experiment. *Agriculture and Forest Meteorology* **95:** 27–51.

GREEN, R. E.; CORNELL, S. J.; SCHARLEMANN, J. P. W. and BALMFORD, A. 2005. Farming and the fate of wild nature. *Science* **307:** 550–555.

GREGORICH, E. G.; ROCHETTE, P.; VAN DEN BYGAART, A. J. and ANGERS, D. A. 2005. Greenhouse gas contributions of agricultural soils and potential mitigation practices in Eastern Canada. *Soil and Tillage Research* **83:** 53–72.

GREGORY, P. J.; JOHNSON, S. N.; NEWTON, A. C. and INGRAM, J. S. I. 2009. Integrating pests and pathogens into the climate change/food security debate. *Journal of Experimental Botany* **60:** 2827–2838.

GROVER, V. I. (EDS.) 2004. *Climate Change: Five years after Kyoto.* Science Publishers Inc., Enfield, USA.

GUO, L. B. and GIFFORD, R. M. 2002. Soil carbon stocks and land use change: A meta analysis. *Global Change Biology* **8:** 345–360.

HALBERG, N.; SULSER, T. B.; ROSEGRANT, M. W. and KNUDSEN, M. T. 2006. The impact of organic farming on food security in a regional and global perspective. In: *Global Development of Organic Agriculture: Challenges and Prospects,* eds. Halberg, N.; Alrød, H. F.; Knudsen, M. T. and Kristensen, E. S. CABI Publishing, Wallingford. pp. 227–322.

HAMELINCK, C. N.; SUURS, R. A. A. and FAAIJ, A. P. C. 2004. Technoeconomic analysis of international bio-energy trade chains. *Biomass & Bioenergy* **29:** 114–134.

HANSEN, J.; RUSSEL, G.; RIND, D.; STONE, P.; LACIS, A.; LEBEDEFF, S.; RUEDY, R. and TRAVIS, L. 1981. Efficient three-dimensional global models for climate studies, models I and II. *Monthly Weather Reviews* **111:** 609–662.

HARDY, R. W. F. and HAVELKA, U. D. 1975. *Photosynthates as Major Factor Limiting Nitrogen Fixing by Field Grown Legumes with Emphasis on Soybeans. Contributions to the Scientific Literatures, Section IV, Biology.* F. I. DuPont de Nemours, Wilmington, Del. pp. 58–76.

HELGASON, B. L.; JANZEN, H. H.; CHANTIGNY, M. H.; DRURY, C. F.; ELLERT, B. H.; GREGORICH, E. G.; LEMKE; PATTEY, E.; ROCHETTE, P. and WAGNER-RIDDLE, C. 2005. Toward improved coefficients for predicting direct N_2O emissions from soil in Canadian agroecosystems. *Nutrient Cycling in Agroecosystems* **71:** 87–99.

HENRY, H. A. L.; CLELAND, E. E.; FIELD, C. B. and VITOUSEK, P. M. 2005. Interactive effects of elevated CO_2, N deposition and climate change on plant litter quality in a California annual grassland. *Oecologia* **142:** 465–473.

HESS, H. D.; TIEMANN, T. T.; NOTO, F.; CARULLA, J. E. and KRUEZER, M. 2006. Strategic use of tannins as means to limit methane emission from ruminant livestock. In: *Greenhouse Gases and Animal Agriculture: An Update,* eds. Soliva C. R.; Takahashi, J. and Kreuzer, M. International Congress Series No. 1293. Elsevier, the Netherlands. pp. 164–167.

HILDÉN, M.; LEHTONEN, H.; BÄRLUND, I.; HAKALA, K.; KAUKORANTA, T. and TATTARI, S. 2005. *The Practice and Process of Adaptation in Finnish Agriculture*. FINADAPT Working Paper 5. Finnish Environment Institute Mimeographs no. 335. Finnish Environment Institute, Helsinki, Finland.

HINDRICHSEN, I. K.; WETTSTEIN, H. R.; MACHMÜLLER, A. and KREUZER, M. 2006. Methane emission, nutrient degradation and nitrogen turnover in dairy cows and their slurry at different production scenarios with and without concentrate supplementation. *Agriculture, Ecosystems and Environment* **113:** 150–161.

HINGANE, L. S.; RUPA KUMAR, K. and RAMANA MURTHY, B. H. V. 1985. Long-term trends of surface air temperature in India. *Journal of Climatology* **5:** 521–528.

HOFREITHER, M. F. and SINABELL, F. 1996. Macroeconomic and agricultural aspects of CO_2 emission. In: *Energy CO_2 Taxation and Agriculture*. 5th Annual Meeting of OGA, Austria. pp. 112.

HOGG, D. 1992. *Agricultural Policies in Response to Climate Change: How the Greenhouse Effect Might Affect Agricultural Development*. Papers in International Development. Center for Development Studies, University College of Swansea. 50p.

HOLE, D. G.; PERKINS, A. J.; WILSON, J. D.; ALEXANDER, I. H.; GRICE, P. V. and EVANS, A. D. 2005. Does organic farming benefit biodiversity? *Biological Conservation* **122:** 113–130.

HOOGWIJK, M. 2004. *On the Global and Regional Potential of Renewable Energy Sources*. Ph.D. Thesis. Copernicus Institute, Utrecht University. 256p. Unpublished.

HOOGWIJK, M.; FAAIJ, A.; EICKHOUT, B.; DE VRIES, B. and TURKENBURG, W. 2005. Potential of biomass energy out to 2100, for four IPCC SRES land-use scenarios. *Biomass & Bioenergy* **29:** 225–257.

HUANG, J.; PRAY, C. and ROZELLE, S. 2002. Enhancing the crops to feed the poor. *Nature* **418:** 678–684.

HUNDAL, S. and KAUR, P. 1996. Crop productivity in the Punjab, India. In: *Climate Variability and Agriculture*, eds., Gadgil, Y. P. and Pant, G. B. Abrol, New Delhi.

HUSTON, M. A. and MARLAND, G. 2003. Carbon management and biodiversity. *Journal of Environmental Management* **67:** 77–86.

IMESON, A.; DUMONT, H. and SEKLIZIOTIS, S. 1987. *Impact Analysis of Climatic Change in the Mediterranean Region*, vol. F. European Workshop on Interrelated Bioclimatic and Landuse Changes. Noordwijkerhout, the Netherlands.

IZAURRALDE, R. C.; MCGILL, W. B.; ROBERTSON, J. A.; JUMA, N. G. and THURSTON, J. J. 2001. Carbon balance of the Breton classical plots over half a century. *Soil Science Society of America Journal* **65:** 431–441.

IZAURRALDE, R. C. and RICE, C. W. 2006. Methods and tools for designing pilot soil carbon sequestration projects. In: *Carbon*

Sequestration in Soils of Latin America, eds. Lal, R.; Cerri, C. C.; Bernoux, M.; Etchvers, J. and Cerri, C. E. CRC Press, Boca Raton, Florida. pp. 457–476.

JACKSON, R. B. IV and GEYER, R. A. 1993. *Greenhouse Gases and Agriculture. A Global Warming Forum: Scientific, Economic and Legal Overview.* pp. 417–444.

JACKSON, R. B.; JOBBÁGY, E. G.; AVISSAR, R.; BAIDYA ROY, S.; BARRETT, D.; COOK, C. W.; FARLEY, K. A.; LE MAITRE, D. C.; MCCARL, B. A. and MURRAY, B. C. 2005 Trading water for carbon with biological carbon sequestration. *Science* **310:** 1944–1947.

JALLOW, B. P. 1995. Emission of the greenhouse gases from agriculture, land use change and forestry in the Gambia. *Environmental Monitoring and Assessment* **38:** 301–312.

JANZEN, H. H. 2004. Carbon cycling in earth systems: A soil science perspective. *Agriculture, Ecosystems and Environment* **104:** 399–417.

JENSEN, B. and CHRISTENSEN, B. T. 2004. Interactions between elevated CO_2 and added N: Effects on water use, biomass, and soil 15N uptake in wheat. *Acta Agriculturae Scandinavica Section B* **54:** 175–184.

JODHA, N. S. 1989. Potential strategies for adapting to greenhouse warming: Perspectives from the developing World. In: *Greenhouse Warming: Abatement and Adaptation* eds. Rosenberg, N. J.; Easterling, W. E.; Crosson, P. R. and Darmstadter, J. Resources for the Future, Washington DC.

JOHNSON, D. E.; WARD, G. M. and TORRENT, J. 1991. The environmental impact of bovine somatotropin (bST) use in dairy cattle. *Journal of Dairy Science* **74S:** 209.

JOHNSON, K. A. and JOHNSON, D. E. 1995. Methane emissions from cattle. *Journal of Animal Science* **73:** 2483–2492.

JONES, C. D.; COX, P. M.; ESSERY, R. L. H.; ROBERTS, D. L. and WOODAGE, M. J. 2003. Strong carbon cycle feedbacks in a climate model with interactive CO_2 and sulphate aerosols. *Geophysical Research Letters* **30:** 32.1–32.4.

JORDAN, E.; KENNY, D.; HAWKINS, M.; MALONE, R.; LOVETT, D. K. and O'MARA, F. P. 2006b. Effect of refined soy oil or whole soybeans on methane output, intake and performance of young bulls. *Journal of Animal Science* **84:** 2418–2425.

JORDAN, E.; LOVETT, D. K.; HAWKINS, M.; CALLAN, J. and O'MARA, F. P. 2006c. The effect of varying levels of coconut oil on intake, digestibility and methane output from continental cross beef heifers. *Animal Science* **82:** 859–865.

JORDAN, E.; LOVETT, D. K.; MONAHAN, F. J. and O'MARA, F. P. 2006a. Effect of refined coconut oil or copra meal on methane output, intake and performance of beef heifers. *Journal of Animal Science* **84:** 162–170.

JUEN, I.; KASER, G. and GEORGES, C. 2007. Modelling observed and future runoff from a glacierized tropical catchment (Cordillera Blanca, Peru). *Global Planetary Change* **59:** 37–48.

KAISER, H. M.; RIHA, S. J.; WILKS, D. S.; ROSSIER, D. G. and SAMPATH, R. 1993. A farm-level analysis of economic and agronomic impacts of gradual warming. *American Journal of Agricultural Economics* **75:** 387–398.

KALRA, N.; CHAKRABORTY, D.; RAMESH, P. R.; JOLLY, M. and SHARMA, P. K. 2007. *Impacts of Climate Change in India: Agricultural Impacts.* Final Report, Joint Indo-UK Programme of Ministry of Environment and Forests, India, and Department for Environment, Food and Rural Affairs (DEFRA), United Kingdom and Unit of Simulation and Informatics, Indian Agricultural Research Institute, New Delhi.

KAMRA, D. N.; AGARWAL, N. and CHAUDHARY, L. C. 2006. Inhibition of ruminal methanogenesis by tropical plants containing secondary compounds. In: *Greenhouse Gases and Animal Agriculture: An Update*, eds. Soliva, C. R.; Takahashi, J. and Kreuzer; M. International Congress Series No. 1293. Elsevier, the Netherlands. pp. 156–163.

KANG, G. D.; CAI, Z. C. and FENG, X. Z. 2002. Importance of water regime during the non-rice growing period in winter in regional variation of CH_4 emissions from rice fields during following rice growing period in China. *Nutrient Cycling in Agroecosystems* **64:** 95–100.

KASIMIR-KLEMEDTSSON, A.; KLEMEDTSSON, L.; BERGLUND, K.; MARTIKAINEN, P.; SILVOLA and OENEMA, O. 1997. Greenhouse gas emissions from farmed organic soils: A review. *Soil Use and Management* **13:** 245–250.

KAUFMANN, R. K. and SNELL, S. E. 1997. A biophysical model of corn yield: Integrating climatic and social determinants. *American Journal of Agronomy and Economics* **79:** 178–190.

KAVI KUMAR, K. S. 2007. Climate change studies in Indian agriculture. *Economic and Political Weekly* **42:** 13–18.

KAVI KUMAR, K. S. 2009. *Climate Sensitivity of Indian Agriculture.* Madras School of Economics, Working Paper No. 43.

KAVI KUMAR, K. S. and PARIKH, J. 1998. Climate change impacts on Indian agriculture: The Ricardian approach. In: *Measuring the Impacts of Climate Change on Indian Agriculture,* eds., Dinar et al. World Bank Technical Paper No. 402. Washington DC.

KAVI KUMAR, K. S. and PARIKH, J. 2001a. Socio-economic impacts of climate change on Indian Agriculture. *International Review for Environmental Strategies* 2: 277–293.

KAVI KUMAR, K. S. and PARIKH, J. 2001b. Indian agriculture and climate sensitivity. *Global Environmental Change* **11:** 147–154.

KEBREAB, E.; CLARK, K.; WAGNER-RIDDLE, C. and FRANCE, J. 2006. Methane and nitrous oxide emissions from Canadian animal agriculture: A review. *Canadian Journal of Animal Science* **86:** 135–158.

KEHRWALD, N. M.; THOMPSON, L. G.; YAO, T. D.; MOSLEY-THOMPSON, E.; SCHOTTERER U. ALFIMOV, V.; BEER, J.; EIKENBERG, J. and DAVIS, M. E. 2008. Mass loss on Himalayan glacier endangers water resources. *Geophysical Research Letters* **35**, L22503.

KELLOGG, W. W. and ZHAO, Z. C. 1988. Sensitivity of soil moisture to doubling CO_2 in climate model experiments. Part I. *North American Journal of Climate* **1**: 348–366.

KENNEDY, P. M. and MILLIGAN, L. P. 1978. Effects of cold exposure on digestion, microbial synthesis and nitrogen transformation in sheep. *British Journal of Nutrition* **39**: 105–117.

KETHUNEN, L.; MUKULA, J.; POHJONEN, V.; RANTANEN, O. and VARJO, U. 1988. The effect of climatic variations on agriculture in Finland. In: *The Impacts of Climatic Variations on Agriculture. Vol. I: Assessment in Cool Temperature and Cold Regions*, eds. Parry, M. L.; Carter, T. R. and Konijn, N. T. Kluwer, Dordrecht, the Netherlands. pp. 550–614.

KHALIL, M. A. K. and SHEARER, M. J. 2006. Decreasing emissions of methane from rice agriculture. In: *Greenhouse Gases and Animal Agriculture: An Update*, eds. Soliva, C. R.; Takahashi, J. and Kreuzer, M. International Congress Series No. 1293. Elsevier, the Netherlands. pp. 33–41.

KIMBALL, B. A. 1983a. *Carbon dioxide and Agricultural Yield. An Assemblage Arid Analysis of 770 Prior Observations*. WCL Report 14. US water Sonserv. Zab. Phoenix, Arizona.

KIMBALL, B. A. 1983b. Carbon dioxide and agricultural yield. An assemblage arid analysis of 430 prior observations. *Agronomy Journal* **75**: 779–788.

KIMBALL, B. A. 1985. Adaptation of vegetative and management practices to a higher CO_2 world. In: *Direct Effects of Increasing CO_2 on Vegetation*, eds. Strain, B. R. and Cure, J. D. National Technical Information Service. US Department of Commerce, Springfield VA.

KIMBALL, B. A. 1986. *Influence of Elevated CO_2 on Crop Yield. CO_2 Enrichment of Greenhouse Crops. Vol. II. Physiology, Yield and Economics.* CRC Press, Boca Raton, Florida. pp. 105–115.

KIMBALL, B. A. and IDSO, S. B. 1983. Increasing atmospheric CO_2: Effects on crop yields, water use and climate. *Agricultural Water Management* **7**: 55–72.

KIM, M. K. and MCCARL, B. A. 2005. *Uncertainty Discounting for Land-Based Carbon Sequestration*. Presented at International Policy Forum on Greenhouse Gas Management. April 2005 Victoria, British Columbia.

KNORR, W.; PRENTICE, I. C.; HOUSE, J. I.; HOLLAND, E. A. 2005. Long-term sensitivity of soil carbon turnover to warming. *Nature* **433**: 298–301.

KOGA, N.; SAWAMOTO, T. and TSURUTA H. 2006. Life cycle inventory-based analysis of greenhouse gas emissions from arable land farming systems in Hokkaido, northern Japan. *Soil Science and Plant Nutrition* **52**: 564–574.

KORONTZI, S.; JUSTICE, C. O. and SCHOLES, R. J. 2003. Influence of timing and spatial extent of savannah fires in southern Africa on atmospheric emissions. *Journal of Arid Environments* **54**: 395–404.

KRAPFENBAUER, A. and WRIESSNIG, K. 1995. Anthropogenic environmental pollution—The share of agriculture. *Bodenkultur* **46:** 269–283.

KREUZER, M. and HINDRICHSEN, I. K. 2006. Methane mitigation in ruminants by dietary means: The role of their methane emission from manure. In: *Greenhouse Gases and Animal Agriculture: An Update*, eds. Soliva, C. R.; Takahashi, J. and Kreuzer, M. International Congress Series No. 1293. Elsevier, the Netherlands. pp. 199–208.

KRIPALANI, R, H.; INAMDAR, S. R. and SONTAKKE, N. A. 1996. Rainfall variability over Bangladesh and Nepal: Comparison and connection with features over India. *International Journal of Climatology* **16:** 689–703.

KÜLLING, D. R.; MENZI, H.; SUTTER, F.; LISCHER, P. and KREUZER, M. 2003. Ammonia, nitrous oxide and methane emissions from differently stored dairy manure derived from grass- and hay-based rations. *Nutrient Cycling in Agroecosystems* **65:** 13–22.

KURKALOVA, L.; KLING, C. L. and ZHAO, J. 2004. Multiple benefits of carbon friendly agricultural practices: Empirical assessment of conservation tillage. *Environmental Management* **33:** 519–527.

LAL, M.; CUBASCH, U.; VOSS, R. and WASZKEWITZ, J. 1995. Effect of transient increases in greenhouse gases and sulphate aerosols on monsoon climate. *Current Science* **69:** 752–763.

LAL, M.; NOZAWA, T.; EMORI, S.; HARASAWA, H.; TAKAHASHI, K.; KIMOTO, M.; ABE-OUCHI, A.; NAKAJIMA, T.; TAKEMURA, T. and NUMAGUTI, A. 2001. Future climate change: Implications for Indian summer monsoon and its variability. *Current Science* **81:** 1196–1207.

LAL, M.; SINGH, S. K.; SRINIVASAN, G.; RATHORE, L. S.; NAIDU, D. and TRIPATHI, C. N. 1999. Growth and yield response of soybean in Madhya Pradesh, India to climate variability and change. *Agricultural and Forest Meteorology* **93:** 53–70.

LAL, M.; SINGH, K. K.; SRINIVASAN, G.; RATHORE, L. S. and SASEENDRAN, A. S. 1998. Vulnerability of rice and wheat yields in NW-India to future change in climate. *Agriculture and Forest Meteorology* **89:** 101–114.

LAL, R., 1999. Soil management and restoration for C sequestration to mitigate the accelerated greenhouse effect. *Progress in Environmental Science* **1:** 307–326.

LAL, R. 2001a. World cropland soils as a source or sink for atmospheric carbon. *Advances in Agronomy* **71:** 145–191.

LAL, R. 2001b. Potential of desertification control to sequester carbon and mitigate the greenhouse effect. *Climate Change* **15:** 35–72.

LAL, R. 2002. Carbon sequestration in dry ecosystems of West Asia and North Africa. *Land Degradation and Management* **13:** 45–59.

LAL, R. 2003. Global potential of soil carbon sequestration to mitigate the greenhouse effect. *Critical Reviews in Plant Sciences* **22:** 151–184.

LAL, R. 2004a. Soil carbon sequestration impacts on global climate change and food security. *Science* **304:** 1623–1627.

LAL, R. 2004b. Soil carbon sequestration to mitigate climate change. *Geoderma* **123:** 1–22.

LAL, R. 2004c. Offsetting China's CO_2 emissions by soil carbon sequestration. *Climatic Change* **65:** 263–275.

LAL, R. 2004d. Carbon sequestration in soils of central Asia. *Land Degradation and Development* **15:** 563–572.

LAL, R., 2004e. Soil carbon sequestration in India. *Climatic Change* **65:** 277–296.

LAL, R. 2005. Soil carbon sequestration for sustaining agricultural production and improving the environment with particular reference to Brazil. *Journal of Sustainable Agriculture* **26:** 23–42.

LAL, R. and BRUCE, J. P. 1999. The potential of world cropland soils to sequester C and mitigate the greenhouse effect. *Environmental Science and Policy* **2:** 177–185.

LAL, R.; FOLLETT, R. F. and KIMBLE, J. M. 2003. Achieving soil carbon sequestration in the United States: A challenge to the policy makers. *Soil Science* **168:** 827–845.

LAMBORG, M. R.; HARDY, R. W. F. and PAUL, E. A. 1983. Microbial effects. In: *CO_2 and Plants*, ed. Lemon, E. R. Westview Press, Boulder, Colorado. pp. 131–176.

LEMON, E. R. 1983. *CO_2 and Plants: The Response of Plants to Rising Levels of Atmospheric Carbon Dioxide*. Westview Press, Boulder, CO, USA.

LENG, R. A. 1991. Improving ruminant production and reducing methane emissions from ruminants by strategic supplementation. EPA Report no. 400/1-91/004, Environmental Protection Agency, Washington DC.

LI, C.; FROLKING, S. and BUTTERBACH-BAHL, K. 2005. Carbon sequestration in arable soils is likely to increase nitrous oxide emissions, offsetting reductions in climate radiative forcing. *Climatic Change* **72:** 321–338.

LIEBIG, M. A.; MORGAN, J. A.; REEDER, J. D.; ELLERT, B. H.; GOLLANY, H. T. and SCHUMAN, G. E. 2005. Greenhouse gas contributions and mitigation potential of agricultural practices in northwestern USA and western Canada. *Soil & Tillage Research* **83Z:** 25–52.

LILA, Z. A.; MOHAMMED, N.; KANDA, S.; KAMADA, T. and ITABASHI, H. 2003. Effect of sarsaponin on ruminal fermentation with particular reference to methane production *in vitro*. *Journal of Dairy Science* **86:** 330–336.

LOBELL, D. B.; BURKE, M. B.; TEBALDI, C.; MASTRANDREA, M. D.; FALCON, W. P. and NAYLON, R. L. 2008. Prioritizing climate change adaptation. Needs for food security in 2030. *Science* **319:** 607–610.

LOBELL, D. B. and FIELD, C. B. 2007. Global scale climate-crop yield relationships and the impacts of recent warming. *Environmental Research Letters* **2:** 1–7.

LONERGAN, S. 1998. Climate warming and India. In: *Measuring the Impact of Climate Change on Indian Agriculture*, eds. Dinar,

A. et al., World Bank Technical Paper No. 402. World Bank, Washington DC.

LONG S. P.; AINSWORTH, E. A.; LEAKEY, A. D. B. and MORGAN, P. B. 2009. Global food insecurity. Treatment of major food crops with elevated carbon dioxide or ozone under large-scale fully open-air conditions suggests recent models may have overestimated future yields. *Philosophical Transactions of Royal Society Biological Sciences* **360:** 2011–2020.

LONG, S. P.; AINSWORTH, E. A.; LEAKEY, A. D. B.; NOSBERGER, J. and ORT, D. R. 2006. Food for thought: Lower-than-expected crop yield stimulation with rising CO_2 concentrations. *Science* **312:** 1918–1921.

LONG, S. P.; AINSWORTH, E. A.; ROGERS, A. and ORT, D. R. 2004. Rising atmospheric carbon dioxide: Plants face the future. *Annual Review of Plant Biology* **55:** 591–628.

LOTTER, D.; SEIDEL, R. and LIEBHARDT, W. 2003. The performance of organic and conventional cropping systems in an extreme climate year. *American Journal of Alternative Agriculture* **18:** 146–154.

LOVETT, D. K. and O'MARA, F. P. 2002. Estimation of enteric methane emissions originating from the national livestock beef herd: A review of the IPCC default emission factors. *Tearmann* **2:** 77–83.

LOVETT, D. K.; SHALLOO, L.; DILLON, P. and O'MARA, F. P. 2006. A systems approach to quantify greenhouse gas fluxes from pastoral dairy production as affected by management regime. *Agricultural Systems* **88:** 156–179.

LOVETT, D.; LOVELL, S.; STACK, L.; CALLAN, J.; FINLAY, M.; CONNOLLY, J. and O'MARA, F. P. 2003. Effect of forage/concentrate ratio and dietary coconut oil level on methane output and performance of finishing beef heifers. *Livestock Production Science* **84:** 135–146.

LUTZ, W.; SANDERSON, W. and. SCHERBOV, S 2001. The end of world population growth. *Nature* **412:** 543–545.

MACHADO, P. L. O. A. and FREITAS, P. L. 2004. No-till farming in Brazil and its impact on food security and environmental quality. In: *Sustainable Agriculture and the International Rice-Wheat System,* eds. Lal, R.; Hobbs, P. R.; Uphoff, N.; Hansen, D. O. Marcel Dekker, New York. pp. 291–310.

MACHADO, P. L. O. A. and SILVA, C. A. 2001. Soil management under no tillage systems in the tropics with special reference to Brazil. *Nutrient Cycling in Agroecosystems* **61:** 119–130.

MACHMÜLLER, A.; OSSOWSKI, D. A. and KREUZER, M. 2000. Comparative evaluation of the effects of coconut oil, oilseeds and crystalline fat on methane release, digestion and energy balance in lambs. *Animal Feed Science and Technology* **85:** 41–60.

MADARI, B.; MACHADO, P. L. O. A.; TORRES, E.; ANDRADE, A. G. and VALENCIA, L. I. O. 2005. No tillage and crop rotation effects on soil aggregation and organic carbon in a

Fhodic Ferralsol from southern Brazil. *Soil and Tillage Research* **80:** 185–200.

MÄDER, P.; FLIESSBACH, A.; DUBOIS, D.; GUNST, L.; FRIED, P. and NIGGLI, U. 2002. Soil fertility and biodiversity in organic farming. *Science* **296:** 1694–1697.

MALL, R. K.; SINGH, R.; GUPTA, A.; SRINIVASAN, G. and RATHORE, L. S. 2006. Impact of climate change on Indian agriculture: A review. *Climatic Change* **78:** 445–478.

MALL, R. K.; SINGH, R.; GUPTA, A.; SRINIVASAN, G. and RATHORE, L. S. 2007. Impact of climate change on Indian agriculture: A review. Erratum, *Climatic Change* **82:** 225–231.

MANDAL, N. 1998. *Simulating the Impact of Climatic Variability and Climate Change on Growth and Yield of Chickpea and Pigeonpea Crops.* M.Sc. Thesis. Indian Agricultural Research Institute, New Delhi. Unpublished.

MANNE, A. S. and RICHELS, R. G. 2004. A multi-gas approach to climate policy. In: *The Global Carbon Cycle. Integrating Humans, Climate, and the Natural World*, eds. Field, C. B. and Raupach, M. R. SCOPE 62. Island Press, Washington DC. pp. 439–452.

MARACCHI, G.; SIROTENKO, O. and BINDI, M. 2005. Impacts of present and future climate variability on agriculture and forestry in the temperate regions. *European Climate Change* **70:** 117–135.

MARLAND, G.; McCARL, B. A. and SCHNEIDER, U. A. 2001. Soil carbon: Policy and economics. *Climatic Change* **51:** 101–117.

MARLAND, G.; PIELKE JR., R. A.; APPS, M.; AVISSAR, R.; BETTS, R. A.; DAVIS, K. J.; FRUMHOFF, P. C. et al. 2003a. The climatic impacts of land surface change and carbon management and the implications for climate-change mitigation policy. *Climate Policy* **3:** 149–157.

MARLAND, G.; WEST, T. O.; SCHLAMADINGER, B. and CANELLA, L. 2003b. Managing soil organic carbon in agriculture: The net effect on greenhouse gas emissions. *Tellus* **55B:** 613–621.

MARRIOTT, E. E. and WANDER, M. M. 2006. Total and labile soil organic matter in organic and conventional farming systems. *Soil Science Society of America Journal* **70:** 950–959.

MAY, W. 2002. Simulated changes of the Indian summer monsoon under enhanced greenhouse gas conditions in a global time-slice experiment. *Geophysical Research Letters* **29:** 221–224.

McCARL, B. A. and SCHNEIDER, U. A. 2001. Greenhouse gas mitigation in U.S. agriculture and forestry. *Science* **294:** 2481–2482.

McCARTHY, J. J.; CANZIANI, O. F.; LEARY, N. A.; DOKKEN, D. J. and WHITE, K. S. (EDS.) 2001. *WG II: Climate Change 2001: Impacts, Adaptation & Vulnerability.* Cambridge University Press, Cambridge.

McCRABB, G. C., 2001. Nutritional options for abatement of methane emissions from beef and dairy systems in Australia. In:

Greenhouse Gases and Animal Agriculture, eds. Takahashi, J. and Young, B. A. Elsevier, Amsterdam. pp. 115–124.

McCrabb, G. J.; Kurihara, M. and Hunter, R. A. 1998. The effect of finishing strategy of lifetime methane production for beef cattle in northern Australia. *Proceedings of the Nutrition Society of Australia* **22:** 55.

McGinn, S. M.; Beauchemin, K. A.; Coates, T. and Colombatto, D. 2004. Methane emissions from beef cattle: Effects of monensin, sunflower oil, enzymes, yeast, and fumaric acid. *Journal of Animal Science* **82:** 3346–3356.

McSwiney, C. P. and Robertson, G. P. 2005. Nonlinear response of N_2O flux to incremental fertilizer addition in a continuous maize (*Zea mays* L.) cropping system. *Global Change Biology* **11:** 1712–1719.

Mendelsohn, R. and Schlesinger, M.E. 1999. Climate response functions. *Ambio* **28:** 362–366.

Mendelsohn, R.; Nordhaus, W. D. and Shaw, D. 1994. The impact of global warming on agriculture: A Ricardian analysis. *American Economic Reviews* **84:** 753–771.

Menon, S.; Hansen, J.; Nazarenko, L. and Luo, Y. 2002. Climate effects of black carbon aerosols in China and India. *Science* **297:** 2250–2253.

Mohandass, S.; Kareem, A. A.; Ranganathan, T. B. and Jeyaraman, S. 1995. Rice production in India under current and future climates. In: *Modeling the Impact of Climate Change on Rice Production in Asia*, eds., Matthews, R. B.; Kropff, M. J.; Bachelet, D. and Laar van, H. H. CAB International, UK.

Monteny, G. J.; Bannink, A. and Chadwick, D. 2006. Greenhouse gas abatement strategies for animal husbandry. *Agriculture, Ecosystems and Environment* **112:** 163–170.

Monteny, G. J.; Groenestein, C. M. and Hilhorst, M. A. 2001. Interactions and coupling between emissions of methane and nitrous oxide from animal husbandry. *Nutrient Cycling in Agroecosystems* **60:** 123–132.

Mooley, D. A. and Parthasarathy, B. 1984. Fluctuations of all India summer monsoon rainfall during 1871–1978. *Climatic Change* **6:** 287–301.

Mooney, S.; Antle, J. M.; Capalbo, S. M. and Paustian, K. 2004. Influence of project scale on the costs of measuring soil C sequestration. *Environmental Management* **33(S1):** S252–S263.

Mooney, H.; Cropper, A. and Reid, W. 2005. Confronting the human dilemma. *Nature* **434:** 561–562.

Morison, J. I. L. 1987. Intercellular CO_2 concentration and stomatal response to CO_2. In: *Stomatal Function*, eds. Zeiger, E.; Cowan, I. R. and Farquhar, G. D. Stanford University Press, USA. pp. 229–251.

Morison, J. I. L. and Gifford, R. M. 1983. Stomatal sensitivity to CO_2 and humidity. *Plant Physiology* **71:** 789–796.

MOSIER, A. and KROEZE, C. 2000. Potential impact on the global atmospheric N₂O budget of the increased nitrogen input required to meet future global food demands. *Chemosphere—Global Change Science* **2:** 465–473.

MOSIER, A. R. 2001. Exchange of gaseous nitrogen compounds between agricultural systems and the atmosphere. *Plant and Soil* **228:** 17–27.

MOSIER, A. R. 2002. Environmental challenges associated with needed increases in global nitrogen fixation. *Nutrient Cycling in Agroecosystems* **63:** 101–116.

MOSIER, A. R.; DUXBURY, J. M.; FRENEY, J. R.; HEINEMEYER, O.; MINAMI, K. and JOHNSON, D. E. 1998. Mitigating agricultural emissions of methane. *Climatic Change* **40:** 39–80.

MOSIER, A. R.; HALVORSON, A. D.; PETERSON, G. A.; ROBERTSON, G. P. and SHERROD, L. 2005. Measurement of net global warming potential in three agroecosystems. *Nutrient Cycling in Agroecosystems* **72:** 67–76.

MÜLLER, A. 2009. *Climate Change Mitigation: Unleashing the Potential of Agriculture*. Presentation made to the UNFCCC Ad Hoc Working Group on Long-Term Cooperative Action, 4 April 2009, Bonn, Germany. http://unfccc.int/meetings/ad_hoc_working_groups/lca/items/4815.php

MURRAY, B. C.; McCARL, B. A. and LEE, H. C. 2004. Estimating leakage from forest carbon sequestration programs. *Land Economics* **80:** 109–124.

MURRAY, R. M.; BRYANT, A. M. and LENG, R. A. 1976. Rate of production of methane in the rumen and the large intestine of sheep. *British Journal of Nutrition* **36:** 1–14.

MUTUO, P. K.; CADISCH, G.; ALBRECHT, A.; PALM, C. A. and VERCHOT, L. 2005. Potential of agroforestry for carbon sequestration and mitigation of greenhouse gas emissions from soils in the tropics. *Nutrient Cycling in Agroecosystems* **71:** 43–54.

NELSON, G. C.; ROSEGRANT, M. W.; JAWOO KOO; ROBERTSON, R.; SULSER, T.; ZHU, T.; RINGLER, C. et al. 2009. *Climate Change: Impact on Agriculture and Costs of Adaptation*. Food Policy Report. International Food Policy Research Institute, Washington DC. 19p.

NEWBOLD, C. J.; LÓPEZ, S.; NELSON, N.; OUDA, J. O.; WALLACE, R. J. and MOSS, A. R. 2005. Proprionate precursors and other metabolic intermediates as possible alternative electron acceptors to methanogenesis in ruminal fermentation in vitro. *British Journal of Nutrition* **94:** 27–35.

NEWBOLD, C. J.; OUDA, J. O.; LÓPEZ, S.; NELSON, N.; OMED, H.; WALLACE, R. J. and MOSS, A. R. 2002. Propionate precursors as possible alternative electron acceptors to methane in ruminal fermentation. In: *Greenhouse Gases and Animal Agriculture*. eds. Takahashi, J. and Young, B. A. Elsevier, Amsterdam. pp. 151–154.

NEWBOLD, C. J. and RODE, L. M. 2006. Dietary additives to control methanogenesis in the rumen. In: *Greenhouse Gases and Animal Agriculture: An Update*, eds. Soliva, C. R.; Takahashi, J. and Kreuzer, M. International Congress Series No. 1293, Elsevier, the Netherlands. pp. 138–147.

NEWMAN, J. A. 2004. Climate change and cereal aphids: The relative effects of increasing CO_2 and temperature on aphid population dynamics. *Global Change Biology* **10:** 5–15.

NEWMAN, J. E. 1980. Climate change impacts on the growing season of the North American 'corn belt'. *Biometeorology* **7:** 128–142.

NICHOLSON, S. E. 1994. Recent rainfall fluctuations in Africa and their relationship to past conditions over the continent. *The Holocene* **4:** 121–131.

NICHOLSON, S. E. 2001. Climatic and environmental change in Africa during the last two centuries. *Climate Research* **17:** 123–144.

NIGGLI, U.; FLIESSBACH, A.; HEPPERLY, P. and SCIALABBA, N. 2009. *Low Greenhouse Gas Agriculture: Mitigation and Adaptation Potential of Sustainable Farming Systems, Rev. 2*. FAO, Rome.

NIGGLI, U.; SLABE, A.; SCHMID, O.; HALBERG, N. and SCHLUETER, M. 2008. *Vision for an Organic Food and Farming Research Agenda to* 2025. Organic Knowledge for the Future. 44p. http://www.tporganics.eu/upload/TPOrganics_VisionResearchAgenda.pdf

NINAN, K. N. and BEDAMATTA, S. 2012. *Climate Change, Agriculture, Poverty and Livelihoods: A Status Report*. Working Paper 277. The Institute for Social and Economic Change, Bangalore. 37p.

NORBY, R. J.; KOBAYASHI, K. and KIMBALL, B. A. 2001. Commentary: Rising CO_2-future ecosystems. *New Phytology* **150:** 215–221.

OELBERMANN, M.; VORONEY, R. P. and GORDON, A. M. 2004. Carbon sequestration in tropical and temperate agroforestry systems: A review with examples from Costa Rica and southern Canada. *Agriculture Ecosystems and Environment* **104:** 359–377.

OENEMA, O., WRAGE, N., VELTHOF, G. L., VAN GROENIGEN, J. W.; DOLFING, J. and KUIKMAN, P. J. 2005. Trends in global nitrous oxide emissions from animal production systems. *Nutrient Cycling in Agroecosystems* **72:** 51–65.

OGLE, S. M.; BREIDT, F. J.; EVE, M. D. and PAUSTIAN, K. 2003. Uncertainty in estimating land use and management impacts on soil organic storage for US agricultural lands between 1982 and 1997. *Global Change Biology* **9:** 1521–1542.

OGLE, S. M.; BREIDT, F. J. and PAUSTIAN, K. 2005. Agricultural management impacts on soil organic carbon storage under moist and dry climatic conditions of temperate and tropical regions. *Biogeochemistry* **72:** 87–121.

OGLE, S. M.; CONANT, R. T. and PAUSTIAN, K. 2004. Deriving grassland management factors for a carbon accounting approach developed by the intergovernmental panel on climate change. *Environmental Management* **33:** 474–484.

OGLE, S. M. and PAUSTIAN, K. 2005. Soil organic carbon as an indicator of environmental quality at the national scale: Monitoring methods and policy relevance. *Canadian Journal of Soil Science* **8:** 531–540.

OLESEN, J. E. and BINDI, M. 2002. Consequences of climate change for European agricultural productivity, land use and policy. *European Journal of Agronomy* **16:** 239–262.

OLESEN, J. E.; SCHELDE, K.; WEISKE, A.; WEISBJERG, M. R.; ASMAN, W. A. H. and DJURHUUS, J. 2006. Modelling greenhouse gas emissions from European conventional and organic dairy farms. *Agriculture, Ecosystems and Environment* **112:** 207–220.

OLSSON, L. and ARDÖ, J. 2002. Soil carbon sequestration in degraded semiarid agro-ecosystems: Perils and potentials. *Ambio* **31:** 471–477.

ORAM, P. A. 1985. Sensitivity of agricultural production to climate change. *Climate Change* **7:** 129–152.

OSMOND, C. B.; OBJORKMAN and ANDERSON, D. J. 1980. *Physiological Processes in Plant Ecology*. Ecological Studies 36. Springer-Verlag, New York. pp. 419–425.

PAN, G. X.; ZHOU, P.; ZHANG, X. H.; LI, L. Q.; ZHENG, J. F.; QIU, D. S. and CHU, Q. H. 2006. Effect of different fertilization practices on crop C assimilation and soil C sequestration: A case of a paddy under a long-term fertilization trial from the Tai Lake region, China. *Acta Ecologica Sinica* **26:** 3704–3710.

PANT, G. B.; RUPA KUMAR, K. and BORGAONKAR, H. P. 1999. Climate and its long-term variability over the Western Himalaya during the past two Centuries. In: *The Himalayan Environment*, eds. Dash, S. K. and Bahadur, J. New Age International (P) Limited, Publishers, New Delhi.

PARK, R.; TREHAN, M. S.; MUVSIL, P. W. and HOWE, R. C. 1988. *The Effects of Sea Level Rise on US Coastal Wetlands. Potential Effects of Global Climate Change on the US*. United States Environmental Protection Agency, Washington. Appendix B 1.1–1.55.

PARRY, M. L. 1990. *Climate Change and World Agriculture*. Earthscan London.

PARRY, M. L.; BLANTRAN, M.; DE ROZARI, A. L.; CHONG, S. and PANICH, S. 1992. *The Potential Socio-Economic Effects of Climate Change in South East Asia*. United Nations Environment Programme, Nairobi.

PARRY, M. L.; CARTER, T. R. and KONIJN, N. 1988. *The Impact of Climatic Variation on Agriculture, Vol. 1. Assessment in Cool Temperature and Cold Regions*. Kluwer, Dordrecht, the Netherlands.

PARRY, M. L.; CARTER, T. R. and PORTER, J. H. 1989. The greenhouse effect and the future of UK agriculture. *Journal of Royal Agricultural Society England* **150:** 120–131.

PARRY, M. L.; PORTER, J. H. and CARTER, D. H. 1990. Climatic change and its implication for agriculture. *Outlook on Agriculture* **19:** 9–15.

PARRY, M. L.; ROSENZWEIG, C. and LIVERMORE, M. 2005. Climate change, global food supply and risk of hunger. *Philosophical Transactions of the Royal Society* **360:** 2125–2136.

PARRY, M. L.; ROSENZWEIG, C.; IGLESIAS, A.; LIVERMORE, M. and FISCHER, G. 2004. Effects of climate change on global food production under SRES emissions and socio-economic scenarios. *Global Environmental Change* **14:** 53–67.

PATRA, A. K.; KAMRA, D. N. and AGARWAL, N. 2006. Effect of spices on rumen fermentation, methanogenesis and protozoa counts in *in vitro* gas production test. In: *Greenhouse Gases and Animal Agriculture: An Update,* eds. Soliva, C. R.; Takahashi, J. and Kreuzer, M. International Congress Series No. 1293. Elsevier, the Netherlands. pp. 176–179.

PATTEY, E.; TRZCINSKI, M. K. and DESJARDINS, R. L. 2005. Quantifying the reduction of greenhouse gas emissions as a result of composting dairy and beef cattle manure. *Nutrient Cycling in Agroecosystems* **72:** 173–187.

PAUL, E. A.; MORRIS, S. J.; SIX, J.; PAUSTIAN, K. and GREGORICH, E. G. 2003. Interpretation of soil carbon and nitrogen dynamics in agricultural and afforested soils. *Soil Science Society of America Journal* **67:** 1620–1628.

PAUSTIAN, K.; BABCOCK, B. A.; HATFIELD, J.; LAL, R.; MCCARL, B. A.; MCLAUGHLIN, S.; MOSIER, A. et al 2004. *Agricultural Mitigation of Greenhouse Gases: Science and Policy Options.* CAST (Council on Agricultural Science and Technology) Report, R141 2004. 120p.

PAUSTIAN, K.; COLE, C. V.; SAUERBECK, D. and SAMPSON, N. 1998. CO_2 mitigation by agriculture: An overview. *Climatic Change* **40:** 135–162.

PEIRIS, D. R.; CRAWFORD, J. W.; GRASHOFF, C.; JEFFERIES, R. A.; PORTER, J. R. and MARSHALL, B. 1996. A simulation study of crop growth and development under climate change. *Agriculture and Forest Meteorology* **79:** 271–287.

PÉREZ-RAMÍREZ, J.; KAPTEIJN, F.; SCHÖFFEL, K. and MOULIJN, J. A. 2003. Formation and control of N_2O in nitric acid production: Where do we stand *today? Applied Catalysis B: Environmental* **44:** 117–151.

PFEFFER, W. T.; HARPER J. T. and O'NEEL, S. 2008. Kinematic constraints on glacier contributions to 21st-century sea-level rise. *Science* **321:** 1340–1343.

PFIFFNER, L. and LUKA, H. 2003. Effects of low-input farming systems on carabids and epigeal spiders—A paired farm approach. *Basic and Applied Ecology* **4:** 117–127.

PFIFFNER, L.; MERKELBACH, L. and LUKA, H. 2003. Do sown wildflower strips enhance the parasitism of lepidopteran pests in cabbage crops? *International Organization for Biological and Integrated Control of Noxious Animals and Plants/West Palaearctic Regional Section Bulletin* **26:** 111–116.

PHETTEPLACE, H. W.; JOHNSON, D. E. and SEIDL, A. F. 2001. Greenhouse gas emissions from simulated beef and dairy livestock systems in the United States. *Nutrient Cycling in Agroecosystems* **60:** 9–102.

PHILLIPS, D. A.; NEWELL, K. D.; HASSALL, S. A. and FILLING, C. E. 1976. The effect of CO_2 enrichment on root nodule development and symbiotic N_2 reduction in *Pisum sativum* L. *American Journal of Botany* **63:** 356–362.

PIMENTEL, D.; HARVEY, C.; RESOSUDARMO, P.; SINCLAIR, K.; KURZ, D.; McNAIR, M.; CRIST, S. et al., 1995. Environmental and economic costs of soil erosion and conservation benefits. *Science* **267:** 1117–1123.

PIMENTEL, D.; HEPPERLY, P.; HANSON, J.; DOUDS, D. and SEIDEL, R. 2005. Environmental, energetic, and economic comparisons of organic and conventional farming systems. *Bioscience* **55:** 573–582.

PINARES-PATIÑO, C. S.; ULYATT, M. J.; WAGHORN, G. C.; HOLMES, C. W.; BARRY, T. N.; LASSEY, K. R. and JOHNSON, D. E. 2003. Methane emission by alpaca and sheep fed on lucerne hay or grazed on pastures of perennial ryegrass/white clover or birdsfoot trefoil. *Journal of Agricultural Science* **140:** 215–226.

PITOVRANOV, S. E.; LAKIMETS, V.; KISELEV, V. I. and SIROTENKO, O. D. 1988. The effects of climatic variations on agriculture in the subarctic zone of the USSR. In: *The Assessment in Cold Temperature and Cold Regions*, eds. Parry, M. L.; Carter, T. R. and Konijn, N. T. Dordrecht, the Netherlands. pp. 617–724.

PROCTOR, P. and COLE, G. 2002. *Grasp the Nettle: Making Biodynamic Farming and Gardening Work.* Random House Publishing, New Zealand.

RATHORE, L. S.; SINGH, K. K.; SASEENDRAN, S. A. and BAXLA, A. K. 2001. Modelling the impact of climate change on rice production in India. *Mausam* **52:** 1.

REAY, D. S.; SMITH, K. A. and EDWARDS, A. C. 2003. Nitrous oxide emission from agricultural drainage waters. *Global Change Biology* **9:** 195–203.

REEDER, J. D.; SCHUMAN, G. E.; MORGAN, J. A. and LECAIN, D. R. 2004. Response of organic and inorganic carbon and nitrogen to long-term grazing of the short grass steppe. *Environmental Management* **33:** 485–495.

REGANOLD, J. P.; ELLIOT, L. F. and UNGER, Y. L. 1987. Long-term effects of organic and conventional farming on soil erosion. *Nature* **330:** 370–372.

REILLY, J.; HOHMANN, N. and KANE, S. 1994. Climate change and agricultural trade: Who benefits and who loses? *Global Environmental Change* **4:** 24–36.

RICE, C. W. and OWENSBY, C. E. 2001. Effects of fire and grazing on soil carbon in rangelands. In: *The Potential of*

U.S. Grazing Lands to Sequester Carbon and Mitigate the Greenhouse Effect, eds. Follet, R.; Kimble, J. M. and Lal, R. Lewis Publishers, Boca Raton, Florida. pp. 323–342.

RICHTER, B. 2004. Using ethanol as an energy source. *Science* **305:** 340.

RICHTER, G. M. and SEMENOV, M. A. 2005. Modelling impacts of climate change on wheat yields in England and Wales: Assessing drought risks. *Agricultural Systems* **84:** 77–97.

ROBERTSON, G. P. 2004. Abatement of nitrous oxide, methane and other non-CO_2 greenhouse gases: The need for a systems approach. In: *The Global Carbon Cycle. Integrating Humans, Climate, and the Natural World*, eds. Field, C. B. and Raupach, M. R. SCOPE 62. Island Press, Washington DC. pp. 493–506.

ROBERTSON, G. P. and GRACE, P. R. 2004. Greenhouse gas fluxes in tropical and temperate agriculture: The need for a full-cost accounting of global warming potentials. *Environment, Development and Sustainability* **6:** 51–63.

ROBERTSON, G. P.; PAUL, E. A. and HARWOOD, R. R. 2000. Greenhouse gases in intensive agriculture: Contributions of individual gases to the radiative forcing of the atmosphere. *Science* **289:** 1922–1925.

ROCHETTE, P. and JANZEN, H. H. 2005. Towards a revised coefficient for estimating N_2O emissions from legumes. *Nutrient Cycling in Agroecosystems* **73:** 171–179.

ROCKSTRÖM, J. 2003. Water for food and nature in drought-prone tropics: Vapour shift in rain-fed agriculture. *Philosophical Transactions of the Royal Society of London B* **358:** 1997–2009.

ROGERS, H. H.; BECK, R. D.; BINGHAM, G. E.; CURE, J. D.; DAVIS, J. M.; HECK, W. W.; RAWLINGS, J. D. et al. 1981. *Response of Vegetation to CO2. Field Studies of Plant Responses to Elevated CO2*. Report no. 5. Botany Department, North Carolina State University Raleigh.

ROGERS, H. H.; BINGHAM, G. E.; CURE, I. E.; HECK, W. W.; HEAGLE, A. S.; ISRAEL, D. W.; SMITH, J. M.; SURANO, K. A. and THOMAS, J. F. 1980. *Response of Vegetation to CO2. Field Studies of Plant Responses to Elevated CO2*. Report no. 1. US Department of Energy and Department of Agriculture, Washington DC.

ROGERS, H. H. and DAHLMAN, R. C. 1993. Crop responses to CO_2 enrichment. *Vegetation* **104–105:** 117–131.

ROGERS, H. H.; SIONIT, N.; CURE, J. D.; SMITH, J. M. and BINGHAM, G. F. 1984. Influence of elevated CO2 on water relations in soybeans. *Plant Physiology* **74:** 233–238.

ROGERS, H. H.; THOMAS, J. F. and BINGHAM, G. E. 1983. Response of agronomic and forest species to elevated atmospheric CO_2. *Science* **220:** 428–429.

ROGNER, H.; CABRERA, M.; FAAIJ, A.; GIROUX, M.; HALL, D.; KAGRAMANIAN, V.; SERGUEI et al. 2000. Energy resources. In: *World Energy Assessment of the United Nations*, eds. Goldemberg, J. et al. UNDP, UNDESA/WEC. UNDP, New York. pp. 135–171.

ROHLING, E. J.; GRANT, K.; HEMLEBEN, C.; SIDDALL, M.; HOOGAKKER, B. A. A.; BOLSHAW, M. and KUCERA, M. 2008. High rates of sea-level rise during the last interglacial period. *Natural Geoscience* **1**: 38–42.

ROSE, S.; AHAMMAD, H.; EICKHOUT, B.; FISHER, B.; KUROSAWA, A.; RAO, S.; RIAHI, K. and VAN VUUREN, D. 2007. *Land in Climate Stabilization Modeling.* Energy Modeling Forum Report, Stanford University. http://www.stanford.edu/group/EMF/projects/group21/Landuse.pdf

ROSEGRANT, M.; PAISNER, M. S. and MEIJER, S. 2001. *Long-Term Prospects for Agriculture and the Resource Base.* The World Bank Rural Development Family. Rural Development Strategy Background Paper #1. The World Bank, Washington.

ROSEGRANT, M. W. and CLINE, S. A. 2003. Global food security: Challenges and policies. *Science* **302**: 1917–1919.

ROSENBERG, N. J. 1992. Adaptation of agriculture to climate change. *Climate Change* **21**: 385–405.

ROSENBERG, N. J.; KIMBALL, B. A.; MARTIN, P. and COOPER, C. F. 1990. From climate and CO_2 enrichment to evapotranspiration. In: *Climate Change and US Water Resources*, ed. Waggonen, P. E., John Wiley, New York. pp. 151–175.

ROSENZWEIG, C. 1985. Potential CO_2 induced effects on North American wheat producing region. *Climate Change* **7**: 367–389.

ROSENZWEIG, C. and HILLEL, D. 1993. Agriculture in a greenhouse world. *Research and Exploration* **9**: 208–221.

ROSENZWEIG, C. and HILLEL, D. 1998. *Climate Change and Global Harvest.* Oxford University Press, Oxford, UK. pp. 135–154.

ROSENZWEIG, C. and PARRY, M. 1994. Potential impacts of climate change on world agriculture. *Nature* **367**: 133–138.

ROY, R. N.; MISRA, R. V. and MONTANEZ, A. 2002. Decreasing reliance on mineral nitrogen: Yet more food. *Ambio* **31**: 177–183.

RUMPLER, W. V.; JOHNSON, D. E. and BATES, D. B. 1986. The effect of high dietary cation concentrations on methanogenesis by steers fed with or without ionophores. *Journal of Animal Science* **62**: 1737–1741.

RUPA KUMAR, K. 2002. Regional climate scenarios. In: *Proc. TERI Workshop on Climate Change: Policy Options for India.* New Delhi, September, 5–6, 2002.

RUPA KUMAR, K. and ASHRIT, R. G. 2001. Regional Aspects of global climate change simulations: Validation and assessment of climate response over Indian monsoon region to transient increase of greenhouse gases and sulphate aerosols. *Mausam* **52**: 229–244.

RUPA KUMAR, K.; KRISHNA KUMAR, K. and PANT, G. B. 1994. Diurnal asymmetry of surface temperature trends over India. *Geophysical Research Letters* **21**: 677–680.

RUPA KUMAR, K.; KUMAR, K.; PRASANNA, V.; KAMALA, K.; DESPHNADE, N. R.; PATWARDHAN, S. K. and PANT, G. B. 2003. Future climate scenario. In: *Climate*

Change and Indian Vulnerability Assessment and Adaptation. Universities Press (India) Pvt. Ltd., Hyderabad.

RUPA KUMAR, K.; PANT, G. B.; PARTHASARATHY, B. and SONTAKKE, N. A. 1992. Spatial and sub-seasonal patterns of the long-term trends of Indian summer monsoon rainfall. *International Journal of Climatology* **12:** 257–268.

SAHOO, S. K. 1999. *Simulating Growth and Yield of Maize in Different Agro-Climatic Regions.* M.Sc. Thesis. Indian Agricultural Research Institute, New Delhi. Unpublished.

SAINI, A. D. and NANDA, R. 1986. Relationship between incident radiation, leaf area and dry matter yield in wheat. *Indian Journal of Agricultural Sciences* **56:** 512–519.

SALINGER, M. J. 1988. Climatic warning: Impact on the New Zealand growing season and implications for temperate Australia. In: *Greenhouse Planning for Climatic Change*, ed. Pearman, G. H. CSIRO, Australia. pp. 564–575.

SANCHEZ, P. A. 2002. Soil fertility and hunger in Africa. *Science* **295:** 2019–2020.

SANCHEZ, P. A. and SWAMINATHAN, M. S. 2005. Cutting world hunger in half. *Science* **307:** 357–359.

SANDERS, J. 2007. *Economic Impact of Agricultural Liberalization Policies on Organic Farming in Switzerland.* Ph.D. thesis, Aberystwyth University.

SANDS, R. D. and McCARL, B. A. 2005. Competitiveness of terrestrial greenhouse gas offsets: Are they a bridge to the future? In: *Abstracts of USDA Symposium on Greenhouse Gases and Carbon Sequestration in Agriculture and Forestry*, March 22–24, USDA, Baltimore, Maryland.

SANGHI, A.; MENDELSOHN, R. and DINAR, A. 1998. The climate sensitivity of Indian agriculture. In: *Measuring the Impact of Climatic Change on Indian Agriculture*, eds. Dinar, A.; Mendelsohn, R.; Evenson, R.; Parikh, J.; Sanghi, A.; Kumar, K.; McKinsey, J. and Lonergan, S. World Bank Technical Report No. 409. World Bank, Washington DC.

SASEENDRAN, A. S. K.; SINGH, K. K.; RATHORE, L. S.; SINGH, S. V. and SINHA, S. K. 2000. Effects of climate change on rice production in the tropical humid climate of Kerala, India. *Climatic Change* **44:** 495–514.

SATHAYE, J.; SHUKLA, P. R. and RAVINDRANATH, N. H. 2006. Climate change, sustainable development and India: Global and national concerns. *Current Science* **90:** 314–325.

SCHILS, R. L. M.; VERHAGEN, A.; AARTS, H. F. M. and SEBEK, L. B. J. 2005. A farm level approach to define successful mitigation strategies for GHG emissions from ruminant livestock systems. *Nutrient Cycling in Agroecosystems* **71:** 163–175.

SCHLESINGER, M. J. and MITCHELL, J. F. B. 1985. Model projections for equilibrium response to increased CO_2 concentration. In: *Projecting the Climatic Effects of Increased CO_2*, eds. MacCracken, M. C. and Luther, F. M. DOE/ER-2037. US Department of Energy, Washington DC. pp. 81–148.

SCHLESINGER, W. H. 1999. Carbon sequestration in soils. *Science* **284:** 2095.

SCHMIDELY, P. 1993. Quantitative review on the use of anabolic hormones in ruminants for meat production. I. Animal performance. *Annales de Zootechie* **42:** 333–359.

SCHNABEL, R. R.; FRANZLUEBBERS, A. J.; STOUT, W. L.; SANDERSON, M. A. and STUEDEMANN, J. A. 2001. The effects of pasture management practices. In: *The Potential of U.S. Grazing Lands to Sequester Carbon and Mitigate the Greenhouse Effect*, eds. Follett, R. F.; Kimble, J. M. and Lal, R. Lewis Publishers, Boca Raton, Florida. pp. 291–322.

SCHNEIDER, U. A. and MCCARL, B. A. 2003. Economic potential of biomass based fuels for greenhouse gas emission mitigation. *Environmental and Resource Economics* **24:** 291–312.

SCHOLES, R. J. and BIGGS, R. 2004. *Ecosystem Services in Southern Africa: A Regional Assessment.* CSIR, Pretoria.

SCHOLES, R. J. and VAN DER MERWE, M. R. 1996. Sequestration of carbon in savannas and woodlands. *The Environmental Professional* **18:** 96–103.

SCHUMAN, G. E.; HERRICK, J. E. and JANZEN, H. H. 2001. The dynamics of soil carbon in rangelands. In: *The Potential of U.S. Grazing Lands to Sequester Carbon and Mitigate the Greenhouse Effect*, eds. Follett, R. F.; Kimble, J. M. and Lal, R. Lewis Publishers, Boca Raton, Florida. pp. 267–290.

SEINO, H. 1995. *Implications of Climate Change for Crop Production in Japan. Climate Change and Agriculture: Analysis of Potential Agricultural Impacts.* Special Publication no. 59. American Society of Agronomy, Madison. pp. 293–306.

SHEEHAN, J.; ADEN, A.; PAUSTIAN, K.; KILLIAN, K.; BRENNER, J.; WALSH, M. and NELSON, R. 2004. Energy and environmental aspects of using corn stover for fuel ethanol. *Journal of Industrial Ecology* **7:** 117–146.

SIEGRIST, S.; STAUB, D.; PFIFFNER, L. and MÄDER, P. 1998. Does organic agriculture reduce soil erodibility? The results of a long-term field study on loess in Switzerland. *Agriculture, Ecosystems and Environment* **69:** 253–264.

SINCLAIR, T. R. ET AL. 2000. Leaf nitrogen concentration of wheat subjected to elevated (CO_2) and either water or N deficits. *Agricultural Ecosystem and Environment* **79:** 53–60.

SINGH, N. and SONTAKKE, N. A. 2002. On climatic fluctuations and environmental changes of the Indo-Gangetic Plains, India. *Climatic Change* **52:** 287–313.

SINGH, R. S.; NARAIN, P. and SHARMA, K. D. 2001. Climate changes in Luni river basin of arid western Rajasthan (India). *Vayu Mandal* **31:** 103–106.

SINHA, S. K. and SWAMINATHAN, M. S. 1991. Deforestation, climate change and sustainable nutrients security. *Climatic Change* **16:** 33–45.

SIROTENKO, O. D.; ABASHINA, H. V.; PARLOVA, V. N. and DIXON. R. K. 1997. Sensitivity of the Russian agriculture to change in climate, CO_2 and tropospheric ozone concentrations and soil fertility. *Climate Change* **36:** 217–232.

SISTI, C. P. J.; SANTOS, H. P.; KOHHANN, R.; ALVES, B. J. R.; URQUIAGA, S. and BODDEY, R. M. 2004. Change in carbon and nitrogen stocks in soil under 13 years of conventional or zero tillage in southern Brazil. *Soil and Tillage Research* **76:** 39–58.

SIX, J.; OGLE, S. M.; BREIDT, F. J.; CONANT, R. T.; MOSIER, A. R. and PAUSTIAN, K. 2004. The potential to mitigate global warming with no-tillage management is only realized when practiced in the long term. *Global Change Biology* **10:**155–160.

SMIT, B.; LUDLOW, L. and BRKLACICH, M. 1989. Implication of a global climatic warming for agriculture a review and appraisal. *Journal of Environmental Quality* **17:** 519–527.

SMITH, J. U.; SMITH, P.; WATTENBACH, M.; ZAEHLE, S.; HIEDERER, R.; JONES, R. J. A.; MONTANARELLA, L.; ROUNSEVELL, M. D. A.; REGINSTER, I. and EWERT, F. 2005b. Projected changes in mineral soil carbon of European croplands and grasslands, 1990–2080. *Global Change Biology* **11:** 2141–2152.

SMITH, K. A. and CONEN, F. 2004. Impacts of land management on fluxes of trace greenhouse gases. *Soil Use and Management* **20:** 255–263.

SMITH, P. 2004a. Carbon sequestration in croplands: The potential in Europe and the global context. *European Journal of Agronomy* **20:** 229–236.

SMITH, P. 2004b. Engineered biological sinks on land. In: *The Global Carbon Cycle. Integrating Humans, Climate, and the Natural World*, eds. Field, C. B. and Raupach, M. R. SCOPE 62. Island Press, Washington DC. pp. 479–491.

SMITH, P. 2004c. Monitoring and verification of soil carbon changes under Article 3.4 of the Kyoto Protocol. *Soil Use and Management* **20:** 264–270.

SMITH, P.; ANDRÉN, O.; KARLSSON, T.; PERÄLÄ, P.; REGINA, K.; ROUNSEVELL, M. and VAN WESEMAEL, B. 2005a. Carbon sequestration potential in European croplands has been overestimated. *Global Change Biology* **11:** 2153–2163.

SMITH, P.; GOULDING, K. W.; SMITH, K. A.; POWLSON, D. S.; SMITH, J. U.; FALLOON, P. D. and COLEMAN, K. 2001. Enhancing the carbon sink in European agricultural soils: Including trace gas fluxes in estimates of carbon mitigation potential. *Nutrient Cycling in Agroecosystems* **60:** 237–252.

SMITH, P.; MARTINO, D.; CAI, Z.; GWARY, D.; JANZEN, H. H.; KUMAR, P.; McCARL, B. et al. 2007b. Greenhouse gas mitigation in agriculture. *Philosophical Transactions of the Royal Society Biological Sciences* **363:** 789–813.

SMITH, P.; MARTINO, D.; CAI, Z.; GWARY, D.; JANZEN, H. H.; KUMAR, P.; McCARL, B. A. et al. 2007c. Policy and technological constraints to implementation of greenhouse gas mitigation options in agriculture. *Agriculture, Ecosystems and Environment* **118:** 6–28.

SMITH, P.; MARTINO, D.; CAI, Z.; GWARY, D.; JANZEN, H.; KUMAR, P.; MCCARL, B. et al. 2007a. Agriculture. In: *Climate Change 2007: Mitigation. Contribution of Working Group III to the Fourth Assessment Report of the Intergovernmental Panel on Climate Change*, eds. Metz, B; Davidson, O. R.; Bosch, P. R.; Dave, R. and Meyer, L. A. Cambridge University Press, Cambridge, United Kingdom and New York, NY, USA.

SOLIVA, C. R.; TAKAHASHI, J. and KREUZER, M. (EDS.) 2006. *Greenhouse Gases and Animal Agriculture: An Update*. International Congress Series No. 1293, Elsevier, the Netherlands 377p.

SOUSSANA, J. F.; LOISEAU, P.; VIUCHARD, N.; CESCHIA, E.; BALESDENT, J.; CHEVALLIER, T. and ARROUAYS, D. 2004. Carbon cycling and sequestration opportunities in temperate grasslands. *Soil Use and Management* **20:** 219–230.

SPATARI, S.; ZHANG, Y. and MACLEAN, H. L. 2005. Life cycle assessment of switchgrass- and corn stover-derived ethanol-fueled automobiles. *Environmental Science and Technology* **39:** 9750–9758.

STALEY, J. T. and JOHNSON, S. N. 2008. Climate change impacts on root herbivores. In: *Root Feeders: An Ecosystem Perspective*, eds. Johnson, S. N., and Murray, P. J. CABI, Wallingford, UK.

STEPHENSON, B. D.; DOUVILLE, H. and RUPA KUMAR, K 2001. Searching for a fingerprint of global warming in the Asian summer monsoon. *Mausam* **52:** 213–220.

STERN, N. 2006. Stern review on the economics of climate change. Her Majesty's Treasury and the Cabinet Office, London. http://www.hmtreasury.gov.uk/independent_reviews/stern_review_economics_climate_change/sternreview_index.cfm

STERN, N. 2007. *The Economics of Climate Change: The Stern Review*. Cambridge University Press, Cambridge.

STOCKLE, O. C.; DYKE, P. T.; WILLIAMS, J. R.; JONES, C. A. and ROSENBERG, N. J. 1992b. A method for estimating the direct and climatic effects of rising atmospheric CO_2 on growth and yield of crops: Part II—Sensitivity analysis at three sites in the Midwestern USA. *Agricultural Systems* **38:** 239–256.

STOCKLE, O. C.; WILLIAMS, J. R.; ROSENBERG, N. J. and JONES, C. A. 1992a. A method for estimating the direct and climatic effects of rising atmospheric CO_2 on growth and yield of crops: Part I—Modification of the EPIC model for climate change analysis. *Agricultural Systems* **38:** 225–238.

STOTT, P. A.; STONE, D. A. and ALLEN, M. R. 2004. Human contribution to the European heat wave of 2003. *Nature* **432:** 610–614.

STRENGERS, B.; LEEMANS, R.; EICKHOUT, B.; DE VRIES, B. and BOUWMAN, L. 2004. The land-use projections and resulting emissions in the IPCC SRES scenarios as simulated by the IMAGE 2.2 model. *GeoJournal* **61:** 381–393.

TAN, G. X. and SHIBASAKI, R. 2006. Global estimation of crop productivity and the impacts of global warming by GIS and EPIC integration. *Ecological Modeling* **168**: 357–370.

TAO, F. L.; YOKOZAWA, M.; HAYASHI, Y. and LIN, E. D. 2003. Future climate change, the agricultural water cycle, and agricultural production in China. *Agricultural Ecosystem and Environment* **97**: 361–361.

TATE, K. R.; ROSS, D. J.; SCOTT, N. A.; RODDA, N. J.; TOWNSEND, J. A. and ARNOLD, G. C. 2006. Post-harvest patterns of carbon dioxide production, methane uptake and nitrous oxide production in a *Pinus radiata* D. Don plantation. *Forest Ecology and Management* **228**: 40–50.

TENGÖ, M. and BELFRAGE, K. 2004. Local management practices for dealing with change and uncertainty: A cross-scale comparison of cases in Sweden and Tanzania. *Ecology and Society* **9**: 4. http://www.ecologyandsociety.org/vol19/iss3/art4/

THAPLIYAL, V. and KULSHRESTHA, S. M. 1991. Climate change and trends over India. *Mausam* **42**: 333–338.

TILMAN, D.; FARGIONE, J.; WOLFF, B.; D'ANTONIO, C.; DOBSON, A.; HOWARTH, R.; SCHINDLER, D.; SCHLESINGER, W. H.; SIMBERLOFF, D. and SWACKHAMER, D. 2001. Forecasting agriculturally driven global environmental change. *Science* **292**: 281–284.

TINKER, P. B.; INGRAM, J. S. I. and STRUWE, S. 1996. Effects of slash and burn agriculture and deforestation on climate change. *Agriculture, Ecosystems and Environment* **58**: 13–22.

TITUS, J. G. 1990. Effect of climate change on sea level rise and the implications for world agriculture. *Horticultural Science* **25**: 1567–1571.

TORBERT, H. A.; PRIOR, S. A.; HOGERS, H. H. and WOOD, C. W. 2000. Review of elevated atmospheric CO_2 effects on agro-ecosystems: Residue decomposition processes and soil C storage. *Plant and Soil* **224**: 59–73.

TREWAVAS, A. 2002 MALTHUS FOILED AGAIN and AGAIN. *Nature* **418**: 668–670.

TUBIELLO, F. N. and EWERT, F. 2002. Simulating the effects of elevated CO_2 on crops: Approaches and applications for climate change. *European Journal of Agronomy* **18**: 57–74.

TUBIELLO, F. N.; ROSENZWEIG, C.; GOLDBERG, R. A.; JAGTAP, S., and JONES, J. W. 2002. Effects of climate change on US crop production: Simulation results using two different GCM scenarios. Part I: wheat, potato, maize, and citrus. *Climate Research* **20**: 259–270.

TUCK, G.; GLENDINING, M. J.; SMITH, P.; HOUSE, J. I. and WATTENBACH, M. 2006. The potential distribution of bioenergy crops in Europe under present and future climate. *Biomass and Bioenergy* **30**: 183–197.

UNKOVICH, M. 2003. Water use, competition, and crop production in low rainfall, alley farming systems of south-eastern Australia. *Australian Journal of Agricultural Research* **54**: 751–762.

UPRETY, D. C.; CHAKRAVARTY, N. V. K.; KATIYAL, R. K. and ABROL, Y. P. 1996. Climate variability and Brassica. In: *Climate Variability and Agriculture*, eds. Abrol, Y. P.; Gadgil, S. and Pant, G. B. Narosa Publishing House, New Delhi.

VAN GROENIGEN, K. J.; GORISSEN, A.; SIX, J.; HARRIS, D.; KUIKMAN, P. J.; VAN GROENIGEN, J. W. and VAN KESSEL, C. 2005. Decomposition of 14C-labeled roots in a pasture soil exposed to 10 years of elevated CO_2. *Soil Biology and Biochemistry* **37**: 497–506.

VAN NEVEL, C. J. and DEMEYER, D. I. 1995. Lipolysis and biohydrogenation of soybean oil in the rumen in vitro: Inhibition by antimicrobials. *Journal of Dairy Science* **78**: 2797–2806.

VAN NEVEL, C. J. and DEMEYER, D. I. 1996. Influence of antibiotics and a deaminase inhibitor on volatile fatty acids and methane production from detergent washed hay and soluble starch by rumen microbes in vitro. *Animal Feed Science and Technology* **37**: 21–31.

VAN OOST, K.; GOVERS, G.; QUINE, T. A. and HECKRATH, G. 2004. Comment on 'Managing soil carbon' (I). *Science* **305**: 1567.

VAN WILGEN, B. W.; GOVENDER, N.; BIGGS, H. C.; NTSALA, D. and FUNDA, X. N. 2004. Response of savanna fire regimes to changing fire-management policies in a large African National Park. *Conservation Biology* **18**: 1533–1540.

VENKATARAMAN, C.; HABIB, G.; EIGUREN-FERNANDEZ, A.; MIGUEL, A. H. and FRIEDLANDER, S. K. 2005. Residential biofuels in south Asia: Carbonaceous aerosol emissions and climate impacts. *Science* **307**: 1454–1456.

VINER, D.; SAYER, M.; UYARRA, M. and HODGSON, N. 2006. *Climate Change and the European Countryside: Impacts on Land Management and Response Strategies*. Report prepared for the Country Land and Business Association, UK. Publ., CLA, UK. 180p.

VON CAEMMERER, S. and FURBANK, R. T. 2003. The C-4 pathway: An efficient CO_2 pump. *Photosynthesis Research* **77**: 191–207.

VU, J. C. V.; ALIEN, L. H. JR.; BROOTE, K. J. and BOWES, G. 1997. Effects of elevated CO_2 and temperature on photosynthesis and RuBisco in rice and soybean. *Plant Cell and Environment* **20**: 68–76.

WAGGONER, P. T. 1992. Preparing US agriculture for global climate change. *Science of Food and Agriculture* **4**: 2–5.

WANG, B., H. NEUE, and H. SAMONTE, 1997. Effect of cultivar difference on methane emissions. *Agriculture, Ecosystems and Environment*, **62**: 31–40.

WANG, M. X. and SHANGGUAN, X. J. 1996. CH_4 emission from various rice fields in PR China. *Theoretical and Applied Climatology* **55**: 129–138.

WASSMANN, R.; LANTIN, R. S.; NEUE, H. U.; BUENDIA, L. V.; CORTON, T. M. and LU, Y. 2000. Characterization of methane emissions from rice fields in Asia. III. Mitigation options and future research needs. *Nutrient Cycling Agroecosystems* **58:** 23–36.

WATSON, R.; ZINYORWERA, M. and MOSS, R. (EDS.) 1996. *Climate Change 1995: Impacts, Adaptation and Mitigation of Climate Change*, Contribution of WG II to the Second Assessment Report of the IPCC, Cambridge University Press, Cambridge, UK.

WEBSTER P. J. 2008. Myanmar's deadly daffodil. *Natural Geoscience* **1:** 488–490.

WEST, T. O. and MARLAND, G. 2003. Net carbon flux from agriculture: Carbon emissions, carbon sequestration, crop yield, and land-use change. *Biogeochemistry* **63:** 73–83.

WEST, T. O. and POST, W. M. 2002. Soil organic carbon sequestration rates by tillage and crop rotation: A global data analysis. *Soil Science Society of America Journal* **66:** 1930–1946.

WIDODO, W.; VU, J. C. V.; BOOTE, K. J.; BAKER, J. T. and ALLEN JR., L. H. 2003. Elevated growth CO_2 delays drought stress and accelerates recovery of rice leaf photosynthesis. *Environment and Experiment Botany* **49:** 259–272.

WILKS, D. S. 1988. Estimating the consequences of CO_2 induced climate change on North American grain agriculture using GIS information. *Climate Change* **13:** 19–42.

WILLER, H. and KILCHER, L. (EDS.) 2009. *The World of Organic Agriculture*. Statistics and Emerging Trends 2009. IFOAM, DE-Bonn and FiBL, CH-Frick.

WILLIAMS, G. V. D. and OAKES, W. T. 1978. Climatic resources for maturing barley and wheat in Canada. In: *Essays on Meteorology and Climatology—In Honour of R. W. Longley*, eds. Haye, K. D. and Reinelt, E. R. Studies in Geography Monograph 3, University of Canada. 367p.

WITTWER, S. H. 1980. CO_2 and climate change: An agricultural perspective. *Journal of Soil Water Conservation* **35:** 116–120.

WITTWER, S. H. 1985. CO_2 levels in the biosphere: Effects on plant productivity. *Critical Review of Plant Science* **2:** 171–198.

WITTWER, S. H. 1986. *Worldwide Status and History of CO_2 Enrichment: An Overview. CO_2 Enrichment of Greenhouse Crops, Vol. I. Status and CO_2 Sources*. CRC Press, Boca Raton, Florida.

WITTWER, S. H. 1990. Implication of the greenhouse effect on crop productivity. *Horticulture Science* **25:** 1560–1567.

WITTWER, S. H. 1995. *Food, Climate and CO_2: The Global Environment and World Food Production*. Michigan State University, Michigan.

WITTWER, S. H. and ROBB, W. 1964. CO_2 enrichment of greenhouse atmospheres for food crop production. *Economic Botany* **18:** 34–56.

WOLIN, E. A.; WOLF, R. S. and WOLIN, M. J. 1964. Microbial formation of methane. *Journal of Bacteriology* **87:** 993–998.

WRIGHT, A. D. G.; KENNEDY, P.; O'NEILL, C. J.; TROOVEY, A. F.; POPOVSKI, S.; REA, S. M.; PIMM, C. L. and KLEIN. L. 2004. Reducing methane emissions in sheep by immunization against rumen methanogens. *Vaccine* **22:** 3976–3985.

WYSS, E.; NIGGLI, U. and NENTWIG, W. 1995. The impact of spiders on aphid populations in a strip-managed apple orchard. *Journal of Applied Entomology* **119**: 473–478.

XIANG, C. G.; ZHANG, P. J.; PAN, G. X.; QIU, D. S. and CHU, Q. H. 2006. Changes in diversity, protein content, and amino acid composition of earthworms from a paddy soil under different long-term fertilizations in the Tai Lake Region, China. *Acta Ecologica Sinica* **26**: 1667–1674.

XU, H.; CAI, Z. C.; JIA, Z. J. and TSURUTA, H. 2000. Effect of land management in winter crop season on CH4 emission during the following flooded and rice-growing period. *Nutrient Cycling in Agroecosystems* **58**: 327–332.

XU, H.; CAI, Z. C. and TSURUTA, H. 2003. Soil moisture between rice-growing seasons affects methane emission, production, and oxidation. *Soil Science Society of America Journal* **67**: 1147–1157.

XU, S.; HAO, X.; STANFORD, K.; MCALLISTER, T.; LARNEY, F. J. and WANG, J. 2007. Greenhouse gas emissions during co-composting of cattle mortalities with manure. *Nutrient Cycling in Agroecosystems* **78**: 177–187.

YAGI, K.; TSURUTA, H. and MINAMI, K. 1997. Possible options for mitigating methane emission from rice cultivation. *Nutrient Cycling in Agroecosystems* **49**: 213–220.

YAN, X.; OHARA, T. and AKIMOTO, H. 2003. Development of region-specific emission factors and estimation of methane emission from rice field in East, Southeast and South Asian countries. *Global Change Biology* **9**: 237–254.

YOSHINO, M. M.; HORIE, T.; SEINO, H.; TSUJII, H; UCHIJIMA, T. and UCHIJIMA, Z. 1988. The impact of climatic variations in agriculture. In: *The Assessment in Cold Temperature and Cold Regions*, eds. Parry, M. L.; Carter, T. R. and Konijn, N. T. Dordrecht, the Netherlands. pp. 725–863.

ZEDDIES, J. 1995. Agricultural policy and climate protection. *Agrar Wirtschaft* **44**: 157–159.

ZEHNDER, G.; GURR, G. M.; KÜHNE, S.; WADE, M. R.; WRATTEN, S. D. and WYSS, E. 2007. Arthropod pest management in organic crops. *Annual Review of Entomology* **52**: 57–80.

ZEMP, M.; ROER, I.; KÄÄB, A.; HOELZLE, M.; PAUL, F. and HAEBERLI, W. 2008. *Global Glacier Changes: Facts and Figures*. UNEP World Glacier Monitoring Service Zurich, Switzerland.

ZHANG, J. C. 1989. THE CO_2 problem in climate and dryness in North China. *Meteorological Magazine* **15**: 3–8.

ZHAO, Z. C. and KELLOGG, W. W. 1988. Sensitivity to soil moisture to doubling of CO2 in climate model experiments. Part II. *The Asian Monsoon Regional Journal of Climate* **1**: 367–378.

ZHOU, X. L.; HARRINGTON, R.; WOIWOD, I. P.; PERRY, J. N.; BALE, J. S. and CLARK, S. J. 1995. Effects of temperature on aphid phenology. *Global Change Biology* **1**: 303–313.

CHAPTER TWO

Climate change and agromet advisory services in Indian agriculture

Rakesh Singh Sengar and H.S. Bhadoria

Contents

Abstract

The improved estimates of global change impacts on global-scale crop yield trends will require several scientific advances. Some, such as predicting the rates of global temperature increase or the behaviour of farmers in the face of gradual trends, are beyond the scope of the traditional plant physiology community. Into mid-century, the growth rates in aggregate crop productivity will continue to be mainly driven by technological and agronomic improvements, just as they have for the past century. Even in the most pessimistic scenarios, it is highly unlikely that climate change would result in a net decline in global yields. Instead, the relevant question at the global scale is how much of a headwind climate change could contribute in the perpetual race to keep productivity growing as fast as demand? Overall, the net effect of climate change and CO_2 on the global average supply of calories is likely to be fairly close to zero over the next few decades, but it could be as large as 20–30% of overall yield trends. Of course, this global picture hides many changes at smaller scales that could be of great relevance to food security, even if global production is maintained.

2.1 Introduction

Climate change is a long-term shift in the statistics of the weather (including its averages). For example, it could show up as a change in climate normals (expected average values for temperature and precipitation) for a given place and time of year, from one decade to the next (Hansen 2002). We know that the global climate is changing. The last decade of the twentieth century and the beginning of the twenty-first century has been the warmest period in the entire global instrumental temperature record, which began in the mid-nineteenth century.

Global climate change

1. Global mean temperatures have increased by 0.74°C in the last 100 years (Rathore et al. 2003).

2. Greenhouse gases (GHG) (CO_2, methane, nitrous oxide) increases are mainly caused by fossil fuel use and a change in land usage (Figure 2.1).

3. Temperatures will increase by 1.8–6.4°C by 2100 AD. Greater increase in rabi.

4. Precipitation is likely to increase in kharif.

5. Snow cover is projected to contract.

6. More frequent hot extremes, heavy precipitations.

7. Sea level rises to be 0.18–0.59 m (Figure 2.2).

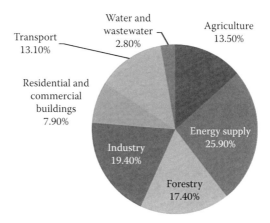

FIGURE 2.1 Contribution of different sectors in the world to climate change. Sources of greenhouse gas emissions.

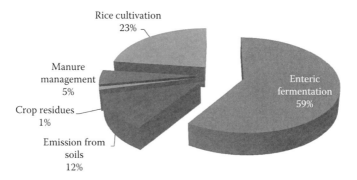

FIGURE 2.2 Sectors of agriculture in India that contribute to climate change.

2.2 Projected impacts of climate change on Indian agriculture

1. Imbalance in food trade due to positive impacts on Europe and North America; however, with negative impacts on the United States (Singh et al. 2009).
2. Increased water, shelter and energy requirement for livestock; implications for milk production.
3. Increasing sea and river water temperatures are likely to affect fish breeding, migration and harvests. Coral reefs would start declining around 2030.
4. Considerable effect on microbes, pathogens and insects.

2.3 Adaptation and mitigation framework: The need to consider the emerging scenario

1. Greater demand for (quality) food; yields need to increase by 30–50% by 2030 (Figure 2.3).
2. Increasing urbanisation and globalisation.
3. Increasing competition from other sectors for land, energy, water and capital.
4. Climate change is a continuous process; greater focus is on short-term actions on adaptation and mitigation (Table 2.1).

2.4 Why is agro-meteorological advisory services required in India?

- About 60% of the people depend on agriculture in India.
- About 43% of India's land is under agricultural use.
- Agriculture plays an important role in the Indian economy.
- To increase awareness about climate changes among farmers (Varshneya et al. 2009).

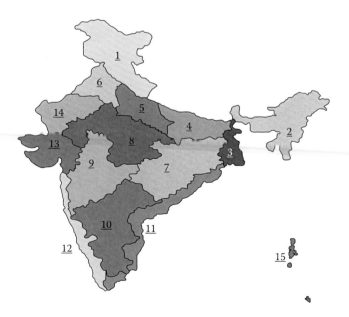

FIGURE 2.3 Agro-climatic zones of India.

Table 2.1   Agro-climatic regions/zones in India

S.no.	Agro-climatic regions/zones	States represented
1	Western Himalayan region	Himachal Pradesh, Jammu and Kashmir, Uttarakhand
2	Eastern Himalayan region	Arunachal Pradesh, Assam, Manipur, Meghalaya, Mizoram, Nagaland, Sikkim, Tripura, West Bengal
3	Lower Gangetic plain region	West Bengal
4	Middle Gangetic plain region	Uttar Pradesh, Bihar
5	Upper Gangetic plain region	Uttar Pradesh
6	Trans-Gangetic plain region	Chandigarh, Delhi, Haryana, Punjab, Rajasthan
7	Eastern plateau and hills region	Chhattisgarh, Jharkhand, Madhya Pradesh, Maharashtra, Orissa, West Bengal
8	Central plateau and hills region	Madhya Pradesh, Rajasthan, Uttar Pradesh
9	Western plateau and hills region	Madhya Pradesh, Maharashtra
10	Southern plateau and hills region	Andhra Pradesh, Karnataka, Tamil Nadu
11	East coast plains and hills region	Andhra Pradesh, Orissa, Pondicherry, Tamil Nadu
12	West coast plains and ghat region	Goa, Karnataka, Kerala, Maharashtra, Tamil Nadu
13	Gujarat plains and hills region	Gujarat, Dadra and Nagar Haveli, Daman and Diu
14	Western dry region	Rajasthan
15	Island region	Andaman and Nicobar Islands, Lakshadweep

Note: Planning Commission has identified 15 resource development regions in the country, 14 in the mainland and remaining one in the islands of Bay of Bengal and Arabian Sea.

- Agro-meteorology may help farmers through weather forecasting in agricultural activities such as seed sowing, irrigation, spraying of chemicals, harvesting and so forth (Figure 2.4).

2.5   Role of weather forecasting in agricultural management

- Selection of crop and varieties (Maji et al. 2008)
- Land preparation and ratio management under crops and varieties
- Deciding seed sowing and harvesting time of crops

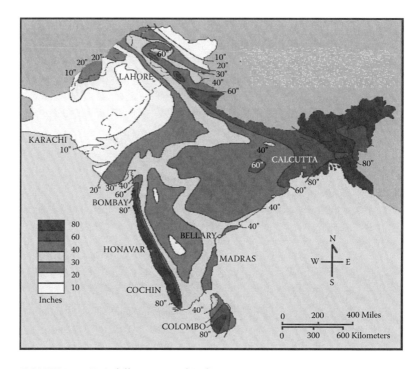

FIGURE 2.4   Rainfall pattern of India.

- Helping in irrigation management according to rainfall forecasting (Surendra et al. 2009)
- Helping in the protection of crops from adverse climatic situation through weather forecasting, such as frost
- Helping in deciding time for the application of fertilisers, weed killer, fungicide, insecticide and so on (Figure 2.5)

2.6 Development history of weather forecasting based agro-advisory service

- Department of Agro-Meteorology established in 1932 at the national level at New Delhi (Marty et al. 2008)
- Farmers' weather bulletin started in 1945
- State-level agro-advisory services started in 1976
- Medium-range weather forecasting-based agro-advisory services for agro-climatic zones of India started in 1991 by National Centre for Medium Range Weather Forecasting, New Delhi (Figure 2.6)

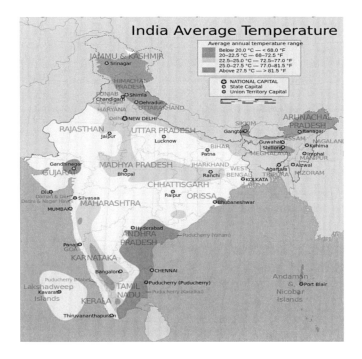

FIGURE 2.5 (**See colour insert.**) Temperature pattern of India.

- National Centre for Medium Range Weather Forecasting merged with Indian Meteorological Department, New Delhi
- District-level agro-advisory services started in 2008
- Block-level agro-advisory services started in 2013

2.7 Major aspects of agro-advisory service

1. Sowing/transplanting of kharif crops according to information about effectiveness of a monsoon (Figure 2.7)
2. Sowing of rabi crops according to moisture availability and temperature suitability (Seth et al. 2009)
3. Advice regarding the spray of insecticide or fungicide according to wind velocity and direction (Figure 2.8)
4. Advice about delaying the application of fertilisers, according to the intensity of rainfall (Figure 2.9)
5. Forecasting of harmful insects and disease severity according to weather parameters

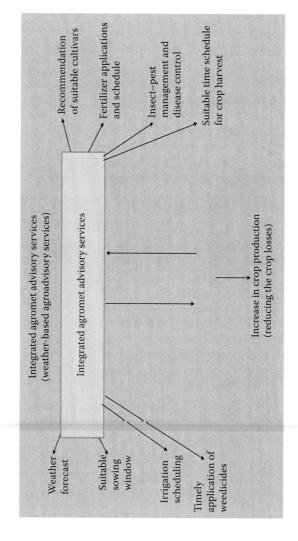

FIGURE 2.6   Major aspects of agro-advisory service.

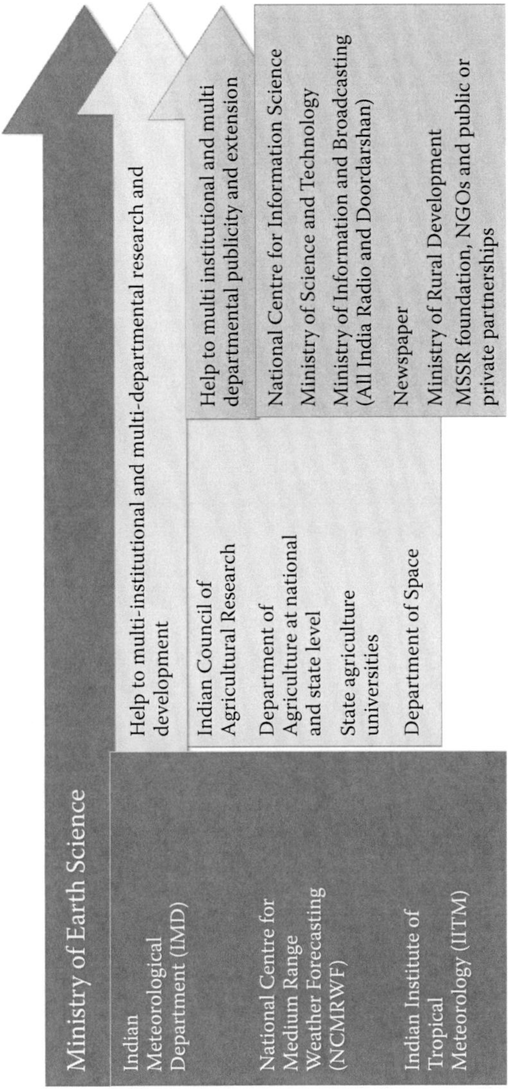

Ministry of Earth Science

Indian Meteorological Department (IMD)

National Centre for Medium Range Weather Forecasting (NCMRWF)

Indian Institute of Tropical Meteorology (IITM)

Help to multi-institutional and multi-departmental research and development

Indian Council of Agricultural Research

Department of Agriculture at national and state level

State agriculture universities

Department of Space

Help to multi institutional and multi departmental publicity and extension

National Centre for Information Science

Ministry of Science and Technology

Ministry of Information and Broadcasting (All India Radio and Doordarshan)

Newspaper

Ministry of Rural Development

MSSR foundation, NGOs and public or private partnerships

FIGURE 2.7 Associated agencies.

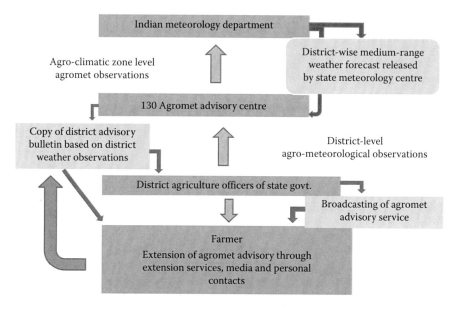

FIGURE 2.8 (See colour insert.) District-level agromet advisory service.

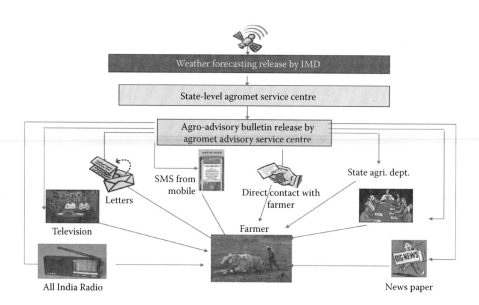

FIGURE 2.9 (See colour insert.) Effective communication mediums for communication of weather forecasting between agromet service centre and beneficiaries.

✓ TV   Existing arrangements
✓ Air
✓ Print
✓ Personal contact

New institutional arrangements

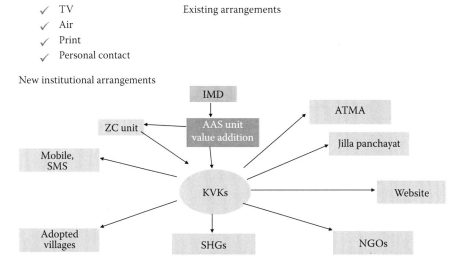

FIGURE 2.10   Dissemination of weather forecast.

6. Advice about the optimum time of irrigation according to pan evaporation (Figure 2.10)

7. Advice regarding the optimum time of crop harvest

2.8 Future strategies

- Agro-advisory services will be started at the tehsil/block level
- Weather forecasting SMS facilities on mobile will be increased
- Quality of weather forecasting will be increased at the block level
- Participation in weather-based crop insurance scheme
- Increase in the accuracy of weather forecasting

References

HANSEN, J.W.B. 2002. Realizing the potential benefits of climate prediction to agriculture: Issue approaches and challenges. *Agricultural System* 74: 309–330.

MAJI, M.D., BANERJEE, R., DAS, N.K., CHAKRABORTY, S. and BAJPAI, A.K. 2008. Role of meterological factors on the incidence of mulberry diseases. *Journal of Agrometeorology* 4: (special issue-Part I): 193–196.

MURTHY, N.S., SHAH, S., and SINGH, R.K. 2008. Climate and its variability over the western Himalaya. *Journal of Agrometeorology* 11(special issues-II): 296–298.

RATHORE, L.S, SINGH, K.K. and GUPTA, A. 2003. National Center for Medium Range Weather Forecasting: Activities, current status and future plans. In *National Seminar on Agrometeorology in the New Millennium: Prospective and Challenges*, October 20–31, PAU, Ludhiana. Abstract/Souvenir, pp. 141–147.

SETH, R., ANSARI, V.A., DATTA, M. 2009. Weather-risk hedging by farmers: An empirical study of willingness-to-pay in Rajasthan, India. *The Journal of Risk Finance*, 10(1): 54–66.

SINGH, P.K., SINGH, K.K., RATHORE, L.S. and BAXLA, A.K. 2009. Climatic variability in Jhansi region of Uttar Pradesh. *Journal of Agrometeorology* 11(special issue-part I): 51–53.

SINGH, S., SINGH, D., and RAO, VUM, 2009. Weather index crop insurance: A climatic risk management option for farmer in rural India. *Journal of Agrometeorology* 11(special issues-I): 245–248.

VARSHNEYA, M.C., VAIDYA, V.B., PANDEY, V., DAMLE, K. 2009. Validation of rainfall forecast predicted by Monsoon Research Almanac-2009. *Journal of Agrometeorology* 11(special issue): 175–179.

CHAPTER THREE

Crop modelling for agriculture production and management

Padmakar Tripathi and Arvind Kumar

Contents

Abstract

Crop modelling can play a significant part in systems approaches by providing a powerful capability for scenario analysis. Crop modelling has developed extensively over the past 30 years and a diverse range of crops models are now available. It is argued, however, that the tendency to distinguish between and separate the so-called 'scientific' and 'engineering' challenges and approaches in crop modelling has constrained the maturation of modelling. It is considered that effective crop modelling must combine a scientific approach to enhance understanding with an application orientation to retain a focus on prediction and problem solving. Greater use of crop simulation models has also been suggested to increase the efficiency of different trials. While simulation models successfully capture the temporal variation, they use a lumped parameter approach that assumes spatial variability of the soils, crops or climate.

3.1 Introduction: Crop weather simulation modelling

Crop is defined as 'aggregation of individual plant species grown in a unit area for economic purpose', whereas irreversible increase in size and volume and the consequences of differentiation and distribution occurring in a plant is known as *growth*. Reproducing the essence of a system without reproducing the system itself is called *simulation*. In simulation, the essential characteristics of the system are reproduced in a model, which is then studied in an abbreviated time scale.

The agricultural region can be considered a collection of individual fields that vary in environmental conditions and management practices. An increase in the population, demands an increase in agricultural production with available resources. Efficient management of available resources with variable weather conditions is essential to increase the productivity of agriculture. In addition to this, the focus of agricultural production is changing from quantity towards quality and sustainability (Aggarwal et al., 1997). Solution of these new challenges requires consideration of how numerous components interact to effect plant growth. These transitions force farmers and agricultural advisors to deal with increasing bulks of information (Aggarwal et al., 2006). They need to analyse vast and sporadically located information resources. The information gathering process is cumbersome and sometimes unreliable. Often the task to select, combine and analyse the information is demanding. As information technology has opened up new challenges to automate data and analysis, computer programmes that simulate the crop growth or yield of crops under different management regimes, help farmers make technical decision to manage their crops better. Since 1960, the large-scale evolution of computers allowed the ability to synthesise detailed knowledge on plant physiological processes in order to explain the functioning of crops as a whole. Insights into various processes were expressed using mathematical equations and integrated in simulation models. Therefore, a model can be defined in different ways by scientists: (a) A model is a schematic representation of the conception of an agricultural system or an act of mimicry or a set of equations, which represents the behaviour of a system. (b) A model is 'a representation of an object, system or idea in some form other than that of the entity itself'. Its purpose is usually to aid in explaining, understanding or improving the performance of a system.

In the beginning, models were meant to increase the understanding of crop behaviour by explaining crop growth and development, in terms of understanding physiological mechanisms (Bachelet et al., 1993). Over the years, new insights and different research questions motivated further development of simulation models. In addition to their explanatory function, the applicability of well-tested models for extrapolation and prediction was quickly recognised and more application-oriented models were developed. For instance, demands for advisory systems for farmers and scenario studies for policy makers resulted in the evolution of models geared towards tactical and strategic decision support, respectively. Now, crop growth modelling and simulation have become accepted tools for agricultural research (Boote and Toolenaar, 1994).

3.2 Types of crop models

Depending on the purpose for which they are designed, models are classified into different groups or types.

Statistical models

These models express the relationship between yield or yield components and weather parameters. In these models, relationships are measured in a system using statistical techniques. In a statistical model approach, one or several variables (representing weather or climate, soil characteristics or a time trend) are related to crop responses such as yield and yield contributing characters. The independent variables are weather parameters derived from agrometeorological variances. The weighting coefficients in these equations are obtained in the statistical manner using standard statistical procedures. Such variables are used as multivariable regression analysis. This statistical approach does not easily lead to an exploration of the cause and effect or relationship, but it is a very practical approach for the assessment or prediction of yield and its related parameters. The coefficients in the statistical model and the validity of the estimates depend to a large extent on the design of the model, as well as on the representations of the input data. If the soil and climate conditions and the cropping practice are fairly homogeneous over a specific area represented by the input data, or if soil and geography are properly weighted in the equations, then it can be expected that the coefficients and the estimates have a practical significance for the assessment of the crop conditions or predictions of yield for any specific area in question. Regression models are attractive because of their simple and straightforward relation between yield and one or more environment factors, but these are not accurate enough to be used for other areas and other crops (Chou and Chen, 1995). Despite this limitation, they are used extensively for the prediction of a single crop yield over a large region with a variety of soils, agronomic practices and insect-disease problems. A combination of such factors is still beyond the success of dynamic simulation models.

The following points may be incorporated to provide the accurate forecast of crop growth and development by a statistical model:

1. In a statistical model, each predictor for the regression equation must have a significant value, and the year must be included in all equations reflecting the impact of technology. Also, the equation predictor for border district must be sorted out.

2. A model for different climatic conditions within the districts and ensemble technique for crop yield forecast should be developed.

3. The ecological level should be included in the crop simulation model. At any altitude, the weather data can be taken by multiplying the lapse rate at this altitude.

4. For validation of forecast, the trail/experimental field should be at a controlled condition and also at different climatic conditions/ecological conditions/district levels so that it represents the farmers' field condition.

5. Statistical significance of the parameter is required for regression equation.

6. Different weather variables, namely, T_{max}, T_{min}, relative humidity (morning and evening) and rainfall having significant value are considered to develop the regression equation for prediction of yield. A regression model is required to isolate the yield dependent on weather only, instead of composite weather and agricultural package fertiliser, irrigation, seed and so on. Thus, a regression model, though economic in time consumption, might contain large error percentages in yield prediction. A case study of a wheat crop yield prediction of 14 districts of eastern Uttar Pradesh during 2012–2013 at pre-harvest stage (on 15 March, 2013) has been shown in Table 3.1, compared with a yield predicted during 2011–2012 on the same date. This shows the error variability in the range of 0–10% on either side in the model.

Mechanistic models	These models explain not only the relationship between weather parameters and yield but also the mechanism of these models (explains the relationship of influencing dependent variables). These models are based on physical selection.
Deterministic models	These models estimate the exact value of the yield or dependent variable. Usually, these are developed by mathematical techniques and have well-defined coefficients.
Stochastic models	For this model, the probability element is attached to each output. For each set of inputs, different outputs are given along with probabilities. These models define the yield or state of dependent variable at a given rate.
Dynamic models	Time is included as a variable in this model. Both dependent and independent variables have values that remain constant over a given period of time.
Static models	In a static model, time is included not as a variable. Dependent and independent variables having values remain constant over a given period of time.
Descriptive model	A descriptive model defines the behaviour of a system in a simple manner. The model reflects little or none of the mechanisms that are the causes of the phenomena, but consists of one or more mathematical equations. An example of such an equation is the one derived from successively measured weights of

Table 3.1 Statistical yield forecast of wheat crop at F_3 stage (pre-harvest) for 2012–2013 of districts of eastern Uttar Pradesh

Crop	District	Equation	Predicted yield (kg/ha) 2012–2013	Average yield (last 20 years) (kg/ha)	Error % as compared to 2011–2012
Wheat	Faizabad	3125.326 + 0.258 (Z251) + 0.6698 (Z131) + 0.258 (Z141)	3048	2482.4	3.85
	Sultanpur	2966.38 + 31.97 (TIME) + 1.118 (Z250) + 25.897 (Z21)	3112	2515.4	2.11
	Barabanki	3008.265 + 1.2153 (Z151) + 1.242(Z131) + (−11.268) (Z31)	3348	2658.7	8.67
	Gonda	2335.224 + 0.1454 (Z341) + 0.158 (Z141) + 0.222 (Z230)	3087	2494.8	5.54
	Basti	2298.172 + 52.89 (TIME) + 0.258 (Z341) + 0.120 (Z230)	2958	2371.5	6.87
	Shravasti	4008.352 + 44.58 (TIME) + 0.141 (Z141) + 0.187 (Z341)	2806	2243.9	4.39
	Balrampur	3589.368 + 51.426 (TIME) + 0.589 (Z241) + 0.158 (Z230)	3154	2362.7	4.32
	Siddharthnagar	3589.125 + 0.1542 (Z451) + 5.14 (Z41)	2882	2305.5	10.39

Wheat				
Maharajganj	$2589.3 + (-52.258) (Z11) + 1587.3$ $(TIME) + 2.8569 (Z121)$	3258	2459.2	−0.32
Gorakhpur	$2789.325 + 1.158 (Z141) + 3.698 (Z451)$	3080	2394.1	1.57
Santkabirnagar	$4285.325 + 30.48 (TIME) + (-0.158) (Z141)$	2936	2462.5	−2.68
Ambedkarnagar	$3896.58 + 1.158 (Z121) + 22.587$ $(TIME) + 14.211 (Z31)$	3681	3095.5	5.40
Bahraich	$2116.358 + 1.0123 (Z241) + 2.258$ $(Z121) + 4.103 (Z20) + (-1.4106) (Z10)$	2911	2345.6	−0.11
Raibareilly	$1124.358 + 32.151 (TIME) + 0.1367$ $(Z131) + 5.5821 (Z10)$	3156	2480.8	8.52
Deoria	$2965.3154 + 89.687 (Z21)$	2509	2363.1	3.34

a crop. The equation is helpful to determine quickly the weight of the crop where no observation was made.

Explanatory models

This model consists of quantitative description of the mechanisms and processes that causes the behaviour of the system. To create this model, a system is analysed and its processes and mechanisms are quantified separately. The model is built by integrating these descriptions for the entire system. It contains descriptions of distinct processes such as leaf area expansion, tiller production and so on. Crop growth is a consequence of these processes.

Simulation models

Computer models, in general, are a mathematical representation of a real-world system. One of the main goals of crop simulation models is to estimate agricultural production as a function of weather and soil conditions as well as crop management. These models use one or more sets of differential equations and calculate both rate and state variables over time, normally from planting until harvest maturity or final harvest.

The Earth's land resources are finite, whereas the number of people that the land must support continues to grow rapidly. This creates a major problem for agriculture. The production/productivity must be increased to meet the rapidly growing demands, while the natural resources must be protected. New agricultural research is needed to supply information to farmers, policy makers and other decision makers on how to accomplish sustainable agriculture over the wide variations in climate around the world. In this direction, explanation and prediction of growth of managed and natural ecosystems in response to climate and soil-related factors are of increasing importance as the objectives of science (Dhaliwal et al., 1997). Quantitative prediction of complex systems, however, depends on integrating information through levels of organisation, and the principal approach for that is through the construction of statistical and simulation models. In simulation of systems, use and balance of carbon, beginning with the input of carbon from canopy assimilation, forms the essential core of most simulations that deal with the growth of vegetation. Systems are webs or cycles of interacting components. Change in one component of a system produces changes in other components because of the interactions. For example, a change in weather to warm and humid may lead to a more rapid development of a plant disease, a loss in the yield of a crop and consequent financial adversity for individual farmers and also for the people of a region. Most natural systems are complex. Many do not have boundaries. The bio-system is composed of a complex interaction among the soil, the atmosphere and the plants that live in it. A change or alteration of one element may yield both desirable and undesirable consequences. Minimising the undesirable, while reaching the desired end result is the principal aim of the agro-meteorologists. In any engineering work related to agricultural meteorology, the use of mathematical modelling is essential. Of the different modelling techniques, mathematical modelling enables one to predict the behaviour of design while keeping the expense at a minimum. Agricultural systems are basically modified ecosystems. Managing these systems is very difficult (Hoogenboom et al., 1999). These systems are influenced by the weather both in length and breadth. So, these have to be managed through systems models, which are possible only through classical engineering expertise. A simple example of a simulation model has been presented in Figure 3.1.

In the mid-1960s, crop simulation models integrated knowledge of physiological processes and morphological traits to help explain yield formation in environments varying in physical, biological and agronomic factors. These simulations can be used to evaluate key interactions quickly and identify traits with the greatest impact on yield potential and for assessing the relationship between crop productivity and environmental factors. They have been shown to be efficient in determining the response of crop plants to changes in weather. Examples of such models include erosion productivity impact calculator (EPIC), CERES and GAPS. The Sites Network for Agrotechnology

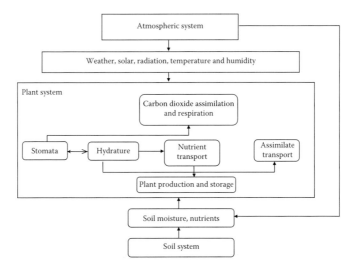

FIGURE 3.1 Simple example of a simulation model.

Transfer (IBSNAT) project proposed many models for soya bean growth—SOYGRO. Other models in the areas of irrigation management, soil physics, nutrition, pests and diseases are already in use. There is even a model used for a long time in biology and agriculture;, the one invented to describe the progeny resulting from crossing plants with one or more characters controlled by dominant-recessive genes. One of the most important uses of models is to forecast the results produced by a given system in response to a given set of inputs. One very important future use of models in agriculture is to forecast the effects of certain environmental conditions and agricultural practices on crop performance. Being useful tools for researchers, models have also been developed and applied to solve complex agricultural problems. There was practically no demand for farm-level models a couple of decades ago, but today, many farmers have mobile phones through which they can access the Internet. Another growing group of model users in the governmental agencies are concerned with developing agricultural and environmental policies.

In planning and analysing agricultural systems, it is essential not only to consider variability but also to think of it in terms directly relevant to components of the system. Such analysis may be a relatively straightforward probabilistic analysis of particular events, such as the start of the crop season in India. The principal effects of weather on crop growth and development are well understood and predictable. Crop simulation models can predict responses to large variations in weather. At every point of the application, weather data are the most important input. The main goal of most applications of crop models is to predict commercial output such as grain yield, fruits, root, biomass for fodder and so on. In general, the management applications of crop simulation models can be defined as (1) strategic applications (crop models are run prior to planting), (2) practical applications (crop models are run prior to and during crop growth) and (3) forecasting applications (models are run to predict yield both prior to and during crop growth). Crop simulation models are used in the United States and in Europe by farmers, private agencies and policy makers, to a great extent, for decision making. Under Indian climatic conditions, these applications have an excellent role to play. The reasons include the dependence on monsoon rains for all agricultural operations in India. Once the arrival of the monsoon is delayed, the policy makers and agricultural scientists are under tremendous pressure. They need to go for contingency plans. These models enable the evaluation of alternative management strategies, quickly, effectively and at no/low cost.

To account for the interaction of the management scenarios with weather conditions and the risk associated with unpredictable weather, the simulations are conducted for at least 20–30 different weather seasons or weather years. Currently, historical weather data is used when available and, if not, weather generators are used.

A few models commonly used for forecasting the yield using agrometeorological, crop inputs and crop characters, are listed next.

DSSAT crop models The decision support system for agrotechnology transfer (DSSAT) has been in use for the last 15 years by researchers worldwide. This package incorporates models of 16 different crops with software that facilitates the evaluation and application of the crop models for different purposes. Over the last few years, it has become increasingly difficult to maintain the DSSAT crop models, partly due to fact that there were different sets of computer code for different crops with little attention given to software design at the levels of the crop models themselves. Thus, the DSSAT crop models have been redesigned and programmed to facilitate a more efficient incorporation of new scientific advances, applications, documentation and maintenance. The DSSAT was originally developed by an international network of scientists, cooperating in the International Benchmark Sites Network for Agrotechnology Transfer project to facilitate the application of crop models in a systems approach to agronomic research. Its initial development was motivated by a need to integrate knowledge about soil, climate, crops and management in order to make better decisions about transferring production technology from one location to others where the soils and climate differed. The systems approach provided a framework in which research is conducted to understand how the system and its components function. This understanding is then integrated into models that allow one to predict the behaviour of the system for given conditions. After one is confident that the models simulate the real world adequately, computer experiments can be performed hundreds or even thousands of times for given environments to determine how to best manage or control the system. The DSSAT helps decision makers by reducing the time and human resources required for analysing complex alternative decisions. It also provides a framework for scientific cooperation through research to integrate new knowledge and apply it to research questions. Prior to the development of the DSSAT, crop models were available, but these were used mostly in labs where they were created. The DSSAT is a collection of independent programmes that operate together; crop simulation models are at its centre. Databases describe weather, soil, experiment conditions and measurements; and genotype information for applying the models to different situations. Software helps users prepare these databases and compare simulated results with observations to give them confidence in the models or to determine if modifications are needed to improve accuracy (Uehara, 1989; Jones et al., 2003). In addition, the programme contained in DSSAT allows users to simulate options for crop management over a number of years in order to assess the risks associated with each option. DSSAT was first released in 1989 with its V 2.1; additional releases were made: V 3.0 in 1994, V 3.5 in 1989 and V 4.5 is the latest version, being used for forecasting. The schematic diagram of the DSSAT model is shown in Figure 3.2.

DSSAT models and crops   The crop and models included in DSSAT are as listed below.

CERES models for cereals   The CERES (Crop Estimation through Resource and Environment Synthesis) family of crop models included rice, wheat, barley, maize, sorghum and millet.

CROPGRO models for legumes   The CROPGRO (CROP GROwth) family of crop models included soya bean, drybean, peanut and chickpea.

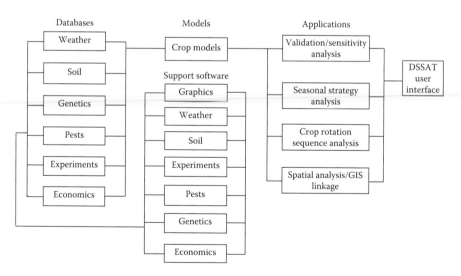

FIGURE 3.2 Schematic drawing of DSSAT components.

SUBSTOR models for root crops The root crop or SUBSTOR (SUBterranean STORage) models included cassava, aroid and potato.

Other crop models Other crop models included in DSSAT are for sugarcane, tomato, sunflower and pasture.

InfoCrop model InfoCrop is a decision support system (DSS) that has been developed by the National Agricultural Technology Project (NATP) by the Indian Council of Agricultural Research (ICAR). It is based on a generic crop model that has been developed to provide a platform to scientists to build their applications around it and to meet the goals of stakeholders need for information. The models in this DSS have a similar structure and are designed to simulate the effects of weather, soils, agronomic management, including planting, nitrogen, residues, irrigation and major pests on crop growth and yield. In particular, it is based on MACROS, WTGROWS, ORYZA 1 and SUCROS models. It is user-friendly and is targeted to increase applications of crop models in research and development, and also has simple and easily available input requirements. InfoCrop has been developed for 12 crops, namely, rice, wheat, sorghum, millet, sugarcane, chickpea, pigeon pea, cotton, mustard, groundnut, potato and, of course, maize. The flowchart of input and output files of the InfoCrop model and other characters have been depicted in Figure 3.3. It is a dynamic crop yield model, developed by Aggarwal and his coworkers from the Centre for Application of Systems Simulation, IARI, New Delhi. It is a mechanistic and dynamic crop simulation model, which can deal with the interaction among weather, crop/variety, soils and management, besides major pests. It has the capacity to evaluate the production of major annual crops, namely, rice, wheat, sorghum, millet, sugarcane, chickpea, pigeon pea, cotton, mustard, groundnut, potato and maize, and has a built-in database of Indian soils.

The InfoCrop model provides several outputs relating to growth and development, water use, N uptake, soil carbon, greenhouse gas emissions and yield losses due to various pests. It can be used to accelerate the application of available knowledge at field, farm and regional levels. This model also has the capability of analyzing experimental data, estimating the potential yield and yield gaps and also assessing the impacts of climatic variability and climate change. The model also works efficiently for management optimisation and assesses the environmental impact study. Thus, this model is most versatile and has many agricultural applications used for DSSAT.

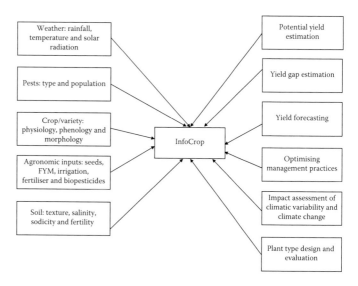

FIGURE 3.3 Context diagram of InfoCrop depicting the input requirement on the left-hand side and its possible application on the right-hand side.

InfoCrop is a DSS based on crop models that has been developed by a network of scientists to provide a platform for scientists and extension workers to build their applications around it and to meet the goals of the stakeholders need for information. The model is designed to simulate the effects of weather, soils, agronomic management, nitrogen, water and major pests on crop growth and yield, water and nitrogen management and greenhouse gas emission. InfoCrop is user-friendly, targeted to increase the applications of crop models in research and development and has simple and easily available input requirements. InfoCrop is developed for 12 crops, including rice and wheat. Crop models in InfoCrop are sensitive to the environment; for example, radiation, temperature, rainfall, wind speed, vapour pressure, flooding, frost; soil (depth of planting/transplanting, seed rates, amount and time of irrigation and N fertilisation (including organic) in different soil depths and pests, for example, population/severity of pests and their timing of appearance).

3.3 Climate change and crop modelling

Climate change is defined as 'any long-term substantial deviation from the present climate because of variations in weather and climatic elements'.

Below are the some of the factors that are responsible for climate change:

1. Natural causes such as changes in the earth's revolution, changes in the area of the continents, variations in the solar system and so forth.

2. Owing to human activities, the concentrations of carbon dioxide and certain other harmful atmospheric gases have been increasing. The present level of carbon dioxide is 325 ppm and it is expected to reach 700 ppm by the end of this century because of the present trend of burning forests, grasslands and fossil fuels. Few models predicted an increase of 2.3–4.6°C in the average temperature and precipitation per day from 10% to 32% in India.

Role of crop modelling in climate change studies

In recent years, there has been a growing concern that changes in climate will lead to a significant damage to both market and no-market sectors. Climate change will have a negative effect in many countries. But the adaptation of farmers to climate change, through changes in farming practices, cropping patterns and use of new technologies will help to ease the impact. The variability of our climate and especially the associated weather extremes is currently one of the concerns of the scientific community as well as the general community. The application of crop models to study the potential impact of climate change and climate variability provides a direct link among models, agrometeorology and the concerns of the society. As climate change deals with future issues, the use of crop simulation models compared to surveys proves to be a more scientific approach to study the impact of climate change on agricultural production and world food security. DSSAT is one of the first packages that modified weather simulation generators or introduced a package to evaluate the performance of models for climate change situations.

3.4 Applications and uses of crop growth models in agriculture

Crop growth models are being developed to meet the demands under the following situations in agricultural meteorology:

1. When farmers face the difficult task of managing their crops on poor soils in harsh and risky climates

2. When scientists and research managers need tools that can assist them in taking an integrated approach to

finding solutions in the complex problem of weather, soil and crop management

3. When policy makers and administrators need simple tools that can assist them in policy management in agricultural meteorology

The potential uses of crop growth models for practical applications are as follows.

On-farm decision making and agronomic management

The models allow the evaluation of one or more options that are available with respect to one or more agronomic decisions such as

- Determining the optimum planting date as shown in Figures 3.4 and 3.5
- Determining the best choice of cultivars
- Evaluating weather risk
- Investment decisions

Crop growth models can be used to predict crop performance in regions where the crop has not been grown before or not grown under optimal conditions. A model can calculate the probabilities of grain yield levels for a given soil type based on rainfall. Investment decisions such as purchase of irrigation systems can be taken with an eye on long-term usage of the equipment thus acquired.

Understanding of research

In agrometeorological research, the crop models basically help in

- Testing scientific hypothesis
- Highlighting when information is missing
- Organising data

Date of transplanting	Simulated	Observed	% of Error
5th July	4271	4110	3.7
15th July	3624	3460	4.5
25th July	3599	3290	8.0

Error percentage increases with delay in transplanting.

FIGURE 3.4 Comparison of observed with simulated values of yield in kg/ha of rice at different dates of transplanting.

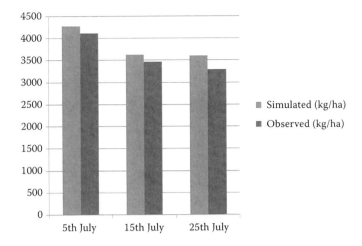

FIGURE 3.5 CERES-Rice model for yield prediction in kg/ha at different dates of transplanting.

- Integrating across disciplines
- Assisting in genetic improvement
  - Evaluate optimum genetic traits for specific environment
  - Evaluate cultivar stability under long-term weather

Policy management

Policy management is one very useful application of crop simulation models. The issues range from global (impacts of climate change on crops) to field-level (effect of crop rotation on soil quality). During 1997, it was shown that in Burkina Faso, crop simulation modelling using satellite and ground-based data could estimate millet production for an early warning of famine which can allow policy makers the time they need to take appropriate steps to ameliorate the effects of global food shortage on vulnerable urban and rural populations. In Australia, it was observed that during November–December when the SOI (Southern Oscillation Index) phase is positive, there is an 80% chance of exceeding average district yields. Conversely, in years when the November–December SOI phase is either negative or rapidly falling, there is only a 5% chance of exceeding average district yields, but 95% chance of below average yields. This information allows the industry to adjust strategically for the expected volume of production. Crop models can be used to understand the effects of climate change such as consequences of elevated CO_2 and changes in temperature and rainfall on

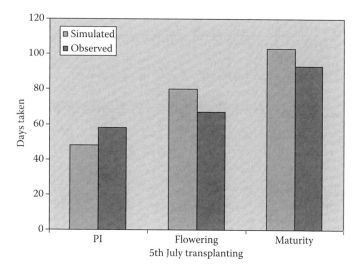

FIGURE 3.6   CERES-Rice model for yield prediction in kg/ha at different phenophases.

crop development, growth and yield. Ultimately, the breeders can anticipate future requirements based on the climate change.

Improved quantitative models for forecasting regional productivity with the various factors will be crucial for evaluating trade-offs associated with potential changes. As optimum crop production estimation becomes more complex, involving several factors such as fertiliser, pest control, genotype, environment and cultural practices, conducting trials with various combinations of these factors becomes very complex and expensive. The influence of soil, water and climatic variables on rice productivity can be effectively estimated through different rice models. The use of rice crop models is very important for suggesting best management practices, forecasting yields, pest and disease incidences, suitable varieties and best sowing dates for optimum crop production with variable climatic conditions. The days taken by rice crop for panicle initiation, flowering and maturity were simulated and compared with observed values. It was found that the model fairly simulated the days taken as shown in Figure 3.6.

3.5   Crop simulation model and agricultural production

Generally, the relation of weather factors with the growth and development of a crop is expressed by an equation that is commonly termed as a model. Thus, a model serves the purpose of predicting the behaviour of crops in response to the weather variations. The model simulates/limitates the behaviour of a real crop by predicting the growth of its components. Crop growth is a very complex phenomenon and a product of a series of complicated interactions of soil, plant and weather. A crop growth model synthesises our insights into the physiological and ecological processes that govern crop growth into mathematical equations.

Several regression models have been developed by many workers to predict the relationship with rice crop productivity and its components. These models, when dealing with multi-year time series, usually include a technology trend factor, thus lumping everything other than climatic factors into one regressor. In addition to climatic factors, there are a large number of edaphic, hydrologic, biotic, agronomic and socio-economic factors that influence crop growth and productivity. Crop models can accelerate inter-disciplinary knowledge utilisation in agricultural research and development. These models present an opportunity for assessing potential production in a region and facilitate analysis of the sustainability options for agricultural development, including planning of resource allocation. These approaches have been used in the recent past for determining the production potential of a location knowing its resources, germplasm and the level of available technology, in matching agrotechnology with the resources of farmers and in analysing the precise reasons for yield gap, in estimating crop yield before the actual harvest and in studying short- and long-term consequences of climatic variability and climatic change on agriculture.

The use of various crop simulation models has been classified into three primary categories: (i) for research knowledge synthesis, (ii) for crop system decision management and (iii) and for policy analysis. Crop models have been used to assist in the genetic improvement of crops by (i) determining optimal genetic traits of plants for specific environment and (ii) predicting the performance of new cultivars for specific environments, thus reducing the number of locations or seasons of multilocation breeding trials. The greater application of crop models in agricultural research and development, however, requires a simple, user-friendly modelling framework, whose inputs are easily available/measurable. In addition, the framework should provide a structure that can be easily integrated in the application and not be very user-friendly. Preliminary results have also indicated that some of them do not perform very well in many tropical environments characterised by limited inputs and semi-arid climate.

The CERES-Rice model is a process-oriented, management-level model of rice crop growth and development that predicts the duration of growth, the average growth rates and the amount of assimilate partitioned to the economic yield components of the plant (Hundal and Kaur, 1999). The simulation processes of the model are dynamic and are affected by environmental and culti-var-specific factors. The duration of growth for a particular culti-var, however, is highly dependent on its thermal environment and to some extent the photoperiod during floral induction. Therefore, the model requires input data such as daily weather data, initial soil conditions, crop management and crop cultivar informa-tion. The daily weather data includes solar radiation, precipita-tion, maximum and minimum temperature. Initial soil conditions involve drainage and runoff coefficients, initial soil water, rooting preference factors, organic nitrogen and carbon contents. The out-put data for each model simulation run encompasses the results of simulated daily growth and development, carbon balance, soil water balance, nitrogen balance and mineral nutrient aspects.

3.6 Conclusion

Crop modelling can play a significant role in system approaches by providing a powerful capability for scenario analysis. Crop modelling has developed extensively over the past 30 years and a diverse range of crops models are now available. It is argued, however, that the tendency to distinguish between and sepa-rate the so-called 'scientific' and 'engineering' challenges and approaches in crop modelling has constrained the maturation of modelling. It is considered that effective crop modelling must combine a scientific approach to enhance understanding with an applications orientation to retain a focus on prediction and problem-solving. Greater use of crop simulation models has also been suggested to increase the efficiency of different tri-als. While simulation models successfully capture the temporal variation, they use a lumped parameter approach that assumes the spatial variability of the soils, crops or climate.

References

AGGARWAL, P.K., BANERJEE, B., DARYAEI, M.G., BHATIA, A., BALA, A., RANI, S., CHANDER, S, PATHAK, H. and KALRA, N. 2006. InfoCrop: A dynamic simulation model for the assessment of crop yield, losses due to pest and environmental impact of agro-ecosystem in tropical environments: IT. Performance of the model, *Agricultural Systems*, 89: 47–67.

AGGARWAL, P.K., KROPTT, M.J., CASMAN, K.G. and TEN BERGE, H.F.M. 1997. Simulating genotypic strategies for increasing rice yield potential in irrigated, tropical environments. *Field Crops Research*, 51: 5–17.

BACHELET, D., VAN SICKLE, J. and GAY, C.A. 1993. The impacts of climate change on rice yield: Evaluation of the efficacy of different modelling approaches. In: F.W.T. Penning de Vries, P. Teng and K. Metselaar (Eds.), *Systems Approaches for Agricultural Development*. Kluwer Academic Publishers, The Netherlands, pp. 145–174.

BOOTE, K.J. and TOOLENAAR, M. 1994. Modelling genetic yield potential. In: K.J. Boote, J.M. Bennett, T.R. Sinclair and G.M. Paulsen (Eds.), *Physiology and Determination of Crop Yield*. ASA, CSSA, SSSA, Madison, WI, pp. 533–566.

CHOU, T.Y. and CHEN, H.Y. 1995. The application of a decision-support system for agricultural land management. *Journal of Agriculture and Forestry*, 44: 75–89.

DHALIWAL, L.K., SINGH, G. and MAHI, G.S. 1997. Dynamic simulation of wheat growth, development and yield with CERES-Wheat model. *Annals of Agricultural Research*, 18: 157–164.

HOOGENBOOM, G., WILKENS, P.W., THORNTON, P.K., JONES, J.W., HUNT, L.A. and IMAMURA, D.T. 1999. Decision support system for agrometeorology transfer V 3.5. In: Hoogenboom, G., Wilkens, P.W., Tsuji, G.Y. (Eds.), *DSSAT Version 3*. vol. 4 (ISBN1-886684-04-9), University of Hawaii, Honolulu, HI, pp. 1–36.

HUNDAL, S.S. and KAUR, P. 1999. Evaluation of agronomic practices for rice using computer simulation model, CERES-Rice. *Oryza*, 36(1): 63–65.

JONES, J.W., HOOGENBOOM, G., PORTER, C.H., BOOTE, K.J., BATCHELOR, W.D., HUNT, L.A., WILKENS, P.W., SINGH, V., GIJSMAN, A.J. and RITCHIE, J.T. 2003. The DSSAT cropping system model. *European Journal of Agronomy* 18: 235–265.

UEHARA, H. 1989. Occlusion of small hepatic veins associated with systemic lupus erythematosus with the lupus anticoagulant and anti-cardiolipin antibody. *Hepatogastroenterology*, 36(5): 393–397.

CHAPTER FOUR

Statistical techniques for studying the impact of climate change on crop production

Seema Jaggi, Eldho Varghese and Arpan Bhowmik

Contents

Abstract

Statistical science plays a major role in any scientific investigation. The use of appropriate statistical techniques for analysing data is very crucial to obtain a meaningful interpretation of the investigation. Statistical analyses play an important role in agro-meteorology, as they provide a means of interrelating series of data from diverse sources, namely, biological data, soil and crop data and atmospheric measurements. Owing to the complexity and multiplicity of the effects of environmental factors on agricultural production, it is necessary to use statistical techniques to detect the interactions of these factors and their practical consequences. Based on the objectives and interests of the research, appropriate statistical models starting from multiple regression analysis to logistic regression to time series modelling have to be employed. The main objective of this chapter is to provide the basic idea behind some statistical tools that can be successfully employed to study the effect of climatic factors on crop production.

4.1   Introduction

Statistics has two major components: descriptive statistics and inferential statistics. Descriptive statistics gives numerical and graphical procedures to summarise a collection of data in a clear and understandable way whereas inferential statistics provides procedures to draw inferences about a population from a sample. Descriptive statistics are used to describe the basic features of the data in a study. They provide simple summaries about the sample and the measures. Together with simple graphics analysis, they form the basis of virtually every quantitative analysis of data. Descriptive statistics are used to present quantitative descriptions in a manageable form. In a research study, there may be lots of measures. Descriptive statistics help us to simplify large amounts of data in a sensible way. Using the numerical approach one might compute statistics such as the mean and standard deviation which are precise and objective. Graphical methods are more suited for identifying patterns or trends in the data. The numerical and graphical approaches complement each other. Some of the graphical approaches are *Box plot*, which is an excellent tool for conveying location and variation information in datasets, particularly for detecting and illustrating location and variation changes between different groups of data; *scatter plot*, which reveals relationships or association between two variables, such relationships manifest themselves by any non-random structure in the plot; *probability plot*, which is a graphical technique for assessing whether or not a dataset follows a given distribution such as the normal; *histogram* and *stem-and-leaf plot*, which is a display that organises data to show its shape and distribution and so on.

Rainfall is an important climatic variable that imposes crop production risks, especially on rain-fed subsistence cultivation

Table 4.1 Total annual rainfall (mm) and production (million tonnes) of wheat in India (1950–1999)

Year	0	1	2	3	4	5	6	7	8	9
195-	1136.1	1051	1108.1	1200.4	1213.8	1321.7	1390.6	1136.3	1268.9	1324.7
	6.46	6.18	7.5	8.02	9.04	8.76	9.4	7.99	9.96	10.32
196-	1241	1366.6	1169.4	1172	1205.9	1008.3	1092.3	1162.3	1078	1158.2
	11	12.07	10.78	9.85	12.26	10.4	11.39	16.54	18.65	20.09
197-	1174	1202.3	1022.6	1207.5	1073.9	1280.8	1156.8	1266.5	1229.8	1101.3
	26.41	24.74	21.78	24.1	28.84	29.01	31.75	35.51	31.83	36.31
198-	1358.9	1184.5	1117.2	1265.7	1143.1	1137.6	1142	1082.2	1298.1	1149
	36.31	37.45	42.79	45.48	44.07	47.05	44.32	46.17	54.11	49.85
199-	1125.2	1168.9	1119.9	1209.4	1277	1198.6	1190	1211.8	1250.9	1181
	55.14	55.69	57.21	59.84	65.77	62.1	69.35	66.35	71.29	76.37

Source: Indiastat.com.

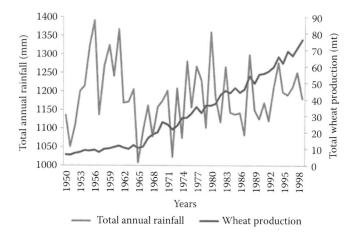

FIGURE 4.1 (**See colour insert.**) Trend in total annual rainfall (mm) and production (million tonnes) of wheat over years.

systems on marginal land. Table 4.1 gives the total annual rainfall (mm) along with the total production (million tonnes) of wheat in India from 1950 to 1999. It can be seen from Figure 4.1 that the trend in total annual rainfall (mm) and production (million tonnes) of wheat over years are not similar. However, the state-wise data may depict some strong relation between annual rainfall and production.

Probability distributions consist of mutually exclusive and exhaustive compilation of all random events that can occur for a particular process and the probability of each event's occurring. It is a mathematical model that represents the distributions of the universe obtained either from a theoretical population or from the actual world; the distribution shows the results we would obtain if we took many probability samples and computed the statistics for each sample. Some well-known probability distributions are uniform distribution, binomial distribution, Poisson distribution, normal distribution, exponential distribution and so on. Fitting these distributions to the data provides some idea regarding the distributional pattern.

4.2 Relationship studies

Climate is a paradigm of a complex system. It has many variables, which act on a wide range of space–time scales. Statistical tools can be employed for studying the relationship between these variables. Some of such tools are discussed as follows.

Correlation analysis

Given a pair of related measures (X and Y) on each of a set of items, the correlation coefficient (r) provides an index of the degree to which the paired measures co-vary in a linear fashion. In general, r will be positive when items with large values of X also tend to have large values of Y, whereas items with small values of X tend to have small values of Y. Correspondingly, r will be negative when items with large values of X tend to have small values of Y whereas items with small values of X tend to have large values of Y. Numerically, r can assume any value between −1 and +1 depending upon the degree of the linear relationship. Plus and minus one indicate perfect positive and negative relationships whereas zero indicates that the X and Y values do not co-vary in any linear fashion. This is also called as *Pearson product-moment correlation coefficient.* The values of the correlation coefficient have no units. A scatter plot provides a picture of the relation, the value of the correlation is the same if you switch the Y (vertical) and X (horizontal) measures.

Let (x_i, y_i), $i = 1,2, ..., n$ denote a random sample of n observations from a bivariate population. The sample correlation coefficient r is estimated by the formula

$$r = \frac{Cov(x,y)}{\sqrt{V(x)V(y)}}$$

Table 4.2 Station-wise maximum and minimum temperature and rainfall in India for the year 2012

Stations	Minimum temperature (°C)	Maximum temperature (°C)	Highest 24-hour rainfall (mm)
Agartala AP	7.0	37.8	102.3
Cherrapunji	0.8	27.2	772.2
Dibrugarh AP	6.6	37.0	137.2
Guwahati AP	6.4	37.4	111.3
Imphal AP	2.0	35.6	84.8
Passighat	8.5	35.9	261.1
Shillong	2.1	28.9	134.0
Tezpur	7.0	36.4	161.3
Baghdogra AP	5.3	36.8	134.0
Berhampore	9.3	43.4	91.0
Kolkata	10.0	40.5	87.4
Cooch Behar AP	5.0	36.5	259.6

Source: Indiastat.com.

or

$$r = \frac{n\sum_{i=1}^{n} x_i y_i - \left(\sum_{i=1}^{n} x_i\right)\left(\sum_{i=1}^{n} y_i\right)}{\sqrt{\left[n\sum_{i=1}^{n} x_i^2 - \left(\sum_{i=1}^{n} x_i\right)^2\right]\left[n\sum_{i=1}^{n} y_i^2 - \left(\sum_{i=1}^{n} y_i\right)^2\right]}}$$

The correlation matrix for the above three variables given in Table 4.2 are as follows:

	Minimum temperature	Maximum temperature	Rainfall
Minimum temperature	1	0.847	−0.508
		(0.001)	(0.092)
Maximum temperature	0.847	1	−0.677
	(0.001)		(0.016)
Rainfall	−0.508	−0.677	1
	(0.092)	(0.016)	

Note: Values in parenthesis are the significance level.

It is seen that the minimum and maximum temperatures show a strong positive correlation whereas maximum temperature and rainfall shows a negative correlation and are significant.

In a multi-variate setup, the partial correlation is the correlation between two variables after eliminating the effects on other variables. Partial correlation coefficient between minimum and maximum temperature after controlling the variability due to rainfall is found to be 0.794 and is significant (0.004), which shows that rainfall has indirect effect on correlation between minimum and maximum temperature.

Regression analysis

The correlation coefficient measures the extent of interrelation between two variables that are simultaneously changing with mutually extended effects. In certain cases, changes in one variable are due to changes in a related variable, but there need not be any mutual dependence. One variable is considered to be dependent on the other as it changes. The relationship between variables of this kind is known as regression. Regression analysis is a statistical tool for the investigation of relationships between variables (Draper 1998; Montgomery 2006). When such relationships are expressed mathematically, it will predict the value of one variable from the knowledge of the other. For instance, the photosynthetic and transpiration rates of trees are found to depend on atmospheric conditions, like temperature or humidity, but it is unusual to expect a reverse relationship. The dependent variable is usually denoted by Y and the independent variable by X. When only two variables are involved in regression, the functional relationship is known as simple regression. If the relationship between the two variables is linear, it is known as simple linear regression; otherwise it is known as non-linear regression. Regression analysis is widely used for prediction and forecasting.

Multiple regression models Suppose Y denotes the yield of a crop over a certain period of time which depends on p explanatory variables X_1, X_2, ..., X_p such as maximum temperature, minimum temperature, atmospheric pressure, rainfall, CO_2 concentration in the atmosphere and so on over that specified period of time, and suppose we have n data points. Then a multiple regression model of crop yield that might describe this relationship can be written as

$$y_i = \beta_0 + \beta_1\, x_{1i} + \beta_2 x_{2i} + \cdots + \beta_p x_{pi} + \varepsilon_i, \text{ for } i = 1,\, 2, ..., n$$

(4.1)

where β_0 is the intercept term. The parameter β_k (for k = 1, 2, ..., p) measures the expected change in Y per unit change in X_k when all other k − 1 variables are held constant. Here, ε is an identical and independently distributed (iid) random variable with zero mean and constant variance. The prediction of the future crop yield is possible once all the parameters are estimated.

Problem of multi-collinearity in multiple regression model The individual regression coefficient in a multiple regression model determines the interpretation of the model. However, some of the climatic factors such as rainfall, relative humidity and so on are not independent of each other and there exists a close relationship between many of the climatic explanatory variables. So, inference based on the usual regression model may be erroneous or misleading. When the explanatory variables are not orthogonal; rather, there exists near-linear dependencies among them, the problem of multi-collinearity is said to exist in the regression setup. One of the major consequences of multi-collinearity is the large variances and co-variances for the least square estimators of the regression coefficients. Owing to this, the estimates of the regression coefficients become unstable which ultimately leads to imprecise interpretation of the regression model. Thus, a multi-collinearity problem in the multiple regression setup must be detected properly before going for the usual multiple regression. Lots of techniques are available in the literature that deals with the detection of multi-collinearity. Some of the most simple and user-friendly techniques are the examination of the correlation matrix and the detection of multi-collinearity based on conditional number and conditional indices, which has been discussed.

Detection of multi-collinearity by the examination of correlation matrix The simplest measure which is available in the literature is the inspection of off-diagonal elements ρ_{ij} in $\mathbf{X'X}$, where $\mathbf{X}$ is an $n \times p$ matrix of the levels of the various climatic explanatory variables (here, one has to consider variables involved in the regression model as unit length scaled variable). If two climatic explanatory variables, say X_i and X_j (for i, j = 1, 2, ..., n), are nearly linear related, then $|\rho_{ij}|$ will be near to unity. So, by examining the correlation matrix, one can easily detect the problem of multi-collinearity.

Detection of multi-collinearity through condition number and condition indices Another simple measure is the measure in which the characteristic roots or eigen values of $\mathbf{X'X}$ can be used for the detection of multi-collinearity among the climatic explanatory variables. Let, $\lambda_1, \lambda_2, ..., \lambda_p$ be the eigen values of $\mathbf{X'X}$. When one or more near linear relationships exist in the data, one or more eigen values will be small. Apart from the eigen values, the *condition number* of $\mathbf{X'X}$ may also be preferred. The conditional number of $\mathbf{X'X}$ is defined as

$$\kappa = \frac{\lambda_{max}}{\lambda_{min}} \tag{4.2}$$

Generally, when the condition number is less than 100, it can be said that there is no serious problem of multi-collinearity. Moderate to strong multi-collinearity exists when the condition number lies between 100 and 1000. But if the condition number exceeds 1000, then one has to give special consideration to the problem of multi-collinearity as the condition number of more than 1000 indicates severe multi-collinearity.

Another measure based on the eigen values of $\mathbf{X'X}$ is the use of *condition indices* of $\mathbf{X'X}$ which are defined as

$$\kappa = \frac{\lambda_{max}}{\lambda_k}, \quad k = 1, 2, \ldots, p \tag{4.3}$$

Clearly, the largest condition index is nothing but the condition number defined in Equation 4.2. The number of condition indices that are large (say >1000) is a useful measure of the number of near-linear dependencies in $\mathbf{X'X}$.

Regression analysis using qualitative climatic explanatory variables

The variables employed in regression analysis are usually quantitative variables, that is, the variables have a well-defined scale of measurements. Sometimes the climatic data available may have nominal or ordinal explanatory variables apart from various quantitative explanatory variables. For example, data related to rainfall over a certain period of time for a particular region are not available; rather, available data indicate the rainfall as above normal or below normal over that specified period of time for all the data points. Furthermore, data related to all other climatic variables such as temperature, atmospheric pressure and so on are available in usual manner. So, in this situation, all other explanatory variables apart from rainfall are quantitative in nature wherein rainfall is a nominal variable with two levels as above normal or below normal. So it is better to mention it as an attribute rather than specifying it as a variable. So a question may arise how to incorporate this attribute into the regression model?

One solution for quantifying such attributes is the use of *indicator variables* or *dummy variables*. A dummy variable is an artificial variable constructed such that it takes the value '1' whenever the qualitative phenomenon it represents occurs (say above normal in the above case), and taking value as '0' otherwise. Once created, dummies are then used in regression analyses just like other explanatory variables, yielding standard ordinary least square results. There is a numeric way in which one can choose a dummy variable. In general, the most useful dummy variable setups are simple in form, employing levels of '0' and '1' or ' − 1' and '1.' Once the qualitative explanatory variables are quantified by the use of dummy variable, one can easily employ multiple linear regression on crop yield based on various explanatory variable in the usual manner. Furthermore, sometimes a particular qualitative variable may have more than two levels. Suppose, in the available dataset, there are p explanatory variables. Out of these p explanatory variables, suppose apart from the first explanatory variable rainfall, all other $(p - 1)$ variables are quantitative climatic explanatory variables wherein rainfall is a categorical variable with three levels mentioned as above normal, normal and below normal. In this situation, to quantify the attribute, one has to use two dummy variables Z_1 and Z_2. One of the possible ways of defining the two dummy variables may be

$Z_1 = 0$ and $Z_2 = 0$	Below normal rainfall
$Z_1 = 0$ and $Z_2 = 1$	Normal rainfall
$Z_1 = 1$ and $Z_2 = 1$	Above normal rainfall

Then a multiple regression model of crop yield that might describe this relationship can be written as

$$y_i = \beta_0 + \beta_1 z_{1i} + \beta_2 z_{2i} + \beta_3 x_{2i} + \cdots + \beta_p x_{pi} + \varepsilon_i, \quad \text{for } i = 1, 2, \ldots, n$$
$$(4.4)$$

where the symbols have their usual meaning as defined earlier. Generally, when there are m levels of qualitative explanatory variables, then one has to employ $m - 1$ dummy variables for that particular qualitative explanatory variable. Sometimes, more than one qualitative climatic explanatory variable exists in the dataset. In such situations, more than one dummy variable may have to be used.

Logistic regression analysis

In all the above cases discussed so far, it is generally assumed that the dependent variable, that is, the crop yield is quantitative in nature. But situations may arise when, in the available dataset, the yield of crop is not quantitative in nature; rather, it is qualitative in nature. For example, suppose data related to the yield of crop are not numeric; rather, in the available dataset, it is expressed as high yield and low yield. In that situation, yield is a qualitative response variable with two levels: high yield and low yield. Further, let high yield be denoted by 1 and low yield be denoted by 0. In such cases, the usual multiple linear regression theory is not appropriate. Rather, the statistical model preferred for the analysis of such binary (dichotomous) responses is the binary logistic regression model, developed primarily by Cox (1958) and Walker and Duncan (1967). Thus, binary logistic regression is a mathematical modelling approach that can be used to describe the relationship of several independent variables to a binary (dichotomous) dependent variable.

Logistic regression allows the prediction of discrete variables by a mix of continuous and discrete predictors. It addresses the same questions that multiple regression does but with no distributional assumptions on the predictors (the predictors do not have to be normally distributed, linearly related or have equal variance in each group).

Suppose Y denotes the crop yield over a certain period of time and it has two levels, namely, high yield $(Y = 1)$ and low yield $(Y = 0)$. Further, let there be p climatic explanatory variables such as rainfall, maximum temperature, minimum temperature, relative humidity, atmospheric pressure and so on over that specified period of time. Suppose, π_i denotes the probability that the ith observation of the dependent variable takes the value 1, that is, $Y_i = 1$, for $i = 1, 2, ..., n$. Then, the simple logistic regression model that best describes the situations can be expressed as

$$\pi_i = P(Y_i = 1 | X_{1i} = x_{1i}, ..., X_{pi} = x_{pi})$$
$$= e^z (1 + e^z) \tag{4.5}$$
$$= 1/(1 + e^{-z})$$

where $z = \beta_0 + \beta_1 x_{1i} + \cdots + \beta_p x_{pi}$ Here, all other symbols have their usual meaning as defined earlier. The parameters can be estimated by fitting the logistic regression. Based on the fitted value, one can predict the different levels of crop yield.

Time series approach

Climate is a very heterogeneous factor. The global economy witnessed a major setback time due to various natural phenomena such as drought, flood, tsunami and so on. All these are the outcomes of climate change. These natural phenomena directly or indirectly affect the yield of various crops over the years. So the alternative approach for studying the impact of climate change is to use time series modelling by considering the yield of the present year as the dependent variable and lagged variable as explanatory variables. The simplest of these kinds are autoregressive (AR) models, moving average (MA) models, autoregressive moving average (ARMA) models and autoregressive integrated moving average (ARIMA) models. The following is a brief description related to these models.

Suppose Y_t is a discrete time series variable that takes different values over a period of time. Then the corresponding pth-order AR model, that is, AR (p) model, is defined as

$$AR(p): Y_t = \beta_0 + \beta_1 Y_{t-1} + \beta_2 Y_{t-2} + \cdots + \beta_p Y_{t-p} + \varepsilon_t \tag{4.6}$$

where y_t is the response variable, say the crop yield at time t; Y_{t-1}, Y_{t-2}, …, Y_{t-p} are the respective variables at different time with lag; β_0, β_1, …, β_p are the regression coefficients; and ε_t is the error term.

Similarly, qth-order MA model, that is, MA (q) model may be specified as

$$\text{MA(q): } Y_t = \theta_0 + \theta_1 \varepsilon_{t-1} + \theta_2 \varepsilon_{t-2} + \cdots + \theta_q \varepsilon_{t-q} + \varepsilon_t \quad (4.7)$$

where θ_0, θ_1, …, θ_q are the coefficients.

AR and MA models can be effectively combined to form a more general and useful class of time series models known as ARMA models. However, they can only be used successfully when the data are stationary. This model class can be extended to non-stationary series by allowing differencing of the data series. These are known as ARIMA models (Box and Jenkins, 1970). There are many varieties of ARIMA models available in the literature. The general no-seasonal model is known as ARIMA (p,d,q), where p is the order of the AR part, d is the degree of first differencing involved and q is the order of the MA part. The value of p and q may be inferred by looking at auto-correlation function (ACF) and partial auto-correlation function (PACF) as given in Table 4.3.

Table 4.4 highlights the production (in million tonnes) of wheat in India from 1949–1950 to 2010–2011.

From Figure 4.2, it is clear that the data are non-stationary. So, in order to make the data stationary, one has to perform a differencing operation. After the first differencing, the data become stationary. Graphical representation of stationary data has been presented in Figure 4.3. Once the data become stationary, one can easily fit the ARIMA model.

In Figures 4.4 and 4.5, ACF and PACF operations have been performed. Based on this figure, one can easily say that for the

Table 4.3 Primary distinguishing characters of theoretical ACFs and PACFs for stationary process

Process	ACF	PACF
AR	Tails off towards zero (exponential decay or damped sign wave)	Cuts off to zero (after lag p)
MA	Cuts off to zero (after lag q)	Tails off towards zero (exponential decay or damped sign wave)

Table 4.4 Production (in million tonnes) of wheat in India from 1949–1950 to 2010–2011

Year	Production (mt)	Year	Production (mt)	Year	Production (mt)
1949–1950	6.40	1970–1971	23.83	1991–1992	55.69
1950–1951	6.46	1971–1972	26.41	1992–1993	57.21
1951–1952	6.18	1972–1973	24.74	1993–1994	59.84
1952–1953	7.50	1973–1974	21.78	1994–1995	65.77
1953–1954	8.02	1974–1975	24.10	1995–1996	62.10
1954–1955	9.04	1975–1976	28.84	1996–1997	69.35
1955–1956	8.76	1976–1977	29.01	1997–1998	66.35
1956–1957	9.40	1977–1978	31.75	1998–1999	71.29
1957–1958	7.99	1978–1979	35.51	1999–2000	76.37
1958–1959	9.96	1979–1980	31.83	2000–2001	69.68
1959–1960	10.32	1980–1981	36.31	2001–2002	72.77
1960–1961	11.00	1981–1982	37.45	2002–2003	65.76
1961–1962	12.07	1982–1983	42.79	2003–2004	72.15
1962–1963	10.78	1983–1984	45.48	2004–2005	68.64
1963–1964	9.85	1984–1985	44.07	2005–2006	69.35
1964–1965	12.26	1985–1986	47.05	2006–2007	75.81
1965–1966	10.40	1986–1987	44.32	2007–2008	78.57
1966–1967	11.39	1987–1988	46.17	2008–2009	80.68
1967–1968	16.54	1988–1989	54.11	2009–2010	80.80
1968–1969	18.65	1989–1990	49.85	2010–2011	86.87
1969–1970	20.09	1990–1991	55.14		

Source: Indiastat.com.

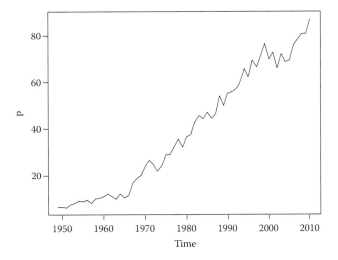

FIGURE 4.2 Graphical representation of wheat production data.

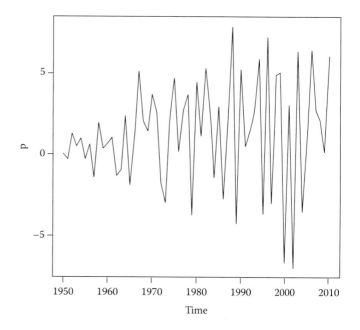

FIGURE 4.3   Graphical representation of stationary data.

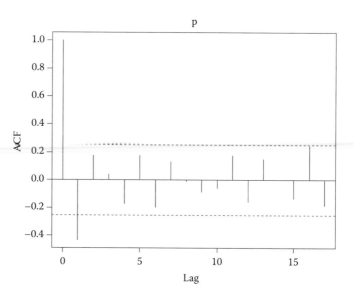

FIGURE 4.4   Auto-correlation function (ACF).

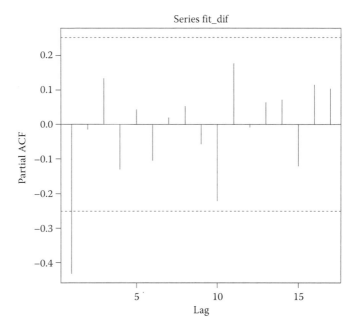

FIGURE 4.5   Partial auto-correlation function (PACF).

aforementioned dataset, ARIMA (1,1,1) can be fitted. The fitted ARIMA (1,1,1) model without intercept is

$$\hat{Y}_t = -0.4435Y_{t-1} - 0.9497\varepsilon_{t-1}$$
$$\quad\quad(0.1182)\quad\quad(0.0615)$$

(4.8)

The number in the bracket indicates the standard error of the parameter estimates.

4.3 Impact studies

Climate change impact studies here refer to research and investigations designed to find out what effect future changes in climate could have on the agriculture and allied sector. It requires some tools and techniques that facilitate proper assessment. A lot of research is going on in different fields of science with respect to climate change. Here, some statistical techniques are discussed which could be used for such studies to a great extent.

A test for structural change

Structural change refers to a long-term shift in the agricultural production. Over the last few decades, the structure of agricultural production around the world has been changing. The factors influencing this change have to be addressed.

Consider a dataset for two periods of time with Period I, say drought, having n_1 observations, and Period II, say normal, having n_2 observations. Here, the objective is to find out whether there is any structural change or shift in the yield pattern between the drought and normal periods. Let there be two variables Y = productivity and X = area under cultivation and data were collected at two periods of time and given as follows:

Period I		Period II	
Productivity	Area	Productivity	Area
1	2	1	2
2	4	3	4
2	6	3	6
4	10	5	8
6	13	6	10
		6	12
		7	14
		9	16
		9	18
		11	20

To investigate this, the following models are defined:

Period I: $Y = \alpha_1 + \beta_1 X + u$ (Drought)
Period II: $Y = \alpha_2 + \beta_2 X + u$ (Normal)

Here, the aim is to test $\alpha_1 = \alpha_2$ and $\beta_1 = \beta_2$.

Test statistic
Let the number of restrictions be s. The test statistic is as follows:

$$F = \frac{(RSS_r - RSS_u)/s}{RSS_u/(n - p)}$$

which follows the F distribution with (s, n − p) d.f. RSS_r is the residual sum of squares under null hypothesis and RSS_u is the residual sum of squares under usual model (Waterman 1974).

For Period I, the fitted model is

$$\hat{Y} = -0.0625 + 0.4375X \text{ and } RSS_1 = 0.6875$$

and for Period II the fitted model is

$$\hat{Y} = 0.4000 + 0.5091X \text{ and } RSS_2 = 2.4727$$

Residual sum of squares $(RSS_u) = RSS_1 + RSS_2 = 3.1602$
Residual sum of squares under null hypothesis (RSS_r)
$= 6.5565$

$$F = \frac{(6.5565 - 3.1602)/2}{3.1602/(15 - 4)} = 5.91$$

It can be inferred that there is structural change in the yield pattern between drought and normal periods as $F_{0.05}(2,11) = 3.98$.

Paired t-test for assessing the impact

When the two samples are not independent, but the sample observations are paired together, then this test is applied. The paired observations are on the same unit or matching units. It is often used to compare 'before' and 'after' scores in experiments to determine whether significant change has occurred; for example, to know the impact of climate change on a yield of perennial crops over years, assuming the rest of the variations as constant. Let (x_i, y_i), $i = 1,...,n$ be the pairs of observations and let $d_i = x_i - y_i$. Our aim is to test $H_0 : \mu_1 = \mu_2$.

Test statistic

$$t = \frac{\bar{d}}{s_d/\sqrt{n}}$$

follows t distribution with $n - 1$ d.f., where $\bar{d} = 1/n \sum_{i=1}^{n} d_i$ and $s_d^2 = 1/(n - 1) \sum_{i=1}^{n} (d_i - \bar{d})^2$.

Cluster and discriminant analysis

Cluster analysis is a technique for grouping individuals or objects into unknown groups. In agriculture, cluster analysis has been used for diversity analysis, which is the classification of genotypes into arbitrary groups on the basis of their characteristics. In agro-meteorology, cluster analysis can be used to analyse historical records of the spatial and temporal variations in pest/insect populations in order to classify regions on the basis of population densities and the frequency and persistence of outbreaks. The analysis can be used to improve regional monitoring and control of pest populations. Clustering techniques require that one define a measure of closeness or similarity between two observations. Clustering algorithms may be hierarchical or non-hierarchical. Hierarchical methods can be either agglomerative or divisive. An agglomerative hierarchical method starts with the individual objects, thus there are as many clusters as objects. The most similar objects are first grouped and these initial groups are merged according to their similarities. Eventually, as the similarity decreases, all sub-groups are fused into a single cluster.

Divisive hierarchical methods work in the opposite direction. An initial single group of objects is divided into two sub-groups such that the objects in one sub-group are far from the objects in the others. These sub-groups are then further divided into dissimilar sub-groups. The process continues until there are as many sub-groups as objects, that is, until each object forms a group. The results of both an agglomerative and divisive method may be displayed in the form of a two-dimensional diagram known as dendrogram, which illustrates the mergers or divisions that have been made at successive levels. K-means clustering is a popular non-hierarchical clustering technique. It begins with user-specified clusters and then reassigns data on the basis of the distance from the centroid of each cluster. See Johnson and Wichern (2006) and Hair et al. (2006) for more detailed explanations.

Discriminant analysis is a multi-variate technique concerned with classifying distinct set of objects (or set of observations) and with allocating new objects or observations to the previously defined groups. It involves deriving variates, which are a combination of two or more independent variables that will discriminate best between *a priori* defined groups. The objectives of discriminant analysis are (i) identifying a set of variables that best discriminates between the groups, (ii) identifying a new axis, Z, such that new variables Z, given by the projection of observations onto this new axis, provides the maximum separation or discrimination between the groups and (iii) classifying future observations into one of the groups.

References

Box, G.E.P. and Jenkins, G.M. 1970. *Time Series Analysis: Forecasting and Control.* Holden-Day, San Francisco.

Cox, D.R. 1958. The regression analysis of binary sequences (with discussion). *Journal of the Royal Statistical Society* B, **20**, 215–242.

DRAPER, N.R. and SMITH, H. 1998. *Applied Regression Analysis.* 3rd ed., Wiley, New York.

HAIR, J.F., ANDERSON R.E., TATHAM, R.L. and BLACK, W.C. 2006. *Multivariate Data Analysis.* 5th ed., Pearson Education Inc, New Delhi, India.

JOHNSON, R.A. and WICHERN, D.W. 2006. *Applied Multivariate Statistical Analysis.* 5th ed., Pearson Prentice Hall Inc., London.

MONTGOMERY, D.C., PECK, E.C. and VINING, G. 2006. *Introduction to Linear Regression Analysis.* 3rd ed., Wiley, New York.

WALKER, S.H. and DUNCAN, D.B. 1967. Estimation of the probability of an event as a function of several independent variables. *Biometrika,* **54**, 167–178.

WATERMAN, M.S. 1974. A restricted least squares problem. *Technometrics,* **16**, 135–138.

CHAPTER FIVE

Nanotechnological interventions for mitigating global warming

Anjali Pande, Madhu Rawat, Rajeev Nayan,
S.K. Guru and Sandeep Arora

Contents

Abstract

The rapidly changing climate, due to global warming, is a major cause of concern for agriculture scientists. Undesirable climatic variability is bound to adversely affect the agricultural productivity, leading to food and nutritional insecurity for the burgeoning population. Therefore, one of the biggest challenges is to augment agricultural productivity, which is decreasing under environmental pressure. The scenario requires an intensive research for increased adaptation and devising novel strategies for mitigating the harmful effects of global warming on crop growth and productivity. Crop adaptation processes can be made more efficient and directed through targeted modifications of the cellular processes. Nanotechnology offers a promising strategy for non-intrusive engineering of crops under changing climatic conditions. Limitations faced by the conventional technologies in agriculture can be easily and efficiently addressed by means of this technology, with respect to the vulnerability to changing climatic conditions. Nanotechnology holds the potential to revolutionise agriculture and allied systems, through the development of efficient nano-delivery systems, nano-fertilisers, nano-biosensors and so on, for increasing the growth and productivity. Direct as well as indirect nano-technological interventions, though still in their nascent stage, are bound to play a pivotal role in sustainable agricultural development under global warming.

5.1 Introduction

During the past century, the average global temperature rose significantly as a consequence of greenhouse effect. This occurs due to the rise in greenhouse gasses, such as carbon dioxide (CO_2) emitted from burning fossil fuels or from deforestation, which traps heat that would otherwise escape from the Earth. The current scientific view is that most of the increase in global temperature is caused due to human activity. As a result, the concentration of greenhouse gasses increases in the atmosphere, which has led to an elevation in the Earth's average surface temperature. Most scientists agree that the planet's rise in temperature will continue at an increasing rate. Under these circumstances, crop productivity will be at stake. On the one hand, the increase in temperature leads to droughts, cracked fields and low rainfall, which adds to the intensity and frequency of dusty storms, ultimately depleting the quality of agricultural land and making it permanently unsuitable for cultivation. On the other, because of the higher temperatures, the seasons are becoming unstable. Consequently, the amount of rainfall will be severely affected and hence the crop productivity. Therefore, under such conditions, crops in many regions will be prone to environmental stresses. These changes not only affect plant growth and yield, but also have an adverse effect on agricultural productivity. Furthermore, the impact of climate change also poses a serious threat to food security and needs to be much better understood. According to a study, wheat yields in recent years marginally decreased in India, France and China compared with what they probably would have been without rising temperatures. Researchers have also claimed that corn yields were off a few percentage points in some of the countries from what would have been expected normally. Most of the mechanisms and two-way interactions between agriculture and climate are known, even though they are not always well understood. It is evident that the relationship between climate change and agriculture is still very much a matter of conjecture with many uncertainties (Rosenzweig and Hillel, 1993).

5.2 Impact of climate change on crop production

Climate change has adversely affected global agriculture in terms of productivity, economy and food security. Crop productivity is vulnerable to decreases or increases in precipitation. Small changes in temperature and rainfall could have significant effects on the quality of cereals, fibre and beverage crops, fruits and some aromatic and medicinal plants with resultant implications on their prices and trade. Agricultural trade has risen, and is expected to increase further because of instability in production quality and quantity due to climate change impacts. Furthermore, not only will climate potentially decrease the amount of land available for agricultural production but there will also be an increase in competition for resources with other developmental needs, such as infrastructure. Climate change will, undeniably, lead to more pressure on an already-volatile economy. Agriculture affects all livelihoods, occupying approximately 40% of the land globally, consuming 70% of the global water resources and affecting biodiversity at all scales; from genetic to the ecosystem. An analysis of the biophysical impact of climate changes associated with global warming shows that higher temperatures generally hasten plant maturity in annual species, thus shortening the growth stages of crop plants. Global warming in the short term is likely to favour agricultural production in temperate regions (largely northern Europe, parts of North America) and negatively impact tropical crop production (South Asia, Africa). According to Rosenzweig and Liverman (1992), the regions differ significantly, both in the biophysical characteristics of their climate and soil and in the vulnerability of their agricultural systems. Tropical areas are generally expected to have the biggest decreases in agricultural production. Another factor affecting agriculture with this background is the increasing rate of microbial decomposition of organic matter, which adversely affects soil fertility in the long run. Also, studies analysing the effects on pests and diseases suggest that temperature increases may extend the geographic range of some insect pests currently limited by temperature.

The effects of increased ultraviolet-B (UV-B) radiation are also of great concern. The increase in UV-B radiation reduces the yield in certain agricultural crops. Olszyk and Ingram (1993) have also discussed the potential of crop-quality reduction due to the 'Effects of UV-B and Global Climate Change on Rice Production' in the Environmental Protection Agency International Rice Research Institute (EPA/IRRI) cooperative research plan (Dai et al., 1992).

In the long run, the outcome of climatic change will have the following effects on agriculture:

- The quality and quantity of agricultural productivity will be adversely affected.
- There will be changes of water use (irrigation) and agricultural inputs such as herbicides, insecticides and fertilisers.
- Several environmental factors will be altered and will have their negative impact, particularly in relation to frequency and intensity of soil drainage (leading to nitrogen leaching), soil erosion and reduction of crop diversity.
- The cultivable land, land speculation, land renunciation and hydraulic amenities will be at stake. There will be less rural space available for cultivation.
- Adaptation in organisms may lead to changes in competition levels, and humans may develop an urgency to develop more competitive organisms, such as flood-resistant or salt-resistant varieties of crops; also implementing various methods and technologies so as to develop new varieties of crops to combat the growing disease and pests.

5.3 Mitigation strategies for reducing crop productivity losses

It is apparent that, in general, agricultural productivity will decrease to a greater or lesser extent under increasing climatic variability, particularly with rising temperatures and fluctuating extreme precipitation. Agricultural resources and their associated supply chains need to be managed sustainably, as climate change is a major challenge to sustainable development. Therefore, one of the biggest challenges is to sustain our agricultural resource that is decreasing under such pressure. Only through scientific agricultural practices will our work with sustaining crop productivity be ready for the challenge to meet the increasing demand against the background of reducing resources in a changing climate scenario, while also minimising further environmental degradation. This would require increased adaptation and mitigation research, capacity building, changes in policies, regional cooperation, support of global adaptation and mitigation funds and other resources. Simple adaptations such as, a change in planting dates and crop varieties could help in reducing the impacts of climate change to some extent. Changing varieties, such as changing the planting date, is a first line of defence for farmers to consider (Wolfe et al., 2008). Losses in wheat production can be reduced from 4 to 5 million tonnes to 1–2 million tonnes if a large percentage of farmers could change to timely planting. This may, however, not be easy to implement due to constraints associated with some crops. Another optimisation strategy is 'deficit irrigation', in which irrigation is applied during drought-sensitive growth stages of crops. Outside these periods, irrigation is limited or even unnecessary if rainfall provides a minimum supply of water. Water restriction is limited to drought-tolerant phenological stages, often the vegetative stages and the late ripening period. The total irrigation application is therefore not proportional to the irrigation requirements throughout the crop cycle. In other words, deficit irrigation aims at stabilising yields and obtaining maximum crop water productivity rather than maximum yields (Zhang and Oweis, 1999). Additional strategies for increasing our adaptive capacity include bridging yield gaps to augment production, development of adverse climate-tolerant genotypes and land use systems, assisting farmers in coping with the current climatic risks by providing weather-linked value-added advisory services and crop/weather insurance and improved land and water use management and policies.

GE (genetically engineered) crops created by recombinant DNA (rDNA) could be an overwhelming aid for increasing the agricultural growth and productivity against rising temperatures. Natural processes can be made more efficient and directed through targeted modifications by means of mutation, classical breeding or rDNA technology. Breeding for new cultivars/varieties that are tolerant to higher temperatures are likely to be advantageous under the changed climatic scenario, possibly through genetic engineering. Varieties with improved tolerance to heat or drought, or adapted to take advantage of a longer-growing season for increased yield, will be available for some crop species.

Another emergent strategy to support agricultural productivity under changing climatic conditions is the application of nanotechnology. Nanotechnology is new to agriculture, a developing field of less than a decade old. It is considered as an enabling technology by which the existing materials, virtually all man-made materials and systems, can acquire different properties rendering them suitable for numerous novel applications. Nanotechnology holds the potential to revolutionise agriculture and food systems. Success has already been achieved for manufacturing nano-pesticides, nano-fertiliser and many other nano-products for increasing the growth and productivity of crops in adverse climatic conditions. Therefore, the possibilities of applying nanotechnology to solve the problems of agriculture with respect to its vulnerability to changing climatic conditions should be worth exploring.

5.4 Genetic engineering as a possible alternative

Increasing CO_2, global mean temperatures, varying rainfall patterns and frequent weather changes are occurring due to climate change. Such factors place a direct impact on the health and well-being of crops, thereby affecting small landholders, subsistence agriculture and food security in the developing world (Howden et al., 2007). Crop modelling shows that climate change will likely reduce agricultural production, thus reducing food availability (Lobell and Field, 2007) and affecting food security. Plant breeding, appropriate crop husbandry, sound natural resource management and agricultural policy interventions will be needed to ensure food availability and reduce poverty in a world affected by climate change (Howden et al., 2007). Persistent efforts in various research fields have been going on to develop new cultivars that can respond to environments with abiotic stresses (Bhatnagar-Mathur et al., 2008). Abiotic stresses aggravated by climate change pose a serious threat to the sustainability of crop yields and account for substantial yield losses. Scientific knowledge of the processes of abiotic stress tolerance in crops continues to develop and guides conventional breeding and genetic engineering of new crop cultivars. The modern tools of cell and molecular biology have shed light on control mechanisms for abiotic stress tolerance, and for engineering stress-tolerant crops based on the expression of specific stress-related genes. Such trait-based approaches to crop genetic enhancement have led to genetic manipulations (through transgenic approaches), thereby resulting in desired genotypes. Hence, research efforts in genetic engineering are advancing so as to keep pace with predicted environmental changes that will be more variable and stressful. Climate change will also be associated with increased water stresses in many regions due to changes in rainfall distribution and because increased temperatures under low relative humidity will result in a greater evaporative demand, thereby reducing water use efficiency, particularly in drought-prone environments. Bennett (2003) summarises options for water productivity enhancement through crop breeding and biotechnology (Bt), whereas Ortiz et al. (2007) provides an overview of transgenic research for drought-prone environments. Innovations in crop genetic enhancement have provided some of the best options for farmers, especially in the developing world, to combat against global warming, water scarcity, flooding and salinity. Genetic enhancement of crops brings innovations to farming systems as a result of new findings and ensuing knowledge from research. Crop improvement has been accelerated by the genetic engineering of new traits, particularly those that are not amenable to conventional breeding. Farmers grew about 114.3 million ha of transgenic crops in 2007 (with a growth rate of 12% vis-à-vis the previous year).

5.5 The Indian scenario

GE crops promise to make a great, possibly indispensable, contribution to reducing mass hunger even in adverse climatic conditions. Yet, the development of GM (genetically modified) crops has recently caused widespread unease in many countries. Controversies over GM crops and GM food in India have summed up many of the issues. Maharashtra, Karnataka and Tamil Nadu had an average of 42% increase in yield with GM cotton in 2002, the first year of commercial GM cotton planting, but due to a severe drought in Andhra Pradesh that year, the parental cotton plant used in the genetic-engineered variant was not well suited to the extreme drought, and hence, no increase in yield. Drought-resistant variants were developed with substantially reduced losses to insect predation. By 2011, 88% of Indian cotton was made GM. However, recently, the cotton bollworm has been developing a resistance to Bt cotton and the Indian Agriculture Ministry linked farmers' suicides in India to the declining performance of Bt cotton for the first time. Consequently, in 2012, Bt cotton was banned in Maharashtra and there were orders for its socio-economic studies to be conducted by independent institutes.

A variety of ethical issues are raised in response to the use of genetic engineering for enhancing agricultural productivity under changing environmental conditions. The ethical principle of beneficence demands action to eliminate hunger and disease. Keeping this in mind, the core ethical values should provide guidance in research, to modify nature in the service of human needs. There should be a balance between the need for technological advancement with the duty to protect and preserve our environment for future generations.

5.6 Nanotechnology as an alternative to GM crops

Food security is one of the biggest challenges in the current scenario, especially in developing countries. To meet this challenge, several methods such as crop management, crop improvement and crop protection from pests and diseases have been employed since time immemorial so as to enhance the growth and productivity of crops. Although these conventional methods and improved technologies are successful in enhancing the productivity of crops, due to certain limitations they are unable to break through some of the bottlenecks in agriculture. The conventional breeding methods under crop improvement for developing high-yielding varieties with desired traits such as stress tolerance (biotic/abiotic), herbicide tolerance, insect resistance and so on, are successful, but they are time consuming because it takes years to develop a new variety. Conversely, the genetic engineering approach is much faster, but not equally accepted worldwide due to various limitations associated with it. For example, the GE insect-resistant plants exert their effect only on being chewed by the insects; this means that the plant has to get damaged unnecessarily. In some cases, the insects may also acquire resistance against the bacterial toxins used in GE crops. Similarly, weed varieties also become resistant to herbicides. Therefore, despite the efforts made by the scientists, productivity could not realise its potential. Recently, the emergence of nanotechnology and its application in agriculture has raised hopes for improving agricultural productivity by overcoming the problems encountered in conventional agricultural practices. Although nanoparticles are comparatively very small in size (Figure 5.1), they have far-reaching consequences on the biological system.

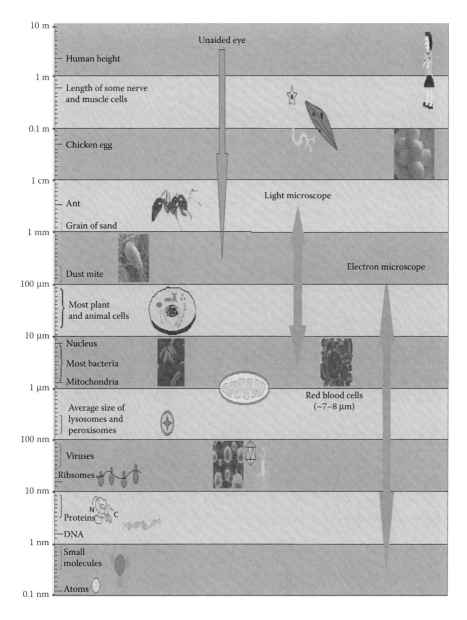

FIGURE 5.1 (**See colour insert.**) Size variations in biomolecules and living organisms.

The Indian Government is also looking forward to its development as a viable alternative to other limiting methods to boost the agricultural productivity. The Planning Commission of India has recommended nanotechnology research and development as one of the six areas of investment (Sreelata, 2008).

5.7 Nanotechnology in agriculture

Earlier, the application of nanotechnology in agriculture was only theoretical, but in recent years its practical knowledge has gained new heights and will continue to have its significant impact on this field. Nanotechnology exploits the properties, processes and phenomena of matter at the nanometer (1 to ~100 nm) scale. To realise their practical application, nanoparticles with different sizes, shapes and composition need to be synthesised. For the synthesis of nanoparticles, researchers routinely practice either 'top-down' or 'bottom-up' approaches (Figure 5.2). In the top-down approach, scientists try to formulate nanoparticles by using larger ones to direct their assembly. The bottom-up approach is a process that builds larger and more complex systems by starting at the molecular level while maintaining a precise control of the molecular structure.

Thus, by controlling nano-scale composition, size and shape we can create new materials with new properties (Figure 5.3). The development of new functional materials and smart delivery systems for agrochemicals such as herbicides, fertilisers and pesticides, smart systems integration for food processing, packaging and so forth (Moraru et al., 2003), are some of its applications in this area. The potential of nanotechnology is also increasing with the identification of suitable techniques and sensors for precision agriculture, natural resource management and early detection of pathogens and contaminants in food products. Some of its applications have been broadly discussed in this chapter.

5.8 Applications of nanotechnology

Crop improvement

Nanoparticles have the ability to work at the cellular level and rearrange the atoms in the DNA of an organism for the

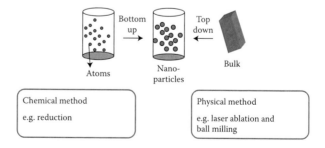

FIGURE 5.2 **(See colour insert.)** Classical approaches for nanoparticle synthesis.

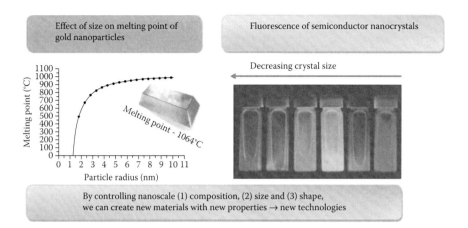

FIGURE 5.3 (**See colour insert.**) The effect of particle size on chemical and physical properties of nanoparticles.

expression of the desired character, thus shortening the time-consuming conventional methods. Nanotechnology has the potential to modify the molecular constitution of the crop plants, making it suitable for cultivation under adverse conditions. Both natural and induced mutations have an important role in crop improvement. Under this background, nanotechnology creates a new dimension without the use of chemical compounds and physical mutagens such as x-ray, γ-ray and so on. In Thailand, Chiang Mai University's Nuclear Physics Laboratory has developed a new white-grained rice variety from a traditional purple-coloured rice variety called Khao Kam. The colour of the leaves and stems of Khao Kam were changed from purple to green and the grain became whitish (ETC, 2004). The principle behind such a change was that a nano-sized hole was drilled through the wall and membrane of a rice cell, to insert a nitrogen atom. The nitrogen atom is passed through the hole and a particle beam was used to stimulate rearrangement of the DNA. This newly derived organism is designated as an atomically modified organism (AMO), because of its evolution through change at the atomic level. Further research work for crop improvement is still going on worldwide.

Gene delivery Non-viral gene-delivery systems have gained considerable attention as compared to viral systems, even though the efficacy of DNA transfection is very low in the former case. Non-viral vectors have several advantages, such as they are relatively easy to prepare, less immunogenic and onco-genic and there is no viral recombination. One such example of non-viral vectors is a functionalised nanoparticle. The functionalised nanoparticles have the ability to incorporate genetic materials such as plasmid DNA (deoxyribonucleic acid), RNA (ribonulceic acid) and siRNA (small interfering ribonuleic acid), but with little toxicity. This demonstrates a new era in pharmacotherapy for delivering genes into specific tissues and cells (Jin et al., 2009).

Nanotechnology in disease diagnostics   Crop productivity is greatly affected by diseases. The detection of the disease at the exact stage is essential to effectively prevent it. Viral diseases are the toughest ones to control as compared to other diseases. To prevent most of the diseases, pesticides are routinely used; this is not only associated with residual toxicity and environmental hazards but also results in crop yield loss, if applied after the appearance of the disease. Nano-based viral diagnostics include a multiplexed diagnostic kit development which plunges for the detection of the exact strain of the virus and its stage of application. Along with the detection power of these nano-based diagnostic kits, they can also increase the speed of detection. The detection and utilisation of biomarkers for accurate indication of the stages of disease with differential protein production in both healthy and diseased states, lead to the identification of several proteins during the infection cycle. These proteins can be used as markers for that particular disease stage.

Nanotechnology in pest control   With the advancement of nanotechnology, nanoparticles are now being used to produce pesticides, insecticides and insect repellants (Owolade et al., 2008). Nanoencapsulation is a process that involves the slow and efficient release of chemicals such as insecticides into a particular host plant for insect pest control. Nanoencapsulation with nanoparticles in the form of pesticides allows proper absorption of the chemical by the plants unlike large particles (Scrinis and Lyons, 2007). Release mechanisms of nanoencapsulation are diffusion, dissolution, biodegradation and osmotic pressure at specific pH. This process is also capable of delivering DNA and other desired chemicals into plant tissues for protection against insect pests. The ongoing research on silkworm, *Bombyx mori* L. race Nistari clearly shows that nanoparticles could stimulate more production of fibroin protein that can help to produce carbon nanotubes in future (Bhattacharyya et al., 2008; Bhattacharyya, 2009). Nanoencapsulation is currently the most promising technology for protection of host plants against insect pests, but the toxicological and ecotoxicological risks linked to this emerging technology have not been examined. Research on nanoparticles and insect control should be geared towards the introduction of faster and ecofriendly pesticides in the future (Bhattacharyya et al., 2007). Therefore, the leading chemical companies focus especially on nanopesticides formulation for targeting the host tissue through nanoencapsulation. Thus, nanotechnology will surely revolutionise agriculture in the near future.

Nanotechnology for mitigating climate change
Nanotechnology is flourishing as one of the newest approaches to combat the climate change. Under sub-optimal conditions, the potential of gold nanoparticles in alleviating the oxidative damage to *Brassica juncea* has already been explored (Arora et al., 2012). Besides this, they have also concluded that the gold nanoparticles improve the redox status of the plants under adverse conditions, thereby facilitating healthy survival of this crop. The significant increase in seed yield was also observed in gold nanoparticle-treated plants. Thus, nanotechnology paves the way for food security even under the unfavourable environmental conditions.

Nanoparticles for environmental remediation
Nanoparticles represent a new category of environmental remedy technologies that provide cost-effective solutions to some of the most challenging environmental problems. Research has shown that iron nanoparticles are very effective for the transformation and detoxification of a wide variety of common environmental contaminants, such as chlorinated organic solvents, organo-chlorine pesticides and so on. Iron nanoparticles have a large surface area, high surface reactivity and also provide enormous flexibility for *in situ* applications. Catalysed and supported nanoparticles have been synthesised to further enhance the speed and efficiency of remediation. Recent research has suggested that in a remediation technique, the use of iron nanoparticles has the following advantages: (1) effective for transforming a large variety of environmental contaminants, (2) cost-effective and (3) non-toxic. Recent laboratory research has largely established iron nanoparticles as effective reductants and catalysts for a variety of common environmental contaminants, including chlorinated organic compounds and metal ions. Using a palladium and iron nanoparticles dose at 6.25 gL^{-1}, all chlorinated compounds were reduced below the detectable limits (Chinnamuthu et al., 2009). Zero-valent iron (ZVI) can be used as a chemical reductant for the removal of

chlorinated and nitroaromatic compounds under anaerobic environmental conditions (O'Hara et al., 2006). Another application of nanoparticles is the sequestration of metallic ion and heavy metal from aqueous solutions (e.g. Ag, Hg, Pb, Cu, Zn, Sb, Mn, Fe, As, Ni, Al, Pt, Pd and Ru). In the presence of magnetic ions such as iron sulphide, heavy metals precipitate onto the bacterial cell wall, making the bacteria sufficiently magnetised for removal from the suspension by magnetic separation procedure. Research has shown that certain bacteria could produce iron sulphonamide. These particles could be used for the removal of harmful agents from the surrounding environment. This new method employs molecular templates to coat nanoparticles of magnetite with mesoporous silica.

5.9   Conclusion

Nanotechnology opens up avenues for new innovations in the various dimensions of biological sciences, especially in agriculture. It not only paves the way to combat the various environmental issues due to global warming but also provides a solution to the major problems in agriculture. Many other limitations faced by conventional and other technologies in agriculture can be easily and efficiently addressed through this technology. Even though it has fruitful applications in every sector, still there are no rules and recommendations for its safe and benign use. Therefore, its handling guidelines should be formulated and should come into effect to realise the optimum benefit of nanotechnology in agriculture and other areas to combat global warming.

Suggested reading

ARORA, S., SHARMA, P., KUMAR, S., NAYAN, R., KHANNA, P.K. and ZAIDI, M.G.H. 2012. Gold-nanoparticle induced enhancement in growth and seed yield of *Brassica juncea*. *Plant Growth Regul.* 66: 303–310.

BARIK, T.K., SAHU, B. AND SWAIN, V. 2008. Nanosilica—From medicine to pest control. *Parasitol. Res.* 103(2): 253–258.

BENNETT, J. 2003. Opportunities for increasing water productivity of CGIAR crops through plant breeding and molecular biology. In: *Water Productivity for Agriculture: Limits and Opportunities from Improvement,* J.W. Kijne, R. Barker and D. Molden, (eds.), Wallingford, CAB, UK. pp. 103–126.

BHATNAGAR-MATHUR, P., VADEZ, V. and SHARMA, K.K. 2008. Transgenic approaches for abiotic stress tolerance in plants: Retrospect and prospects. *Plant Cell Rep.* 27: 411–424.

BHATTACHARYYA, A. 2009. Nanoparticles—From drug delivery to insect pest control. *Akshar* 1(1): 1–7.

BHATTACHARYYA, A. and DEBNATH, N. 2008. Nano particles— A futuristic approach in insect population. In: *Proceedings on UGC Sponsored National Seminar on Recent Advances in Genetics and Molecular Biology, Biotechnology and Bioinformatics,* 21st and 22nd November, Jointly organized by Department of Zoology and Botany, Vidyasagar College, Kolkata—700006. West Bengal, India.

BHATTACHARYYA, A., GOSH, M., CHINNASWAMY, K.P., SEN, P., BARIK, B., KUNDU, P. and MANDAL, S. 2008. Nano-particle (allelochemicals) and silkworm physiology. In: *Recent Trends in Seribiotechnology,* K.P. Chinnaswamy and A. Vijaya Bhaskar Rao, (eds.), Bangalore, India. pp. 58–63.

BHATTACHARYYA, A. et al. 2007. Bioactivity of nanoparticles and allelochemicals on stored grain pest—*Sitophilus oryzae* (L.) (Coleoptera: Curculionidae). Accepted in *27th Annual Session of the Academy of Environmental Biology and National Symposium of Biomarkers of Environmental Problems,* 26–28 October, 2007, Jointly organized by Department of Zoology and Department of Environmental Sciences, Charan Singh University, Meerut, UP, India.

BHATTACHARYYA, A. et al. 2010. Nano-particles—A recent approach to insect pest control. *Afr. J. Biotech.* 9(24): 3489–3493.

CHINNAMUTHU, C.R. and MURUGESA BOOPATHI, P. 2009. Nanotechnology and agroecosystem. *Madras Agric. J.* 96(1–6): 17–31.

DAI, Q., CORONEL, V.P., VERGARA, B.S., BARNES, P. W. and QUINTOS, A.T. 1992. Ultraviolet-B radiation effects growth and physiology of four rice cultivars. *Crop Sci.* 32: 1269–1274.

DING, W.K. and SHAH, N.P. 2009. Effect of various encapsulating materials on the stability of probiotic bacteria. *J. Food Sci.* 74(2): M100–M107.

ETC. 2004. Atomically modified rice in Asia: www.etcgroup.org/article.asp?newsid=444.

HOWDEN, S.M., SOUSSANA, J.-F., TUBIELLO, F.N., CHHETRI, N., DUNLOP, M. and MEINKE, H. 2007. Adapting agriculture to climate change. *Proc. Natl. Acad. Sci.* 104: 19691–19696, doi:10.1073/pnas.0701890104.

JIN, S., LEACH, J.C. and YE, K. 2009. Micro and nano technologies in bioanalysis: Nanoparticle-mediated gene delivery. In: *Methods in Molecular Biology,* R.S. Foote and J.W. Lee, (eds.), Humana Press, LLC, USA.

LOBELL, D.B. and FIELD, C.B. 2007. Global scale climate–crop yield relationships and the impacts of recent warming. *Environ. Res. Lett.* 2: 014002.

MORARU, C.I., PANCHAPAKESAN, C.P., QINGRONG, H., TAKHISTOV, P., SEAN, L. and KOKINI, J.L. 2003. Nanotechnology: A new frontier in food science. *Food Technol.* 57: 24–29.

OAKDENE, H. 2007. Environmentally beneficial nanotechnologies, 9. http://www.defra.gov.uk/environment/nanotech/policy/pdf/ envbeneficialreport. Pdf

O'HARA, S., KRUG, T., QUINN, J., CLAUSEN, C. and GEIGER, C. 2006. *Field and Laboratory Evaluation of the Treatment of DNAPL Source Zones Using Emulsified Zero-Valent Iron. Remediation Journal*, San Francisco, USA: Wiley Periodicals, Inc., Spring 2006.

OLSZYK, D.M. and INGRAM, K.T. 1993. *Effect of UV-B and Global Climate Change on Rice Production*. EPA/IRRI cooperative research plan. IRRI, Philippines.

ORTIZ, R., IWANAGA, M., REYNOLDS, M.P., WU, H. and CROUCH, J. 2007. Overview on crop genetic engineering for drought prone environments. *J. Semi-Arid Trop. Agricu. Res.* 4(1): 1–30.

OWOLADE, O.F., OGUNLETI, D.O. and ADENEKAN, M.O. 2008. Titanium dioxide affects disease development and yield of edible cowpea. *EJEAF Chem.* 7(50): 2942–2947.

ROSENZWEIG, C. and HILLEL, D.1993. Agriculture in a greenhouse world: Potential consequences of climate change. *National Geographic Research and Exploration* 9: 208–221.

ROSENZWEIG, C. and LIVERMAN, D. 1992. Predicted effects of climate change on agriculture: A comparison of temperate and tropical regions. In: *Global Climate Change: Implications, Challenges, and Mitigation Measures*, S.K. Majumdar, (ed.), PA: The Pennsylvania Academy of Sciences. pp. 342–361.

SCRINIS, G. and LYONS, K. 2007. The emerging nano-corporate paradigm nanotechnology and the transformation of nature, food and agri-food systems. *Int. J. Sociol. Food Agric.* 15(2): 22–44.

SREELATA, M. 2008. India looks to nanotechnology to boost agriculture. http://www.scidev.net/en/news/india-looks-to-nanotechnology-to boostagriculture. Html

SUKUL, N.C., SINGH, R.K., SUKUL (CHUNARI), S., SEN, P., BHATTACHARYYA, A., SUKUL, A. and CHAKRABARTY, R. 2009. Potentized drugs enhance growth of pigeon pea. *Environ. Ecol.* 26(3): 1115–1118.

TORNEY, F. 2009. Nanoparticle mediated plant transformation. Emerging technologies in plant science research. *Interdepartmental Plant Physiology Major Fall Seminar Series. Physics*. p. 696.

WOLFE, B.B., HALL, R.I., EDWARDS, T.W.D., JARVIS, S.R., SINNATAMBY, R.N., YI, Y. and JOHNSTON, J.W. 2008. Climate-driven shifts in quantity and seasonality of river discharge over the past 1000 years from the hydrographic apex of North America. *Geophys. Res. Lett.* 35: 10.1029/2008GL036125.

ZHANG, H. AND OWEIS, T. 1999. Water-yield relations and optimal irrigation scheduling of wheat in the Mediterranean region. *Agric. Water Manage.* 38: 195–211.

CHAPTER SIX

Role of biotechnology in climate resilient agriculture

Pradeep K. Jain, Pooja Choudhary and Dinesh K. Sharma

Contents

Abstract

Change of climate over time has led to decrease in crop yield due to inadequate rainfall, various abiotic stresses, potential weeds, pests and diseases caused by fungi, bacteria and viruses. Biotechnology and the application of advanced techniques in agriculture will help in creating plants that will adapt to these new climatic conditions. One of the important ways of adapting to such changes is to apply agricultural biotechnological strategies that counter the effects of such changes by improving crop productivities per unit area of land cultivars. The increasing demand for food crops worldwide can be satisfied in two ways: first is to increase the area under production and the second is to improve productivity on existing arable land.

6.1 Introduction

Climate change is a significant and lasting change in the statistical properties of the climate system when considered over long periods of time (Mtui, 2011). The main reason behind climate change is either the Earth's natural forces, which basically include solar radiation and continental drift, or human activities. Climate change has obvious and direct effects on the agricultural sector, and if the global state is taken into account, the reverse, that is, the impact of agriculture on climate is also increasingly evident. Agricultural activities result in the large-scale emission of greenhouse gases through the use of

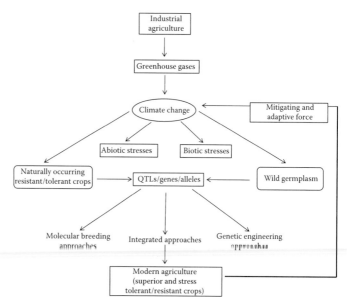

FIGURE 6.1 Agricultural production acts both as the contributor and the potential mitigating and adaptive force towards climate change. Germplasm collection including naturally occurring tolerant/resistant crops and wild relatives can be used to isolate QTL(s), gene(s) or allele(s) conferring tolerance/resistance to various biotic and abiotic stresses by employing modern genomics approaches (molecular breeding, genetic engineering and integrated breeding). These QTL(s), gene(s) and allele(s) can be used to develop modified crops that are better adapted to the various stresses.

fossil fuel-based inputs, livestock production, soil erosion, land conversion and deforestation. Agriculture, the second largest industrial sector, directly accounts for approximately 14% of global greenhouse gas emissions and indirectly for an additional 17% of emissions (IPCC, 2007). Hence, the agricultural sector acts both as the contributor to the climate change and,through adjustment in practices, as a potential mitigating and adaptive force (Figure 6.1). Research and development of new technologies have always been important to agricultural production to achieve the goal of increasing output per unit of land, labour and other input. The need of agricultural innovation has become even more apparent with the emergence of the issue of climate change. Here, we describe the potential role that biotechnology and other advanced agricultural practices can play in climate change mitigation and adaptation.

6.2 Industrial agriculture and the climate change

In the twenty-first century, climate change is one of the most serious and extensive challenges faced by the modern world and, as per the Environmental Protection Agency (EPA), it is mainly caused by the upsurge of the greenhouse gases in the Earth's atmosphere. Greenhouse gases are constituents of the atmosphere (both natural and anthropogenic) that absorb and emit radiations at specific wavelengths within the spectrum of infrared radiation emitted by the Earth's surface, the atmosphere and clouds (IPCC, 2007). The primary greenhouse gases in the Earth's atmosphere include water vapour (H_2O), carbon dioxide (CO_2), nitrous oxide (N_2O), methane (CH_4) and ozone (O_3). Certain man-made greenhouse gases such as the halocarbons and other chlorine and bromine containing substances are also present in the atmosphere besides sulphur hexafluoride (SF_6), hydrofluorocarbons and perfluorocarbons. Infrared opacity of the atmosphere increases with increased levels of the greenhouse gases, an imbalance that can only be remunerated by an increase in the temperature of the surface–troposphere system. This phenomenon is termed as the greenhouse effect (IPCC, 2007). With the modernisation of the society, the level of greenhouse gases released into the atmosphere has increased, leading to an increase in anthropogenic changes in the climate. According to Yohe and Tol (2007), due to the increase in greenhouse gas emissions, global temperatures could rise by 2–3°C by 2050, resulting in the rise of sea levels and a change in the prototype of vegetation and animal migration.

Agriculture, the world's largest industry, is one of the biggest contributors in these greenhouse gas emissions and subsequently the changes in the climate, with maximum impact coming from the use of industrialised inputs such as machinery and fertilisers. Hence, before examining the effect of climatic change on agriculture, it is imperative to understand the current industrial agricultural system and its effect on climate change. 'Industrial agriculture' describes the agricultural methods used post-green revolution and the term 'green revolution' refers to the introduction of scientific technology into agriculture, especially hybrid seeds and chemical inputs such as fertilisers, pesticides and herbicides. Green revolution changed the scenario of world agriculture from a primary ecological process to one of the technological developments, revolutionising the world's food system. Before the 1900s, animals and human power were used instead of machinery to manage agricultural crops, and fertilisers comprised animal waste, crop residue and local organic matter. Agricultural yields obtained from these low-input and labour-intensive methods were low but stable. Pest outburst or severe weather was avoided by growing more than one crop or variety in the field, with farmers relying more prominently on natural process of earth instead of industrial inputs. Hence, in this system the relations between the agriculture and ecology were very strong and the farmer's understanding of the ecological process played a major role in the success of the crops. This early agrarian system soon started shifting away from the ecological methods toward mechanised farming due to the industrial revolution, which formed a part of green revolution. Because of an increase in population and subsequently the fear of food shortages in the future, alternative systems of agriculture based on modern machinery and technology became a vital part of government policy by reducing the human input and increasing the technological input. Hence, this industrial technology boom changed the agrarian system and the face of society.

In the 1960s and 1970s, green revolution based on increased use of technology further revolutionised the agricultural system. Norman Borlaug, the father of green revolution allowed Mexico's green revolution to spread worldwide by developing high yielding hybrid semi-dwarf wheat in 1940s, which were able to produce higher yields when combined with chemical inputs such as pesticides and fertilisers. Soon with the help of various funding agencies like Food and Agriculture Organization (FAO) and the United States Agency for International Development (USAID), hybrid technology succeeded in making its way to India, Asia and across Europe. Hence, with the green revolution came the miracle of the twentieth century, which was the huge amount (doubled and tripled) of food produced from the same amount of land.

To produce high yields using hybrid seeds, chemical fertilisers were required, which were specific to a single crop and, hence, encouraged monocultures giving further rise to pest problems, which were then tackled with another chemical, that is, pesticides. Hence, with the increase of hybrid seeds, dependence on chemical inputs grew and the technological progress in agriculture, which appeared to be favourable at first glance, resulted in the explosion of many problems and complications over time, particularly greenhouse gas emissions. Agricultural practices, including deforestation, cattle feed lots, chemical use (fertilisers, pesticides and herbicides), use of fuels and manufacturing of on-farm machines and harvesting methods accounted for 25% of greenhouse gas emission (FAO, 2007), making agriculture the second largest industrial sector contributing to greenhouse gases. Looking at such a large impact of technological advancement in agriculture on climate change, it becomes imperative to limit all the aspects of agricultural greenhouse gas emissions. The agricultural aspect of climate change is primarily a technological problem, but is also influenced by political and social factors. However, despite political and social limitations, there are immediate benefits of biotechnology in agriculture that can be seen working in the current agricultural system. It is these benefits that hold the promise for reducing the immediate impact of agriculture on climate change and addressing the urgent problem of greenhouse gas emissions. While it is important for alternate movements, like the Polyface and similar sustainable farms, to continue growing and supporting the entire systematic agricultural change, it is also essential to immediately change the current system of industrial agriculture which accounts for the majority of the agricultural causes of climate change. It is therefore essential to find ways to immediately tackle the greenhouse gas emissions from large-scale industrial farms, and biotechnology holds one such immediate solution. The FAO says 'agriculture can be part of the solution by contributing to climate change mitigation, through carbon conservation, sequestration and substitutions and establishing agricultural systems that can buffer extreme events' (FAO, 2007).

Current and forecasted climatic conditions such as temperature extremes (hot and cold), drought, heat waves and the changing pattern of rainfall pose a serious challenge for agricultural production worldwide, affecting plant growth and yield, and causing billions of dollars in losses (Boyer, 1982). Hence, the global climate change is associated with the problem of food insecurity, hunger and malnutrition, particularly in South Asia and sub-Saharan Africa (Nelson et al., 2009; Parry et al., 2009). For example, the global temperature increased between 1981 and 2002, reducing the yields of major cereals by costing as much as $5 billion per year (Lobell and Field, 2007). The productivity of maize was drastically reduced by heat waves and drought in Italy (Ciais et al., 2005). Heat waves also affected wheat production in Central Asia and extreme flooding in South Asia in 2009–2010. In addition to the challenges associated with the climate change like extreme temperatures, drought and flooding, the biotic stresses such as pests, diseases and alien weed species also affect the current cropping systems (Hyman et al., 2008; Wassmann et al., 2009).

6.3 Biotechnology for climate change mitigation and crop adaptation

Agricultural biotechnology can play a positive role in addressing the problems associated with climate change by mitigating the impact of climate change, creating adaptation techniques and reduced-impact agricultural methods. Climate change mitigation refers to human interventions to reduce the sources or decrease the intensity of negative effects of climate change. Generally the climate change mitigation strategies involve reductions in the concentration of atmospheric greenhouses gases either by keeping a check on their sources or by increasing their sinks (IPCC, 2007). Climate change adaptation strategy involves the reduction in the vulnerability of natural and human systems to climate change effects (IPCC, 2007).

6.4 Biotechnology for climate change mitigation

Biotechnology plays a great role in reducing the on-farm fuel consumption by reducing the usage of chemical inputs and employing low-till or no-till agricultural methods. The use of chemical inputs such as fertilisers, herbicides and pesticides has become a common practice in the industrial agriculture worldwide, especially marginal landscapes, and has resulted in the global scale contamination of the environment with toxins that change the course of biogeochemical cycles. While these chemical inputs help to accelerate crop growth and increase yields, their effects on the climate change through the emissions of greenhouse gases are becoming increasingly evident.

Biotechnology provides a valuable solution for reducing the amount of chemical fertilisers used in conventional farming, finally leading to a reduction in the amount of greenhouse gases released into the atmosphere. This has been made possible by the development and use of modern biotechnology such as genetically modified organisms (GMOs) that have low fertiliser input needs. For example, the rice and canola developed by Arcadia Biosciences are genetically modified (GM) to use nitrogen more efficiently, resulting in reduced fertiliser needs. This technology, which is referred to as nitrogen use efficiency (NUE), allows farmers to produce yields equivalent to conventional agriculture without a significant requirement for nitrogen fertilisers. Artificial inorganic nitrogenous fertilisers like ammonium sulphate, ammonium chloride, ammonium phosphate, sodium nitrate and calcium nitrate are responsible for the formation and release of greenhouse gases (especially N_2O) from the soil to atmosphere when they interact with common soil bacteria (Brookes and Barfoot, 2009). Additionally, improved NUE in crops leads to the lower emission of greenhouse gases in the atmosphere through reduced fertiliser application. The reduced input of nitrogen fertilisers also means less nitrogen pollution of ground and surface waters. The GMOs and other related technologies like organic farming also reduce on-farm fuel usage, leading to reduction in CO_2 emissions, by decreasing the necessity and frequencies of spraying with fertilisers, pesticides and herbicides. Additionally, the GM crops will continue to reduce greenhouse gas emissions through reduced fertiliser application by combining the initial CO_2 reduction with further improvements in biotechnology research.

The use of environment-friendly biotechnology-based fertilisers (composted humus and animal manure) should be encouraged to reduce the negative effects of artificial fertilisers. Organic farming based on biofertilisers, crop rotation and intercropping with leguminous plants having nitrogen-fixing abilities are among some of the conventional biotechnological strategies for reducing artificial fertilisers use (Varshney et al., 2011). The use of genetically engineered techniques to improve *Rhizobium* inoculants led to the development of strains with improved nitrogen-fixing characteristics. The non-leguminous cereal crops, such as rice and wheat, can be made to fix nitrogen in the soil by inducing nodular structures on their roots using biotechnological approaches (Yan et al., 2008). Additionally, manipulation of animal diet and manure management can reduce CH_4 and N_2O emissions from animal husbandry (Johnsona et al., 2007). Agricultural biotechnology should provide for solutions to fight against climate change. In this context, biofuels produced both from traditional and GMO crops, such as sugarcane, oilseed, rapeseed and jatropha, will play a crucial role in reducing the adverse effects of greenhouse gases emission, particularly CO_2 by the transport sector. Hence, energy-efficient farming will depend on machines that use bio-ethanol and biodiesel instead of the conventional fossil fuels. A plantation of perennial non-edible oil seed producing plants will help in clearing the atmosphere and producing biodiesel fuel for direct use in the energy sector or in blending biofuels with fossil fuel in certain proportion, thereby minimising the use of fossil fuels to some extent (Jain and Sharma, 2010).

The capture or uptake of the carbon-containing substances, particularly carbon dioxide, is often referred to as carbon sequestration. It is commonly used to describe any increase in soil organic carbon content caused by the change in land management (Powlson et al., 2011). The soil carbon sequestration is one of the important strategies to limit the increase of the atmospheric CO_2 concentration. One way to enhance carbon sequestration is by reducing conventional tillage. Conventional tillage means to completely turn the soil to reduce the need for weed control and receive higher yield. However, tillage causes high erosion rates, resulting in the release of CO_2 into the atmosphere and the loss of other nutrients from soil. Tillage also increases the speed of decomposition of organic matter in the soil by increasing the availability of oxygen in the soil. An alternate approach to conventional tillage is the conservation tillage, which leaves approximately 30% of crop residue on the land to help reduce soil erosion from wind and rain. In this way, it reduces the loss of CO_2 from the agricultural systems and also plays a vital role in reducing water loss through evaporation, increasing soil stability and in maintaining cool soil microclimate. Conservation tillage is considered the superior option to conventional tillage as it reduces erosion and sedimentation in nearby waterways and allows for more natural soil cycles. Biotechnology takes conservation tillage a step further by creating GM crops like herbicide-tolerant (HT) seeds that reduce the need for tillage and allow farmers to adopt 'no-till' farming practices. In no-till farming, crops are specifically designed to reduce the impacts of soil preparation through plowing, ripping or turning the soil. HT crops allow farmers to apply herbicides to the emerging weeds rather than incorporating into the soil through tillage. This strategy has been made possible only by the use of biotechnology, which allowed the development of GM seeds, in the absence of which herbicides would have killed both

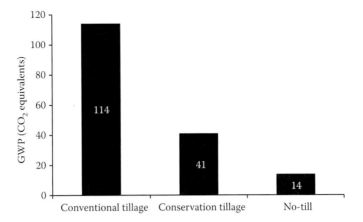

FIGURE 6.2   The gross global warming potential (GWP) for conventional, conservation and no-till agriculture.

the crops and the weeds. The main purpose of the herbicide-resistant plants is to reduce the need for tillage, finally providing protection to nearby environments through reduced erosion and enhanced soil sequestration. For example, the GM herbicide-resistant Round up Ready™ soybean accounted for up to 95% of no-till areas in the United States of America (USA) and in Argentina, leading to the sequestration of 63,859 million tonnes of CO_2 (Kleter et al., 2008). HT crops allow farmers to kill only the weeds avoiding the greenhouse gas intensive process of weed control by traditional tillage, finally leading to more soil carbon sequestration. No-till agriculture, in addition to carbon sequestration, reduces the consumption of fuel to operate equipment, thereby reducing CO_2 emissions. The gross global warming potential (GWP) for no-till agriculture is drastically lower than both traditional and conservation tillage (Figure 6.2). Reduction of fuel usage due to the application of biotechnology amounted to savings of about 962 million kg of CO_2 emitted in 2005, while the adoption of reduced tillage or no-tillage practices led to the reduction of 40.43 kg/ha CO_2 emissions due to less fuel usage, respectively. Therefore, in terms of carbon sequestration and reduced greenhouse gas emissions, it is clear that GM HT crops are beneficial for climate change mitigation.

6.5   Biotechnology for crop adaptation

In addition to the climate change mitigation strategies, it is equally important to give attention to the various adaptation strategies for the effects of climate change that are already in progress. Change of climate over time has led to a decrease in crop yield due to inadequate rainfall, various abiotic stresses, potential weeds, pests and diseases caused by fungi, bacteria and viruses. Biotechnology and the application of advanced techniques in agriculture will help in creating plants that will adapt to these new climatic conditions. One of the important ways of adapting to such changes is to apply agricultural bio-technological strategies that counter the effects by improving crop productivities per unit area of land cultivars.

The increasing demand for food crops worldwide can be satisfied in two ways: first is to increase the area under production and the second is to improve productivity on existing arable land. Given the limited amount of land for cultivation and a continuously changing climate, the second option seems to be more lucrative. Some of the conventional biotechnological options that organic farming technologies using biofertilisers include good agronomical practices such as land management, crop rotation, mixed farming, intercropping with leguminous plants with nitrogen fixing abilities and application of traditional and indigenous knowledge on known chemical pests and disease control methods (Bianchi et al., 2006). In this way, agricultural biotechnology and other advanced breeding strategies may help to further achieve higher yields and meet the needs of an expanding population with limited land and water resources.

Adaptation to abiotic stresses

Climate change poses an enormous intimidation in terms of the available agricultural land and fresh water use. Abiotic stress conditions such as salinity, drought, extreme temperatures, chemical toxicity and oxidative stress impose negative effects on agriculture and the natural environment (Bartels and Sunkar, 2005). Rising sea levels increase water salinity and force migration, resulting in greater population density with reduced viable crop land and fresh water for irrigation. About 25 million acres of arable land is lost each year due to salinity caused by indefensible irrigation techniques (Ruane et al., 2008). It is estimated that if the increase in salinity continues with this speed, it will lead to 30% loss of arable land within 25 years and 50% by the year 2050 (Wang et al., 2003; Valliyodan and Nguyen, 2006). Seventy percent of the available fresh water consumed is accounted by the agriculture sector (Brookes and Barfoot, 2008), which is likely to increase with the increasing temperature associated with climate change. Increasing harsh conditions will force the plants to use more energy and, hence, more water to grow. The problem is aggravated when the rising sea levels decrease available arable land and fresh water sources. This condition requires the need for an agriculture that truly conserves both water and land, and still gives a higher yield to feed the growing population. Biotechnology can be employed to generate an agricultural system that will be more water-efficient in the large-scale production methods.

Keeping the above information in view, the solutions that facilitate the adaptation of crops to these abiotic stresses (drought, salinity, etc.) need to be developed. The conventional approaches to reduce the effects of these abiotic stresses involve selecting and growing stress-resistant crops that can tolerate harsh conditions on marginal lands. Examples of such crops include cassava, millet and sunflower (Manavalan et al., 2009). Tissue culture and breeding are also being used to cross stress-tolerant crops with high-yielding species, generating stress-tolerant high-yielding hybrids (Ruane et al., 2008). Although the biotechnology community generally focuses on either molecular breeding or genetic engineering approaches, it is evident that there is a need to target complex problems caused by different stresses using integrated biotechnology approaches. As the whole genome sequence of plant, physical maps, genetics and functional genomics tools are becoming increasingly available, integrated approaches using molecular breeding and genetic engineering offer new opportunities for improving stress resistance (Manavalan et al., 2009). Hence, an outline for breeding a plant for the abiotic stress should incorporate conventional breeding and germplasm selection, elucidation of specific molecular control mechanisms in tolerant and sensitive genotypes, biotechnology-oriented improvement of selection and breeding procedures (functional analysis, marker probes and transformation with specific genes) and improvement and adaptation of current agricultural practices (Wang et al., 2003).

Activation and regulation of specific stress-related genes form the basis of the control mechanisms for abiotic stress tolerance (Table 6.1). Genetically engineered plants are based on different stress mechanisms, like metabolism, regulatory controls, ion transport, antioxidants and detoxification, late embryogenesis abundance, heat shock process and heat proteins (Wang et al., 2003). A number of high-yielding GM crops tolerant to abiotic stress have already been made available, some of which include tobacco (Hong et al., 2000), *Arabidopsis thaliana* and *Brassica napus* (Jaglo et al., 2001), tomato (Hsieh et al., 2002; Zhang and Blumwald, 2002), rice (Yamanouchi et al., 2002), maize, cotton, wheat and oilseed rape (Yamaguchi and Blumwald, 2005). As drought and water

Table 6.1   List of representative genes conferring stress tolerance in plants

S. no.	Name of gene	Full form	Trait
1.	DREB	Dehydration responsive element binding factor	Improved drought and salt tolerance
2.	SUB	Submergence (ethylene response factor like gene)	Submergence tolerance
3.	HSP	Heat shock protein	Improved drought and salt tolerance
4.	NAC	NAM/ATAF/CUC	Improved drought tolerance
5.	ERF	Ethylene response factor	Improved drought tolerance
6.	HARDY	AP2/ERF gene	Improved drought and salt tolerance
7.	HSF	Heat shock factor	Improved temperature tolerance
8.	MYC	—	Improved drought tolerance
9.	MYB	—	Improved drought tolerance
10.	ABF	Abscisic acid responsive factor	Improved drought tolerance
11.	P5CS	Pyrroline-5-carboxylate synthase	Improved drought and salt tolerance
12.	TPS	Trehalose-6-phosphate synthetase	Improved drought tolerance
13.	IMT	Myo-inositol-O-methyltransferase	Improved drought and salt tolerance
14.	CodA	Choline oxidase	Improved cold and salt tolerance
15.	ProDH	Proline dehydrogenase	Improved salt tolerance
16.	OAT	Ornithine amino transferase	Improved NaCl or mannitol tolerance
17.	BADH	Betaine aldehyde dehydrogenase	Improved salt tolerance
18.	Cu/ZnSOD	Superoxide dismutase	Improved cold and oxidative stress tolerance
19.	ALDH	Aldehyde dehydrogenase	Improved drought, salt and oxidative stress tolerance
20.	CDPK	Calcium-dependent protein kinase	Improved drought and salt tolerance
21.	NDPK	Nucleotide diphosphate kinase	Improved cold and salt tolerance
22.	NHX	Na^+/H^+ antiporter	Improved salt tolerance

Table 6.1   (continued) List of representative genes conferring stress tolerance in plants

S. no.	Name of gene	Full form	Trait
23.	SOS	Salt overly sensitive	Improved salt tolerance
24.	Glyoxylase	—	Improved salt tolerance
25.	NCED	9-*cis*-epoxycarotenoid dioxygenase	Improved drought tolerance
26.	Invertase	—	Improved salt tolerance

Source: Adapted from Bartels B and Sunkar R 2005. *Crit. Rev. Plant Sci.*, 24: 23–58.

scarcity are becoming more prevalent, biotechnology will help create plants that can withstand these harsh conditions. There are examples where plants are engineered to reduce the levels of poly (ADP ribose) polymerase, an important stress-related enzyme, resulting in GM plants that are able to survive drought and showed 44% increase in yield compared to their non-GM counterparts (Brookes and Barfoot, 2008). The United Kingdom Agricultural Biotechnology Council (ABC) is working on another technology, which involves the use of transcription factors and stress genes that act as genetic switches. This technology has resulted in a twofold increase in productivity for Arabidopsis and a 30% increase in yield for maize during severe water stress. Additionally, new areas of research in biotechnology are working toward creating plants that are resistant to salt by introducing a gene from salt-tolerant mangroves into food crops. With this technology, the available water sources can be used more efficiently and the lands near rising oceans that are subject to ground water salination will become fertile for these salt-tolerant seeds. Creating plants with increased yields means less land will be needed to plant and grow food. With growing populations and climate-induced land loss, producing higher yields on less land will become an essential component of agriculture. In this context, in addition to hardier and more water-efficient plants, biotechnology is also creating more space-efficient plants.

Adaptation to biotic stresses

Strains, resistant to biotic stresses such as insects, fungi, bacteria and virus have been developed through conventional landscape-management practices and breeding initiatives, leading to crop adaptation. For example, agricultural pest control strategies have been significantly benefited by the ability of the soil bacteria (*Bacillus thuringiensis, Bt*) gene to be transferred into maize, cotton and other crops to import protection against insects. *Bt* crops proved to be highly beneficial tools for the integrated pest management program by providing farmers with new pest control strategies (Zhe and Mitchell, 2011). For example, transgenic canola (oilseed rape) and soybean have been modified to be resistant to specific herbicides (Bonny, 2008). Also, GM cassava, potatoes, bananas and other crops that are resistant to fungi, bacteria and viruses are in development; some have already been commercialised while others are undergoing field trials (Van Camp, 2005). Studies carried out between 2002 and 2005 found that biotic stress-resistant GM crops account for an increase in the average yield of 11–12% for canola and maize compared to conventional crops (Brookes and Barfoot, 2008, 2009; Gomez-Barbero et al., 2008).

6.6 Conclusion

The development and application of plant biotechnology can contribute optimistically towards climate change adaptation and mitigation through reduction of CO_2 emissions, carbon sequestration, reduced fuel use, adoption of environmentally friendly fuels and reduced artificial fertiliser use, employing biofuels for improved soil fertility and crop adaptability. These procedures, on the one hand, help in improving agricultural productivity and food security, whereas on the other hand protect our environment from adverse effects of climate change. There is harmony among scientific groups that climate variability is a result of direct and indirect anthropogenic activities. An integrated approach combining both the conventional and modern agricultural biotechnology approaches will not only contribute to increased yield and food security, but also significantly contribute to climate change adaptation and mitigation initiatives.

Acknowledgements

PKJ and PC sincerely acknowledge the financial support provided by ICAR and the support provided by project director, NRC on Plant Biotechnology. DKS acknowledges the support and encouragement provided by the director, IARI.

References

BARTELS B, SUNKAR R 2005. Drought and salt tolerance in plants. *Crit. Rev. Plant Sci.*, 24: 23–58.

BIANCHI FJJA, BOOIJ CJH, TSCHARNTKE T 2006. Sustainable pest regulation in agricultural landscapes: A review on landscape composition, biodiversity and natural pest control. *Proc. Royal Soc.*, 273(B): 1715–1727.

BONNY S 2008. Genetically modified glyphosate-tolerant soybean in USA: Adoption factors, impacts and prospects. A review. *Agro. Sustain. Dev.*, 28: 21–32.

BOYER JS 1982. Plant productivity and environment. *Science*, 218: 443–48.

BROOKES G, BARFOOT P 2008. GM Crops: Global socio-economic and environmental impacts 1996–2006. *J. AgBio. Forum*, 11(1): 21–38.

BROOKES G, BARFOOT P 2009. Global impact of biotech crops: Income and production effects, 1996–2007. *J. AgBio. Forum*, 12(2): 184–208.

CIAIS P, REICHSTEIN M, VIOVY N, GRANIER A, OGEE J, ALLARD V, AUBINET M et al. 2005. Europe-wide reduction in primary productivity caused by the heat and drought in 2003. *Nature*, 437: 529–533.

Food and Agriculture Organization 2007. http://www.fao.org/.

GOMEZ-BARBERO G, BERBEL J, RODRIGUEZ-CEREZO E 2008. BT corn in Spain—the performance of the EU's first GM crop. *Nature Biotechnol.*, 26: 384–386.

HONG Z, LAKKINENI K, ZHANG K, VERMA DPS 2000. Removal of feedback inhibition of *delta*-pyrroline-5-carboxylate synthase results in increased proline accumulation and protection of plants from osmotic stress. *Plant Physiol.*, 122: 1129–1136.

HSIEH TH, LEE JT, YANG PT, CHIU LH, CHARNG YY, WANG YC, CHAN MT 2002. Heterology expression of *Arabidopsis* C-repeat/dehydration response element binding factor I gene confers elevated tolerance to chilling and oxidative stresses in transgenic tomato. *Plant Physiol.*, 129: 1086–1094.

HYMAN G, FUJISAKA S, JONES P, WOOD S, DE VICENTE MC, DIXON J 2008. Strategic approaches to targeting technology generation: assessing the coincidence of poverty and drought-prone crop production. *Agric. Syst.*, 98: 50–61.

IPCC 2007. Climate Change 2007. Impacts, adaptation and vulnerability. Working Group II Contribution to the IPCC Fourth Assessment Report. Summary to Policymakers. Available online: (http://www.ipcc.ch).

JAGLO KR, KLEFF S, AMUNSEN KL, ZHANG X, HAAKE V, ZHANG JZ, DEITS T, THOMASHOW MF 2001. Components of Arabidopsis C-repeat/dehydration response element binding factor or cold-response pathway are conserved in *Brassica napus* and other plant species. *Plant Physiol.*, 127: 910–917.

JAIN S, SHARMA MP 2010. Prospects of biodiesel from Jatropha in India: A review. *Renew. Sustain. Energy Rev.*, 14(2): 763–771.

JOHNSONA JMF, FRANZLUEBBERSB AJ, WEYERSA SL, REICOSKYA DC 2007. Agricultural opportunities to mitigate greenhouse gas emissions. *Environ. Poll.*, 150(1): 107–124.

KLETER GA, HARRIS C, STEPHENSON G, UNSWORTH J 2008. Comparison of herbicide regimes and the associated potential environmental effects of glyphosate-resistant crops versus what they replace in Europe. *Pest Manage. Sci.*, 64: 479–488.

LOBELL DB, FIELD CB 2007. Global scale climate-crop yield relationships and the impacts of recent warming. *Environ. Res. Lett.*, 2: 014002.

MANAVALAN LP, GUTTIKONDA SC, TRAN LP, NGUYEN HT 2009. Physiological and molecular approached to improve drought resistance in soybean. *Plant Cell Physiol.*, 50(7): 1260–1276.

MTUI GYS 2011. Involvement of biotechnology in climate change adaptation and mitigation: Improving agricultural yield and food security. *Int. J. Biotechnol. Mol. Bio. Res.*, 2: 222–231.

NELSON GC, ROSEGRANT MW, KOO J, ROBERTSON R, SULSER T, ZHU T, RINGLER C et al. 2009. Climate change: Impact on agriculture and costs of adaptation. *International Food Policy Research Institute*, (Washington, D.C.).

PARRY M, EVANS A, ROSEGRANT MW, WHEELER T 2009. Climate change and hunger: Responding to the challenge. *World Food Programme*, (Rome).

POWLSON DS, WHITMORE AP, GOULDING KWT 2011. Soil carbon sequestration to mitigate climate change: A critical re-examination to identify the true and false. *Eur. J. Soil Sci.*, 62: 42–55.

RUANE J, SONNINO F, STEDURO R, DEANE C 2008. Coping with water scarcity in developing countries: What role for agricultural biotechnologies? Land and water Discussion Paper No. 7. *Food and Agricultural Organization (FAO)*, p. 33.

VALLIYODAN B, NGUYEN HT 2006. Understanding regulatory networks and engineering for enhanced drought tolerance in plants. *Curr. Opin. Plant Biol.*, 9(2):189–195.

VAN CAMP W 2005. Yield enhancing genes: seeds for growth. *Curr. Opin. Biotechnol.*, 16: 147–153.

VARSHNEY RK, BANSAL KC, AGGARWAL PK, DATTA SK, CRAUFURD PQ 2011. Agricultural biotechnology for crop improvement in a variable climate: Hope or hype? *Trends Plant Sci.*, 16:363–371.

WANG W, VINOCUR B, ALTMAN A 2003. Plant responses to drought, salinity and extreme temperatures: Towards genetic engineering for stress tolerance. *Planta*, 218: 1–14.

WASSMANN R, JAGADISH SVK, SUMFLETH K, PATHAK H, HOWELL G, ISMAIL A, SERRAJ R, REDOÑA E, SINGH RK, HEUER S 2009. Regional vulnerability of climate change impacts on Asian rice production and scope for adaptation. *Adv. Agron.*, 102: 91–133.

YAMAGUCHI T, BLUMWALD E 2005. Developing salt tolerant crop plants: Challenges and opportunities. *Trends Plant Sci.*, 10: 615–620.

YAMANOUCHI U, YANO M, LIN H, ASHIKARI M, YAMADA K 2002. A rice spotted leaf gene Sp17 encodes a heat stress transcription factor protein. *Proc. Natl. Acad. Sci. USA*, 99: 7530–7535.

YAN Y, YANG J, DOU Y, CHEN M, PING S, PENG J, LU W et al. 2008. Nitrogen fixation island and rhizosphere competence traits in the genome of root-associated *Pseudomonas stutzeri* A1501. *Proc. Natl. Acad. Sci. USA*, 105 (21): 7564–7569.

YOHE G, TOL R 2007. Report on Reports—The Stern Review: Implications for Climate Change, *Environment* http://www.environmentmagazine.org/Archives/Back%20Issues/March%.

ZHANG HX, BLUMWALD E 2002. Transgenic salt-tolerant tomato plants accumulate salt in foliage but not in fruit. *Nat. Biotechnol.*, 19: 765–768.

ZHE D, MITHCELL PD 2011. Can conventional crop producers also benefit from *Bt* technology? *Agricultural and Applied Association Series*. Paper No. 103584.

CHAPTER SEVEN

Climate change effect on sugarcane productivity

Kalpana Sengar, Rakesh Singh Sengar,
Kanhaiya Lal and Vivekanand Pratap Rao

Contents

Abstract

Sugarcane (*Saccharum* spp.) is main cash crop through-out the tropical regions of the world. It represents an important food and bioenergy source, being cultivated in many tropical and subtropical countries, and covering more than 23 million hectares worldwide, with a production of 1.6 billion metric tons of crushable stems. This crop is responsible for almost two thirds of the global sugar production.

The predicted outcomes of climate change in the Indian context include a rise in temperature, decreased rainfall, altered rainfall patterns, floods, drought, waterlogging and increased CO_2. Increased temperature during the maturity period of sugarcane may adversely affect the juice quality, especially juice content; increased summer drying of the crop could result in substantial yield loss. It is also possible that there will be an autocatalytic component to global warming. Photosynthesis and respiration of plants and microbes increase with temperature, especially in temperate latitudes. As respiration increases with increasing temperatures more than photosynthesis, global warming is likely to increase the flux of carbon dioxide to the atmosphere, which would constitute a positive feedback to global warming.

7.1 Introduction

To optimise sugarcane improvement, it is necessary to know the impact a selected trait will have on the general physiology of the plant. However, this is not yet possible as there are too many gaps in our knowledge of the unique development and physiology of sugarcane. Such gaps impair our ability to enhance desired agronomical traits. For example, selection of sugarcane varieties with increased photosynthetic capacity may be useless if sugar accumulation is constrained by temperature, water deficit or nutrient availability (Inman-Bamber et al. 2002). It may prove difficult to consistently increase sucrose levels in the culm without first knowing the factors that affect sugarcane yield and carbon partitioning.

A key aspect to increase sugarcane yield is the regulation of its photosynthetic apparatus. Sugarcane C4 metabolism concentrates CO_2 in photosynthetically active tissues, a strategy that has an energy cost that may be offset by the reduction in photorespiration rates. There are at least three distinctive forms of C4 metabolism that can be identified by the decarboxylation enzymes they use: NADP$^+$–malic enzyme (NADP-ME), NAD$^+$–malic enzyme (NAD-ME) and phosphoenolpyruvate carboxykinase (PCK). There is evidence that sugarcane has both NADP-ME and PCK (Calsa and Figueira 2007), which suggest the two types of C4 metabolism might complement each other (Christin et al. 2007). The physiological implication of the presence of both pathways and how they could be explored to increase sugarcane yield is still unknown.

It is also important to detail how carbon demands in the culm affect photosynthetic rates. Photosynthetic rates decrease with plant age, which could be a result of physiological limitations to sucrose accumulation in the culms (McCormick et al. 2006). This regulation is mediated by hexose, but little is known about the downstream pathways of this signal (McCormick et al. 2008a).

The relationship between sink and source is a key step in the identification of targets that can be changed in order to improve sucrose accumulation. Sucrose production and storage is associated with the demand imposed by sink organs (McCormick et al. 2008b). For example, when the leaf growth is reduced, sucrose content tends to increase in culm (Inman-Bamber and Smith 2005). Furthermore, transgenic varieties that express an enzyme that converts sucrose into isomaltulose showed increased photosynthesis, probably due to the introduction of this new carbon sink (Wu and Birch 2007). Finally, the reduction of leaf elongation induced by water deficit redirects the carbon partition and provides an increase in sucrose content (Inman-Bamber et al. 2004). Experiments showed that water stress reduced the whole plant photosynthesis by 18% and the plant extension rate by 41%, resulting in a 19% reduction in total biomass.

However, water stress increased the sucrose mass gain by 27% and increased sucrose content of the dry mass by 37%, confirming that water deficit reduced the demand for photo assimilation for producing fibre and tops so that excess assimilate was allowed to accumulate in the form of sucrose (Inman-Bamber et al. 2008).

The impact of water deficit on the physiology or developmental process and on gene expression is also under study on six different sugarcane varieties in four regions of Brazil. As expected, the preliminary physiological measures showed that different cultivars utilise different mechanisms to survive water stress (Paros et al. 1989). For example, one cultivar utilised leaf rolling to reduce water loss, whereas a different variety increased root to shoot growth to reduce water loss and to increase water uptake (L. Endres, personal communication).

Over the next decades, climate change and increased CO_2 levels are projected to impact the productivity of all crops. The CO_2 levels are predicted to increase from about 379 ppm in 2005 to 730–1020 ppm by the end of the century (IPCC 2007). To design sugarcane crops for maximum productivity in such a changing environment, it is necessary to study how the increase of CO_2 levels affects sugarcane physiology. An increase in the levels of CO_2 will reduce the rate of photorespiration in all plants, but considerably more in C3 plants than C4 plants. Nevertheless, C4 plants do increase their biomass when CO_2 levels are increased from 370 to 720 ppm. This increase in biomass of C4 plants is associated more with the increase in water use efficiency than in the reduction of photorespiration (Vu et al. 2006; de Souza et al. 2008). An efficient use of water leads to a lower rate of water depletion in the soil, which increases resistance to drought (Vu and Allen 2009). Higher CO_2 levels change both the metabolites and transcript level of a number of sugarcane genes (Vu et al. 2006; de Souza et al. 2008), but how each change impacts sugarcane physiology remains unknown. Yield increases of 60% were observed on sugarcane grown in open top chamber under 720 ppm CO_2, which indicates that yields potential may increase under those conditions (de Souza et al. 2008).

Many other physiological traits need to be detailed before a strategy can be designed to improve them. For example, numerous details of sugarcane C4 photosynthesis and other metabolic pathways are needed to detect which steps constrain sugarcane yield. Understanding the mechanisms regulating the transition from vegetative to reproductive growth would allow the control of flowering for breeding and reduce the loss of fixed carbon for reproduction. In addition, little is known about what limits the capacity of sugarcane to store high concentrations of sucrose in the parenchyma tissue of the stalk (McCormick et al. 2008a). Sucrose content variation depends on the morphology of the plant, such as size of the canopy and responses to ripening stimuli, such as mild water stress, and how these traits influence the supply and demand for photo assimilation (Inman-Bamber et al. 2009). The photomorphogenic control of sugarcane development can be modified by treatment with gibberellic acid (GA_3). This phytohormone induces a significant increase in stem cell elongation, which increases the capacity for sucrose storage in sugarcane seedlings.

In the next few years, many physiological puzzles have to be solved. Initially, the results obtained in more controlled greenhouse conditions under varying field conditions will need to be confirmed. Sugarcane transgenics, either overexpressing or silencing specific candidate genes will allow the testing of many hypotheses, while physiological experiments will help in identifying new candidate genes. System biology coupled with yet-to-be developed models will integrate physiology data with massive amounts of proteomic, metabolomic and transcriptomic data, to allow a more targeted approach toward understanding the limits of sugarcane productivity.

7.2 Present scenario

Indian sugar industry, the second largest after the textile industry, plays a vital role in the socio-economic transformation of rural India. India is the second largest producer of sugar after Brazil.

About 4 million growers are involved in the cultivation of sugarcane. Sugarcane is cultivated under a number of biotic and abiotic stresses, resulting in yield stagnation/fluctuation and low recovery. Sugarcane cultivation has assumed five dimensions, namely, maximisation of productivity, minimisation of cost production, sustainability, value addition and competitiveness.

In India, the total area under sugarcane cultivation was 41 lac hectare with productivity of 70 t/ha. Also, 10–20% of the sugar recovery was done during 2009–2012. Being a C4 plant, sugarcane is physiologically one of the most efficient solar energy harvesting plant. As per the agro-biological calculation and considering 50% use of solar radiation and 30% transpiration loss in sugarcane, it is possible to harvest 600 t/ha of total biomass. Some of the progressive growers have achieved 350 t/ha of cane yield; and therefore, there is a great scope to bridge the gap between potential and actual cane yields. Sugarcane agriculture in the country is associated with inherent inconsistencies in area and production due to various factors like climate, cane and sugar pricing, pricing of other commodities, cost of inputs and labour, labour availability and so on. The last decade saw the widest fluctuation in sugarcane production ranging from 12.7 million tonnes in 2004–2005 to the record production of 28.4 million tonnes in 2006–2007, leading to either a deficit or surplus situation. The present requirement of sugar in the country is 23 million tonnes, which is the highest in the world. The current production can meet the domestic requirement with an occasional surplus. Domestic sugar price in India is among the lowest in the world. The production cost of Indian sugar is estimated to be in the medium range—costlier than Australia and Brazil but lower than that of the United States. In the future, we may face stiff competition from African countries, whose production costs are lower than India.

The growth of the sugarcane agriculture in the country has been spectacular: From 1.17 m hectare in 1930–1931, the cane area increased to 5.1 million hectare by 2006–2007; almost a fourfold increase. During this period, the productivity went up from 31 to 68 t/ha, sugarcane production increased from 37 million tonnes to 355 million tonnes and sugar production had gone up from 0.12 million tonnes to 28.4 million tonnes. Sugar recovery also showed an improvement from 9.0% to 10.27%. The number of sugar factories went up from 29 to over 500 at present. The growth in cane and sugar production was contributed by two factors; a fourfold increase in cane area and improvement in productivity by more than 100%. Both these factors were possible because of the development of new,

well-adapted varieties, efficient crop production and crop production technologies. Owing to this, between 1961 and 2009,
the production of sugarcane increased at a compound annual
growth rate (CAGR) of 2.42%.

7.3 The sugarcane cycle

The Indian sugar industry is cyclical in nature. One or two
years of excess sugar production is followed by a few years of
shortage. Surplus cane and sugar production leads to low sugar
prices and accumulated stocks, resulting in losses to the sugar
factories. Consequently, the cane price is not paid in time to
the farmer. Then they switch over to other crops, leading to a
reduction in cane area and sugar production and thereby bringing down sugar production. The rainfall pattern seriously influences the cane area and production as well.

7.4 Contribution to economy

The contribution of sugarcane to India's GDP is 1.1%, which
is significant considering that the crop is grown only in 2.55%
of the gross-cropped area. The contribution of sugarcane to
the agricultural GDP has steadily increased from about 5% in
1990–1991 to 10% in 2010–2011. During the last two decades,
the average annual growth of sugarcane in the agriculture sector was about 2.6% against the overall growth of 3% in the
agriculture sector of the country. In India, sugar is an essential
item of mass consumption and the cheapest source of energy,
supplying around 10% of the daily calorie intake. Apart from
sugar, sugarcane also supplements the energy sector through
ethanol and electricity production.

7.5 Crop profile

Sugarcane cultivation in the south (South- and Southeast Asia)
extent from 7 to 32 N latitude covering both tropics and subtropics. The regions located south of 23 N latitude are ideally suited
for growing sugarcane in view of long sunshine hours throughout the year, which facilitates continued growth. The yield levels
in subtropical areas are significantly less (−56 t/ha) compared to
the tropics (−82 t/ha), since sprouting and growth during winter
months are severely affected. Development of varieties capable
of winter sprouting and growth is most essential if the current

yield levels of the subtropics are to be improved. A large number of improved high-yielding, well-adapted, location-specific varieties are currently available. Uttar Pradesh ranks first in cane area, accounting for over 40% of the total. The crop is grown under a wide range of agro-climatic condition, and the productivity ranges from 25 t/ha in Chhattisgarh to 105 t/ha in Tamil Nadu, while the average productivity in the country is about 65 t/ha. The average sugar recovery in the country is about 10.2%. Sixty to seventy percent of cane produced in the country is used for sugar production, while the rest is used for seed, chewing, juice and for the production of jaggery and khandsari. Sugarcane has now emerged as a multiproduct crop used for food, fuel, energy and fibre. Sugarcane is photosynthetically one of the most efficient crop, fixing 2.3% of solar radiation. One hectare of sugarcane may produce 100 tonnes of green matter every year, which is more than twice the agriculture yield of most other commercial crops. One hectare of sugarcane land with a yield of 82 t/ha produces about 7000 L of ethanol. Effective conversion of bagasse to ethanol is a major research focus today. The global availability of bagasse is estimated to be about 425 million tonnes annually. This huge biomass can be an important feedstock for the production of bio-ethanol. This is particularly important for countries like India, where the scope for increasing the production of ethanol from molasses or sugarcane juice is very limited. However, the limitation so far had been the lack of cost-effective technologies to convert bagasse to ethanol.

Yield gaps

The experimental maximum yield in sugarcane is 325 t/ha which is hardly achieved, though individual farmers have reported yields close to this. There is a wide gap in productivity between the tropical and the subtropical regions of the country; the farmers averaging about 82 t/ha in tropical and the later 56t/ha in subtropical respectively (Yasuda et al. 1982).

Production constraint

Sugarcane cultivation in the country falls under five agro-climatic regions: peninsular, east coast, northwest, north central and northeast zones. The productivity in each zone is affected by a varied number of factors. The yield in subtropical India is affected by the prolonged winter, which reduces the growing period. Drought, waterlogging, salinity and alkalinity affect cane production significantly in many states. Among the sugarcane disease, red rot is prevalent throughout the country and this disease can cause substantial loss. The emergence and spread of yellow leaf disease across the country is a major cause of concern. Continued mono-cropping of sugarcane without crop rotation and organic recycling for several decades have depleted the soil fertility considerably. Sugarcane is a labour- and input-intensive crop that remains in the field for more than a year. The cost of sugarcane cultivation has gone up significantly due to the increase in cost of labour and inputs. Labour availability for major operations such as harvest has also become scarce due to the migration of labourers seeking urban employment. The cost of harvest is Rs. 450–650/tonne in the tropics which is more than 25% of the sale value of the products.

7.6   Demand for sugar and allied products

Sugarcane is the basic raw material for sugar production, while molasses bagasse from the feedstock are the by-products of the sugar industry for ethanol production and cogeneration, respectively. The projected requirement of sugar in 2030 is 36 million tonnes, which is about 50% higher than at present. To achieve this target, the sugarcane production should be about 500 million tonnes from the current 350 million tonnes for which the production has to be increased by 7–8 million tonnes annually. The increased production has to be achieved from the existing cane area through improved productivity and sugar recovery since further expansion in cane area is not feasible.

International trends

The current global sugarcane production is about 165 m tonnes, nearly 80% of which is contributed by cane sugar. The global demand for sugar at present is close to 167 m tonnes, and it is estimated that by 2030 the requirement will go up by over 50% to 257 m tonnes. The major producers are Brazil, India, China, Thailand, the United States, Mexico and Australia. The new sugar industries are coming up in many African countries such as Uganda, Tanzania, Zambia, Nigeria, Ethiopia, Guinea and so on, while sugar production in some of the traditional sugar-producing countries such as Fiji and Mauritius has come down significantly for various reasons.

7.7   Challenge of climate change

The success of agriculture in India much depends on normal monsoons as the favourable weather condition. In the National Action Plan on Climate Change (NAPCC) the government has listed eight missions to combat climate change and bring down the country's emission level in the long term. A mission on sustainable agriculture and a mission on strategic knowledge on climate science are the most important.

Consequent to changes in the soil temperature, soil moisture and composition of gases in the root zone, there are likely changes with respect to root growth, composition of root exudates, soil processes, nutrient dynamics, decomposition and so on. The increase in CO_2 content will have a beneficial effect on C_3 crop and the dicot weeds may compete with the sugarcane crop. There could be higher incidence of pests and disease under the altered temperature regime. If the ambient temperature remains within the favourable range for pests, insect species will complete more generations, thereby leading to a larger population than normal. Climate change is likely to affect the pathogen, host or the host–pathogen interaction. The change in climatic conditions will have an impact on the pathogen variability. Given the changing climatic conditions, the overall disease scenario in sugarcane suggests that besides resistant varieties, other approaches are needed to be taken care of for disease management.

7.8 Conclusion

The overall scenario that is emerging is that climate change will affect sugarcane productivity through reduced growth, increased weed competition and increased incidence of pests and disease. Studies have estimated that for every 2°C rise in temperature, the sucrose yield will be reduced by about 30%. The decrease in yield is attributed to increased moisture stress caused by the warmer climate.

References

CALSA T Jr, FIGUEIRA A. 2007. Serial analysis of gene expression in sugarcane (*Saccharum* spp.) leaves revealed alternative C4 metabolism and putative antisense transcripts. *Plant Molecular Biology* **63**:745–762.

CHRISTIN PA, SALAMIN N, SAVOLAINEN V, DUVALL MR, BESNARD G. 2007. C4 photosynthesis evolved in grasses via parallel adaptive genetic changes. *Current Biology* **17**:1241–1247.

DE SOUZA AP, GASPAR M, DA SILVA EA, ULIAN EC, WACLAWOVSKY AJ, NISHIYAMA MY Jr, DOS SANTOS RTMM, SOUZA G, BUCKERIDGE MS. 2008. Elevated CO_2 increases photosynthesis, biomass and productivity, and modifies gene expression in sugarcane. *Plant, Cell and Environment* **31**:1116–1127.

Intergovernmental Panel on Climate Change (IPCC). 2007. Fourth Assessment Report. Cambridge University Press.

INMAN-BAMBER NG, SMITH DM. 2005. Water relations in sugarcane and response to water deficits. *Field Crops Research* **92**:185–202.

INMAN-BAMBER NG, BAILLIE C, WILLCOX J. 2002. Tools for improving efficiency of limited water use in sugarcane. *Proc. Irrigation Assoc. Aust., National Conference*, Sydney, 21–23 May, 2002, 251–259.

INMAN-BAMBER NG, ATTARD SJ, SPILLMAN MF. 2004. Can lodging be controlled through irrigation? *Proceedings of the Australian Society Sugar Cane Technology* **26**: (CD-ROM.) 11 p.

INMAN-BAMBER NG, BONNETT GD, SPILLMAN MF, HEWITT ML, JACKSON J. 2008. Increasing sucrose accumulation in sugarcane by manipulating leaf extension and photosynthesis with irrigation. *Crop and Pasture Science* **59**(1):13–26.

INMAN-BAMBER NG, BONNETT GD, SPILLMAN MF, HEWITT ML, XU J. 2009. Source-sink differences in genotypes and water regimes infl uencing sucrose accumulation in sugarcane stalks. *Crop and Pasture Science* **60**:316–327.

MCCORMICK AJ, CRAMER MD, WATT DA. 2006. Sink strength regulates photosynthesis in sugarcane. *New Phytologist* **171**:759–770.

MCCORMICK AJ, CRAMER MD, WATT DA. 2008a. Regulation of photosynthesis by sugars in sugarcane leaves. *Journal of Plant Physiology* **165**:1817–1829.

MCCORMICK AJ, CRAMER MD, WATT DA. 2008b. Changes in photosynthetic rates and gene expression of leaves during a source–sink perturbation in sugarcane. *Annals of Botany* **101**:89–102.

PAROS JP, BONNEL E FEREROL, MAUBUSSIAN JC. 1989. Altered rust resistance and yield components arising from leaf tissue and bud culture in sugarcane. *Proceedings of 20th Congress ISSCT*, Sao Paula, Brazil, 732–738.

VU JC, ALLEN LH Jr, GESCH RW. 2006. Up-regulation of photosynthesis and sucrose metabolism enzymes in young expanding leaves of sugarcane under elevated growth CO_2. *Plant Science* **171**:123–131.

WU L, BIRCH R. 2007. Doubled sugar content in sugarcane plants modified to produce a sucrose isomer. *Plant Biotechnology Journal* **5**:109–117.

YASUDA T, MAEGAWA H, YAMAGUCHI T. 1982. The selection for the tolerance of mineral stress in tropical plant tissue culture. In: *Plant Tissue Culture* (ed. A. Fujiwara), pp. 491–492. Maruzen, Tokyo.

CHAPTER EIGHT

Global warming impact on rice crop productivity

D.P. Singh

Contents

Abstract

Rice is a major food crop of Asia and of the world. The impact of global warming on rice may be due to a rise in sea level, thus resulting in inundated lands with sea water which makes these lands unsuitable for rice cultivation. Climate change is estimated to affect 20 million hectares of the world's rice-growing area adversely, mainly in India and Bangladesh. It is forecasted by the International Food Policy Research Institute that by 2050, the rice prices will increase between 32% and 37% as a result of climate change due to the reduction in rice productivity by 14% in South Asia, 10% in East Asia and the Pacific and 15% in sub-Saharan Africa. The rise in carbon dioxide levels in the environment may result in higher biomass in rice, which, depending on the type of cultivars, may or may not increase the grain yield. Climate change may be tackled by adopting proper strategies in research and policies of different countries. A proactive approach to this may save the rice production, as well as help in reducing emissions of greenhouse gas 'methane' from rice cultivation.

8.1 Introduction

The world population is growing steadily on the one hand, whereas land and water resources are declining on the other. Rice is the primary staple food for more than half the world's population. Asia represents the largest producing and consuming region. A total of 651 million tonnes of rice (paddy) was harvested in Asia in 2012 (FAOSTAT, 2012). Rice production is rising in South Asia but falling in the east. It is also a staple food in sub-Saharan Africa, preferred in China and the only available domestic staple in many countries in Asia (FAOSTAT, 2010). An increase in rice production by ≈1% annually is estimated to meet the growing demand for food that will result from population growth and economic development (Rosegrant et al., 1995). Global population is predicted to rise to over 9 billion by 2050, which will lead to a 25% increase in the demand for rice. Most of this increase must come from greater yields on existing cropland to avoid environmental degradation, destruction of natural ecosystems and loss of biodiversity (Cassman, 1999; Tilman et al., 2002). During this period, a warmer climate may decrease rice yields by 8%. Fresh global water supply will be needed to accommodate increased rice production and an additional 57,280,000,000 L (1432 L × 40,000,000 kg) of fresh water will be required. Fresh water demands will be more in highly populated countries like India and China, which are the main producers of rice in the world.

8.2 Global warming

'Global warming' refers to the rise in the average temperature of the earth's atmosphere and oceans. The greenhouse gases (carbon dioxide, water vapour, nitrous oxide and methane) trap heat and light from the sun in the Earth's atmosphere and lead to an increase in the temperature. Huge quantities of greenhouse gases are released into the atmosphere due to mining and combustion of fossil fuels, deforestation and maintenance of livestock herds and also due to rice production. The increase in temperature harms people, animals and plants, including rice. The higher the concentration of greenhouse gases, the more the trapping of heat in the atmosphere and the reduced escape of heat back into space. The higher heat results in a change in climate and altered weather patterns. In 2001, the 'UN-sponsored Intergovernmental Panel on Climate Change' reported that worldwide temperatures have increased by more than 0.6°C in the past century and estimated that by 2100, average temperatures will increase by between 1.4°C and 5.8°C (Nguyen, 2005).

8.3 Global warming and rice productivity

High temperatures or global warming would decrease the rice production globally (Furuya and Koyama, 2005). There is a need to plan for appropriate strategies to adapt to and mitigate the global warming for achieving long-term food security. Both lowland rice cultivation and upland rice production under slash-and-burn shifting cultivation results in the emission of methane and nitrous oxide gases and, thus, contributes to global warming. Increased concentration of carbon dioxide in the atmosphere along with rising temperatures are two major factors making rice agriculture a larger source of greenhouse gas 'methane' which may double by the end of this century. Methane is produced from carbon and hydrogen by bacteria in the soil. Some carbon enters the soil from the roots of rice plants, which have taken it from the atmosphere via photosynthesis. The rice plants grow faster under higher carbon dioxide concentration. This growth, in turn, pumps up the metabolism of methane-producing microorganisms in soil in rice field, thus generating more methane. Rice farming is responsible for approximately 10% of the methane released. Researchers at Northern Arizona University gathered published research from 63 different experiments on rice paddies, mostly from Asia and North America. The meta-analysis was used and found two strong patterns: first, more carbon dioxide boosted emissions of methane from rice paddies, and second, higher temperatures caused a decline in rice yields. According to the study, in the future the amount of methane emitted from rice paddies is likely to increase. Together, higher carbon dioxide concentrations and warmer temperatures predicted by the end of this century will double the amount of methane emitted per kilogram of rice produced (NAU, 2013). Since half of the worlds' human population is highly dependent on rice, the production systems for this crop are, thus, vital for the reduction of hunger and poverty. The cultivation of rice extends from dry lands to wet-lands and from the banks of the Amur River at 53° north latitude to Central Argentina at 40° south latitude. Rice is also grown in cool climates at altitudes of over 2600 m above sea level in the mountains of Nepal, as well as in the hot deserts of Egypt. However, most of the annual rice production comes from tropical climate areas. In 2004, more than 75% of the global rice harvested area (about 114 million out of 153 million ha) came from the tropical region whose boundaries are formed by the Tropic of Cancer in the Northern Hemisphere and the Tropic of Capricorn in the Southern Hemisphere (Nguyen, 2005). The temperature increases, which results in rising seas and changes in rainfall patterns and distribution, and may affect the land and water resources required for rice production and achieving the desired productivity of rice crops. The highest limit of temperature for growth of rice is 45°C and temperatures above this will be adverse for yields. The optimum temperature range for rice at different stages after germination is 35–31°C whereas for ripening it is 20–29°C (Table 8.1). The temperature may affect and produce abnormal symptoms in rice (Table 8.2). Such a rapid increase during the crop growth stages, particularly during extremely sensitive reproductive and early grain-filling stages of rice (*Oryza sativa* L.), leads to reduced biomass, grain yield and quality. Hence, increasing diurnal temperature tolerance in

Table 8.1 Critical temperatures for the development of rice plant at different growth stages

Growth stages	Critical temperature (°C)		
	Low	High	Optimum
Germination	16–19	45	18–40
Seedling emergence	12	35	25–30
Rooting	16	35	25–28
Leaf elongation	7–12	45	31
Tillering	9–16	33	25–31
Initiation of panicle primordia	15	–	–
Panicle differentiation	15–20	30	–
Anthesis	22	35–36	30–33
Ripening	12–18	>30	20–29

Source: From Yoshida, S. 1978. *Tropical Climate and Its Influence on Rice*. IRRI Research Paper Series 20. Los Baños, Philippines, IRRI.

Table 8.2 Symptoms of heat stress in rice

Growth stage	Symptoms
Vegetative	White leaf tip, chlorotic bands and blotches, white bands and specks, reduced tillering, reduced height
Reproductive anthesis	Reduce spikelet number, sterility
Ripening	Reduced grain-filling

Source: From Yoshida, S. 1981. *Fundamentals of Rice Crop Science.* Los Baños, Philippines, IRRI. 269 pp.

rice is a more sustainable approach than altering well-established cropping patterns, which will inevitably lead to yield penalties (Nagarajan et al., 2010). The current temperatures are already approaching critical levels during the susceptible stages of the rice plant, namely, Pakistan/North India (October), South India (April, August), East India/Bangladesh (March–June), Myanmar/Thailand/Laos/Cambodia (March–June), Vietnam (April/August), Philippines (April/June), Indonesia (August) and China (July/August) (Wassmann et al., 2009b).

8.4 Land and water resources for rice

The increase in temperature will create more land and water for growing rice in areas outside the tropical region (Darwin et al., 2005). The areas of coastal regions in the United States (Florida, much of Louisiana), the Nile Delta and Bangladesh will become unsuitable for rice with the rise of sea level by 88 cm (Kluger and Lemonick, 2001).

During the last two decades, night temperatures have increased at a much faster rate than day temperatures and global climate models predict a further increase in its frequency and intensity. The rice crop is affected both at the vegetative and reproductive stage due to a rise in temperature and, hence, productivity is also affected. The temperatures required at different crop growth phases are given in Table 8.1. High temperatures may result in various possible injuries to rice crops (Table 8.2). High temperatures for 1–2 h at anthesis may result in sterility of the rice crop. Mohandrass et al. (1995) predicted a decline in yield by 14.5% in summer rice in India by 2005 based on experiments at multi-locations. In the Philippines too, yields of dry-season rice declined by 10% for every 1°C increase in growing-season minimum temperatures, whereas the effect of maximum temperature on crop yields was insignificant. Peng et al. (2004) provided evidence in support of statements that decreased rice yields from increased nighttime temperature was associated with global warming. Temperature and radiation had statistically significant impacts during both the vegetative and ripening phases of the rice plant. Welch et al. (2010) concluded that diurnal temperature variation must be considered when investigating the impacts of climate change on irrigated rice in Asia. Higher temperatures can adversely affect rice yields through two principal pathways, namely (i) high maximum temperatures that cause—in combination with high humidity—spikelet sterility and adversely affect grain quality, and (ii) increased nighttime temperatures that may reduce assimilate accumulation. On the other hand, some rice cultivars are grown in extremely hot environments, so that the development of rice germplasm with improved heat resistance can capture an enormous genetic pool for this trait. The results show that high night temperature compared with high day temperature reduced the final grain weight by a reduction in grain growth rate in the early or middle stages of grain filling, and also reduced cell size midway between the central point and the surface of the endosperm (Morita et al., 2005). In the Philippines, rice production may decline up to 75% by 2100 because of global warming and Filipinos will have to settle for meals with little or no rice unless the government aggressively implements climate change adaptation programmes (Antiporda, 2013). Transpiration from rice panicles can help lower the temperature of the panicle, which is the susceptive organ for high-temperature-induced spikelet sterility. By increasing the transpiration, the heat damage to the panicle predicted to occur due to global warming may be avoided. To examine the possibility of genetic improvement in transpiration conductance of intact rice panicles (gpI), we measured gpI at the time of flowering in the open field in 21 rice varieties of widely different origins. Thus, the target of improvement in gpI against high-temperature-induced spikelet sterility should be set at the level of the existing varieties with the highest gpI (Fukuoka et al., 2012). Tao et al. (2008) studied the impact of global warming on rice production and water use in China, against a global mean temperature. They found the median values of rice yield decrease ranged from 6.1% to 18.6%, 13.5% to 31.9% and 23.6% to 40.2% for GMT changes of 1°C, 2°C and 3°C, respectively. Yoshimoto et al. (2010) synthesised a process-based model study in tandem with FACE experiments for studing the effects of climate change on rice yields in Japan. They found that it not only contributes to reducing the evaluation uncertainties, but also validates the adapting or avoiding studies of heat stress or negative influence on rice under projected climate change. The biomass production in rice will be more, which may or may not increase the grain yield. The higher temperatures can result in sterility in flowers, which will then adversely affect yields. The higher respiration losses due to a rise in temperature will also make rice less productive. IRRI research indicated that a rise in nighttime temperature by 1°C may result in losses in rice yields by about 10%.

8.5 Salinity, flooding and rice

Rice is highly sensitive to salinity. Salinity often coincides with other stresses in rice production, namely drought in inland areas or submergence in coastal areas. Submergent tolerance of rice plants has substantially been improved by introgressing the Sub1 gene into popular rice cultivars in many rice-growing areas in Asia. The rice crop has many unique features in terms of susceptibility and adaptation to climate change impacts due to its semi-aquatic phylogenetic origin. The bulk of global rice supply originates from irrigated systems, which are to some extent shielded from immediate drought effects. The buffer effect of irrigation against climate change impacts, however, will depend on the nature and state of the respective irrigation system (Wassmann et al., 2009b). Although rice can grow in water fields, submerged crops under water for long periods of time are not tolerated by rice plants. Flooding due to sea level rises in coastal areas and tropical storms will hinder rice production. At present, about 20 million hectares of the world's rice-growing area is at risk of occasionally being flooded to submergence level in India and Bangladesh. Wassmann et al. (2009b) in his review paper mentioned that the mega-deltas in Vietnam, Myanmar and Bangladesh are the backbone of the rice economy in the respective country and will experience specific climate change impacts due to sea level rise. Significant improvements of the rice production systems, that is, higher resilience to flooding and salinity, are crucial for maintaining or even increasing yield levels in these very productive deltaic regions. The other 'hotspot' with especially high climate change risks in Asia is the Indo-Gangetic Plains (IGP), which will be affected by the melting of the Himalayan glaciers. The dominant land use type in the IGP is rice–wheat rotation. The geo-spatial vulnerability assessments may become crucial for planning targeted adaptation programmes, but policy frameworks are needed for their implementation (Wassmann et al., 2009a).

8.6   Water shortage and rice productivity

Rice cultivation needs plenty of water. The changes in climate leading to a week without rain in upland rice-growing areas and 2 weeks in shallow lowland rice-growing areas can cause reduction in rice yields in the range of 17–40%. The intensity and frequency of droughts are predicted to increase in rain-fed rice-growing areas. Such changes are also expected in reduce water-short irrigated areas for rice cultivation. It affects rain-fed rice production in an area of 23 million hectares in South and Southeast Asia and about 20 million hectares of rain-fed lowland rice in Africa. The scarcity of water may also affect rice production in Australia, China, the United States and other countries. Drought stress is also expected to aggravate through climate change; a map superimposing the distribution of rain-fed rice and precipitation anomalies in Asia highlights especially vulnerable areas in East India/Bangladesh and Myanmar/Thailand (Wassmann et al., 2009a,b).

8.7   Global warming and its impact on pests, diseases and weeds

The IRRI experiments over the last 10 years at farmers' fields indicated that rice diseases and pests are influenced greatly by climate change. The incidence of diseases like brown spot and blast increases due to shortages of water, irregular rainfall patterns and related water stresses. On the other hand, the incidence of sheath blight or whorl maggots or cutworms reduces due to a change in the environmental conditions and shifts in production practices adopted by farmers to reduce the impact of climate change. It, thus, results in an emergence of new crop health dynamics. Global warming will enhance rice–weed competition. Rodent population outbreaks in Asia may increase due to unseasonal and asynchronous cropping as a result of extreme weather events. A combined simulation model (CERES-Rice coupled with BLASTSIM) was used to study the effects of global temperature change on rice leaf blast epidemics in several agro-ecological zones in Asia. At least 5 years of historical daily weather data were collected from each of 53 locations in five Asian countries (Japan, Korea, China, Thailand and

Philippines). Two weather generators, WGEN and WMAK, from the Decision Support System for Agrotechnology Transfer, were utilised to produce estimated daily weather data for each location. Thirty years of daily weather data produced by one of the generators for each location were used as input to the combined model to simulate blast epidemics for each temperature change. Maximum blast severity and the area under the disease progress curve caused by leaf blast which resulted from 30-year simulations were statistically analysed for each temperature change and for each location. Simulations suggest that temperature changes had significant effects on disease development at most locations. However, the effect varied in different agro-ecological zones. In the cool subtropics such as Japan and northern China, elevation of temperature above normal temperature resulted in more severe blast epidemics. In warm/cool humid subtropics, elevation of temperature caused significantly less blast epidemics. However, lower temperature caused insignificant difference in disease epidemics compared with that in normal temperature. Conditions in the humid tropics were opposite to those in cool areas, where daily temperature changes by $-1°C$ and $-3°C$ resulted in significantly more severe blast epidemics, and temperature changes by $+1°C$ and $+3°C$ caused less severe blast. Scenarios showing blast intensity affected by temperature change in different agro-ecological zones were generated with a geographic information system (GIS) (Luo et al., 1998).

8.8 Strategies for mitigating effects of global warming on rice production

The paddy experiments carried out at UC Davis and Trinity College Dublin indicated that increased carbon dioxide in the atmosphere boosted rice yields by 24.5% and methane emissions by 42.2%, increasing the amount of methane emitted per kilo of rice (Soos, 2012). There are several options available to reduce methane emissions from rice agriculture. The management practices such as mid-season drainage and using alternative fertilisers as well as switching to more heat-tolerant rice varieties and adjusting sowing dates are some of the measures suggested to reduce the methane emissions (NAU, 2013).

The following are few of the strategies which may be adopted to counter the effect of global warming on rice:

1. To breed and release new rice cultivars with better adaptation to high temperature and other climatic stresses

2. Develop new agro-ecosystem models which may be capable of predicting more correctly the consequences of climate change and land use change at different scales

3. Deployment of new management strategies for an eco-logical intensification of rice landscapes in Asia for increasing resource use efficiency, enhanced ecosystem resilience and a reduction in global warming potential

4. Create national and regional adaptation and mitigation policies for climate change on rice-based agriculture and net contributions of rice systems to global warming

References

ANTIPORDA, J. 2013. Decline in rice production due to global warming seen. *The Manila Times.net* (www.manilatimes.net/index.php/news/top-stories/38680-decline - 77 k).

CASSMAN, K.G. 1999. Ecological intensification of cereal production systems: Yield potential, soil quality, and precision agriculture. *Proc. Natl. Acad. Sci. USA* 96: 5952–5959.

DARWIN, R., TSIGAS, M., LEWANDROWSKI, J. and RANESES, A. 2005. World agriculture and climate change: Economic adaptation. USDA Agricultural Economic Report No. 703. 86 pp.

FAOSTAT. 2010. FAO statistical database (www.fao.org).

FAOSTAT. 2012. FAO statistical database (www.fao.org).

FUKUOKA, M., YOSHIMOTO, M. and HASEGAWA, T. 2012. Varietal range in transpiration conductance of flowering rice panicle and its impact on panicle temperature. *Plant Prod. Sci.* 15: 258–264.

FURUYA, J. and KOYAMA, O. 2005. Impacts of climatic change on world agricultural product markets: Estimation of macro yield functions. *JARQ* 39: 121–134.

KLUGER, J. and LEMONICK, M.D. 2001. Global warming. *Time Magazine*, 23 April 2001: 51–59.

LUO, Y., TENG, P.S., FABELLAR, N.G. and TeBEEST, D.O. 1998. The effects of global temperature change on rice leaf blast epidemics: A simulation study in three agroecological zones. *Agriculture Ecosystems & Environment* 68: 187–196.

MOHANDRASS, S., KAREEM, A.A., RANGANATHAN, T.B. and JEYARAMAN, S. 1995. Rice production in India under the current and future climate. In: *Modeling the Impact of Climate Change on Rice Production in Asia*, R.B. Mathews, M.J. Kroff, D. Bachelet and H.H. van Laar (eds), pp. 165–181. United Kingdom, CAB International.

MORITA, S., YONEMARU, J.-I. and TAKANASHI, J.-I. 2005. Grain growth and endosperm cell size under high night temperatures in rice (Oryza sativa L.). *Annals of Botany* 95: 695–701.

NAGARAJAN, S., JAGADISH, S.V.K., HARI PRASAD, A.S., THOMAR, A.K., ANJALI, A., MADAN P. and AGARWAL, P.K. 2010. Local climate affects growth, yield and grain quality of aromatic and non-aromatic rice in northwestern India. *Agriculture, Ecosystems & Environment* 138: 274–281.

NAU. 2013. Research Links Rice Agriculture to Global Warming (http://nau.edu/Research/News-Updates/Research-Links-Rice-Agriculture-to-Global-Warming).

NGUYEN, N.V. 2005. Global climate changes and rice food security. *International Rice Commission Newsletter (FAO)* 54: 24–30.

PENG, S., JIANLIANG, H., SHEEHY, J.E., LAZA, R.C., VISPERAS, R.M., XUHUA, Z., CENTENO, G.S., KHUSH, G.S. and CASSMAN, K.G. 2004. Rice yields decline with higher night temperature from global warming. *PNAS* 101: 9971–9975.

ROSEGRANT, M.W., SOMBILLA, M.A. and PEREZ, N. 1995. *Food, Agriculture and the Environment Discussion Paper No. 5* (International Food Policy Research Institute, Washington, DC).

SOOS, A. 2012. Rice and global warming. *Environmental News Network* (http://www.enn.com/pollution/article/45133).

TAO, F., HAYASHI, Y., ZHANG, Z., SAKAMOTO, T. and YOKOZAWA, M. 2008. Global warming, rice production, and wateruse in China: Developing a probabilistic assessment. *Agricultural and Forest Meteorology* 148: 94–110.

TILMAN, D., CASSMAN, K.G., MATSON, P.A., NAYLOR, R. and POLASKY, S. 2002. Agricultural sustainability and intensive production practices. *Nature* 418: 671–677.

WASSMANN, R., JAGADISH, S.V.K., HEUER, S., ISMAIL, A., REDONA, E., SERRAJ, R., SINGH, R.K., HOWELL, G., PATHAK, H. AND SUMFLETH, K. 2009a. Climate change affecting rice production: The physiological and agronomic basis for possible adaptation strategies. *Advances in Agronomy* 101: 59–122.

WASSMANN, R., JAGADISH, S.V.K., SUMFLETH, K., PATHAK, H., HOWELL, G., ISMAIL, A., SERRAJ, R., REDONA, E., SINGH, R.K. and HEUER, S. 2009b. Regional vulnerability of climate change impacts on Asian rice production and scope for adaptation. *Advances in Agronomy* 102: 91–133.

WELCH, J.R., VINCENT, J.R., AUFFHAMMER, M., MOYA, P.F., DOBERMANN A. and DAWE, D. 2010. Rice yields in tropical/subtropical Asia exhibit large but opposing sensitivities to minimum and maximum temperatures. *PNAS* 107: 14,562–14,567.

YOSHIDA, S. 1978. *Tropical Climate and Its Influence on Rice.* IRRI Research Paper Series 20. Los Baños, Philippines, IRRI.

YOSHIDA, S. 1981. *Fundamentals of Rice Crop Science.* Los Baños, Philippines, IRRI. 269 pp.

YOSHIMOTO, M., YOKOZAWA, M., IIZUMI, T., OKADA, M., NISHIMORI, M., MASAKI, Y., ISHIGOOKA, Y. et al. 2010. Projection of effects of climate change on rice yield and keys to reduce its uncertainties. *Crop, Environment & Bioinformatics* 7: 260–268.

CHAPTER NINE

Unfolding the climate change impact on medicinal and aromatic plants

Manish Das

Contents

Abstract

The United Nations declared 2010 as the Year of Biodiversity to encourage nations to conserve their plant and animal species. The Convention on Biological Diversity (CBD) was signed more than a decade ago, but species continue to disappear worldwide at a rapid rate. Local communities who have used medicinal plants for generations say that these species are becoming hard to find. They say climate change is a factor. The CBD embedded three main goals into the national biodiversity strategies of all countries who signed it. These were sustainable use and the fair and equitable sharing of benefits of genetic resources. In respect of wild Indian medicinal plants, Article 8 of the CBD relates to *in situ* conservation. In particular clause 8(j) and 8(d) are very relevant to medicinal plants. However, as far as implementation is concerned much remains to be done.

9.1 Introduction

As a direct consequence of CBD, the Biological Diversity Act was enacted in 2002. A National Biodiversity Strategy and Action Plan (NBSAP) was prepared and subsequently a National Biodiversity Authority (NBA) was constituted. However, one is not aware of any significant initiative under this Act by NBA as far as the conservation and sustainable use of medicinal plants is concerned or, more importantly, as far as the preservation of traditional knowledge, innovations and practices of indigenous and local communities or their wider application is concerned. There appears to be some restrictions with regard to regulatory functions.

9.2 Is there a big loss of medicinal plant species in India?

An institutional mechanism needs to be put in place to systematically assess and enlist the decline and loss of medicinal plant species and to monitor and assess threats to wild populations of prioritised species. Article 8d of CBD specifically states: 'Promote the protection of ecosystems, natural habitats and the maintenance of viable populations of species in natural surroundings'. The Ministry of Environment and Forest (MoEF), however, has to have long-term programme, strategy or dedicated funding for monitoring viable populations and undertaking assessment of medicinal plants. National Medicinal Plants Board located at New Delhi and Indian Council of Agriculture Research may have to take the lead in this direction.

On a relatively small scale, some efforts have been undertaken by NGOs using International Union for Conservation of Nature (IUCN) Red List Categories and Criteria. According to such studies, 335 wild medicinal plants of India have been identified as being under various categories of threat of extinction ranging from near threatened, vulnerable, endangered to critically endangered. Eighty-four of these species of conservation concern have been recorded in high volume trade.

There is a need to put in place a long-term systematic threat assessment programme for important biota, including medicinal plants, which can be anchored in a network of institutions that have the competence to study different groups of plants and animals.

9.3 What are the most significant losses?

For example, there are six plant species of high conservation concern. These are *Aconitum heterophyllum*, *Coscinium fenestratum*, *Decalepis hamiltonii*, *Picrorhiza kurroa*, *Saraca asoca* and *Taxus wallichiana*.

These six plant species are valuable medicinal plant species that are currently being used in high quantities by India's herbal industry. They are of extreme conservation concern because of the rapid decline of their wild populations. The plant materials of these species are obtained entirely from the wild. They represent different life forms ranging from herbs to shrubs, climbers and trees. Their medicinal uses are described in the codified Indian systems of medicine, namely Ayurveda, Siddha and Unani.

9.4 What impact will the decline of these plants have?

These six species are used to treat many disease conditions, namely as anti-inflammatory, analgesic, anti-diarrhoeal, antipyretic, anti-diabetic, anti-cancer, in liver diseases as well as gynaecological disorders.

Their decline will adversely affect the current usage for health care and treatment of disease conditions. The extinction of such plant species will be an irreparable loss of the wild gene pool that has evolved over several millennia. Once lost, these species will not be reproducible through any synthetic means. It will be a huge loss for our future generations.

9.5 Are these losses because of climate change or because of over-extraction?

The decline and loss of wild populations of valuable wild Indian medicinal plants is due to the combined impact of habitat loss and degradation as well as over-exploitation. Climate change is also cited as a reason but there are no serious studies that have exposed this relationship. A few recent studies, outside India, have speculated about the fragmentation and decline of wild populations of some plant species in the mountains ecosystems due to climate change.

Medicinal plants constitute around 40% of the known diversity of vascular plant species of India. Conservation of Indian flora merits high priority. A national agenda for conservation of medicinal plants should be made.

9.6 Climate change

Although the terms 'global warming' and 'climate change' are often used interchangeably, 'climate change' is often the preferred term of many environmental organisations and government agencies. Climate change refers to any significant change in measures of climate (such as temperature, precipitation or wind) over an extended period of time (decades or longer). Global warming refers to an increase in the temperature of the atmosphere that can contribute to change in global climate patterns. The Intergovernmental Panel on Climate Change considers 'climate change' to mean any change in climate over time, whether due to natural variability or as a result of human activity. The United Nations Framework Convention on Climate Change defines 'climate change' as a change in climate that is attributable directly or indirectly to human activity that alters atmospheric composition.

The success of mankind's ability to meet the challenges of climate change will depend on how well it conserves the world's plants. Governments must act now, if plants are to continue to provide the resources and ecosystem services upon which all other species depend.

The climate change challenge for plants

Wild plant conservation has three mutually dependent aims:

- Maintaining plant species and their genetic diversity.
- Achieving sustainable use of wild plant resources.
- Securing plants and natural vegetation as providers of ecosystem services.

These aims are most likely to be achieved where efforts are focused on maintaining plants within robust ecosystems. However, the ability of national governments to achieve these aims is under increasing pressure because of climate change; the impact of which is seen at all levels of species' survival, including

- A continuing shift in the potential ranges of many plant species, causing them to become extinct in their existing locations. Many will find it difficult to 'follow the climate', lacking adequate means of dispersal and finding their paths impeded by human destruction of wild habitats.
- Increasing scarcity of food, fuel, forage, medicines and many other resources derived from wild plants. This will be a serious problem for the billions of people, especially in developing countries, who rely on such resources for their subsistence and livelihoods.
- The necessity to maintain water supplies, flood control and soil stability, all of which rely on natural vegetation in both river catchments and coastal margins. Water supplies, already under stress globally, will come under even greater pressure, further exacerbating potential resource conflicts.

Meeting the challenge: Important plant areas

Wild plants play a fundamental role in enabling national governments to sustain delivery of social and economic development and climate change magnifies the significance of this role. The critical factor in securing sustainable management of national plant resources is how governments involve the people and groups for whom the resources have most value.

Climate change is affecting medicinal and aromatic plants around the world and could ultimately lead to losses of some key species. This conclusion is based on the research, observations and opinions of multiple medicinal plant researchers and conservationists, as reported in the cover article of the latest issue of Herbal Gram (Vol. 81), the quarterly journal of the American Botanical Council (ABC).

Endangering medicinal plants advertisement

A 14-page article, based on recent climate change research and the perspectives of 15 scientific researchers, medicinal plant conservation experts and others, explores the current and potential effects of climate change on medicinal and aromatic plants. The article notes that species endemic to regions or ecosystems that are especially vulnerable to climate change, such as Arctic and alpine regions, could be most at risk. *Rhodiola rosea* of the Canadian Arctic and snow lotus (*Saussurea laniceps*) of the Tibetan mountains are specifically identified as medicinal species that could face significant threats from climate change. Researchers who have studied medicinal plants of Arctic and alpine areas and discovered potential threats posed by climate change provide information on their findings.

The article further explores effects of climate change that appear to be impacting plants including medicinal throughout the world. For example, climate change has led to shifts in seasonal timing and/or ranges for many plants, which could ultimately endanger some wild medicinal populations. Extreme weather events, meanwhile, have begun to impact the production and harvesting of various medicinal plants around the world. For instance, recent abnormally hot summers have prevented reseeding of medicinal plants such as chamomile (*Matricaria recutita*) in Germany and Poland, and increasingly severe flooding in Hungary has reduced harvests of fennel (*Foeniculum vulgare*) and anise (*Pimpinella anisum*) in that country.

Climate change recognised as one of the greatest challenges

Climate change has become increasingly recognised as one of the greatest challenges to humankind and all other life on Earth. Worldwide changes in seasonal patterns, weather events, temperature ranges and other related phenomena have all been reported and attributed to global climate change. Numerous experts in a wide range of scientific disciplines have warned that the negative impacts of climate change will become much more intense and frequent in the future—particularly if environmentally destructive human activities continue unabated.

Medicinal and aromatic plants and climate change

Like all living members of the biosphere, medicinal and aromatic plants (MAPs) are not immune to the effects of climate change. Climate change is causing noticeable effects on the life cycles and distributions of the world's vegetation, including wild MAPs. Some MAPs are endemic to geographic regions or ecosystems particularly vulnerable to climate change, which could put them at risk. Concerns regarding the survival and genetic integrity of some MAPs in the face of such challenges are increasingly being discussed within various forums.

Although scientists do not know whether climate change poses a more prominent or immediate threat to MAP species than other threats, it does have the potential to exert increasing pressures upon MAP species and populations in the coming years. The possible effects on MAPs may be particularly significant due to their value within traditional systems of medicine and as economically useful plants. The future effects of climate change are largely uncertain, but current evidence suggests that these phenomena are having an impact on MAPs and that there are some potential threats worthy of concern and discussion.

Some studies have demonstrated that temperature stress can affect the secondary metabolites and other compounds that plants produce, which are usually the basis for their medicinal activity. But few studies have been conducted *in situ* (in natural settings) or *ex situ* (in a controlled non-natural setting) to mimic conditions of global warming.

The taste and medicinal effectiveness of some Arctic plants could possibly be affected by climate change. It was noted that such changes could either be positive or negative, although it seems more likely that the effects would be negative since secondary metabolites are produced in larger quantities under stressed conditions and—for Arctic plants—warmer temperatures would likely alleviate environmental stress. However, that the production of plants' secondary metabolites are influenced by multiple factors—including diseases, competition between plants, animal grazing, light exposure, soil moisture and so on—and that these other factors may mitigate the effects of climate change on plants' secondary metabolites.

Recently NordGen, an organisation based in Alnarp, Sweden, collected samples of four medicinal plant species from Greenland for preservation and evaluation: angelica (*Angelica archangelica*, Apiaceae), yarrow (*Achillea millefolium*, Asteraceae), *Rhodiola rosea* (aka golden root, Crassulaceae) and thyme (*Thymus vulgaris*, Lamiaceae). These four MAPs are not currently endangered in Greenland, nor are they currently listed on the Convention in Trade in Endangered Species (CITES) appendices. However, collectors interested in preserving current plant genotypes from rapidly warming areas, such as Greenland, must do so before new genotypes arrive in response to climate change. Moreover, plant populations in Greenland are often isolated by the territory's many huge ice sheets, and this can limit the populations' available gene pools and subsequent abilities for genetically adapting to new climatic conditions. Capturing genetic diversity becomes increasingly important since it is possible that populations will lose genetic diversity in response to the changing environment.

Researchers have found that some cold-adapted plant species in alpine environments have begun to gradually climb higher up mountain summits—a phenomenon correlated with warming temperatures (Held et al. 2005). In some cases, these plants migrate upward until there are no higher areas to inhabit, at which point they may be faced with extinction. Additionally, the upward migration of plant species can lead to increased competition for space and resources, causing further stress among alpine plant populations.

A Global team found that useful Tibetan plants (predominantly medicinal plants) accounted for 62% of all plant species in the alpine Himalayan sites that they examined. Further, although overall species richness was found to decline with elevation from the lowest summits to the highest, the proportion of useful plants stayed approximately constant. This high percentage of useful plants confirms the importance of the Himalayas for Tibetan medicine and reflects the dangers posed by potential plant losses from climate change.

However, a few medicinal alpine species are restricted to the upper alpine zone, such as *Artemisia genipi* (Asteraceae) and *Primula glutinosa* (Primulaceae). These species may experience greater impacts from warming temperatures, possibly leading to local endangerment.

Medicinal and aromatic plants in other threatened regions

Although Arctic and alpine areas are experiencing some of the most rapid changes from global warming, other ecosystems are also considered particularly threatened by the ongoing effects of climate change. Among these ecosystems are islands and rainforests. Islands are considered especially at risk from rising ocean levels, in addition to changing temperatures and weather patterns. The world's oceans also absorb excess heat from the atmosphere, and as water warms it expands in volume (a process known as thermal expansion), which will similarly contribute to global sea level rise.

Despite these threats, experts have indicated that island MAPs may not be significantly affected by conditions related to climate change. Many of the plants used by island communities are common species that are widespread and highly adaptable.

Common medicinal plants of the Pacific islands include noni (*Morinda citrifolia*, Rubiaceae), naupaka (*Scaevola* spp,

Goodeniaceae), kukui (*Aleurites moluccana*, Euphorbiaceae) and milo (*Thespesia populnea*, Malvaceae). These and other medicinal plant species of the area grow relatively fast, have high reproduction rates and are typically resistant to salt water and wind, making them more resilient to some of the predicted effects of global climate change.

Similarly medicinal plants of the Mediterranean islands do not appear to be under any considerable threat from conditions of climate change. According to de Montmollin, most wild collected MAPs, such as thyme (*Thymus* spp, Lamiaceae) and rosemary (*Rosmarinus* spp, Lamiaceae), are rather widespread and located at lower altitudes, making them less vulnerable to climate change than plants with narrower ecological requirements.

Rainforest ecosystems are also considered to be particularly threatened by climate change. Climate modelling studies have indicated that these regions are likely to become warmer and drier, with a substantial decrease in precipitation over much of the Amazon.

There is not much, if any, published evidence on MAPs that could be at risk in the rainforest from climate change, and experts are unable to comment on specific MAPs that may be vulnerable to climate change in rainforests. However, the expected loss of general biodiversity in the Amazon, as noted in the IPCC report, indicates the potential to lose both known and undiscovered MAP species.

Widespread effects of climate change on medicinal and aromatic plants

Some effects of climate change appear to be impacting plants worldwide. For instance, evidence has shown that climate change has been affecting vegetation patterns such as phenology (the timing of life cycle events in plants and animals, especially in relation to climate) and distribution. Some wild plants, including MAPs, have begun to flower earlier and shift their ranges in response to changing temperatures and weather patterns. Shifting phenologies and ranges may seem of little importance at first glance, but they have the potential to cause great challenges to species' survival. They further serve as harbingers of future environmental conditions from climate change. Increased weather extremes are also predicted to accompany climate change, and plant species' resilience in the face of these weather events may also factor into their abilities to adapt and survive.

Shifts in phenology

The life cycles of plants correspond to seasonal cues, so shifts in the timing of such cycles provide some of the most compelling evidence that global climate change is affecting species and ecosystems. Available evidence indicates that spring emergence has generally been occurring progressively earlier since the 1960s. Such accelerated spring onset has generated noticeable changes in the phenological events of many plant species, such as the timing of plants' bud bursts, first leafings, first flowerings, first seed or fruit dispersal and so on. Records indicate that many plants—including MAPs—have started blooming earlier in response to the earlier occurrences of spring temperatures and weather.

There is a lot of variability between species, and it can be difficult to predict how climate change will affect the phenologies of different plants. In one finding it is reported that phenological shifts of medicinal plants are not significantly affecting wild harvesting practices. It was noted that there was always variations in the timing of the seasons, and collectors of wild medicinals are accustomed to adjusting their harvesting schedules accordingly.

It was noted that early blooming can become detrimental if an area is prone to cold spells late in the spring season. If a cold spell occurs a few days or weeks after early blooming has commenced, then those early buds or fruits could freeze, potentially killing or affecting the production of some economically useful plants. Apple orchards of North Carolina suffered severely from this type of scenario, and the medicinal plant bloodroot (*Sanguinaria canadensis*, Papaveraceae) has also been susceptible to frost following early blooming.

The impact of extreme weather events

Mounting evidence indicates that extreme weather events such as storms, droughts and floods have become more prevalent and intense across the globe in recent years. The frequency and severity of these events are expected to increase in the future as a result of continued warming, having negative effects on human health, infrastructure and ecosystems. Extreme weather events have been known to affect harvesters' and cultivators' abilities to grow and/or collect medicinal plant species, and such difficulties have certainly been reported in recent years.

Effects in European countries

Extreme weather conditions throughout Europe are impacting medicinal plant production from seeding to harvesting, such as chamomile in Germany and Poland. In the first year fennel (*Foeniculum vulgare*, Apiaceae) was recorded as having no yield at all in Bulgaria, due to drought conditions during the spring in that country. Due to long and dry summers in Serbia, accompanied by other extreme weather conditions such as strong rains and winds, have sometimes made it impossible for harvesters to perform second cuttings of the aerial parts of cultivated herbs such as peppermint.

Effects in African countries

Medicinal plants on other continents have also been impacted by severe weather conditions. Africa's Sahel region experienced one of the most severe droughts of the twentieth century. In Africa, medicinal plants of the Sahel include hibiscus (*Hibiscus sabdariffa*, Malvaceae), myrrh (*Commiphora africana*, Burseraceae), frankincense (*Boswellia* spp, Burseraceae), baobab (*Adansonia digitata*, Malvaceae), moringa (*Moringa oleifera*, Moringaceae) and various aloes (*Aloe* spp, Liliaceae) and were affected. Future drought from climate change could have devastating effects on the region's already suffering ecosystems and harvesting capabilities.

Effects in India

India, whose climate is largely controlled by an annual monsoon, appears to be experiencing increasingly severe and erratic precipitation. A recent study found that the overall amount of monsoon rainfall across Central India has remained relatively stable over the past century; however, moderate rainfall events during monsoon seasons have significantly decreased while extreme rainfall events have greatly increased since the early 1980s. This increase in extreme rainfall events could indicate greater potential for future natural disasters. Experts have claimed that the frequency and intensity of flooding has likewise been increasing in India in recent years as well as hailstorms that have caused huge agricultural losses across areas of India.

States like Gujarat and Rajasthan experienced hailstorms and rains in 2006, 2007 and 2008, at times when such events traditionally have not occurred within the past 50 years. Hail and rainstorms have also damaged psyllium (*Plantago ovata*, Plantaginaceae), wheat (*Triticum aestivum*, Poaceae) and cumin (*Cuminum cyminum*, Apiaceae) crops in the area. The destruction of Indian psyllium crops from hail and rainstorms resulted in a smaller than usual annual yield for 2008. Similarly, it was noted that the availability of menthol crystals was affected by heavy monsoon rainfall, which occurred earlier than usual in Northern India and reportedly damaged wild mint (*Mentha arvensis*, Lamiaceae) crops in 2008.

Hurricane seasons could also be affected by climate change, although experts do not agree on the possible effects. Some experts believe that hurricanes will increase in frequency, duration and intensity; others predict that hurricanes will either not be significantly affected or might even be inhibited by factors related to warming. Regardless, shifts (whether increasing or decreasing) in hurricane activity have the potential to affect the availability of medicinal plants.

Increasing evidence and studies have thus shown that at least some types of extreme weather events have been striking more frequently and with greater force throughout the world. Although particular weather events cannot be definitively blamed on climate change, the negative effects of some recent droughts, storms and floods on herbal crops demonstrate the threat that increased extreme weather could pose to the availability and supply of MAPs.

9.7 Conclusion

The effects of climate change are apparent within ecosystems around the world, including medicinal and aromatic plant populations. Medicinal and aromatic plants (MAPs) in Arctic and alpine areas face challenges associated with their rapidly changing environments, and some researchers have raised concerns regarding the possible losses of local plant populations and genetic diversity in those areas. Shifting phenologies and distributions of plants have been recorded worldwide, and these factors could ultimately endanger wild MAP species by disrupting synchronised phenologies of interdependent species, exposing some early blooming MAP species to the dangers of late cold spells, allowing invasives to enter MAP species' habitats and compete for resources and initiating migratory challenges, among other threats. Extreme weather events already impact the availability and supply of MAPs on the global market, and projected future increases in extreme weather are likely to negatively affect MAP yields even further.

Climate change may not currently represent the biggest threat to MAPs, but it has the potential to become a much greater threat in future decades. Many of the world's poorest people rely on medicinal plants not only as their primary healthcare option, but also as a significant source of income. The potential loss of MAP species from effects of climate change is likely to have major ramifications on the livelihoods of large numbers of vulnerable populations across the world. Further, the problems associated with climate change are likely to be much more difficult to combat than other threats to MAPs. The problems posed by warming temperatures, disrupted seasonal events, extreme weather and other effects of climate change, on the other hand, cannot be so quickly and easily resolved.

Climate change and its effects will certainly increase in the near future, although the extent to which they do so cannot presently be determined. The effects of climate change on medicinal plants, in particular, has not been well-studied and is not fully understood. As the situation unfolds, climate change may become a more pressing issue for the herbal community, potentially affecting users, harvesters and manufacturers of MAP species.

Bibliography

BHARDWAJ J, SINGH S and SINGH D. 2007. Hailstorm induced crop losses in India: Some case studies. *4th European Conference on Severe Storms*, Trieste, Italy; abstract, September 10–14, 2007.

CAVALIERE C. 2008. Drought reduces 2007 saw palmetto harvest. *HerbalGram*, 77:56–57.

CLELAND E E, CHUINE I, MENZEL A, MOONEY H A and SCHWARTZ M D. 2007. Shifting plant phenology in response to global change. *Trends in Ecology and Evolution*, 72(7):357–364.

DEAN C. 2007. Will warming lead to a rise in hurricanes? *New York Times*, May 29, 2007; F5.

DENYER S. 2007. Floods find India wanting as climate change looms. *Hindustan Times*, August 8, 2007.

GORE A. 2006. *An Inconvenient Truth.* New York: Rodale.

GOSWAMI B N, VENUGOPAL V, SENGUPTA D, MADHUSOODANAN M S and XAVIER P K. 2006. Increasing trend of extreme rain events over India in a warming environment. *Science*, 314:1442–1445.

HAWKINS B, SHARROCK S and HAVENS K. 2008. *Plants and Climate Change: Which Future?* Richmond, UK: Botanic Gardens Conservation International.

HELD I M, DELWORTH T L, LU J, FINDELL K L and KNUTSON T R. 2005. Simulation of Sahel drought in the 20th and 21st centuries. *Proceedings of the National Academy of Sciences,* USA, 105(50):17891–17896.

Intergovernmental Panel on Climate Change. *Climate Change 2007: Synthesis Report.* November 2007.

LAW W and SALICK J. 2005. Human-induced dwarfing of Himalayan snow lotus, *Saussurea laniceps* (Asteraceae). *Proceedings of the National Academy of Sciences, USA,* 102(29):10, 218–10, 220.

MALCOLM J R, LIU C, NEILSON R P, HANSEN L and HANNAH L. 2006. Global warming and extinctions of endemic species from biodiversity hotspots. *Conservation Biology*, 20(2):538–548.

NEILSON R P, PITELKA L F and SOLOMON A M. 2005. Forecasting regional to global plant migration in response to climate change. *Bio Science*, 55(9):749–759.

NICKENS T E. 2007. Walden warming. *National Wildlife*, 14:36–41.

PAL J S, GIORGI F and BI X. 2004. Consistency of recent European summer precipitation trends and extremes with future regional climate projections. *Geophysical Research Letters*, 31:L13202.

PARMESAN C and YOHE G. 2003. A globally coherent fingerprint of climate change impacts across natural systems. *Nature*, 421:37–42.

POMPE S, HANSPACH J, BADECK F, KLOTZ S, THUILLER W and KUHN I. 2008. Climate and land use change impacts on plant distributions in Germany. *Biology Letters*, 4:564–567.

SCHAR C, VIDALE P L and LUTHI D. 2004. The role of increasing temperature variability in European summer heatwaves. *Nature*, 427:332–336.

SHEA J. 2008. Apple growers hopeful after freeze. *Times-News*, April 6.

THOMAS C D, CAMERON A and GREEN R E. 2004. Extinction risk from climate change. *Nature*, 427:145–148.

WALTHER G R, POST E, and CONVEY P. 2002. Ecological responses to recent climate change. *Nature*, 416:389–395.

YOON C K. 1994. Warming moves plants up peaks, threatening extinction. *New York Times*, June 21, C4.

ZOBAYED S M A, AFREEN F and KOZAI T. 2005. Temperature stress can alter the photosynthetic efficiency and secondary metabolite concentrations in St. John's wort. *Plant Physiology and Biochemistry*, 43:977–984.

CHAPTER TEN

Impact of climate change on Indian agriculture

Samarendra Mahapatra

Contents

Abstract

Agriculture is an economic activity that is highly dependent upon weather and climate in order to produce the food

and fibre necessary to sustain human life. Not surprisingly, agriculture is deemed to be an economic activity that is expected to be vulnerable to climate variability and change. The vulnerability of agriculture to climate variability and change is an issue of major importance to the international scientific community ... stabilisation of greenhouse gas concentrations in the atmosphere at a level that would prevent serious anthropogenic interference with the climate system. Agriculture in developed countries may actually benefit where technology is readily available, along with the employment of appropriate adaptive adjustments.

10.1   Introduction

The atmosphere surrounding the Earth is made up of nitrogen (78%), oxygen (21%) and the remaining 1% is made up of trace gases (called so because they are present in very small quantities) that include carbon dioxide, methane and nitrous oxide. These gases, also called greenhouse gases, act as a blanket and trap the heat radiating from the Earth, thus making the atmosphere warm. Beginning with the industrial revolution, global atmospheric concentrations of these greenhouse gases have increased markedly as a result of human activities. The global increase in carbon dioxide concentration is primarily due to fossil fuel use and land use change, while those of methane and nitrous oxide are primarily due to agriculture. As a result, we are witnessing global warming. The increasing greenhouse gases (GHG) resulted in global warming by 0.74°C over the past 100 years and 11 of the 12 warmest years were recorded during 1995–2006. The Intergovernmental Panel for Climate Change (IPCC) projections on temperature predicts an increase of 1.8–4.0°C by the end of this century. Some changes will affect agriculture through their direct and indirect effect on crops, soils, livestock, fisheries and pests. Tropical countries are likely to be affected more compared to the countries situated in temperate regions. The brunt of environmental changes is expected to be very high in India due to a greater dependence on agriculture, limited natural resources, an alarming increase in human and livestock population, a changing pattern in land use and socio-economic factors that pose a great threat in meeting the food, fibre, fuel and fodder requirement. There is a likelihood of a considerable impact on agricultural land use due to snow melt, availability of irrigation, frequency and intensity of inter- and intra-seasonal droughts and floods, soil organic transformation matters, soil erosion and availability of energy as a consequence of global warming, impacting agricultural production and hence the nations' food security. Global warming due to the greenhouse effect is expected to impact the hydrological cycle, namely, precipitation, evapotranspiration, soil moisture and so forth, which would pose new challenges for agriculture.

10.2 Climate change impacting agriculture in India: As per Dr. Swaminathan

The impact of climate change on the farm sector would be profound and 1°C rise in temperature could lead to wheat yield losses of around 6 million tonnes per year in India. 'Climate change is impacting agriculture in countries like India. For one degree rise in temperature in areas like Uttar Pradesh, Punjab and Haryana could amount to a loss of about 6 million tonnes of wheat annually', said Swaminathan, while delivering a lecture on sustainable development. Climate change has a different meaning for different parts of the globe. Canada may benefit as the rise in temperature will allow them to grow more crops, but it would have an opposite effect in India. In India, climate change impact could be in terms of high temperature and rise in sea levels. Temperature not only affects the grain output but is also critical in terms of grain filling and pest attack. Suggesting steps to counter climate change, India should utilise its panchayati raj system effectively. 'One woman and one male member of every panchayat should be trained to become Climate Risk Managers who would be taught about various aspects related to the phenomenon', he said, adding that every farm should have a biogas plant and pond to check emission and ensure energy and water security. Crops to be classified into those that are climate resilient and those that are climate sensitive, for instance, wheat is a climate-sensitive crop, while rice shows a wide range of adaptation in terms of growing conditions. Sharing similar concerns, the Centre for Media Studies (CMS) chairman Bhaskar Rao said that agriculture is as important as corruption and no nation can be at peace if the disparities between the haves and the have-nots are huge. Swaminathan's lecture was part of the CMS–Nehru Memorial Museum and Library national lecture series on challenging issues in contemporary India. The vulnerability of Indian agriculture to climate change is well acknowledged. But what is not fully appreciated is the impact this will have on rain-fed (non-irrigated) agriculture, practiced mostly by small and marginal farmers who will suffer the most. The crops that may be hit include pulses and oilseeds, among others. These are already in short supply and are consequently high-priced. Nearly 80 million hectares, out of the country's net sown area of around 143 million hectares, lack irrigation facilities and, hence, rely wholly on rain water for crop growth. Over 85% of the pulses and coarse cereals, more than 75% of the oilseeds and nearly 65% of cotton are produced from such lands. The crop yields are quite low. The available records indicated that the predominantly rain-fed tracts experience three to four droughts every 10 years. Of these, two to three droughts are generally of moderate intensity and one is severe. Most of the rain-fed lands, moreover, are in arid and semi-arid zones where annual rainfall is meagre and prolonged dry spells are quite usual even during the monsoon season. This makes crop cultivation highly risk-prone. If the quantum of rainfall in these areas drops further or its pattern undergoes any distinct, albeit unforeseeable, change in the coming years, which seems quite likely in view of climate change, crop productivity may dwindle further, adding to the woes of rain-fed farmers.

According to the Indian Council of Agricultural Research (ICAR), medium-term climate change predictions have pro-jected the likely reduction in crop yields due to a climate change between 4.5% and 9% by 2039. The long-run predic-tions paint a scarier picture with the crop yields anticipated to fall by 25% or more by 2099. This will have a detrimental effect on farmers' income and purchasing power, with obvious down-the-line repercussions. Though the rainfall records avail-able with the India Meteorological Department do not indicate any perceptible trend of change in overall annual monsoon rainfall in the country, noticeable changes have been observed within certain distinct regions. At least three meteorological sub-divisions—Jharkhand, Chhattisgarh and Kerala—have shown significant decreases in seasonal rainfall though some others have recorded an uptrend in precipitation as well. Since rain-fed crops, such as coarse grains, pulses and oilseeds, are grown mostly during the kharif season, these are impacted by both low as well as excess rainfall. The groundnut crop in the Rayalaseema area of Andhra Pradesh in 2008 can be a case in point. It suffered substantial damage because of high as well as low rainfall at different stages of crop growth. While heavy rainfall early in the season adversely affected the development of pegs (which bear groundnut pods below the soil), the rela-tively drier spell at the later stage hit the development of pods. This aside, climate change is also reflected in the increasingly fluctuating weather cycle with unpredictable cold waves, heat waves, floods and exceptionally heavy single-day downpours.

The most noticeable of such events in recent years included the country-wide drought in 2002, the heat wave in Andhra Pradesh in May 2003, extremely cold winters in 2002 and 2003, and prolonged dry spell in July 2004 and January 2005 in the North, unusual floods in the Rajasthan desert in 2005, drought in the North-East in 2006, abnormal temperature in January and February in 2007, and 23% rainfall deficiency in the 2009 monsoon. All these events took a heavy toll on crop output. Indeed, the silver lining in this dismal scenario is the National Action Plan on Climate Change, launched in 2008, which aims at developing technologies to help rain-fed agriculture adapt to the changing climate patterns. At least four of the eight 'national missions' started under this programme will have direct or indirect bearing on rain-fed farming. These are the missions on sustainable agriculture, water, green India and strategic knowledge. The ICAR-led national agricultural research system is also conducting research on specific projects under the umbrella programme on climate change. 'Apart from the use of technological advances to combat climate change, there has to be sound policy framework and strong political will to achieve this objective', maintains ICAR scientists. State agricultural universities and regional farm research centres, too, will have to play a role in developing local situation-specific strategies for adapting the rain-fed farming to emerging climate patterns.

10.3 Global scenario of climate change

Current scenario

The global atmospheric concentration of carbon dioxide, a GHG largely responsible for global warming, has increased from a pre-industrial value of about 280–379 ppm in 2005. Similarly, the global atmospheric concentration of methane and nitrous oxides, other important GHGs, has also increased considerably. The increase in GHGs was 70% between 1970 and 2004. Eleven of the last 12 years rank among the 12 warmest years in the instrumental record of global surface temperature since 1850. The mean earth temperature has changed by 0.74°C during 1906–2005. Most of the observed increase in global average temperatures since the mid-twentieth century is very likely due to the observed increase in anthropogenic greenhouse gas concentrations. During the last 50 years, cold days, cold nights and frost have become less frequent, while hot days, hot nights and heat waves have become more frequent. The frequency of heavy precipitation events has increased over most land areas. Global average sea level rose at an average rate of 1.8 mm per year over 1961–2003. This rate was faster over 1993–2003, about 3.1 mm per year.

Future projections

The projected temperature increase by the end of this century is likely to be in the range 2–4.5°C with a best estimate of about 3°C, and is very unlikely to be less than 1.5°C. Values substantially higher than 4.5°C cannot be excluded. It is likely that future tropical cyclones will become more intense, with larger peak wind speeds and heavier precipitation. For the next two decades, a warming of about 0.2°C per decade is projected. Even if all future emissions were stopped now, a further warming of about 0.1°C per decade would be expected. Himalayan glaciers and snow cover are projected to contract. It is very likely that hot extremes, heat waves and heavy precipitation events will continue to become more frequent. Increases in the amount of precipitation are very likely in high latitudes, while decreases are likely in most subtropical land regions, continuing observed patterns in recent trends. The projected sea level rise by the end of this century is likely to be 0.18–0.59 m. Average global surface ocean pH is projected to reduce between 0.14 and 0.35 units over the twenty-first century.

10.4 Indian scenario of climate change

Current scenario

Analyses done by the Indian Meteorology Department and the Indian Institute of Tropical Meteorology, Pune (MS), generally show the same trends for temperature, heat waves, glaciers, droughts and floods, and sea level rise as by the Intergovernmental Panel on Climate Change of United Nations. The magnitude of the change varies in some cases. At the national level, there is no trend in monsoon rainfall during the last 100 years, but there are some regional patterns. Areas of an increasing trend in monsoon rainfall are found along the west coast, North Andhra Pradesh and North-West India, and those of a decreasing trend over East Madhya Pradesh and adjoining areas, North-East India and parts of Gujarat and Kerala (−6% to −8% of normal over 100 years). Surface air temperature for the period 1901–2000 indicates a significant warming of 0.4°C for 100 years. The spatial distribution of temperature changes indicated that a significant warming trend has been observed along the west coast, Central India and the interior peninsula and over North-East India. However, a cooling trend has been observed in the North-West and some parts in southern India. Instrumental records over the past 130 years do not show any significant long-term trend in the frequencies of large-scale droughts or floods in the summer monsoon season. The total frequency of cyclonic storms that form over the Bay of Bengal has remained almost constant over the period 1887–1997. There is evidence that glaciers in the Himalayas are receding at a rapid pace.

Future projections

It is projected that by the end of the twenty-first century, rainfall will increase by 15–31%, and the mean annual temperature will increase by 3–6°C. The warming is more pronounced over land areas, with the maximum increase over northern India. The warming is also projected to be relatively greater in winter and post-monsoon seasons.

- Although an increase in carbon dioxide is likely to be beneficial to several crops, the associated increase in temperatures and increased variability of rainfall would considerably impact food production.

- There are a few Indian studies on this theme and they generally confirm a similar trend of agricultural decline with climate change. Recent studies done at the Indian Agricultural Research Institute indicate the possibility of a loss of 4–5 million tonnes in wheat production in the future with every 1°C rise in temperature throughout the growing period (but no adaptation benefits). It also assumes that irrigation would remain available in the future at today's levels. Losses for other crops are still uncertain, but they are expected to be relatively smaller, especially for *kharif* crops.

- It is, however, possible for farmers and other stakeholders to adapt to a limited extent and reduce the losses (possible adaptation options are described later in this document). Simple adaptations such as a change in planting dates and crop varieties could help in reducing the impacts of climate change to some extent. For example, the Indian Agricultural Research Institute study, as quoted above, indicates that losses in wheat production in future can be reduced from 4–5 million tonnes to 1–2 million tonnes if a large percentage of farmers could change to timely planting and to better-adapted varieties. This change of planting would, however, need to be examined from a cropping systems perspective.

- Increasing climatic variability associated with global warming will, nevertheless, result in considerable seasonal/annual fluctuations in food production. Even today, all agricultural commodities are sensitive to such variability. Droughts, floods, tropical cyclones, heavy precipitation events, hot extremes and heat waves are known to negatively impact agricultural production and farmers' livelihood. The projected increase in these events will result in greater instability in food production and threaten the security of farmers' livelihoods.

- Increasing glacier melt in the Himalayas will affect the availability of irrigation, especially in the Indo-Gangetic plains, which, in turn, could create negative consequences for our food production.

- Global warming in the short term is likely to favour agricultural production in temperate regions (largely northern Europe, North America) and negatively impact tropical crop production (South Asia, Africa). This is likely to have consequences on international food prices and trade and, hence, our food security.

- Small changes in temperature and rainfall could have a significant effect on the quality of cereals, fruits, aromatic, and medicinal plants with resultant implications on their prices and trade.

- Pathogens and insect populations are strongly dependent upon temperature and humidity. An increase in these parameters will change their population dynamics resulting in a loss of yield.

- Global warming could increase water, shelter and energy requirement of livestock for meeting the projected milk demands. Climate change is likely to aggravate the heat stress in dairy animals, adversely affecting their productive and reproductive performance. A preliminary estimate indicates that global warming is likely to lead to a loss of 1.6 million tonnes of milk production in India by 2020.

- Increasing sea and river water temperature is likely to affect fish breeding, migration and harvests. A rise in temperature as small as 1°C could have important and rapid effects on the mortality of fish and their geographical distributions. Oil sardine fishery did not exist before 1976 in the northern latitudes and along the east coast as the resource was not available and sea surface temperature were not congenial. With the warming of sea surface, the oil sardine is able to find temperature to its preference, especially in the northern latitudes and eastern longitudes, thereby extending the distributional boundaries and establishing fisheries in larger coastal areas.

- Corals in the Indian Ocean will be soon exposed to summer temperatures that will exceed the thermal thresholds observed over the last 20 years. Annual bleaching of corals will almost become a certainty from 2050.

Organic agriculture, synonymous for biological agriculture, seems to be the feasible solution to the most debated topic 'climate change'. The climate of our world is undergoing a dramatic change. Global warming is rapidly increasing and there is a widespread consensus that the current trend is caused by increased emissions of various greenhouse gases such as carbon dioxide, hydrofluorocarbons, perfluorocarbons, sulphur hexafluoride, methane and nitrous oxide. Greenhouse gases allow short-wave solar radiation to pass into the Earth's atmosphere. They absorb some of the long-wave thermal radiation that is otherwise emitted back out to space, which results in the warming effect on our atmosphere. The emission of greenhouse gases into the atmosphere comes with industrialisation, through deforestation, shifting cultivation and the expansion of intensive agriculture. Present-day agriculture is no longer sustainable in most parts of the country, we can no longer deny chemical fertilisers and pesticides for the sake of susceptibility as defined by the west. The powerful message that distills from all thoughts and dialogues is the move toward Fukuoka's natural farming and Vinoba Bhave's Sarvodaya method of 'Rishi Kheti'. The logic to these naturalists is aimed at a reduced dependency on non-renewable resources, purchased inputs and population control to achieve a higher efficiency of inputs and economic maximisation of yield along with environmental safety. Biological agriculture can be defined as a system that attempts to provide a balanced environment, in which the maintenance of soil fertility and the control of pests and diseases are achieved by the enhancement of natural processes and cycles, with moderate inputs of energy resources, while maintaining an optimum productivity. The chemical agricultural (conventional agriculture) techniques have resulted in a great increase of productivity; however, they have greater negative impacts that include soil erosion or degradation, effects of pesticides, detention of soil health and environment, environmental pollution and so on.

The concentration of CO_2 and other greenhouse gases (GHGs) in the atmosphere is increasing as a result of land use change, besides fossil fuel combustion and cement production. The increase in GHGs in the atmosphere is leading to climate change and global warming. There is a need to reduce GHGs emissions and to increase carbon sinks. Currently, the biosphere is considered to be a carbon sink absorbing about 2.8 gigatonnes of C a year, which represents 30% of fossil fuel emissions.

Increase in water-holding capacity due to organic agriculture

Ecozones	% Increase over chemical farming	
	Range	Mean
Arid	2–9	5.3
Semi-arid	3–16	7.2
Sub-humid	4–15	6.9
Humid	3–17	6.8

Increase in C buildup due to organic agriculture

Ecozones	Additional increase ($\mu g\ g^{-1}$)	
	Range	Mean
Arid	49–83	62.5
Semi-arid	57–98	71.9
Sub-humid	61–101	75.5
Humid	68–102	83.0

Changes in microbial biomass due to organic agriculture

Agro ecosystems	% Increase in microbial biomass due to organic agriculture	
	Range	Mean
Arid	2–33	15.8
Semi-arid	5–25	17.5
Sub-humid	7–28	16.9
Humid	8–30	17.6

Carbon sequestration by organic agriculture as compared to chemical farming

Agro ecosystems	% Increase in C sequestration	
	Range	Mean
Arid	12–25	17.2
Semi-arid	8–19	13.5
Sub-humid	6–15	10.2
Humid	5–14	7.4

Improvement in microbial activity in organic versus chemical farming

Agro ecosystems	% Increase over chemical farming	
	Range	Mean
Arid	4–59	21.5
Semi-arid	6–61	23.7
Sub-humid	8–62	25.9
Humid	9–63	26.5

10.5 Climate variability and food production

Climatic variability and occurrence of extreme events are major concerns for the Indian subcontinent. There is a need to quantify the growth and yield responses of important crops and also identify suitable land use options to sustain agricultural productivity under this large range of climatic variations. In India, the analysis of seasonal and annual surface air temperatures (Pant and Kumar, 1997) has shown a significant warming trend of 0.57°C per hundred years. The warming is found to be mainly contributed by the post-monsoon and winter seasons. The monsoon temperatures do not show a significant trend in any major part of the country. Similar warming trends have also been noticed in Pakistan, Nepal, Sri Lanka and Bangladesh. The rainfall fluctuations in India have been largely random over a century, with no systematic change detectable in the summer monsoon season. However, during recent years areas of increasing trend in the seasonal rainfall have been found along the west coast, north Andhra Pradesh and North-West India and areas of decreasing trend over East Madhya Pradesh, Orissa and North-East India.

Extreme weather conditions, such as floods, droughts, heat and cold waves, flashfloods, cyclones and hailstorms are direct hazards to crops. More subtle fluctuations in weather during critical phases of crop development can also have a substantial impact on yields. Cultivated areas are subject to a broader range of influences, including changes in commodity prices, costs of inputs and availability of irrigation water. Climate may have indirect and possibly lagged influences on harvested areas.

10.6 Projected climate change scenarios over the Indian subcontinent

Climate change is no longer a distant scientific prognosis but is becoming a reality. The anthropogenic increases in emissions of greenhouse gases and aerosols in the atmosphere result in a change in the radiative forcing and a rise in the Earth's temperature. The bottom-line conclusion of the Third Assessment Report of the Intergovernmental Panel on Climate Change (IPCC, 2001) is that the average global surface temperature will increase between 1.4°C and 3°C above 1990 levels by 2100 for low emission scenarios and between 2.5°C and 5.8°C for higher emission scenarios of greenhouse gases and aerosols in the atmosphere.

10.7 Vulnerability of crop production

Estimating the effect of a changing climate on crop production in India is difficult due to the variety of cropping systems and levels of technology used. However, the use of crop growth models is one way in which these effects can be studied, and probably represents the best method we have at present for doing so. Although a large number of simplifying assumptions must necessarily be made, these models allow the complex interaction between the main environmental variables influencing crop yields to be understood.

10.8 Uncertainties due to scenarios and crop models on impact assessment

Estimates of the impact of climate change on crop production could be biased depending upon the uncertainties in climate change scenarios, region of study, crop models used for impact assessment and the level of management. So it is very important to give these uncertainties due importance while assessing the impacts of possible climate change on crop productivity for formulating response strategies.

10.9 Mitigation and adaptation strategies

Examination of relatively recent weather of the last century at many parts of the country indicates warming trends, although they may not be statistically significant, but there are enough indicators to suggest a modest increase in CO_2 and temperature. In spite of the uncertainties about the precise magnitude of climate change on regional scales due to scenarios and crop models on impact assessment, an assessment of the possible impacts of climate change on India's agricultural production under varying socio-economic conditions is important for formulating response strategies, which should be practical, affordable and acceptable to farmers. The identification of suitable response strategies is the key to sustainable agriculture. The important mitigation and adaptation strategies required to cope with anticipated climate change impacts include adjustment in sowing dates, breeding of plants that are more resilient to the variability of climate and improvement in agronomic practices.

10.10 Predicted climate change impacts on agriculture

The predicted changes to agriculture vary greatly by region and crop. The findings for wheat and rice are reported here.

Wheat production

- The study found that increases in temperature (by about 2°C) reduced potential grain yields in most places. Regions with higher potential productivity (such as northern India) were relatively less impacted by climate change than areas with lower potential productivity (the reduction in yields was much smaller).

- Climate change is also predicted to lead to boundary changes in areas suitable for growing certain crops as evidence has shown for wheat.

- Reductions in yields as a result of climate change are predicted to be more pronounced for rain-fed crops (as opposed to irrigated crops) and under limited water supply situations because there are no coping mechanisms for rainfall variability.

- The difference in yield is influenced by baseline climate. In sub-tropical environments, the decrease in potential wheat yields ranged from 1.5% to 5.8%, while in tropical areas, the decrease was relatively higher, suggesting that warmer regions can expect greater crop losses.

Rice production

- Overall, temperature increases are predicted to reduce rice yields. An increase of 2–4°C is predicted to result in a reduction in yields.

- Eastern regions are predicted to be impacted the most by increased temperatures and decreased radiation, resulting in relatively fewer grains and shorter grain filling durations.

- By contrast, potential reductions in yield due to increased temperatures in northern India are predicted to be offset by higher radiation, lessening the impacts of climate change.

- Although additional CO_2 can benefit crops, this effect was nullified by an increase of temperature.

Policy implications of climate change

The policy implications for climate change impacts in agriculture are multi-disciplinary, and include possible adaptations to

- Food security policy: To account for changing crop yields (increasing in some areas and decreasing in others) as well as shifting boundaries for crops, and the impact that this can have on food supply.

- Trade policy: Changes in certain crops can affect imports/exports depending on the crop (this is particularly relevant for cash crops such as chillies).

- Livelihoods: With agriculture contributing significantly to GNP, it is critical that policy addresses issues of loss of livelihood with changes in crops, as well as the need to shift some regions to new crops, and the associated skills training required.

- Water policy: Impacts vary significantly according to whether crops are rain-fed or irrigated, so the water policy would need to consider the implications for water demand of agricultural change due to climate change.

- Adaptive measures: Policy makers will also need to consider adaptive measures to cope with changing agricultural patterns. Measures may include the introduction of alternative crops, changes to cropping patterns and promotion of water conservation and irrigation techniques.

Need for further research

Owing to the complex interaction of climate impacts, combined with varying irrigation techniques, regional factors, and differences in crops, the detailed impacts of these factors need to be investigated further. Specific recommendations for further research include

- Precision in climate change prediction with higher resolution on spatial and temporal scales
- Linking of predictions with agricultural production systems to suggest suitable options for sustaining agricultural production
- Preparation of a database on climate change impacts on agriculture
- Evaluation of the impacts of climate change in selected locations

10.11 Conclusion

There was an asymmetry in the temperature trends in terms of day and night temperatures over India; the observed warming was predominantly due to an increase in maximum temperatures while minimum temperatures remained practically constant during the past century. There is likely to be a substantial increase in extreme maximum and minimum temperatures all over the country due to an increase in greenhouse gas concentrations. This is a very important finding from the agricultural point of view, as the *mid-day high temperature* increases the saturation deficit of the plants. It accelerates photosynthesis and the ripening of fruits. When high temperature occurs in combination with high humidity, it favours the development of many plant diseases. High temperature also affects plant metabolism. However, *high night temperature* increases respiration. Increases in food grain production during the last three decades made India self-sufficient and contributed tremendously to their food security. The latter, however, is now at risk due to the increased demand of a continuously increasing population. Also, the situation is grim due to the decline in soil fertility, the decline in groundwater level, rising salinity, resistance to many pesticides, degradation of irrigation water quality and a more rapid decline in the genetic diversity of the popular varieties in the farmers field. It is however of paramount importance to sustain the natural resource. Enhancing the organic matter content of soils will ensure a better soil fertility; irrigation pricing in the western Indo-Gangetic plains will ensure the efforts to increase the efficiency of water use and improve other associated environmental impact. However, since this adversely affects income from the rice–wheat system, there is a considerable socio-political resistance to its implementation. In recent years, the prospect of climate changes has stimulated considerable research interest in attempting to predict how the production of crops will be effected. The purpose of this review was to provide an overview of the likely effect of climate change on food production in India. Several studies projected an increase or decrease in yields of cereal crops (rice, wheat, maize and sorghum), oilseed and pulses crops (soybean, groundnut, chickpea, brassica (mustard) and pigeon pea) depending on interaction of temperature and CO_2 changes, production environment, season and location in India. Still climate change impact studies have not been conducted on several important crops in India such as sugarcane, cotton, jute, sunflower, potato and onion and so on, which may be done in the near future for better assessment of Indian agriculture's vulnerability to climate change.

References and Further Reading

AGGARWAL, P.K. and others in ICAR. 2009. *Vulnerability of Indian Agriculture to Climate Change: Current State of Knowledge*, Indian Agricultural Research Institute, October 2009.

AGRICULTURE AND AGRI-FOOD CANADA. 1996. Strategy for Environmentally Sustainable Agriculture and Agri-Food Development in Canada. Draft document, October 1, 1996.

ALBERTA AGRICULTURE, FOOD AND RURAL DEVELOPMENT (AAFRD). 1996. *Agriculture Statistics Yearbook, 1994*. Edmonton, AB: Market Analysis and Statistics Branch, Alberta Agriculture, Food and Rural Development.

ALLABY, M. 1989. *Dictionary of the Environment*, 3rd edition. New York: University Press.

ARTHUR, L.M. 1988. *The Implication of Climate Change for Agriculture in the Prairie Provinces, Climate Change Digest 88-01*. Downsview, ON: Atmospheric Environment Service.

BAYDACK, R., PATTERSON, J.H., RUBEC, C.D., TYRCHNIEWICZ, A.J. and WEINS, T.W. 1996. Management challenges for prairie grasslands in the twenty-first century. In Samson, F. and Knopf, F. (eds). *Prairie Conservation*. Washington DC: Island Press.

GRID-Arendal, 2003. IPCC Third Assessment Report—Climate Change 2001.

KALRA, N. 2007. *Climate Change Impacts on Agriculture in India*, New Delhi: IARI.

PANT, G. B. and KUMAR, R. 1997. *Climates of South Asia*. Chichester, UK: John Wiley & Sons Ltd.

SHARMA, A. 2008. *Impact of Climate Change on Indian Agriculture*, New Delhi: JNU, October 2008.

CHAPTER ELEVEN

Crop adaptation to climate change

An insight

Ashu Singh, Rakesh Singh Sengar,
Netra Chhetri and R.S. Kureel

Contents

Abstract

Many factors will shape global food security over the next few decades, including changes in rates of human population growth, income growth and distribution, dietary preferences, disease incidence, increased demand for land and water resources for other uses (i.e. bioenergy production, carbon sequestration and urban development) and rates of improvement in agricultural productivity. This latter factor, which we define here simply as crop yield (i.e. metric tons of grain production per hectare of land), is a particular emphasis of the plant science community, as researchers and farmers seek to sustain the impressive historical gains associated with improved genetics and agronomic management of major food crops. Much of what is known about the process of technological innovation in agriculture has yet to be captured in the discussions of climate change adaptation. The development of technological solutions to minimise risks of current climate can lead to two possible outcomes: increase in agricultural productivity and insights about adaptation to future climate change. Research efforts about the role of technological change, driven by climatic constraints, are pivotal in making any assertion about the likely adaptation of agriculture to climate change.

11.1 Introduction

Despite the evidence that technological innovation has been fundamental to growth and development of agriculture around the world, there is a dearth of research on the role of climate as a stimulus for innovation of technologies. It has been argued that lessons about climate adaptation come from our ability to understand the process of the existing technological innovation and its role in enabling farmers to cope with climatic challenges. Observing India's district-level time-series data over 12 years, I study the extent to which technological innovations have provided farmers with options to substitute for climate deficiencies to stabilise and enhance rice (*Oryza sativa* L.) productivity in regions with a sub-optimal climate. Drawing upon the hypothesis of induced innovation, which states that the direction of technological innovation in agriculture is induced by differences in relative resource endowments, my goal is to investigate whether spatial variations in climatic resources prompted the development of location-specific technologies that substituted for climatic limitations in the rice-based cropping system of India.

The main thrust of this study is to investigate if climatic limitations have been factored into the research and development of agricultural technologies, so that rice yields of climatically constrained regions could be sustained through innovations, thereby ensuring productivity convergence across India. Insights from this research will provide informed choices about the possible adaptation strategies that could be considered to ensure food security in the face of deleterious climate change. Easterling (1996) suggests that one way to draw insights about the adaptation of the agricultural system to a changing climate is the use of retrospective analysis to understand how the previous technological innovations have been targeted to address climatic constraints in specific locations. Building upon this notion of retrospective analysis, I investigate the process by which India's agricultural research and development systems have addressed specific climatic challenges and opportunities in the past. Logically, the core organizing questions then become: (a) do local climatic limitations provide incentives for farmers and public institutions to invest in research and development of technologies to overcome these limitations?; and (b) can the efforts of the past few decades to put in place a national agricultural research be a reasonable guide for adaptation to climate change?

The aim of this review is to offer guidance and priorities to federal agencies and private foundations funding research and development, policy makers, the scientific community, and economic sectors as they determine the avenues to best address the pressures facing agriculture today while also developing plans to optimize tomorrow's cropping systems.

11.2 Background

Research and technological innovations in agriculture have enabled farmers to cope with various challenges and have been fundamental to the growth and development of agriculture around the world (Rosenberg, 1992; IFPRI, 2009). They are extremely rich and diverse in nature. One of the notable successes came from a global effort to fight wheat rust—a plague that has been known to humanity for thousands of years, but had never been effectively controlled (Dubin and Brennon, 2009). The wheat rust success evolved into a much larger, more multidimensional series of successes and came to be known as Green revolution. Likewise, successes in Sub-Saharan Africa were not less important in addressing the persistent threat and hunger in the region. For example, in East and South Africa, technological innovation in maize led to a growth in maize yields among the regions, primarily smallholder agriculturists (IFPRI, 2009). The introduction of zero-tillage rice–wheat cultivation techniques in Gangetic plains provided benefit to some 620,000 farmers. In this practice, the seeds are planted in unplowed fields to conserve soil fertility, economise on scarce water, reduce land degradation, and lower production costs (Erenstein, 2009).

Although a future climate caused by global warming may be very different from the one that society has experienced in the past, insights for agricultural adaptation that confront us today may well be found in the experience of how climatic challenges were handled in the past. Historical analogues about climate adaptation includes deliberate translocation of crops across different agro-climatic zones, substitutions of new crops for old crops and innovation of technology in response to scarcity of resources (Easterling, 1996). An example of crop translocation includes expansion of hard-red winter wheat across climatic gradients of the North American Great Plains. From 1920 to 1999, the northern boundary of hard-red winter wheat expanded into a climatic region that was about 4.5°C cooler and 20% drier than the climate for the wheat zone in the 1920s. Interestingly, the southward expansion of hard-red winter wheat has not been as extensive where annual average temperatures at the southern boundary are 2°C greater than those of the 1920 southern boundary (Rosenberg, 1992; Easterling et al., 2004). Thus, hard-red winter wheat has been adapted to cooler and drier climates in the last 80 years. In China, winter wheat planting has shifted from 38°54′N to 41°46′N. This shift was aided by the introduction of freeze-resistant winter-wheat variety from high-latitude countries (Chen and Libi, 1997).

The growth of soybean in Ontario, Canada illustrates the examples of the substitutions of new crops for old ones. Although soybean was cultivated in Ontario throughout the twentieth century, it was not a prominent field crop until the 1970s. Between 1970 and 1997, the total acreage planted to soybean increased by over 500%, with the expansion being attributed to a series of technological innovations made in response to the climatic condition of Ontario (Smithers and Blay-Palmer, 2001). A fundamental climatic constraint to soybean cultivation in Ontario was the prevalence of cold night temperature during flowering, confining soybean cultivation to the extreme southwestern portion of the province. A key innovation to address this constraint was the introduction of cold-tolerant genetic material (*Fiskeby63*) from Sweden that led to the development of *Maple Arrow* cultivar, which played a vital role in the eastward spread of soybean crop. According to Smithers and Blay-Palmer (2001), technological innovations were not only confined to the development of cultivars but also to a range of agronomic activities, including modification of planting time and crop rotation-interrupted pest cycle, enhancing the cultivation of soybean.

The development of cowpea cultivars in the African Sahel illustrates the examples of technological substitutions in response to the existing variability in climatic resources. To escape the effects of drought, scientists in the African Sahel have developed early-maturing cowpea cultivars with different phenological characteristics. For example, to avoid the effects of late-season drought, they have developed cowpea varieties (*Ein El Gazal* and *Melakh*) that mature between 55 and 64 days after planting (Elawad and Hall, 2002). Similarly, to avoid mid-season drought, scientists have also developed a cowpea variety (*Mouride*) that matures between 70 and 75 days after planting (Cisse et al., 1997). Unlike *Ein El Gazal* and *Melakh*, which begin flowering between 30 and 35 days from sowing and have synchronous flowering characteristics, the *Mouride* variety starts flowering in about 38 days after planting and spreads out over an extended period of time, thereby escaping the mid-season drought. To enhance the chances of significant grain production, agriculturists in this region have developed cropping techniques where both types of cowpea (short and medium maturing) are planted together so that variable climatic input is optimised (Hall, 2004). Management of climatic risks is a critical aspect of economic survival. Farmers are understandably risk averse in their adoption of new technology. An interesting example is seen in the adoption of canola in southern Australia where Sadras et al. (2003) offer an analysis of a dynamic cropping strategy based on a putative association between start-of-season rain (April and May) and total seasonal rainfall. The study shows the advantage of long-term income when switching from a cereal-only strategy in a year of low rainfall to a more risky strategy of canola and cereal-based cropping systems in a year of high rainfall.

Yet, notwithstanding this recognition, there is a dearth of research that unravels the role of climate as a stimulus for innovation of appropriate technologies (Ausubel, 1995; Ruttan, 1996; NRC, 1999; Smithers and Blay-Palmer, 2001). Little is known about the manner in which technology has altered the relationship between climate and crop production and the roles that climate has played in the development of the new innovations. This research is a response to the challenge of developing a conceptual model that can be used as a basis for understanding agricultural adaptation to future climate change. More specifically, by using the hypothesis of induced innovation, this research extends the boundaries of climate change research to take into consideration the environmental inducements of technology in developing countries. It is an important area of investigation for at least three reasons. First, a productive and sustainable agricultural system is necessary for providing food security to an ever-growing population of developing countries.

Second, traditionally, agriculture occupies an important role in the local and national economies of most developing countries and is an important source of rural employment. Finally, the relation between climate change and agricultural adaptation represents a classic example of the human environment interface, an area of long-standing interest within geography.

11.3 Theoretical framework

As stated earlier, this research will utilise the theoretical framework of the induced innovation hypothesis to examine the interaction between climate and technology as a foundation for understanding the potential future agricultural adaptation to climate change and variability in India. Induced innovation refers to the process by which societies develop technologies that facilitate the substitution of relatively abundant (hence, cheap) factors of production for relatively scarce (hence, expensive) factors in the economy. Although the hypothesis of induced innovation was originally based on the experience of agricultural development in the United States and Japan (Hayami and Ruttan, 1985), lately, it has been used to explain the complex process of technological and institutional change, which represents a major perspective on international agricultural development (Koppel, 1995). The most fundamental insight of this hypothesis is that investment in innovation of new technology is the function of change (or difference) in resource endowment and the price of the resources that enters into the agricultural production function. This has spawned a conceptual infrastructure that addresses the broader issues of how farmers and their supporting institutions determine priorities for agricultural production.

As shown in Figure 11.1, climate change may alter these climatic resources by changing the growing season length and soil moisture regimes, and by adding heat stress to the plant. Such changes, following the hypothesis of induced innovation, will provide appropriate signals to farmers and public institutions to induce technologies suitable for the new environment. Translating this argument, as presented in the conceptual model, the induced innovation hypothesis suggests an important pathway for the interaction of climate and technology and for the study of the agricultural adaptation to climate change. The strength of this simple framework lies in its ability to highlight the central role of climate as a motivator of technological innovation and ultimately as a source of adaptation. Within this

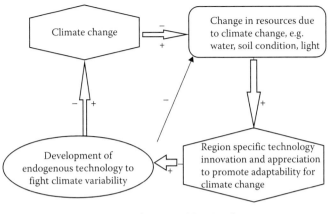

+ = Positive change, – = Negative change

FIGURE 11.1 Conceptual framework: climate–technology interaction.

conceptual framework, I will examine the role of spatial variability in climate as an incentive to the innovation of technologies in the Indiaese agricultural system.

One of the assumptions made by the induced innovation
hypothesis is that when agents of production (e.g. farmers, public institutions) experience problems with change in resource
endowments such as that, perhaps, brought about by climate
change, they are likely to seek new knowledge that will help
overcome these constraints. The change in resource endowment
(see Figure 11.1), therefore, may solicit an adaptive response
whereby farmers and their supportive institutions may adjust
management techniques and the allocation of resources to offset the effect of climate change. More specifically, in a society
(e.g. India) where land is already a scarce resource due to the
combined effect of population growth and unfavorable climate
for crop growth and development, as the pressure to grow food
on climatically less-favoured areas continues, the marginal cost
of production increases relative to the marginal cost of production via the application of technologies. Eventually, societies
will reach a stage where land augmentation will become an
appropriate means of increasing agricultural output. This will
ultimately lead to the development of technologies based on climatic resources of an area. This may be through the adoption
of location-specific crop varieties combined with other management strategies, such as efficient irrigation or application of
chemical fertilisers.

11.4   Climate-induced innovation in agriculture

Climate-induced innovation occurs when location-specific climatic constraints produce new demands on technology. Two outcomes are likely in this process. First, climatic constraints may induce the development of new knowledge to optimise the use of available climatic resources, resulting in increased production. Second, such new knowledge has the potential to enhance the ability of a region to compensate for the constraints imposed by climate and become self-sufficient in agricultural production (Evenson and Gollin, 2000; McCunn and Huffman, 2000). Logically, in the case of India's rice production, climate-induced innovation may provide opportunities for farmers to substitute for climate allowing for increased productivity in climatically less-favorable regions, leading to a convergence of productivity across climatically different regions of the country. The potential for convergence of productivity across different climatic regions can only be realised if and when farmers and research establishments devise and adapt technologies appropriate to the existing region-specific human environment conditions.

Following the thrust of the hypothesis of induced innovation, *a priori*, it can be argued that the innovation of technology in response to scarcity of climatic resources provides potential for rice productivity to grow faster in districts with marginal climate relative to the districts with more favorable climate. This process is asserted ultimately to lead to a convergence in the yield of rice over time. Such processes of targeted technological innovation may be reflected through the development of higher-yielding location-specific rice varieties, the enhancement of land development activities (e.g. irrigation), the development of climate-specific agronomic practices or a combination of all. The adoption of short-season rice varieties, for example, allows farmers to escape the late-season drought that occurs in some areas of the country. Similarly, the presence of irrigation alleviates the scarcity of water, a major constraint in the adoption of improved varieties of rice in India.

The hypothesis of induced innovation has been used to explain the relationship between resource endowment and the development of new technologies. Over time, it has been substantiated through many examples involving technical change in food production. The premise of the hypothesis of induced innovation concerning the role of climate as a stimulant for technological innovation has gone largely unquestioned because it is a difficult assumption to test. The critical question with regard to agricultural adaptation to climate variability and change, therefore, is whether substitution of technologies for climate would be employed in the future? Advances in knowledge can permit the substitution of more abundant resources for increasingly scarce resources to reduce the constraints for agricultural production. For example, innovation of early-maturing cultivars has the greater potential of escaping the effects of drought, which would be increasingly important to address the limitation of water scarcity due to a change in rainfall pattern. In light of this discussion, reorientation of the way society institutes the agricultural research will be necessary to adapt and/or realise the opportunities for technical change provided by new climate. Therefore, a research effort along the path induced by climatic stress is an essential step if meaningful insights are to be obtained with regard to agricultural adaptation to climate change.

11.5 India's biophysical and climatic characteristics

India is gifted with heterogeneous landforms and variety of climatic conditions such as the lofty mountains, the raverine deltas, high altitude forests, peninsular plateaus, variety of geological formations endowed with temperature varying from arctic cold to equatorial hot, and rainfall from extreme aridity with a few cms (<10 cm) to pre humid with world's maximum rainfall (1120 cm) of several hundred centimetre. This provides macro relief of high plateau, open valleys, rolling upland, plains, swampy low lands and barren deserts. These varying environmental situations in the country have resulted in a greater variety of soils. Therefore, the systematic appraisal of agro-ecological regions has tremendous scope in grouping relatively homogenous regions in terms of soil, climate and physiography and conducive moisture availability periods (length of growing season) in planning appropriate land use (see Figure 11.2). The mountain region that lies above the altitudes of 5000 m comprises 35% of India's 147,181 square kilometers of land. The Hills lie between altitudes of 600 and 5000 m, and accounts for 42% of the total land area. The flat Terai region, a northern extension of the Gangetic plain, is located below the 600-m elevation and comprises 23% of the total land area. Each of these regions represents a well-defined geographic area with distinct biophysical characteristics that are significantly different from each other, demanding location-specific technological innovations.

FIGURE 11.2 (**See colour insert.**) Map of India showing three ecological zones (Mountains, Hills and Terai) in the region.

The most outstanding feature of India's climate is the monsoon precipitation, which is characterised by two distinct phases: the 'wet' and the 'dry'. The wet phase (June–September) occurs in the summer season when the country receives over 75% of the annual precipitation (Shrestha et al., 2000). The monsoon, which is highly variable across space and time, is first experienced in the eastern part of the country. The monsoon gradually moves westward with diminishing intensity. The amount of summer monsoon and the number of days with rainfall decrease substantially as it moves to the west and northwestern part of the country (Chalise and Khanal, 1996) and the precipitation pattern becomes more varied with the diverse terrain within each physiographic belt (Chalise, 1994). While the temporal and spatial variability of monsoon rainfall and its social relation of rice production are well recognised, the specific role it plays in the innovation of technology remains understudied. The risks and impacts arising from monsoon variability are site specific and require technological innovations that reflect local conditions.

Although no discernible long-term change in climate has been observed, a study by the Department of Hydrology and Meteorology (DOHM) reveals that the average temperature in India is increasing at a rate of approximately 0.06°C per year. The temperature differences are most pronounced during winter season, and least after the summer monsoon begins (Shrestha et al., 1999). Consistent with the global trend, temperature is increasing at a faster rate in the higher elevations compared to the lower elevations. Notably, the rate of warming is greater in the western half of the country compared to the eastern half. The former is also significantly drier than the latter. Unlike temperature trends, no evidence of change in aggregate precipitation has been noted although studies do point to an increased variability and intensity of rainfall in some regions of the country.

If the observed trends of temperature change are overlain on the prevailing patterns of rainfall of the country, they reveal a negative association between the amount of rainfall and general trends of warming. For example, the Hills and Mountain regions of the western part of the country, which receive lower average rainfall, exhibit a higher degree of warming compared to the central and eastern Hills and Mountain, which are comparatively wetter. Theoretically, if this trend continues in the foreseeable future, the drier regions of the country will become even more so due to projected increases in temperature. For farmers, such a prognosis poses a further challenge in their effort to ensure better rice productivity. A recent study using general circulation models (GCMs) also projects a consistent warming of the Himalaya region (Agrawala et al., 2003). While the study also estimates an overall increase in precipitation, mostly during the monsoon season, it is not clear how these changes will affect the timing and period of monsoon rainfall.

Along with maize, millet, wheat, and barley, rice is an important staple crop, accounting for about 50% of both the total agricultural area and production in the country (Pokhrel, 1997). Rice is grown in all agro-ecological zones, from the subtropical climatic region of the Terai and the valleys to the higher altitudes of 1500 and 3050 m above sea level—the highest elevations in the world known to grow rice. Most rice-growing areas of the country have relatively optimal temperature for rice cultivation except the high Hills and the Mountain. In India, the total area under rice is estimated to be about 1.55 million hectares (HMGN/MFSC, 2002). The two major rice cultivation practices found in India are irrigated and rain-fed wetland (lowland). Both of these practices are common in all three ecological regions. Where there are irrigation facilities, rice fields may be irrigated during the rice-growing season to supplement the rainfall. However, areas under irrigated rice are extremely limited; so, rain-fed cultivation is the dominant practice for about 66% of the rice area (Pokhrel, 1997). The consequences of an adverse climate change could, therefore, have a significant negative effect on rice production.

11.6   Data and methodological approach

The districts for which data on rice productivity and irrigation acreage are available are the primary spatial units of analysis in this study. Of the total 75 districts of India (39 in the Hills, 20 in the Terai and 16 in the Mountains), 73 districts are included. Two districts in the Mountain (Mustang and Manang) are not included since rice is not grown there. The choice of the district as the unit of spatial analysis is further justified because it is the smallest administrative unit that contains the full complement of government services. For example, in the agriculture sector, every district has a government-run Agricultural Development Office (ADO) that employs agricultural extension workers responsible for promoting improved technologies. Each district is also supplemented by the office of the Agricultural Input Corporation (AIC) and the Agricultural Development Bank (ADB); government subsidiaries established to market agro-technologies to the farmers. In addition, the Department of Irrigation (DOI) has its offices at the district level, which are responsible for developing irrigation infrastructure. All these agencies are pivotal in the development of specific agricultural technologies needed in various agro-climatic regions of India.

This study is based on secondary data obtained from the various agencies of the government of India. The data concerning rice yield (productivity and yield are used interchangeably to indicate mean output per unit of land) and irrigation were obtained from the India Agricultural Database (NAD) of the Ministry of Agriculture and Co-operatives (MOAC). The average monthly rainfall data were obtained from the DOHM. The DOHM has compiled the average monthly precipitation for the period between 1968 through 1997 from the records of various meteorological stations throughout the country, and has used the data to represent the monsoon rainfall in this analysis.

Methodologically, following Barro and Sala-I-Martin (1992), convergence can be understood in two ways: convergence in terms of the level of productivity across time, that is, sigma (σ) convergence and the rates of productivity growth across space and time, that is, beta (β) convergence. Conceptually, the two measures used in the literature to test for convergence are related and provide alternative ways to examine similar phenomenon. In this chapter, convergence in terms of the level of productivity across time, that is, sigma (σ) has been examined. An approach to observe the occurrence of sigma (σ) convergence is to plot the evolution of standard deviations over time. For example, McErlean and Wu (2003) show the evolution of σ-convergence by plotting the standard deviations of productivity across the three geographic regions of China from 1985 to 2000. It occurs when the dispersion of rice productivity across 73 districts of India tends to decrease over time.

That is, if

$$\sigma it + T < \sigma it \tag{11.1}$$

where σit is the dispersion of rice yield (yit) across districts i at the initial period and $\sigma it + T$ is the dispersion of rice yield across districts at subsequent periods.

11.7 Results and discussions

I analyse the σ-convergence by examining changes in the evolution of the coefficient of variation (CV) over time at aggregate (national) and disaggregate scales (ecological regions). Analysis of the evolution of σ-convergence across the different scales is especially informative to see whether the pattern of σ-convergence observed at the aggregate scale is also found within the disaggregate scale (e.g. ecological regions). The analysis of σ-convergence provides a measure of variability of rice productivity. It occurs when the variability of rice productivity across the districts decreases over time. Conversely, the increase of variability implies in σ-divergence.

In general, there have been no distinct patterns to suggest the occurrence of σ-convergence in rice productivity because the CVs have not reduced substantially across the districts of India during the 12 years. The CV in 1991/1992 was about 32% and continued to decline until 1996/1997, with a record low in 1995/1996 (23%). The trend reversed thereafter, reaching an all-time high of 40% in 2002/2003. The finding at the aggregate level does not preclude the fact that σ-convergence may not have occurred within a specific ecological region. In the next section, I analyse the evolution of CVs at the scale of ecological regions to see if the pattern observed at the national (coarse-scale) level is also found within the ecological (finer-scale) regions.

Figure 11.2 presents the general trends of the evolution of CVs over time in the three ecological regions. Although exhibiting fluctuations, the CVs have declined in 20 districts of the

Terai region. From a high of 26.30% in 1991/1992, they have declined to a low of 7.60% in 1996/1997. With the exception of 2002/2003, all other years show relatively lower CVs, hovering around 10%. At the same time, CVs in 36 districts of the Hills have remained constant (at around 30%), and present no evidence of either σ-convergence or σ-divergence. In the Mountains, however, the evolution of the CVs, from a low of 5.30% in 1996/1997 to a high of 29.80% in 2001/2002, shows no apparent sign of σ-convergence.

There are several factors that farmers have to consider to make a crop production decision, and their ability to interact with factors such as market risk, varying costs and availability of critical inputs and other environmental risks, makes some farmers (as well as regions) more productive than others. There may be several factors at play in the apparent lack of σ-convergence. The push toward implementing the goals of adoption of climate-appropriate management practices (APP) may have been constrained by widespread inaccessibility due to the difficult geographic terrain, especially in the Hills and the Mountains. At this level of abstraction, however, the trend does provide insights about the unfolding of climate–technology interaction, hence the need for further analysis.

Rice has always been the focus of technological innovation in India. Over time, researchers have developed location-specific technologies (agronomic and cultivars) that reflect local conditions (NRC, 1999). The Coordinated Rice Research Program (CRRP), which functions under NARC, coordinates with the International Rice Research Institute (IRRI) for new genetic materials. In the last 30 years, CRRP has released and recommended more than 40 improved varieties of rice to a wide range of climatic conditions of the country (HMG/MOAC, 2001). Although farmers are selective in accepting them, owing to risks associated with rainfall variability and grain quality, the area covered by these improved varieties has increased steadily over time. The new cultivars recommended for different climatic conditions have extended the technological choices for farmers even in areas with marginal climate. In the early 1990s, the area covered by improved varieties of rice was estimated to be about 46%, but had increased to 71% by the end of the decade (Goletti et al., 2001). It is believed that greater rice productivity in climatically marginal areas can be linked to three overlapping interventions of (a) introduction and adoption of location-specific rice varieties, (b) adoption of climate-appropriate management practices and (c) institutional changes that led to technological innovation in marginal areas.

a. Introduction and adoption of location-specific rice varieties: Agricultural research institutions in India have released over 40 new varieties of rice since it started its formal research and development programme in the country. While most of these varieties were developed for high-potential irrigated land (e.g. 13 for the Terai and the fertile valleys and 11 for the Hills), a number of them were also developed for climatically marginal areas. About 25% of these 40 varieties were specifically recommended for rain-fed regions having intermittent drought periods, of which three were for the rain-fed condition of the mid- and far-western Terai region, four were for the drought-prone regions of the Hills and three were developed as cold-tolerant varieties for high-altitude regions of the Mountains. In the very poor rain-fed rice-growing area of the Mountains and the Hills, participatory plant breeding has led to a successful intervention and adoption of improved rice varieties.

A good example of this was the release of two high-altitude rice varieties, *Machhapuchre-3* and *Machhapuchre-9,* in the mid-1990s. Studies show that *Machhapuchre-3* was significantly superior to local varieties, producing 42% higher yield in rice-growing areas situated between 1500 and 2200 m above sea level (Sthapit et al., 1996; Joshi et al., 2001). Similarly, *Machhapuchre-9* was found to be doing well in areas located at altitudes >2200 m above sea level. Likewise, *Rampur Masuli*, another improved rice variety, has been replacing local low-yielding varieties due to its ability to mature 10–15 days earlier, an important consideration for farmers in regions with intermittent drought. The additional features that have led to a wider adoption of this variety include better tilling, high-yielding capacity and tolerance against foliar diseases (Joshi et al., 2001).

In India, centralised research and development policies of the past may also imply that technological innovation policy could be assessed and planned without much consideration of the particular climatic or other conditions at the local level. It is at the local level that availability of technology and other information determines the production choices of farmers. Understanding how location-specific needs are addressed by farmers and their supporting institutions is the first step towards identifying options for potential agricultural adaptation in a changing context.

b. Adoption of climate-appropriate management practices: Varietal improvement alone will have limited impacts on rice productivity, especially in marginal climatic areas. Low soil fertility and lack of water are other major constraints that are difficult to overcome. Researchers in India have been engaged in devising improved agronomic management practices that alleviate constraints posed by climatic factors. For example, to address the constant dilemma associated with the uncertainty of the onset of monsoon, researchers have been improvising traditional methods of 'direct seeding' often practiced in risk-prone environments (Pandey and Velasco, 2002). According to Pandey and Velasco, the development of suitable varieties, availability of modern tools (e.g. power tiller drill) and increased access to herbicides have made this traditional technology more profitable in risk-prone environments of many Asian countries including India. This method has not only reduced the demand on labour but has thrived in areas of erratic rainfall, especially during the early stages of crop development. According to Tripathi et al. (2004), economic analysis of direct seeding yielded an additional net return of 33% compared to the conventional method of transplanting.

In another example, researchers working with farmers have helped them maximise the yield potential of high yield varieties (HYVs). To do so, farmers were required to follow a set of recommendations, one of which was adhering to specified timing of planting because delayed action could result in substantial loss of yield. A study also shows that improved varieties of rice must be transplanted from the seed bed to the main field between 24 and 28 days to achieve maximum yield potential (Sah et al., 2004). In a country where the timing and intensity of monsoon precipitation is highly variable, such a stringent condition can be problematic. So, it becomes mutually beneficial for both researchers and farmers to understand and implement agronomic practices that will result in higher production. Evidence also suggests that farmers are quite capable of adopting complex technological interventions as long as there is reciprocal relationship between them and the researchers (Witcombe et al., 1996; Joshi et al., 2001).

c. Institutional changes that led to technological innovation in marginal areas: Parallel to the government's effort in developing technologies for improving production in agriculture, there has been a significant policy change that may have contributed to the observed growth in rice productivity. One of the most important policy changes with regard to rice productivity has been the decision by the government to deregulate the fertiliser policy in 1997. This change in policy (i) allowed the private sectors to import and distribute fertilisers, (ii) phased out a fertiliser subsidy and (iii) deregulated fertiliser prices. In the absence of detailed data, it is difficult to precisely assess the impacts of the deregulation policy on the fertiliser use by the farmers. Nonetheless, a study based on the analysis of household-level data collected from 986 farmers indicated a significant growth in the application of fertiliser by the farmers of India. According to this study, 81% of the farmers applied both inorganic and organic fertilisers during the 2001/2002 crop year and reported an increased supply of the fertiliser, something they had not experienced previously.

In the early 1990s, NARC instituted a significant change in agricultural research and development. One of the outcomes was the setting up of Participatory Technology Development (PTD), a programme that focussed on the development of technologies that are appropriate to the climatically marginalised regions of the country. This was achieved through collaborative efforts among all stakeholders in agricultural development including farmers (Witcombe et al., 1996; Sperling and Ashby, 1999). The PTD approach also incorporates indigenous knowledge so that new technologies are best adapted to local social and environmental conditions. The PTD also provided a clearer strategy for coordination of new players (e.g. private enterprises and non-governmental organisations, [NGOs]) involved in the innovation of agricultural technologies in India (Biggs and Gauchan, 2001; Gauchan et al., 2003), an unlikely configuration a decade ago.

A new institutional setting for technological innovation is no doubt complex involving plural systems and multiple sources of innovation. Nevertheless, such environment provides space for a wide range of actors in technological innovation including farmers, private sectors and NGOs (Sthapit et al., 1996), and allows for better interaction and learning. While the earlier work on varietal development lay only within the governmental research institutions, this new institutional arrangement has been able to seek wider partnership among the various stakeholders who are focussed on agricultural development on marginal areas. This partnership has encouraged NGOs and other organisations to become stronger research institutions significantly contributing to innovations of technologies in agriculture. The role of farmers in technological innovation has also grown significantly whereby they are now able to set their agendas based on their own resource endowments, which are facilitated by NARC and NGOs. This new institutional approach has not only improved relationship between farmers and researchers but has also created an environment of dialogue that has benefitted both partners. The impact of PTD is reported to be especially positive in rice production in climatically marginal regions (Sthapit et al., 1996).

11.8   Conclusion

It is a challenge to make a compelling case for technological innovation as being driven solely by climatic factors because India's rice production is framed within the context of other changes that are part of its agricultural development. Yet, this study recognises that climate is one of the most important factors that farmers in the country have to adjust to for their rice production system. More importantly, this research uncovers recent changes in rice production technology made at the local level that is familiar with climatic constraints and local knowledge that signifies the thrust of location-specific innovation. Lack of data has been a major shortcoming in the effort to establish an unambiguous empirical relationship between climate and technologies—this is an open-research issue that can be addressed with time. To partially compensate for this shortcoming, a detailed review of case studies was provided as a qualitative assessment of the development of climate-induced innovations over the period of the study.

The findings from both the empirical and the qualitative assessments indicate that India's research establishment is engaged in and committed to the development of location-specific technologies that address the constraints of climate. Higher rice productivity is not only seen in climatically favorable regions but is also surprisingly observed in areas that are climatically sub-optimal for rice production. The empirical analysis of productivity convergence, even indirectly, implies that technological changes can be represented by examining the direction of productivity over time and is an attempt to approximate the ultimate impacts of climate-induced innovation in agriculture. The development of technological innovations accompanied by changes in agricultural policies may have been responsible for higher rice productivity among the districts with marginal climate. This assertion is supported by both the results of the empirical analysis, showing evidence of productivity convergence, and by the assessment of policies related to research and development. The empirical analysis of productivity convergence, even indirectly, implies that technological changes can be represented by examining the direction of productivity over time and is an attempt to approximate the *ultimate* impacts of climate-induced innovation in agriculture. With respect to policy assessment, the new institutional framework exhibits change in policies that facilitated greater engagement of relevant stakeholders in the development and application of new technologies in rice cultivation (e.g. PTD approach).

References

AGRAWALA, S., V. RAKSAKULTHAI, M. VAN AALST, P. LARSEN, J. SMITH and J. REYNOLDS, 2003. *Development and Climate Change in India: Focus in Water Resources and Hydro Power*. COM/ENV/ EPOC/DCD/DAC 1/FINAL, OECD Paris.

AUSUBEL, J.H., 1995. Technical progress and climatic change. *Energy Policy*, **23**:411–416.

BARRO, R. and X. SALA-I-MARTIN, 1992. Convergence. *Journal of Political Economy*, **100**:223–251.

BIGGS, S. and D. GAUCHAN, 2001. The broader institutional context of participatory plant breeding in the changing natural resources R&D system in India. *An Exchange of Experiences from South and South East Asia: Proceedings of the International Symposium on Participatory Plant Genetic Resource Exchange*, Pokhara, India—CGIAR/PRGA, pp. 61–74.

CHALISE, S.R., 1994. Mountain environments and climate change in the Hindu Kush-Himalayas. In *Mountain Environments in Changing Climates*, ed., Beniston, M., pp. 382–404. London, Routledge.

CHALISE, S.R. and N.R. KHANAL, 1996. *Hydrology of the Hindu Kush-Himalayas*. Report of a Regional Workshop, March 23–24, Kathmandu, UNESCO/ICIMOD.

CHEN, H. and H. LIBAI, 1997. Investigating about varieties filtering in winter wheat northward shifting. *Journal of Chenyang Agricultural University*, **28**:175–179.

CISSE, N., M. NDIAYE, S. THIAW and A.E. HALL, 1997. Registration of Mouride cowpea. *Crop Science*, **35**:1215–1216.

DUBIN, H.J. and J.P. BRENNON, 2009. Fighting a shifting enemy: The international collaboration to combat wheat rusts. In *Millions Fed: Proven Success in Agriultural Development*. David J. Spielman and Rajul Pandy-Lorch (eds.). International Food Policy Research Institute. DOI: http://dx.doi.org/10.2499/0896296598.

EASTERLING, W.E., 1996. Adapting North American agriculture to climate change in review. *Agricultural and Forest Meteorology*, **80**:1–53.

EASTERLING, W.E., H.H. BRIAN and J.B. SMITH, 2004. *Coping with Global Climate Change: The Role of Adaptation in the United States*. Pew Center on Global Climate Change, Arlington, VA, pp. 40.

ELAWAD, H.O.A. and A.E. HALL, 2002. Registration of Ein El Gazal cowpea. *Crop Science*, **42**:1745–1746.

ERENSTEIN, O., 2009. Leaving the plow behind: Zero-tillage rice–wheat cultivation in the Indo-Gangetic Plains. In *Millions Fed: Proven Success in Agriultural Development*. David J. Spielman and Rajul Pandy-Lorch (eds.). International Food Policy Research Institute. DOI: http://dx.doi.org/10.2499/0896296598.

EVENSON, R.E. and D. GOLLIN, 2000. *The Green Revolution: End of Century Perspective*. Paper Presented for the Standing Project on Impact Assessment of the Technical Advisory Committee of the Consultative Group on International Agricultural Research (CGIAR). New Haven, Connecticut, U.S. Economic Growth Center, Yale University.

GAUCHAN, D., M. JOSHI and S. BIGGS, 2003. A strategy for strengthening participatory technology development in agriculture and natural resources innovations systems: The case of India. *International Journal of Technology Management and Sustainable Development*, **2**:39–52.

GOLETTI, F., A. BHATTA and P. GRUHN, 2001. Crop production and productivity growth in India. Discussion Paper No. 2, *Agricultural Sector Performance Review, TA 3536-NEP*, Kathmandu, India, pp. 83.

HALL, A.E., 2004. Breeding for adaptation to drought and heat in cowpea. *European Journal of Agronomy*. Available online @ www.sciencedirect.com.

HAYAMI, Y. and V.W. RUTTAN, 1985. *Agricultural Development: An International Perspective*. The John Hopkins University Press, Baltimore, pp. 506.

HMGN/MFSC, 2002. *Statistical Information on Indiaese Agriculture, 2001/2002*. HMG/MOAC, Kathmandu, India.

IFPRI, 2009. *Millions Fed: Proven Success in Agriultural Development*. In David J. Spielman and Rajul Pandy-Lorch (eds.). International Food Policy Research Institute. DOI: http://dx.doi.org/10.2499/0896296598.

JOSHI, K.D., B.R. STHAPIT and J.R. WITCOMBE, 2001. The impact of participatory plant breeding in landrace diversity: A case of high altitude rice in India. *An Exchange of Experiences from South and South East Asia: Proceedings of the International Symposium on Participatory Plant Genetic Resource Exchange*, Pokhara, India—CGIAR/PRGA, pp. 303–310.

McCUNN, A. and W.E. HUFFMAN, 2000. Convergence in U.S. productivity growth for agriculture: Implications of interstate research spillovers for funding agricultural research. *American Journal of Agricultural Economics*, **82**:370–388.

McERLEAN, S.A. and Z. WU, 2003. Regional agricultural labour productivity convergence in China. *Food Policy*, **28**:237–252.

NATIONAL RESEARCH COUNCIL (NRC), 1999. *Human Dimensions of Global Environmental Change: Research Pathways for the Next Decade*. National Academy Press, Washington DC, pp. 83.

PANDEY, S. and L. VELASCO, 2002. Economic of direct seeding in Asia: Patterns of adoption and research priorities. In Pandey, S. (Ed.), *Direct Seeding: Research Strategies and Opportunities. Proceedings of the International Workshop on Direst Seeding in Asian Rice Systems: Strategic Research Issues and Opportunities*, IRRI, Philippines, pp. 3–14.

POKHREL, T.P., 1997. Rice development program in India. *International Rice Commission Newsletter*, **46**: online at: http://www.fao.org/document

ROSENBERG, N.J., 1992. Adaptation of agriculture to climate change. *Climatic Change*, **21**:385–405.

RUTTAN, V.W., 1996. Research to achieve sustainable growth in agricultural production into the 21st century. *Canadian Journal of Plant Pathology*, **18**:123–132.

SADRAS, V., R. DAVID and K. MIKE, 2003. Dynamic cropping strategies for risk management in dry-land farming systems. *Agricultural Systems*,**76**:929–948.

SAH, R.P., K.B. KOIRALA, K.H. GHIMIRE, H.K. PRASAI, H.S. BHANDARI, R. POUDEL and M. BHATTARAI, 2004. Enhancing yield potential of rice under rainfed environment of western Hills. *Rice Research Proceedings*, NARC, Islamabad, pp. 114–125.

SHRESTHA, A.B., C.P. WAKE, P.A. MAYEWSKI and J.E. DIBB, 1999. Maximum temperature trends in the Himalaya and its vicinity: An analysis based on the temperature record from India for the period 1971–1994. *Journal of Climate*, **12**:2775–2789.

SHRESTHA, A.B., C.P. WAKE, J.E. DIBB, and P.A. MAYEWSKI, 2000. Precipitation fluctuation in the India Himalaya and its vicinity and relationship with some large scale climatological parameters. *Journal of Climatology*, **20**:317–127.

SMITHERS, J. and A. BLAY-PALMER, 2001. Technology innovation as a strategy for climate adaptation in agriculture. *Applied Geography*, **21**:175–197.

SPERLING, L. and J.A. ASHBY, 1999. *Moving Participatory Breeding Forward: The Next Step.* In Collinson, M. (Ed.), History of Farming Systems Research, UK. CAB International.

STHAPIT, B.R., K.D. JOSHI, and J.R. WITCOMBE, 1996. Farmer participatory crop improvement, III: Participatory plant breeding, case of high altitude rice from India. *Experimental Agriculture*, **32**:479–496.

TRIPATHI, J., M.R. BHATTA, S. JUSTICE, and N.K. SHAKYA, 2004. Direct seeding: An emerging resource conserving technology for rice cultivation in the rice–wheat system. *Rice Research Proceedings of the 24th Summer Crop Workshop*, Kathmandu, Nepal, pp. 273–281.

WITCOMBE, J.R., A. JOSHI, K.D. JOSHI, and B.R. STHAPIT, 1996. Farmers participatory crop improvement, I: Methods for varietal selection and breeding and their impacts on bio-diversity. *Experimental Agriculture*, **32**:453–468.

CHAPTER TWELVE

Influence of biotic and abiotic factors on yield and quality of medicinal and aromatic plants

Amit Chauhan, Ram Swaroop Verma and Rajendra Chandra Padalia

Contents

Abstract

According to the World Health Organization (WHO, 1977), 'a medicinal plant' is any plant in which one or

more of its organs contain substances that can be used for
therapeutic purposes or which are precursors for the synthesis of useful drugs. This definition distinguishes those
plants whose therapeutic properties and constituents have
been established scientifically and plants that are regarded
as medicinal but which have not yet been subjected
to thorough investigation. Furthermore, WHO (2001)
defines medicinal plant as herbal preparations produced
by subjecting plant materials to extraction, fractionation,
purification, concentration or other physical or biological processes which may be produced for immediate
consumption or as a basis for herbal products. Aromatic
plants have a pleasant, characteristic fragrant smell.

12.1 Introduction

The origin of the medicinal and aromatic plants (MAPs) is as
old as agriculture, as is its essences and extracts. There utilization begins from wild harvest of plants than selection and
cultivation of useful plants and finally extending them as crop.
MAPs are important in economic, social, cultural and ecological aspects of local communities around the globe. In one form
or another, they benefit practically everyone on Earth through
nutrition, toiletry, body care, incense, and so on. Worldwide, it
is estimated that up to 70,000 species are used in folk medicine
(Farnsworth and Soejarto, 1991). The WHO reports over 21,000
plant taxa used for medicinal purposes (Groombridge, 1992).
In India, which is said to have probably the oldest, richest and
most diverse cultural traditions in the use of medicinal plants;
about 7500 species are used in ethno-medicines (Shankar and
Majumdar, 1997) which is almost half of the country's 17,000
native plant species. In China, the total number of medicinal
plants used in different parts of the country add up to some
6000 species according to Xiao (1991) and to over 10,000
according to He and Sheng (1997). Of these, approximately
1000 plant species are commonly used in Chinese medicine,
and about half of these are considered as the main medicinal
plants (He and Sheng, 1997). In Africa, over 5000 plant species are known to be used for medicinal purposes (Iwu, 1993).
In Europe, with its long tradition in the use of botanicals,
about 2000 medicinal and aromatic plant species are used on
a commercial basis (Lange, 1998). In Germany, Lange (1996)
identified not less than 1500 taxa as sources of medicinal and

aromatic plant material. In Spain, it is estimated that about 800 medicinal and aromatic plant species are used of which 450 species are associated with commercial use (Blanco and Breaux, 1997; Lange, 1998).

The fragrance of these plants is carried in the essential oil fraction. Many aromatic plants are spices. Chandarana et al. (2005) defined spices as any dried, fragrant, aromatic or pungent vegetables or plant substances in whole, broken or ground forms that contribute, to the piquancy of foods and beverages.

Medicinal and aromatic plants contain biologically active chemical substances such as saponins, tannins, essential oils, flavonoids, alkaloids and other chemical compounds (Harborne, 1973; Sofowora, 1993), which have curative properties. These phytochemicals are chemical compounds formed during the normal metabolic processes of plants. These chemicals are often referred to as 'secondary metabolites' of which there are several classes, including alkaloids, flavonoids, coumarins, glycosides, gums, polysaccharides, phenols, tannins, terpenes and terpenoids (Harborne, 1973; Okwu, 2004).

12.2 Selected phytochemical classes

Alkaloids

The term 'alkaloid' was first proposed by Meissner in 1819 to characterise these 'alkali-like' compounds found in plants (Trier, 1931; Pelletier, 1970). Alkaloids rank among the most efficient and therapeutically significant plant substances (Okwu, 2005). Some 5500 alkaloids are known and they comprise the largest single class of secondary plant substances which contain one or more nitrogen atoms, usually in combination as part of a cyclic structure (Harborne, 1973). They are usually organic bases and form salts with acids and when soluble give alkaline solutions. Examples include nicotine (**1**), cocaine (**2**), morphine (**3**) and codeine (**4**) (*Papaver sominferum*), quinine (**5**) (*Cinchona succirubra*), reserpine and (**6**) (*Rauwolfia vomitoria*), all of which has a large demand worldwide. Alkaloid production is a characteristic of all plant organs. They exhibit marked physiological activity when administered to animals (Okwu and Okwu, 2004). Furthermore, alkaloids are often toxic to man and many have dramatic physiological activities, hence their wide use in medicine for the development of drugs (Harborne, 1973; Okwu, 2005).

Phenolic compounds

Phenolics are compounds possessing one or more aromatic rings with one or more hydroxyl groups. They are broadly distributed in the plant kingdom and are the most abundant secondary metabolites of plants, with more than 8000 phenolic structures currently known, ranging from simple molecules such as phenolic acids to highly polymerised substances such as tannins. Plant phenolics are generally involved in defence against ultraviolet radiation or aggression by pathogens, parasites and predators; they also are a contributing factor in plant colours. They are ubiquitous in all plant organs and are therefore an integral part of the human diet. Phenolics are widespread constituents of plant foods (fruits, vegetables, cereals, olive, legumes, chocolate, etc.) and beverages (tea, coffee, beer, wine, etc.), and partially responsible for the overall organoleptic properties of plant foods (Dai and Mumper, 2010).

Plant phenolics include phenolic acids, flavonoids, tannins and the less common stilbenes and lignans. Flavonoids are the most abundant polyphenols in our diets. Some of the most common flavonoids include quercetin (**7**), a flavonol abundant in onion, broccoli, and apple; catechin (**8**), a flavanol found in tea and several fruits; naringenin (**9**), the main flavanone in grape-fruit; cyanidin-glycoside, an anthocyanin abundant in berry fruits (black currant, raspberry, blackberry, etc.); and daidzein, genistein and glycitein, the main isoflavones in soybean (D'Archivio et al., 2007).

7

8

9

Saponins

Saponins are glycosides of both triterpenes and steroids that are characterised by their bitter or astringent taste, foaming property, haemolytic effect on red blood cells and cholesterol-binding properties (Okwu, 2005). Saponins have been shown to possess both beneficial (lowering cholesterol) and deleterious (cytotoxic and permeabilisation of intestinal epithelium) properties and to exhibit structure-dependent biological activity. In medicine, it is used to some extent as an expectorant and an emulsifying agent (Harborne, 1973).

Quinones

Quinones are a large class of compounds endowed with a rich and fascinating chemistry (Patai and Rappaport, 1988). 1,4-Benzoquinone or *p*-benzoquinone is the basic structure of quinonoid compounds. The natural quinone pigments range in colour from pale yellow to almost black and there are over 450 known structures (Harborne, 1973). These compounds are also responsible for the browning reaction in cut or damaged fruits and vegetables, and are an intermediate in the melanin synthesis pathway in human skin. Hypercin (**10**), an anthroquinone which is an example of quinine obtained from St. John's wort (*Hypericum perforatum*), has received much attention as an antidepressant and an antiviral. It also has several antimicrobial properties (Aarts, 1998).

10

Essential oils

Essential oils are generally a complex mixture of terpenoids (terpenes and their oxygenated derivatives). However, other compounds, namely, phenyl propanoids, fatty acids, and their ester, aliphatics, and phenolics also occur in essential oils. Essential oils are the main compounds found in the volatile steam distillation fraction responsible for the characteristic scent, odour or smell found in many plants. Some essential oils possess medicating properties and are used in the pharmaceutical industry. They are commercially important as the basis of natural perfumes and also of spices and are used for flavouring purposes in the food industry. Plant families particularly rich in essential oils include the Compositae, Lamiaceae, Liliaceae, Myrtaceae and others. The terpene essential oils can be divided into two classes: the monoterpenes and sesquiterpenes, C_{10} and C_{15} isoprenoids, which differ in their boiling points (monoterpenes = 140–180°C, sesquiterpenes > 200°C) (Harborne, 1973). Industrially

important essential oil constituents are myrcene (**11**), limonene
(**12**), 1,8-cineole (**13**), linalool (**14**), camphor (**15**), menthol (**16**),
carvone (**17**), methyl chavicol (**18**), thymol (**19**), carvacrol (**20**),
eugenol (**21**), methyl eugenol (**22**), geraniol (**23**), citronellol (**24**),
(*E*)-nerolidol (**25**), patchouli alcohol (**26**) and so on.

12.3   Factors influencing yield and quality of maps

Ecosystems are composed of the nonliving (abiotic) and living
(biotic) components. Ecosystems provide a platform in which
plant evolution occurs, presenting stresses, but also opportunities

to which plants must acclimatise and adapt in order to thrive. Abiotic components of an ecosystem include temperature, soil, water, humidity, light and wind, while biotic components of an ecosystem include parasitic and herbivorous pests, competition from other plants, and favourable (symbiotic) relationships with other organisms along with human (agricultural) operations. All these factors (abiotic and biotic) influencing the yield and quality parameters of MAPs are discussed in detail in the following sections of this chapter.

12.4 Abiotic factors

Temperature

Mean surface air temperatures of the Earth have increased 0.68°C during the last century, and global circulation models project a global warming of 1.4–5.88°C by 2100 (Houghton et al., 2001). Numerous studies have examined the effects of elevated temperatures on plants (Berry and Bjorkman, 1980; Morison and Lawlor, 1999; Rustad et al., 2001). Elevated temperatures reduce stomatal conductance and, subsequently, reduce photosynthesis and growth of many plant species (Berry and Bjorkman, 1980). The photochemical efficiency of photosystem II also decreases at elevated temperatures, indicating increased stress (Gamon and Pearcy, 1989; Maxwell and Johnson, 2000). When plants are stressed, secondary metabolite production may increase because growth is often inhibited more than photosynthesis, and the carbon fixed not allocated to growth is instead allocated to secondary metabolites (Mooney et al., 1991). Several studies have examined the effects of increased temperatures on secondary metabolite production of plants, but most of these studies have contradictory results. Some report that secondary metabolites increase in response to elevated temperatures (Litvak et al., 2002), while others report that they decrease (Snow et al., 2003).

For the last 5–10 years, climate change in Europe has caused an abrupt transition from winter to summer with temperatures in April and May that are more typical for summer temperatures. This has resulted in severe damages; for example, throughout Europe the spring planting of chamomile (*Matricaria recutita*, Asteraceae) has been disastrous with an average yield loss of 80% (Cavaliere, 2009).

Total alkaloid content of *Datura metel* L. has been shown to peak in the hot dry season and is at its lowest during the rainy season in Ibadan (Cavaliere, 2009). To model the effects of the environmental factors, especially temperature, on yield and chemical composition alteration, numerous studies have been conducted.. Couceiro et al. (2006) cultivated *Hypericum perforatum* plants in a growth chamber under controlled environments to determine the effect of two different temperatures, 25°C and 30°C on concentrations of the major bioactive compounds, hypericins and hyperforin. According to the results, hyperforin concentrations were generally 20% lower at 30°C than at 25°C, while pseudohypericin and hypericin concentrations were, respectively, 25% and 30% higher at 30°C than at 25°C. The authors noted that the increases in secondary metabolite levels may be due to the biochemical pathway of a given metabolite that could be stimulated by stress factors such as high temperatures. Zobayed et al. (2005) subjected 70-day-old *H. perforatum* plants, grown under controlled environment to different temperature treatments of 15°C, 20°C, 25°C, 30°C and 35°C before harvesting for 15 days. They observed that high-temperature (35°C) treatment increased the hypericin, pseudohypericin and hyperforin concentrations in the shoot tissues. The total hypericin yield per plant (hypericin + pseudohypericin) was the highest in plants grown under 25°C, and then followed by plants grown at 30°C. The best treatment for hyperforin content per plant was at 30°C. Jensen et al. (1995) reported that a gradual increase in temperature (8°C, 18°C and 28°C) for 2 weeks resulted in a significant increase in hypericin biosynthesis in *H. perforatum* plants cultivated in Canada.

Scutellaria baicalensis Georgi is a traditional Chinese medicinal plant, whose active compounds comprise baicalin, baicalein, wogonoside, wogonin, neobaicalein, visidulin I and oroxylin A. These compounds have anti-burning, anti-tumour and anti-HIV activity (Blach-Olszewska et al., 2008). Li (2008) reported that temperature is an important environmental factor that may affect the medicinal quality of *S. baicalensis*. Extending the studies, Yuan et al. (2011) demonstrated that protracted heat treatment inhibited the accumulation of baicalin and baicalein; however, cells continued growing during the protracted heat stress.

Different seasons are marked with temperature fluctuations showing marked effects on essential oil content and composition of various aromatic plants. The yield of the essential oil of *Thymus serpyllum* during different seasons varied from 0.07% to 0.28% with the highest in summer season (0.28%) and the least in winter (0.07%) in the Kumaon region of the western Himalayas. The major components of the oils were thymol, *p*-cymene, γ-terpinene, 1-octen-3-ol, thymol methyl ether, carvacrol methyl ether, borneol and *p*-cymen-8-ol. Thymol reached the highest values during the autumn (60.1%) season while during the rest of the year, its concentration ranged from 19.4% to 56.4% with the lowest in winter (Verma et al., 2011). In similar studies by the authors in *Origanum vulgare*, it was found that the moderate temperature favoured the conversion of *p*-cymene to carvacrol over extreme temperatures (Verma et al., 2010).

Light

Irradiance is known to regulate not only plant growth and development but also the biosynthesis of both primary and secondary metabolites. Secondary metabolite production might increase (Chauser-Volfson and Gutterman, 1998) or decrease (Gershenzon, 1994) under low light intensity conditions, depending on the type of plant. Low light intensity, for example, increased the methylxanthine content in the leaves of *Ilex paraguariensis* (Coelho et al., 2007), but decreased resin content in the leaves of *Grindelia chiloensis* (Zavala and Ravetta, 2001).

The chemical profile of plants and the accumulation level of a special metabolite in plant tissues can be influenced by several environmental factors such as light quality (Upadhyaya and Furness, 1994) and light intensity (Yamamaura et al., 1989). It was reported that simultaneous irradiation with blue and UV-B light stimulates and affects the generation of phenolic compounds in basil (Nitz and Schnitzler, 2004).

Research showed that photoperiodic treatment itself is an important determinant of monoterpene composition. With regard to terpene composition, the shading of peppermint and changes in the day length has been found to affect the chemical composition of oil (Clark and Menary, 1979, 1980). The highest level of essential oil and concentration of thymol and myrcene in thyme occurred in full sunlight. Leaf length, width, and density of peltate hair decreased with a decrease in light levels (Letchamo and Gosselin, 1995).

Hypericum perforatum L., also known as St. John's wort, has been one of the most important medicinal plants. The most important compound is hypericin which is affected by environmental factors (Zobayed et al., 1995; Zou et al., 2004). Studies showed that hypericin synthesis increased significantly under conditions of high light intensity treatment (400 μmol m^{-2} s^{-1}) (Briskin and Gawienowski, 2001).

Moisture

Plants experience water stress either when the water supply to their roots becomes limited or when the transpiration rate becomes intense. Water stress is primarily caused by water deficit, that is, drought or high soil salinity. In cases of high soil salinity and also in other conditions such as flooding and low soil temperature, water exists in the soil solution, but plants cannot take it up—a situation commonly known as 'physiological drought'. Drought occurs in many parts of the world every year, frequently experienced in the field-grown plants under arid and semi-arid climates. Regions with adequate but non-uniform precipitation also experience water-limiting environments. Since the dawn of agriculture, mild to severe drought has been one of the major production-limiting factors. Consequently, the ability of plants to withstand such stress is of immense economic importance. The general effects of drought on plant growth are fairly well known. However, the primary effect of water deficit at the biochemical and molecular levels are not considerably understood yet and such understanding is crucial. All plants have tolerance to water stress, but the extent varies from species to species. Knowledge of the biochemical and molecular responses to drought is essential for a holistic perception of plant resistance mechanisms to water-limited conditions in higher plants.

Water stress is one of the most important environmental stresses that can depress growth and alter the biochemical properties of plants (Zobayed, 2005). According to Franz (1983), Palevitch (1987) and Marchese and Figueira (2005), one of the most important factors affecting secondary metabolism is soil water capacity. Usually, limited availability of water has a negative effect on plant growth and development. However, a non-severe water deficit has sometimes proved beneficial for the accumulation of biologically active compounds in medicinal and aromatic plants (Palevitch, 1987). Ghershenzon (1984) demonstrated that in herbaceous plants and shrubs, terpenes tend to increase under stress, mainly under severe water deficit conditions. This type of stress is known to increase the amount of secondary metabolites in a variety of medicinal plants, for example, artemisinin in *Artemisia annua* L. (Charles et al., 1993), ajmalicine in *Catharanthus roseus* (Jaleel et al., 2008) and hyperforin in *Hypericum perforatum* (Zobayed et al., 2005). The response of essential oil yield and composition to water stress varies with the duration and severity of stress. Putievesky et al. (1990) also reported that as irrigation intervals became more extended, herbage yield and essential oil yield were reduced in *Pelargonium graveolens*. Similarly, Rajeswara Rao et al. (1996) reported that a wet season encouraged vegetative growth of rose-scented geranium and resulted in a higher essential oil yield.

Similarly, Simon et al. (1992) reported that moderate water stress imposed on sweet basil resulted in higher oil content and greater total oil yield. Furthermore, the authors indicated that water stress changed essential oil composition: water stress increased linalool and methyl chavicol and reduced sesquiterpenes. Contrary to the previously mentioned report, short-term stress (withholding irrigation for 8 days) did not change essential oil yield and oil composition of *Melaleuca alternifolia* (List et al., 1999). In *Isatis indigotica*, extreme water stress has been found to reduce the production of indirubin. However, superior yield and quality could both be obtained at 45–70% of field capacity (Tan et al., 2008). A lot of investigations have showed that water stress increased secondary metabolite accumulation in medical plants, such as *Salvia miltiorrhiza* (Liu et al., 2011), *Bupleuri radix* (Zhu et al., 2009), *Catharanthus roseus* (Abdul et al., 2007), and *Rehmannia glutinosa* (Gaertn.) (Chung et al., 2006). The content of total flavonoids in *Tribulus terrestris* under high levels of water treatment is higher than that occurring in low-water treatments (Yang et al., 2010). In *Ginkgo biloba*, drought stress has been found to promote the growth of quercetin content and to inhibit the increase of rutin in the leaves (He and Zhong, 2003). In *S. baicalensis* Georgi, baicalin increased steadily in the stems and leaves under lower water stress, and it decreased sharply under heavy water stress (Liu et al., 2010).

Water stress reduced fresh and dry weights of *Satureja hortensis* L. (savoury) plants. Severe water stress increased essential oil content more than moderate water stress. The main constituents, such as carvacrol, increased under moderate water stress, while α-terpinene content decreased under moderate and severe water stress of *Satureja hortensis* L. (Baher et al., 2002). Essential oil, total carbohydrate and proline contents were pronouncedly increased with increasing stress levels of *Salvia officinalis* L. (Sage) plants (Hendawy and Khalid, 2005). Sahu (1972) reported that overall growth of *Rauvolfia serpentina* plantation diminished with increasing water stress, but root growth was less influenced than shoot growth. Root yield of the crop grown without irrigation was less than the irrigated ones.

Soil

The edaphic factor pertains to the substratum upon which the plant grows and from which it derives its mineral nutrients and much of its water supply. It involves physical, chemical and biological properties of soils (Mason, 1946a,b). These properties of soil, namely physical, chemical and biological, significantly influence plant growth leading to yield and com-position of medicinal and aromatic plants. According to Janzen (1974), sandy soils, which are poor in nutrients, provide a higher production of secondary metabolites compared with clay soils, which are richer in nutrients. Studies carried out in the Amazonas State, with authentic oil resins of *Copaifera mul-tijuga*, show that there is a dependence between soil texture (sandy or clay) and productivity of the oil resin, with production being higher in clay soils than in sandy soils (Alencar, 1982). However, there was no relationship observed between produc-tion and soil type, but it was observed that the fractions of non-oxygenated sesquiterpenes and diterpene acids are slightly higher in clay soils, while the fraction of oxygenated sesquiter-penes was higher in sandy soils (Medeiros and Vieira, 2008). Moqbeli et al. (2011) reported that soil conditions have a signifi-cant effect on yield and the essential oil of *Melissa officinalis*. It was further reported that in comparison to sand and clay, the loam texture of soil recorded the highest biomass and essential oil content in the plant.

The pH value has an impact on the production of secondary metabolites (Yan et al., 2004; Medentsev et al., 2005; Babula et al., 2006), for example, out of many soil parameters anal-ysed, the highest correlation of the production of the glycoside salidroside in *Rhodiola sachalinensis* was observed to the soil pH (Yan et al., 2004).

There are contradictory reports in the literature concerning the response of essential oil to salt stress. Salt stress decreased essential oil yield in *Trachyspermum ammi* (Ashraf and Orooj, 2006). This negative effect of salt stress in oil yield was also reported for other medicinal plants, for example, *Mentha piperita* (Tabatabaie and Nazari, 2007); peppermint, penny-royal, and apple mint (Aziz et al., 2008); *Thymus maroccanus* (Belaqziz et al., 2009); and basil (Said-Al Ahl and Mahmoud, 2010). Besides, salinity decreased the essential oil yield (Abd El-Wahab, 2006) of fennel. It was also observed that the ane-thole percentage was reduced with saline water. In marjo-ram, the proportions of the main compounds were differently affected by salt (Baatour et al., 2010), while, in *Matricaria recutita*, the main essential oil constituents (α-bisabololoxide B, α-bisabolonoxide A, chamazulene, α-bisabolol oxide A, α-bisabolol, trans-β-farnesene) showed an increase under saline conditions (Baghalian et al., 2008). Also, in *Origanum vulgare*, it was found that the content of the main essential oil constitu-ent (carvacrol) decreased under salt stress, while *p*-cymene and γ-terpinene contents increased under non-salt stress treatments

(Said-Al Ahl and Hussein, 2010). Similar results of an inhibitory effect of high level of salinity were also found on lemon balm (Ozturk et al., 2004), *Majorana hortensis* (Shalan et al., 2006), *Matricaria chamomile* (Razmjoo et al., 2008), *Salvia officinalis* (Ben Taarit et al., 2010) and basil (Said-Al Ahl et al., 2010). On the contrary, an increase of essential oil yield due to lower levels of salinity has been reported in other plant species, for example, *Satureja hortensis* (Baher et al., 2002) and *Salvia officinalis* (Hendawy and Khalid, 2005). It was also shown that essential oil yield of coriander leaves was stimulated only under low and moderate stress, while it decreased at the high salinity level. At low stress, (*E*)-2-decenal, (*E*)-2-dodecenal and dodecanal contents increased (Neffati and Marzouk, 2008).

12.5 Biotic factors

Plants produce diversity of natural products or secondary metabolites with a prominent function of protection against predators and microbial pathogens and are important for the communication of the plants with other organisms. They also serve as defence against abiotic stress (e.g. UV-B exposure) (Schafer and Wink 2009), and are insignificant for growth and developmental processes (Rosenthal et al., 1991).

A vast majority of the different structures of terpenes produced by plants as secondary metabolites are presumed to be involved in defence as toxins and feeding deterrents to a large number of plant-feeding insects and mammals (Gershenzon and Croteau, 1991).

For example, the pyrethroids (monoterpenes esters) that occur in the leaves and flowers of *Chrysanthemum* species show strong insecticidal responses (neurotoxin) to insects such as beetles, wasps, moths, bees, and so on, and is also a popular ingredient in commercial insecticides because of low persistence in the environment and low mammalian toxicity (Turlings et al., 1995).

Abietic acid is a diterpene found in pines and leguminous trees. It is present in or along with resins in resin canals of the tree trunk. When these canals are pierced by feeding insects, the outflow of resin may physically block feeding and serve as a chemical deterrent to continued predation (Bradley et al., 1992). The milkweeds produce several better-tasting glucosides (sterols) that protect them against herbivory by most insects and even cattle (Lewis and Elvin-Lewis, 1977). Halogenated coumarin derivatives work very effectively *in vitro* to inhibit fungal growth. For example, 7-hydroxylated simple coumarins may play a defensive role against parasitism of *Orobanche cernua*, by preventing successful germination, penetration and connection to the host vascular system (Serghini et al., 2001).

Weeds play one of the major roles in the loss of yield and composition of medicinal and aromatic plants. The presence of weeds during different periods of crop growth of cultivar 'Bourbon' resulted in decreases in linalool (9.6–11.4%), isomenthone (7.3–7.9%), citonellol (18.9–21.8%) and citronellyl formate (5.5–6.6%), and increases in geraniol (23.5–26.5%) and geranyl formate (2.2–3.0%). Rose oxides and 10-epi-γ-eudesmol were not affected (Rao and Bhattacharya 1997). Rao et al. (2005) also reported the loss in yield and changes in composition when the oil of three cultivars of rose-scented geranium, namely, Algerian, Bourbon and Kelkar, was co-distilled with companion weeds growing in the field. Water requirement for the growth of weeds is primarily of interest from the stand-point of competition with the crop plant for the available moisture (Gibson, 2000). It has been reported that wild mustard transpires about four times more water than a crop plant (Thakur, 1984). Studies show that weed and canopy architecture, especially plant height, location of branches and height of maximum leaf area determine the impact of competition for light and, thus, have a major influence on crop yield (Cudney et al., 1991). Members of the family Brassicaceae (such as *Coronopus didymus*, a notorious weed in wheat crop) generally produce sulphur compounds such as glucosinolates. Allyl glucosinolate is one of the predominant glucosinolates in many brassicaceous species. In soil, this compound is hydrolysed into allyl isothiocyanate, a volatile compound (Mayton et al., 1996), which may be responsible for allelopathic interference.

Like other crops, medicinal and aromatic crops are susceptible to several pests and diseases which in turn drastically reduce and deteriorate crop yield and composition. *Rhizoctonia solani* has been reported to cause the leaf blight of *C. forskohlii* (Shukla et al., 1993).

Fusarium solani, causing root rot of *C. forskohlii,* has also been reported by Bhattacharya and Bhattacharya (2008). Leaf rust, leaf spot, leaf blight and powdery mildew decreased the concentrations of menthone (from 8.3% to 1.1–2.4%) and isomenthone (from 4.2% to 2.0–3.4%), and increased the content of menthol (from 84.1% to 87.0–90.8%), neomenthol from (1.8% to 2.1–2.8%) and menthyl acetate (from 0.1 to 2.0–4.3%) (Shukla et al., 2000). Wilt disease caused by *Fusarium oxysporum* var. *redolens* reduced biomass yields by 73.6–88.2% and essential oil yields by 69.1–87.1% in Bourbon and Algerian cultivars of rose-scented geranium (*Pelargonium graveolens*) (Rao, 2002).

Finally, human beings are the most significant biotic factor shaping ecosystems. Human beings who manage these factors in terms of irrigation, nutrient input, pest control, land preparation, mixed/relay cropping and other practices are also a biotic component of agroecosystems. Chemical composition in plants is dependent on biosynthetic pathways controlled by several enzymes which are produced by numerous genes, abiotic and biotic factors, including human management that exerts considerable influence on the same. Several agronomic studies taken up on medicinal and aromatic plants invariably examine the quality in relation to agronomic management. A number of studies reported the effects of agricultural practices on the secondary metabolites in medicinal and aromatic plants (Verma et al., 2009, 2011, 2012, 2013).

Planting dates influence growth yield and secondary metabolite production as reported in *Artemisia annua* (Singh et al., 2009b), rose-scented geranium (Kalra et al., 1992) and *Silybum marianum* (Rahimi and Kamali, 2012). In studies conducted in the temperate region of the Himalayas, artemisinin yield in the dried leaves of *Artemisia annua* was found to be maximum among the plants that were transplanted in March (24.39 kg ha^{-1}) and minimum to those transplanted in November (3.39 kg ha^{-1}) (Verma et al., 2011).

Plant density and irrigation methods are the two important factors that directly affect the yield and flower number, amount of essential oils and yield components (Marisol et al., 2003; Saif et al., 2003; Tiwari et al., 2003; Verma et al., 2008; Singh et al., 2009a).

Gengaihi and Abdallah (1978) reported that the number of umbel per plant, seed yield per plant and plant height increased at wider spacing. According to the results of Verzalova et al. (1988), row spacing did not affect the plant height but the number of umbel and seed yield per plant was increased at wider spacing. Naghdibadi et al. (2002) who studied the effect of different plant densities on the yield of dry material of thyme (*Thymus vulgaris*) showed that a higher yield of dry material was obtained with 15 cm densities of planting.

Sharifi and Abbaszadeh (2003) investigated the effect of N fertiliser on essential oil content and composition of the aerial parts of fennel (*Foeniculum vulgare* Mill) and found that N application increased the essential oil content. Ram et al. (2005) evaluated two variables of organic mulch (control and sugarcane trash at 7 t ha^{-1}) and three levels of nitrogen (0, 100 and 200 kg ha^{-1}). Application of N at 200 kg ha^{-1} in the mulched plots significantly enhanced the N uptake by the crop and essential oil content of mint (*Mentha arvensis* L.) over the control; with 100 kg N ha^{-1} being applied to the mulched or unmulched plots and 200 kg N ha^{-1} applied to the unmulched plots. A study was carried out with lemongrass (*Cymbopogon flexuosus*) during 1993–1995 under four rates of applied nitrogen (0, 50, 100 and 150 kg ha^{-1}). Nitrogen application significantly increased crop growth values such as plant height, leaf area index (LAI), herbage and essential oil contents. An application of 100 kg N ha^{-1} was found to be optimal for crop yield (Singh, 1999). The percentage of essential oil and fresh and dry matter of marjoram plants positively responded to increased levels of composted manure compared with chemical fertiliser (Edris et al., 2003).

As discussed earlier, water stress can create significant changes in the yield and composition of MAPs. Irrigation is the prerequisite demand for better yield and quality of crop. Moreover, some interesting studies of irrigation with waste water, saline water, and so on need further research. The essential oil yield of mint increased 14% under irrigation by secondary drainage water as compared to irrigation by agronomical water (Aghayari and Darvishi, 2011). Salinity is a major problem that negatively impacts agricultural productions in many regions of the world. Generally, salinity problems increase with increasing salt concentration in irrigation water (Ayman, 2003). For instance, it was found that the increase of salinity stress decreased almost all growth parameters in *Nigella sativa*, some growth parameters and essential oil amount in Chamomile (Razmjoo et al., 2008) and essential oil yield in lemon balm (Tabatabaie and Nazari, 2007). Also, the effect of salinity parameter on essential oil quality in lemon verbena showed the increased amount of geranial as the salinity level was increased (Ozturk et al., 2004). However, the research showed that chamomile is able to maintain all of its medical properties, under saline conditions and could be cultivated economically in such conditions (Baghalian et al., 2008).

12.6 Conclusion

The effects of climate change are apparent within ecosystems around the world, including medicinal and aromatic plant populations. Changing phenologies and distributions of plants have been recorded worldwide, and these factors could ultimately endanger wild MAP species by disrupting synchronised phenologies of interdependent species, exposing some early-blooming MAP species to the dangers of late cold spells, allowing invasives to enter MAP species habitats and compete for resources, and initiating migratory challenges, among other threats. Extreme weather events already impact the availability and supply of MAPs on the global market, and projected future increases in extreme weather are likely to negatively affect MAP yields even further. What makes medicinal plants unique from other flora is the fact that they, along with other economically useful plants, are collected for human use. Therefore, there is a need for more research into the effects of climate fluctuations on plants in general and MAPs in particular. Climate change may not currently represent the biggest threat to MAPs, but it has the potential to become a much greater threat in the future decades. Many of the world's poorest people rely on medicinal plants not only as their primary healthcare option, but also as a significant source of income. The potential loss of MAP species from effects of climate change is likely to have major ramifications on the livelihoods of large numbers of vulnerable populations across the world.

References

AARTS T. 1998. The dietary supplements industry: A market analysis. Dietary Supplements Conference, Nutritional Business International.

ABD EL-WAHAB M.A. 2006. The efficiency of using saline and fresh water irrigation as alternating methods of irrigation on the productivity of *Foeniculum vulgare* Mill subsp. *vulgare* var. *vulgare* under North Sinai conditions. *Research Journal of Agriculture and Biological Sciences* 2(6): 571–7.

ABDUL J.A., MANIVANNAN P., KISHOREKUMAR A., SANKAR B., GOPI R., SOMASUNDARAM R. and PANNEERSELVAM R. 2007. Alterations in osmoregulation, antioxidant enzymes and indole alkaloid levels in *Catharanthus roseus* exposed to water deficit. *Colloid Surface B* 59: 150–7.

AGHAYARI F. and DARVISHI H.H. 2011. Investigation of irrigation influence with domestic wastewater on quantity and quality features in different mint's species. *Advances in Applied Science Research* 2(5): 557–60.

ALENCAR J.C. 1982. Silvicultural studies of a natural population of *Copaifera multijuga* Hayne (Leguminosae) in Central Amazonia. II. Production of oil-resin. *Amazon Acta* 12(1): 75–89.

ASHRAF M. and OROOJ A. 2006. Salt stress effects on growth, ion accumulation and seed oil concentration in an arid zone traditional medicinal plant ajwain (*Trachyspermum ammi* [L.] Sprague). *Journal of Arid Environment* 64(2): 209–20.

AYMAN A.F. 2003. The use of saline water in agriculture in the Near East and North Africa region: Present and future. *Journal of Crop Production* 7(1–2): 299–323.

AZIZ E.E., AL-AMIER H. and CRAKER L.E. 2008. Influence of salt stress on growth and essential oil production in peppermint, pennyroyal, and apple mint. *Journal of Herbs Spices and Medicinal Plants* 14(1 and 2): 77–87.

BAATOUR O.R., KADDOUR W., WANNES A., LACHAAL M. and MARZOUK B. 2010. Salt effects on the growth, mineral nutrition, essential oil yield and composition of marjoram (*Origanum majorana*). *Acta Physiologiae Plantae* 32: 45–51.

BABULA P., MIKELOVA R., ADAM V., KIZEK R., HAVEL L. and SLADKY Z. 2006. Using of liquid chromatography coupled with diode array detector for determination of naphthoquinones in plants and for investigation of influence of pH of cultivation medium on content of plumbagin in *Dionaea muscipula*. *Journal of Chromatography B* 842(1): 28–35.

BAGHALIAN K., HAGHIRY A., NAGHAVI M.R. and MOHAMMADI A. 2008. Effect of saline irrigation water on agronomical and phytochemical characteristics of chamomile (*Matricaria recutita* L.). *Scientia Horticulturae* 116(4): 437–41.

BAHER Z.F., MIRZA M., GHORBANLI M. and REZAII M.B. 2002. The influence of water stress on plant height, herbal and essential oil yield and composition in *Satureja hortensis* L. *Flavour and Fragrance Journal* 17: 275–7.

BELAQZIZ R., ROMANE A. and ABBAD A. 2009. Salt stress effects on germination, growth and essential oil content of an endemic thyme species in Morocco (*Thymus maroccanus* Ball.). *Journal of Applied Sciences Research* 5(7): 858–63.

BEN TAARIT M.K., MSAADA K., HOSNI K. and MARZOUK B. 2010. Changes in fatty acid and essential oil composition of sage (*Salvia officinalis* L.) leaves under NaCl stress. *Food Chemistry* 9(3): 951–6.

BERRY J. and BJORKMAN O. 1980. Photosynthetic response and adaptation to temperature in higher plants. *Annual Review of Plant Physiology* 31: 491–543.

BHATTACHARYA A. and BHATTACHARYA S. 2008. A study on root rot disease of *Coleus forskohlii* Briq. occurring in Gangetic West Bengal. *Journal of Botanical Society of Bengal* 62: 43–7.

BLACH-OLSZEWSKA Z., JATCZAK B., RAK A., LORENC M., GULANOWSKI B., DROBNA A. and LAMER-ZARAWSKA E. 2008. Production of cytokines and stimulation of resistance to viral infection in human leukocytes by *Scutellaria baicalensis* flavones. *Journal of Interferon and Cytokine Research* 28: 571–81.

BLANCO E. and BREAUX J. 1997. Results of the study of commercialisation, exploitation and conservation of medicinal and aromatic plants in Spain. Unpublished report for TRAFFIC Europe.

BRADLEY D.J., KJELLBORN P. and LAMB C.J. 1992. Elicitor and wound induced oxidative cross linking of a proline rich plant cell protein: A novel rapid defence response. *Cell* 70: 21–30.

BRISKIN D.P. and GAWIENOWSKI M.C. 2001. Differential effects of light and nitrogen on hypericin and leaf glands in *Hypericum perforatum* L. *Plant Physiology* 39: 1075–81.

CAVALIERE C. 2009. The effects of climate change on medicinal and aromatic plants. *Herbal Gram* 81: 44–57.

CHANDARANA H., BALUJA S. and CHAND S.V. 2005. Comparison of antibacterial activities of selected species of Zingiberaceae family and some synthetic compounds. *Turkish Journal of Biology* 29: 83–97.

CHARLES D.I., SIMON J.E., SHOCK C.C., FEIBERT E.B.G. and SMITH R.M. 1993. Effect of water stress and post-harvest handling on artemisinin content in the leaves of *Artemisia annua* L. In: *Proceedings of the Second International Symposium: New Crops, Exploration, Research and Commercialization, Indianapolis, USA, October 1991*; Janick J. and Simon J.E. (eds.). John Wiley and Sons Inc., New York, NY, USA, pp. 640–3.

CHAUSER-VOLFSON E. and GUTTERMAN Y. 1998. Content and distribution of anthrone C-glycosides in the South African arid plant species *Aloe mutabilis* growing in the direct sunlight and the shade in the Negev Desert of Israel. *Journal of Arid Environment* 40: 441–51.

CHUNG I.M., KIM J.J., LIM J.D., YU C.Y., KIM S.H. and HAHN S.J. 2006. Comparison of resveratrol SOD activity, phenolic compounds and free amino acids in *Rehmannia glutinosa* under temperature and water stress. *Environmental and Experimental Botany* 56: 44–53.

CLARK R.J. and MENARY R.C. 1979. The importance of harvest data and plant density on the yield and quality of Tasmanian peppermint oil. *Journal of the American Society for Horticultural Sciences* 104: 702–6.

CLARK R.J. and MENARY R.C. 1980. Environmental effects on peppermint (*Mentha piperita* L.). I. Effect of day length, photon flux density, night temperature and day temperature on the yield and composition of peppermint oil, *Australian Journal of Plant Physiology* 7: 685–92.

COELHO G.C., RACHWAL M.F.G., DEDECEK R.A., CURCIO G.R., NIETSCHE K. and SCHENKEL E.P. 2007. Effect of light intensity on methylxanthine contents of *Ilex paraguariensis* A. St. Hil. *Biochemical Systematics and Ecology* 35: 75–80.

COUCEIRO M.A., AFREEN F., ZOBAYED S.M.A. and KOZAI T. 2006. Variation in concentrations of major bioactive compounds of St. John's wort, Effects of harvesting time, temperature and germplasm. *Plant Science* 170: 128–34.

CUDNEY D.W., JORDAN L.S. and HALL. A.E. 1991. Effect of wild oat (*Avena fatua*) infestations on light interception and growth rate of wheat (*Triticum aestivum*). *Weed Science* 39: 175–79.

DAI J. and MUMPER, R.J. 2010. Plant phenolics: Extraction, analysis and their antioxidant and anticancer properties. *Molecules* 15: 7313–52.

D'ARCHIVIO M., FILESI C., DI BENEDETTO R., GARGIULO R., GIOVANNINI C. and MASELLA R. 2007. Polyphenols, dietary sources and bioavailability. *Annali dell'Istituto Superiore di Sanità* 43: 348–61.

EDRIS A.E., AHMAD S. and FADEL H.M. 2003. Effect of organic agriculture practices on the volatile aroma components of some essential oil plants growing in Egypt II: sweet marjoram (*Origanum marjorana* L.) essential oil. *Flavour and Fragrance Journal* 4: 345–51.

FARNSWORTH N.R. and SOEJARTO D.D. 1991. Global importance of medicinal plants. pp. 25–51. In: *The Conservation of Medicinal Plants*. Akerele O., Heywood V. and Synge H. (eds.). Cambridge University Press, Cambridge.

FRANZ C. 1983. Nutrient and water management for medicinal and aromatic plants. *Acta Horticulturae* 132: 203–15.

GAMON J.A. and PEARCY R.W. 1989. Leaf movement, stress avoidance and photosynthesis in *Vitis californica*. *Oecologia* 79: 475–81.

GENGAIHI E.L. and ABDALLA N. 1978. The effect of date of sowing and plant spacing on yield of seed and volatile oil of fennel. *Journal Pharmazie* 33(9): 605–06.

GERSHENZON J. and CROTEAU R. 1991. Terpenoids. In: *Herbivores Their Interaction with Secondary Plant Metabolites Vol I: The Chemical Participants*, 2nd ed., Rosenthal G.A. and Berenbaum M.R. (eds.). Academic Press, San Diego, pp. 165–219.

GERSHENZON J. 1994. Metabolic cost of terpenoid accumulation in higher plants. *Journal of Chemical Ecology* 20: 1281–2981.

GHERSHENZON J. 1984. Changes in levels of plant secondary metabolites under water and nutrient stress. pp. 273–320. In: *Recent Advances in Phytochemistry—Phytochemical Adaptations to Stress* Timmermann, N., Steelin, C. and Loewus, F.A. (eds). Plenum Press, New York, NY, USA.

GIBSON L.R. 2000. *Plant Competition*. Agronomy Department, Iowa State University.

GROOMBRIDGE B. (ed.). 1992. Global biodiversity. *Status of the Earth's Living Resources*. Chapman and Hall, London, Glasgow, New York.

HARBORNE J.B. 1973. *Phytochemical Methods: A Guide to Modern Techniques of Plant Analysis*. Chapman and Hall Ltd, London, p. 279.

HE B.H. and ZHONG Z.C. 2003. Study on variation dynamics of modular population of *Ginkgo biloba* under different conditions of environmental stress. *Journal of Southwest Agricultural University* 25: 7–10.

HE SHAN-AN and SHENG NING. 1997. Utilization and conservation of medicinal plants in China with special reference to *Atractylodes lancea*. pp. 109–115. In: *Medicinal Plants for Forest Conservation and Health Care. Non-Wood Forest Products 11*. Bodeker, G., Bhat, K.K.S., Burley, J. and Vantomme, P. (eds.). FAO, Rome.

HENDAWY S.F. and KHALID KH.A. 2005. Response of sage (*Salvia officinalis* L.) plants to zinc application under different salinity levels. *Journal of Applied Sciences Research* 1(2): 147–55.

HOUGHTON J.T., DING Y., GRIGGS D., NOGUER J.M., LINDEN P.J.V., DAI X., MASKELL K. and JOHNSON C.A. (EDS.). 2001. *Climate Change 2001: The Scientific Basis*. Cambridge University Press, Cambridge, U.K., 881 pp.

IWU M.M. 1993. *Handbook of African Medicinal Plants*. CRC Press, Boca Raton, Ann Arbor, London, Tokyo.

JALEEL C.A., SANKAR B., MURALI P.V., GOMATHINAYAGAM M., LAKSHMANAN G.M.A. and PANNEERSELVAM R. 2008. Water deficit stress effects on reactive oxygen metabolism in *Catharanthus roseus*; Impacts on ajmalicine accumulation. *Colloids and Surfaces B* 62: 105–11.

JANZEN D.H. 1974. Tropical blackwater rivers, animals and mast fruiting by the Dipterocarpaceae. *Biotropica* 6: 69–103.

JENSEN K.I.N., GAUL O.S., SPECTH E.G. and DOOHAN D.J. 1995. Hypericin content of Nova Scotia genotypes of *Hypericum perforatum* L. *Canadian Journal of Plant Sciences* 75: 923–26.

KALRA A., PARAMESWARAN T.N. and RAVINDRA N.S. 1992. Influence of planting date on plant losses and yield responses of geranium (*Pelargonium graveolens*) to root rot and wilt. *The Journal of Agricultural Science* 118: 309–14.

LANGE D. 1996. Untersuchungen zum Heilpflanzenhandel in Deutschland. Ein Beitrag zum internationalen Artenschutz. German Federal Agency for Nature Conservation, Bonn-Bad Godesberg.

LANGE D. 1998. Europe's medicinal and aromatic plants: Their use, trade and conservation. TRAFFIC International, Cambridge.

LETCHAMO W. and GOSSELIN A. 1995. Effects of HPS supplemental lighting and soil water levels on growth, essential oil content and composition of two thyme (*Thymus vulgaris* L.) clonal selections. *Canadian Journal of Plant Sciences* 75: 231–38.

LEWIS W.H. and ELVIN-LEWIS M.P.F. 1977. *Medical Botany; Plants Affecting Mans Health*. Wiley, New York.

LI H. 2008. Content change of active compounds in different growth and development periods. *Chinese Traditional and Herbal Drugs* 39: 604–7.

LIST S., BROWN P.H. and WALSH K.B. 1999. Functional anatomy of oil glands of Melaleuca. *Austrian Journal of Botany* 43: 629–41.

LITVAK M.E., CONSTABLE J.V.H. and MONSON R.K. 2002. Supply and demand processes as controls over needle monoterpene synthesis and concentration in Douglas fir [*Pseudotsuga menziesii* (Mirb.) Franco]. *Oecologia* 132: 382–91.

LIU H.Y., WANG X.D., WANG D.H., ZOU Z.R. and LIANG Z.S. 2011. Effect of drought stress on growth and accumulation of active constituents in *Salvia miltiorrhiza* Bunge. *Industrial Crops and Products* 33: 84–8.

LIU J.H, ZHANG Y.Q, LI J., HU J.H. and LI Z.H. 2010. Influence of water stress on the physiological and biochemical characteristics of *Scutellaria baicalensis* Georgi. *Plant Physiology and Biochem*istry 11: 22–5.

MARCHESE J.A. and FIGUEIRA G.M. 2005. The use of pre and post-harvest technologies and good agricultural practices in the production of medicinal and aromatic plants. *Brazilian Journal of Medicinal Plants* 7: 86–96.

MARISOL B.D., ROSEMARIE W.E., FELICITAS H.H. and ALEJANDRO M.Y. 2003. Influence of sowing date and seed origin on the of capitul *Calendula officinalis* L. during two growing seasons in Chili. *Agriculture Technology* 63(1): 3–9.

MASON H. 1946a. The edaphic factor in narrow endemism. *The Nature of Environmental Influences Madroño* 8: 209–26.

MASON H. 1946b. The geographic occurrence of plants in highly restricted patterns of distribution. *Madroño* 8: 241–57

MAXWELL K. and JOHNSON G.N. 2000. Chlorophyll fluorescence—A practical guide. *Journal of Experimental Botany* 51: 659–68.

MAYTON H.S., OLIVIER C., VAUGHN S.F. and LORIA R. 1996. Correlation of fungicidal activity of *Brassica* species with allyl isothiocyanate production in macerated leaf tissue. *Phytopathology* 86: 267–71.

MEDEIROS R.S. and VIEIRA G. 2008. Sustainability of extraction and production of copaiba (*Copaifera multijuga* Hayne) oleoresin in Manaus, AM, Brazil. *Forest Ecology and Management* 256: 282–88.

MEDENTSEV A.G., ARINBASAROVA A.Y. and AKIMENKO V.K. 2005. Biosynthesis of naphthoquinone pigments by fungi of the genus *Fusarium. Applied Biochemistry and Microbiology* 41(5): 503–7.

MOONEY H.A., WINNER W.E. and PELL E.J. 1991. *Response of Plants to Multiple Stresses.* Academic Press, San Diego, California, USA.

MOQBELI E., FATHOLLAHI S., OLFATI J.A., PEYVAST G.A., HAMIDOQLI Y. and BAKHSHI D. 2011. Investigation of soil condition on yield and essential oil in lemon balm. *South Western Journal of Horticulture Biology and Environment* 2(1): 87–93.

MORISON J.I.L. and LAWLOR D.W. 1999. Interactions between increasing CO_2 concentration and temperature on plant growth. *Plant, Cell and Environment* 22: 659–82.

NAGHDIBADI H., YAZDANI D., NAZARI F. and MOHAMMADALI S. 2002. Effect of different densities of planting on essential oil component of *Thymus vulgaris. Revive Researches of Medicinal and Aromatic Plant of Iran* 1: 51–7.

NEFFATI M. and MARZOUK B. 2008. Changes in essential oil and fatty acid composition in coriander (*Coriandrum sativum* L.)

leaves under saline conditions. *Industrial Crops and Products* 28: 137–42.

NITZ G.M. and SCHNITZLER W.H. 2004. Effect of PAR and UV-B radiation on the quality and quantity of the essential oil in sweet basil (*Ocimum basilicum* L.), *Acta Horticulturae* 659: 375–81.

OKWU D.E. 2004. Phytochemicals and vitamin content of indigenous spices of South Eastern Nigeria. *Journal of Sustainable Agriculture and Environment.* 6: 30–34.

OKWU D.E. 2005. Phytochemicals, vitamins and mineral contents of two Nigeria medicinal plants. *International Journal of Molecular Medicine and Advance Sciences* 1(4): 375–81.

OKWU D.E. and OKWU M.E. 2004. Chemical composition of *Spondias mombin* Linn. plant parts. *Journal of Sustainable Agriculture and the Environment* 6(2): 140–47.

OZTURK A., UNLUKARA A., IPEKL A. and GURBUZ B. 2004. Effect of salt stress and water deficit on plant growth and essential oil content of lemon balm (*Melissa officinalis* L.). *Pakistan Journal of Botany* 36(4): 787–92.

PALEVITCH D. 1987. Recent advances in the cultivation of medicinal plants. *Acta Horticulturae* 208: 29–35.

PATAI S. and RAPPAPORT Z. 1988. *The Chemistry of Quinonoid Compounds*, Vol II, Wiley, New York.

PELLETIER S.W. 1970. *Chemistry of the Alkaloids*; Van Nostrand Reinhold, New York, NY, USA, p. 1.

PUTIEVESKY E., RAVID U. and DUDAI N. 1990. The effect of water stress on yield components and essential oil of *Pelargonium graveolens* L. *Journal of Essential Oil Research* 2: 111–14.

RAHIMI A. and KAMALI M. 2012. Different planting date and fertilizing system effects on the seed yield, essential oil and nutrition uptake of milk thistle (*Silybum marianum* (L.) Gaertn). *Advances in Environmental Biology* 6(5): 1789–96.

RAJESWARA RAO B.R. and BHATTACHARYA A.K. 1997. Yield and chemical composition of the essential oil of rose-scented geranium (*Pelargonium* species) grown in the presence and absence of weeds. *Flavour and Fragrance Journal* 12: 201–4.

RAJESWARA RAO B.R. 2002. Effect of row spacings and intercropping with cornmint (*Mentha arvensis* L. f. *piperascens* Malinv. Ex Holmes) on the biomass and essential oil yields of rose-scented geranium (*Pelargonium* species). *Industrial Crops and Products* 16: 133–44.

RAJESWARA RAO B.R., KAUL P.N., MALLAVARAPU G.R. and RAMESH S. 1996. Effect of seasonal changes on biomass yield and terpenoid composition of rose-scent geranium (*Pelargonium* species). *Biochemical Systematics and Ecology* 24: 627–35.

RAM D., RAM M. and SINGH R. 2005. Optimization of water and nitrogen application to menthol mint (*Mentha arvensis* L.) through sugarcane trash mulch in a sandy loam soil of semi-arid subtropical climate. *Bio. Tech.* 97(7): 886–93.

RAO B.R.R., KAUL P.N., SINGH K., MALLAVARAPU G.R. and RAMESH S. 2005. Influence of co-distillation with weed biomass on yield and chemical composition of rose-scented geranium (*Pelargonium* species) oil. *Journal of Essential Oil Research* 17(1): 41–43.

RAZMJOO K., HEYDARIZADEH P. and SABZALIAN M.R. 2008. Effect of salinity and drought stresses on growth parameters and essential oil content of *Matricaria chamomilla*. *International Journal of Agriculture Biology* 10(4): 451–54.

ROSENTHAL G.A. 1991. The biochemical basis for the deleterious effects of L-canavanine. *Phytochemistry* 30: 1055–58.

RUSTAD L.E., CAMPBELL J.L. MARION G.M. NORBY R.J. MITCHELL M.J. HARTLEY A.E. CORNELISSEN J.H.C. and GUREVITCH J. 2001. A meta-analysis of the response of soil respiration, net nitrogen mineralization, and aboveground plant growth to experimental ecosystem warming. *Oecologia* 126: 543–62.

SAHU B.N. 1972. Response of *Raovolfia serpentina* to irrigation, nitrogen and phosphate application. *Indian Forester* 18: 312–16.

SAID-AL AHL H.A.H. and HUSSEIN M.S. 2010. Effect of water stress and potassium humate on the productivity of oregano plant using saline and fresh water irrigation. *Ozean Journal of Applied Sciences* 3(1): 125–41.

SAID-AL AHL H.A.H. and MAHMOUD A.A. 2010. Effect of zinc and/or iron foliar application on growth and essential oil of sweet basil (*Ocimum basilicum* L.) under salt stress. *Ozean Journal of Applied Sciences* 3(1): 97–111.

SAID-AL AHL H.A.H., MEAWAD A.A., ABOU-ZEID E.N. and ALI M.S. 2010. Response of different basil varieties to soil salinity. *International Agrophysics* 24: 183–88.

SAIF U., MAQSOOD M., FAROOQ M., HUSSAIN S. and HABIB A. 2003. Effect of planting patterns and different irrigation levels on yield and yield component of maize (*Zea mays* L.). *International Journal of Agriculture and Biology* 1: 64–66.

SCHAFER H. and WINK M. 2009. Medicinally important secondary metabolites in recombinant microorganisms or plants: Progress in alkaloid biosynthesis. *Biotechnology Journal* 4(12): 1684–1703.

SERGHINI K., PEREZ DE LUGUE A., CASTEJON M.M., GARCIA T.L. and JORRIN J.V. 2001. Sunflower (*Helianthus annuus* L.) response to broomraoe (*Orobanche cernua* loefl.) parasitism: Induced synthesis and excretion of 7-hydroxylated simple coumarins. *Journal of Experimental Botany* 52: 227–34.

SHALAN M.N., ABDEL-LATIF T.A.T., and GHADBAN E.A.E.EL. 2006. Effect of water salinity and some nutritional compounds of the growth and production of sweet marjoram plants (*Majorana hortensis* L.). *Egyptian Journal of Agricultural Research* 84(3): 959.

SHANKAR D. and MAJUMDAR B. 1997. Beyond the biodiversity convention: The challenge facing the biocultural heritage of India's medicinal plants. pp. 87–99. In: *Medicinal Plants for*

Forest Conservation and Health Care. Non-Wood Forest Products 11, Bodeker, G., Bhat, K.K.S., Burley, J. and Vantomme, P. (eds.). FAO, Rome.

SHARIFI A.E. and ABBASZADEH B. 2003. Effects of manure and fertilizers in nitrogen efficiency in fennel (*Foeniculum vulgare* Mill). *Iranian Journal of Medicinal and Aromatic Plant Research* 19(3): 133–40.

SHUKLA R.S., CHAUHAN S.S., GUPTA M.L., SINGH V.P., NAQVI A.A. and PATRA N.K. 2000. Foliar diseases of *Mentha arvensis*: Their impact on yield and major constituents of oil. *Journal of Medicinal and Aromatic Plant Sciences* 22: 453–55.

SHUKLA R.S., KUMAR S., SINGH H.N. and SINGH K.P. 1993. First report of aerial blight of *Coleus forskohlii* caused by *Rhizoctonia solani* in India. *Plant Disease* 77: 429.

SIMON J.E., RIESS-BUBENHEIM D., JOLY R.J. and CHARLES D.J. 1992. Water stress induced alterations in essential oil contents and composition of sweet basil. *Journal of Essential Oil Research* 4: 71–75.

SINGH A., RAHMAN L., VERMA R.S., VERMA R.K., SINGH U.B., SINGH S.K., CHAUHAN A. and KUKREJA A.K. 2009a. Effect of plant geometry on growth and yield of lemongrass (*Cymbopogon flexuosus* Ness ex. Stued) cultivars from Uttarakhand hills. *Journal of Medicinal and Aromatic Plant Sciences* 31(1): 10–12.

SINGH M. 1999. Effect of irrigation and nitrogen on herbage, oil yield and water use of lemongrass (*Cymbopogon flexuosus*) on alfisols. *Journal of Agricultural Science* 132: 201–06.

SINGH R., PUTTANNA K., PRAKASA RAO E.V.S., GUPTA A.K., GUPTA M.M. and KHANUJA S.P.S. 2009b. Evaluation of influence of planting date on growth, artemisinin yield and seed yield of *Artemisia annua* under Bangalore agroclimatic conditions. *Archives of Agronomy and Soil Science* 55(5): 569–77.

SNOW M.D., BARD R.R., OLSZYK D.M., MINSTER L.M., HAGER A.N. and TINGEY D.T. 2003. Monoterpene levels in needles of Douglas fir exposed to elevated CO_2 and temperature. *Physiologia Plantarum* 117: 352–58.

SOFOWORA A.E. 1993. *Medicinal Plants and Traditional Medicines in Africa*. 2nd edition. Spectrum Books, Ibadan, Nigeria. p. 289.

TABATABAIE J. and NAZARI J. 2007. Influence of nutrient concentrations and NaCl salinity on the growth, photosynthesis and essential oil content of peppermint and lemon verbena. *Turkish Journal of Agriculture and Forestry* 31: 245–53.

TAN Y., LIANG Z.S., DONG J.E., HAO H.Y. and YE Q. 2008. Effect of water stress on growth and accumulation of active components of *Isatis indigotica*. *China Journal of Chinese Materia Medica* 33: 18–22.

THAKUR C. 1984. *Weed Science*. Metropolitan Book Co. Pvt. Ltd., New Delhi, India, pp. 37–47.

TIWARI K.N., SINGH A. and MAL P.K. 2003. Effect of drip irrigation on yield of cabbage (*Brassica oleracea* L. var. *capitata*) under mulch and mono-mulch condition. *Agricultural Water Management* 58: 19–28.

TRIER G. *Die Alkaloide*; Verlag von Gebrüder, Borntraeger, Berlin, Germany, 1931, pp. 1–10.

TURLINGS T.C.J., LOUGHRIN J.H., MCCALL P.J., ROESE U.S.R., LEWIS W.J. and TUMLINSON J.H. 1995. How caterpillar-damaged plants protect themselves by attracting parasitic wasps. *Proceedings of the National Academy of Sciences of the USA*, 92: 4169–74.

UPADHYAYA M.K. and FURNESS N.H. 1994. Influence of light intensity and water stress on leaf surface characteristics of *Cynoglossum officinale*, *Centaurea* spp. and *Tragopogon* spp. *Canadian Journal of Botany* 72: 1379–86.

VERMA R.S., RAHMAN L., VERMA R.K., CHANOTIYA C.S., CHAUHAN A., YADAV A., YADAV A.K. and SINGH A. 2010. Changes in the essential oil content and composition of *Origanum vulgare* L. during annual growth from Kumaon Himalaya. *Current Science* 98(8): 1010–12.

VERMA R.S., VERMA R.K., CHAUHAN A. and YADAV A.K. Seasonal variation in essential oil content and composition of thyme, *Thymus serpyllum* L. cultivated in Uttarakhand hills *Indian Journal of Pharmaceutical Sciences* 73(2): 233–35.

VERMA R.K., CHAUHAN A., VERMA R.S. and GUPTA A.K. 2011. Influence of planting date on growth, artemisinin yield, seed and oil yield of *Artemisia annua* L. under temperate climatic conditions. *Industrial Crops and Products* 34: 860–64.

VERMA R.K., CHAUHAN A., VERMA R.S., RAHMAN L. and BISHT A. 2013. Improving production potential and resources use efficiency of peppermint (*Mentha piperita* L.) intercropped with geranium (*Pelargonium graveolens* L. Herit ex Ait) under different plant density *Industrial Crops and Products* 44: 577–82.

VERMA R.K., RAHMAN L., KUKREJA A.K., VERMA R.S., SINGH A., CHAUHAN A. and KHANUJA S.P.S. 2008. Effect of nitrogen and plant population density on herb and essential oil yield of Indian basil (*Ocimum basilicum*). *Journal of Medicinal and Aromatic Plant Sciences* 30: 34–39.

VERMA R.K., RAHMAN L., VERMA R.S., YADAV A., MISHRA S., CHAUHAN A., SINGH A., KUKREJA A.K. and KHANUJA S.P.S. 2009. Biomass yield, essential oil yield and resource use efficiency in geranium (*P. greveolens* L. Her. ex. Ait), intercropped with fodder crops. *Archives of Agronomy and Soil Science* 55(5): 557–67.

VERMA R.S., PADALIA R.C. and CHAUHAN A. 2012. Variation in the volatile terpenoids of two industrially important basil (*Ocimum basilicum* L.) cultivars during plant ontogeny in two different cropping seasons from India. *Journal of the Science of Food and Agriculture* 92: 626–31.

VERZALOVA I., KOCURKOVA B. and STAVKOVA I. 1988. The response of two cultivars of fennel (*Foeniculum vulgare* var. vulgare Mill) to row spacing. *Acta Horticulture* 15(2): 101–06.

WHO. 1977. Resolution–Promotion and Development of Training and Research in Traditional Medicine. WHO document No: 30–49.

WHO. 2001. Legal Status of Traditional Medicine and Complementary/Alternative Medicine: A World Wide Review. WHO Publishing 1.

XIAO, PEI-GEN 1991. The Chinese approach to medicinal plants—Their utilization and conservation. pp. 305–313. In: *The Conservation of Medicinal Plants*. Akerele, O., Heywood, V. and Synge, H. (eds.). Cambridge University Press, Cambridge.

YAMAMAURA T., TANAKA S. and TABATA M. 1989. Light dependent formation of glandular trichomes and monoterpenes in thyme seedlings. *Phytochemistry* 28: 741–44.

YAN X., WU S., WANG Y., SHANG X. and DAI S. 2004. Soil nutrient factors related to salidroside production of *Rhodiola sachalinensis* distributed in Chang Bai Mountain. *Environmental and Experimental Botany* 52(3): 267–76.

YANG L., HAN Z.M., YANG L.M. and HAN M. 2010. Effects of water stress on photosynthesis, biomass, and medicinal material quality of *Tribulus terrestris*. *Chinese Journal of Applied Ecology* 21: 2523–2528.

YUAN Y., LIU Y., LUO Y., HUANG L., CHEN S., YANG Z. and QIN S. 2011. High temperature effects on flavones accumulation and antioxidant system in *Scutellaria baicalensis* Georgi cells *African Journal of Biotechnology* 10(26): 5182–92.

ZAVALA J.A. and RAVETTA D.A. 2001. Allocation of photoassimilates to biomass, resin and carbohydrates in *Grindelia chiloensis* as affected by light intensity. *Field Crops Research* 69: 143–49.

ZHU Z.B., LIANG Z.S. and HAN R.L. 2009. Saikosaponin accumulation and antioxidative protection in drought-stressed *Bupleurum chinense* DC. *Environmental and Experimental Botany* 66: 326–33.

ZOBAYED S.M.A., AFREEN F. and KOZAI T. 2005. Temperature stress can alter the photosynthetic efficiency and secondary metabolite concentrations in St. John's wort. *Plant Physiology and Biochemistry* 43: 977–84.

ZOBAYED S.M.A., AFREEN F. and KOZAI T. 2007. Phytochemical and physiological changes in the leaves of St. John's wort plants under a water stress condition. *Environmental and Experimental Botany* 59: 109–16.

ZOU Y., LU Y. and WEI D. 2004. Antioxidant activity of a flavonoid-rich extract of *Hypericum perforatum* L. *in vitro*. *Journal of Agriculture and Food Chemistry* 52: 5032–39.

CHAPTER THIRTEEN

Understanding the patterns of gene expression during climate change

For enhancing crop productivity

Rakesh Srivastava, Rashmi Srivastava and Uma Maheshwar Singh

Contents

Abstract

The extreme challenge for the future is to increase the plant productivity and yield with respect to climate change. In the current global status, aberrant climate change leads to the immense impact on the loss of million tons of crop productivity in agriculture. The consequence is becoming more pathetic by the current and imminent global changes in climate, world population increases, industrialisation toxicity, deterioration of cultivated land and freshwater insufficiency. All these stresses emphasising the development of stress-resistant plants are those that have the ability to adapt and endure the growth and productivity in stressful and harsh environmental changes. It is significant to understand plants stress response mechanism to augment the crop productivity under unfavourable or stressful environmental conditions.

13.1   Introduction

Plants, as a sessile organism, must contend and thrive under multiple climate change threats or environmental stresses. These stresses not only affect the plant crops productivity, but also lead to changes in the plant architecture, growth and development (Walter et al., 2009). During the onset and development of environmental stress within a plant, all the major processes are directly affected such as photosynthesis, transpiration, respiration, energy and lipid metabolism, carbohydrate metabolism, protein synthesis, stomatal conductance and pigment concentrations; then there are secondary stresses such as ion uptake and nutrition stress (affecting the availability, transport and partitioning of nutrients) and oxidative stress which together affect the plant development and growth (Ryan, 1991; Scandalios, 1997; Grene, 2002; Flexas et al., 2004; Saher et al., 2005; Garrett et al., 2006; Chaves et al., 2009; Singh et al., 2009; Walter et al., 2009; Compant et al., 2010; Cramer et al., 2011; Walbot, 2011; Dinakar et al., 2012; Tullus et al., 2012). Plants have to respond and adapt to environmental stresses at the cellular, biochemical, physiological, and molecular levels.

Genetic technologies have revealed many plant genes and their downstream gene activation which are involved in the stress gene expression, stress signal transduction pathways and stress tolerance. These approaches elucidated different mechanisms to fine tune the plant gene expression and the ability to cope in an appropriate manner to environmental stresses. During the change in environmental conditions, the defence against these stresses is a large reprogramming of gene expression through regulation of transcription (Figure 13.1). The development of microarray and high-throughput technologies led to the discovery of several hundred to thousands of genes in plants with altered expression in response to climate changes. Genes with altered expression during climate change stresses are often important for adaptation to stress; transgenic plants overexpress such genes that can have increased change stress tolerance.

Climate change or environmental stresses have a profound influence on the plant growth and productivity in a variety of ways. The potential impacts of climate change have been examined in many crops such as groundnut, rice, wheat, soybean, maize, many vegetables and fruits (Schlenker and Roberts, 2009; Singh et al., 2009, 2014; Hao et al., 2010; Mirade Orduna, 2010; Moretti et al., 2010; Waterer et al., 2010;

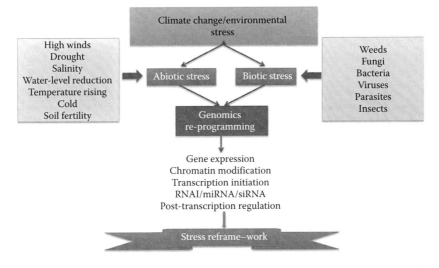

FIGURE 13.1 (**See colour insert.**) A model for abiotic and biotic stress signalling leads to genomics reprogramming during climate change or environmental stress.

Madan et al., 2012; Zhang and Huang, 2012). The future consequences of climate change have been observed in locations such as Africa, Australia, China, Europe, India, Japan, United States and other countries (Olesen and Bindi, 2002; Desch et al., 2007; Schlenker and Roberts, 2009; Hanna et al., 2011; Olesen et al., 2011; Chang et al., 2012; Elsgaard et al., 2012; Mubaya et al., 2012; Zhang and Huang, 2012; Beck, 2013).

Environmental stresses, such as drought, salinity, chilling, freezing, and high temperatures cause adverse effects on the growth of plants and reduce crop productivity, and, in extreme cases, lead to plant death. Abiotic stress is the major cause of crop loss worldwide, reducing average yields for most major crop plants by more than 50% (Bray et al., 2000). Water stress in its extended sense covers both drought and salt stress. Drought and salinity are prevalent in several provinces, and may create serious salinisation issues on more than 50% of all arable lands by the year 2050 (Wang et al., 2003).

Since sequencing of the first plant genome *Arabidopsis thaliana* to the recent high-throughput genomic and proteomic approaches, an enormous quantity of information has been produced to advance our apprehension of how the eukaryotic genome works to induce and synchronise specific programmes of gene expression. Many studies have revealed several molecules, for instance, transcription factors, cofactors, kinases, activators and so on, as promising candidates for common players that are involved in cross talk between abiotic and biotic stress-signalling pathways during the changing environmental conditions. The response to the changing environmental stress in plant development and physiology can be prone to dramatic genomic-reprogramming processes that lead to alternative gene programmes and expression patterns (Figure 13.1).

13.2 Harmonisation of a stress-signalling pathway and gene expression

Plants respond to a multitude of biotic and abiotic environmental stress signals that influence growth and development. Plants are subject to biotic and abiotic stresses, and have developed strategies to protect themselves against these environmental stress attacks. These environmental stress signals are converted into appropriate signalling cascades to endure adverse conditions. Knowledge of these signalling pathways, role and regulation will lead to the design of stress-tolerant plants and a reduction in the loss of crop productivity. To understand the mechanisms of an environmental stress-signalling pathway during climate change, they extensively investigated *A. thaliana, Oryza sativa, Zea mays, Solanum lycopersicum, Triticum*, legumes and many other plants (Zhu, 2001; Bruce et al., 2002; Seki et al., 2003; Baena-Gonzalez and Sheen, 2008; Becana et al., 2010; Walley and Dehesh, 2010; Chew and Halliday, 2011; Dolferus et al., 2011; Santos et al., 2011; Sun et al., 2011; Zhang et al., 2011; Krasensky and Jonak, 2012). The variety of signalling pathways that play a central role during plant stress response are auxin (Popko et al., 2010), abscisic acid (ABA) (Guo et al., 2011; Qin et al., 2011), brassinosteroids (BRs) (Mussig et al., 2006), cytokinin (Ha et al., 2012), ethylene (Etheridge et al., 2005; Zhu and Guo, 2008), gibberellic acid (GA) (Gao et al., 2011), jasmonic acid (JA) (Balbi and Devoto, 2008; Wasternack and Kombrink, 2010), salicylic acid (SA) (Yuan and Lin, 2008; An and Mou, 2011) and nitric oxide (NO) (Gechev et al., 2006; Grun et al., 2006; Qiao and Fan, 2008). Auxin, cytokinin, GA, ABA and ethylene are well accepted as five classes of classic plant hormones. More recently, evidence has accumulated to extend this concept to include BRs, JA, SA and NO (Santner and Estelle, 2009). These signalling molecules not only maintain the development and growth of the plant but are also helpful in providing the tolerance to the plant during a change in environmental conditions.

Auxin

Auxin is a phytohormone, which regulates plant developmental and physiological processes, such as tropic responses to light and gravity, general root and shoot architecture, embryogenesis, vascular differentiation and organogenesis (Woodward and Bartel, 2005). Genome-wide analysis studies show that the transcriptional response to an auxin is rapid and broad, influencing the gene expression of a large and different sets of genes (Goda et al., 2004; Okushima et al., 2005; Overvoorde et al., 2005; Nemhauser et al., 2006). Recent studies manifest that auxin homeostasis directly links growth regulation with stress adaptation responses. Auxin signalling was controlled at the level of gene expression by three gene families, *Aux/IAAs (indole-3-acetic acid), GH3s* and small auxin-up RNAs (*SAURs*). WES1 (a GH3 *protein*) controlled the endogenous auxin (IAA) content through feedback regulation. *WES1* is the gene expression increased by environmental stresses as well as by SA and ABA, causing the reduction of endogenous IAA and resultant growth retardation under stress conditions, which may provide an adaptive strategy on stressed plants (Park, 2007; Park et al., 2007). Auxin, in conjoining with ABA, regulates the gene expression R2R3-type MYB transcription factor, MYB96. MYB96, in turn, regulates lateral root development under drought conditions. Recent studies demonstrate the importance of auxin in cold-stress-mediated plant gene expression and trafficking (Shibasaki et al., 2009; Rahman, 2012). An auxin-resistant mutant, *axr1-24*, shows more resistant salt concentrations, providing a link between auxin and salt stress (Tiryaki, 2007).

Abscisic acid   ABA, a terpenoid phytohormone, signal transduction pathway leads to changes in plant gene expression in many ways, which involves changes in the transcription, transcript processing, chromatin modification and RNA stability (Guo et al., 2011). There are many studies showing that ABA is a key regulatory molecule in the control of gene expression in abiotic and biotic stress such as drought, dehydration, cold, salinity as well as pathogen interaction (Wasilewska et al., 2008). ABA has been essential in transcriptional and post-transcription regulation of stomata aperture, lateral root growth, seed germination, antioxidant response and pathogen defence. Both biotic and abiotic stress genes are induced and controlled by ABA-dependent and ABA-independent pathways. Comparisons of transcriptomes for *Arabidopsis* and rice exposed to ABA and various abiotic stresses have shown changes affecting 5–10% of the genome; more than half of these changes were common to drought, salinity and ABA treatments (Shinozaki et al., 2003; Nakashima et al., 2009). The ABA-regulated genes in *Arabidopsis* seedlings include slightly over 10% of the genome. This is 2–6 times as many genes are regulated by most of the other plant hormones (Nemhauser et al., 2006; Nakashima et al., 2009). ABA passes its signalling and controls gene expression using a different set of receptors, such as, flowering time control protein A (FCA), ChlH/ABA R (abscisic acid receptor)/CCH (conditional chlorina)/GUN5 (genome uncoupled 5), GTGs (GPCR type G proteins) and PYR/PYL/RCARs (pyrabactin/PYR-like/regulatory component of ABA receptors) (Guo et al., 2011). The ABA-induced genes are enriched for those encoding proteins involved in stress tolerance, such as dehydrins and enzymes that detoxify reactive oxygen species (ROS), enzymes of compatible solute metabolism, a variety of transporters, regulatory proteins such as transcription factors, protein kinases and phosphatases and enzymes involved in phospholipid signalling. ABA-repressed gene products are enriched for proteins associated with growth, including cell wall, ribosomal, plasma membrane and chloroplast proteins. *NFYA5* (nuclear factor Y A5), *OCP3* (overexpresser of cationic peroxidase 3), *MYB96* (MYB transcription factor 96), *FTA* (α-farnesyltransferase),

SAL1(3′(2′), 5′-biphosphate nucleocidase, *MSI1*(multi-copy suppressor of Ira1), *TOC1* (timing of CAB expression 1) and many other genes are transcriptionally regulated by ABA-mediated pathway and are crucial in controlling the stomatal aperture and drought resistance. ABA signal transduction has characterised the cell biological and genetic mechanisms upstream and downstream of ROS production. ABA regulated the ROS level in *itn1* (increased tolerance to NaCl), *abi1-1* (ABA insensitive1) and *ost1 (open stomata 1) mutants.* ABA induces the gene expression of ROS-producing enzyme nicotinamide adenine dinucleotide phosphate (NADPH) oxidase (RBOHC and RBOHD) *itn1* mutant. In response to cold, ABA increases the cold-responsive gene expression SCOF-1 DNA-binding activity to the bZIP SGBF-1 in ABRE sequence, suggesting a cooperative role of the two proteins to induce cold tolerance of plants (Kim et al., 2001). Besides these, many genes are regulated by transcription factors responding to drought, salt and cold *via* ABA-dependent or ABA-independent pathways.

Brassinosteroids   BRs are plant steroidal hormones that regulate various aspects of plant growth and development, including cell elongation, photomorphogenesis, xylem differentiation, seed germination and fruit ripening (Clouse and Sasse, 1998). BRs induce gene expression and the ability to enhance resistance in plants in the changing environmental condition or stresses, such as heavy metal stress, water stress, salt stress, high- and low-temperature stress and pathogen attack (Bajguz and Hayat, 2009). BRs treatment induces the expression of three regulatory genes, such as *RBOH, MAPK1* and *MAPK3*, and genes involved in defence and antioxidant responses (Xia et al., 2009). BR induces the expression of genes-encoding transcription factors, such as WRKY6, WRKY30, MYB and MYC. The 24-epibrassinolide (bioactive BRs) induces the expressions of genes-encoding proteins involved in the heat-shock response (HSP and DnaJ), defence (PR-1, PAL and HPL), detoxification (GST, GPX and POD) and antioxidant (CAT, cAPX and MDAR) (Xia et al., 2009). These gene expression results suggest that the involvement of BR induces plant stress tolerance to a variety of environmental stresses. Analysis of T-DNA insertion mutants of four BR response genes indicated that WRKY17, WRKY33, ACP5 and BRRLK have stress-related functions (Divi et al., 2010). For osmotic permeability, BR-deficient *cpd* and BR-insensitive *bri1* mutants showed that BR treatment produced an increase in the osmotic permeability of protoplasts prepared from *cpd* plants (Morillon et al., 2001). The 24-epibrassinolide treatment also leads to significant increases in the levels of hsps90 during thermal stress in *Brassica napus* (Dhaubhadel et al., 2002). BR (24-epibrassinolide) treatments increase the gene expression of arginine decarboxylase (ADC) and improve IAA and ABA levels under heavy metal chromium stress in *Raphanus sativus* (Choudhary et al., 2012). ADC has an important role in response to cold, oxidative stress, salt stress and seed development. T-DNA knockout mutants of *OsGSK1*, a rice orthologue of the BR-negative regulator BIN2, showed greater tolerance to cold, heat, salt and drought stresses as compared with non-transgenic plants (Koh et al., 2007).

Cytokinins

Cytokinins are plant hormones which are implicated in nearly all aspects of plant growth and development, including apical dominance, leaf senescence, nutrient signalling, shoot differentiation, cell division and light responses (Kiba and Mizuno, 2003). Cytokinins homeostasis and signalling are changed and regulated under various biotic and abiotic stresses. For example, *Pseudomonas syringae* pv. *tomato* DC3000 (*Pst*), a bacterial pathogen has a direct effect on cytokinins. The endogenous cytokinins perceived by AHK2 and AHK3 receptors promote SA signalling through ARR2 activation and association with the promoters of pathogenesis-related (PR) genes, which lead to enhanced plant immunity (Choi et al., 2010). Another instance from a recent study has shown that cytokinin level and transport are reduced by drought and/or salinity in various plant species (Argueso et al., 2009; Perilli et al., 2010). Recent characterisation of *Arabidopsis CKX* overexpression and the *ipt1, 3,5,7* mutant plants with reduced endogenous cytokinin levels revealed a strong stress-tolerant phenotype that was associated with increased cell membrane integrity and ABA hypersensitivity (Nishiyama et al., 2011). The cold stress induces *ARR5, 6,7,15* expression, and the *arr5, arr6* and *arr7* mutants exhibit enhanced freezing tolerance and ABA sensitivity similar to *ahk2, 3* and *ahk3, 4* mutants (Jeon et al., 2010). In other recent studies, genome-wide microarray analyses show that more than 10% transcriptional changes in *Arabidopsis* genes in the cytokinin-deficient *ipt1, 3,5,7* mutant were compared to the wild type under both normal and salt stress conditions (Nishiyama et al., 2012).

Ethylene

Ethylene, as a gaseous plant hormone, is involved in many aspects of the plant life cycle, including seed germination, fruit ripening, organ abscission, pathogen response, senescence, root hair development, root nodulation, abscission and so on (Johnson and Ecker, 1998). The production of ethylene is tightly regulated by internal signals during development and in response to environmental stimuli from biotic (e.g. fungal and bacterial disease) and abiotic stresses, such as wounding, hypoxia, ozone, chilling or freezing. Signalling responses to ethylene in *Arabidopsis* are regulated and transferred by five receptors called ethylene receptor1 (ETR1), ETR2, ethylene response sensor1 (ERS1), ERS2 and ethylene insensitive4 (EIN4) (Chang et al., 1993; Hua et al., 1998). The receptors activate Raf-like protein kinase, CTR1, which negatively regulates downstream ethylene-signalling events (Kieber et al., 1993). During either biotic or abiotic stress condition, plants produce increased levels of ethylene, called stress ethylene, which is able to initiate various ethylene-regulated responses (Abeles et al., 1992). Plants deficient in ethylene signalling may show either increased susceptibility or increased resistance during biotic stress. For example, in soybean, mutants with reduced ethylene sensitivity produce less-severe chlorotic symptoms when challenged with the virulent strains *P. syringae* pv. *glycinea* and *Phytophthora sojae*, whereas virulent strains of the fungi *Septoria glycines* and *Rhizoctonia solani* cause more severe symptoms (Hoffman et al., 1999). *Arabidopsis* plants with defects in ethylene perception (*ein2*) or JA signalling (*coi1*) fail to induce a subset of *PR* gene expression, including the plant defensin gene *PDF1.2*, a basic chitinase (*PR-3*) and an acidic have-in-like protein (*PR-4*), resulting in enhanced susceptibility to certain pathogens (Penninckx et al., 1998). Peng et al. (2005) investigated the responses of *Arabidopsis ACS* genes to hypoxia stress and they found that *ACS2, ACS6, ACS7* and *ACS9* were specifically induced during hypoxia. As an environmental signal integrator, ethylene transmits salt stress signalling and enhances plant survival in a DELLA protein-dependent manner (Achard et al., 2006). *Arabidopsis* NAC-type transcription factor gene AtNAC2 incorporates the environmental and endogenous stimuli into the process of plant lateral root development. It has been shown that in the *ein2-1* and *ein3-1* mutant, the salt induction of AtNAC2 was affected (He et al., 2005). Ethylene signal pathways are also essential for defence against ultraviolet B (UV-B) damage. In *etr1-1* (insensitive to ethylene) mutant plants, the UV-B-induced up-regulation of *PR-1* and *PDF1·2* transcript levels was considerably reduced compared with wild-type plants, indicating a role of ethylene in the up-regulation of these genes in response to UV-B exposure (Mackerness, 1999). Another protein in the ethylene-signalling pathway, ethylene response factor protein (JERF3) activates the gene expression through transcription, resulting in decreased accumulation of ROS and, in turn, enhanced adaptation to drought, freezing and salt in tobacco. JERF3 induced the gene expression of *NtCA (carbonic anhydrase), NtSOD (encodes SOD), NtAPX1 (ascorbate peroxidase), NtAPX2 (chloroplas-tic ascorbate peroxidase), NtGPX (glutathione peroxidase), NtSAM1* (S-adenosyl-l-Met synthetase), *TOBLTP* (lipid trans-fer protein), *NtERD10C* (early response to dehydration 10 C), and *NtSPS (Suc-P synthase)* in responding to osmotic stress in tobacco (Wu et al., 2008).

Jasmonic acids   The signalling molecules of JA are important for many aspects of gene expression in plant growth, development and defence. The importance of JA for wound signalling and its role in the defence against insect attack was discovered in solana-ceous plants. In *Arabidopsis*, several mutants were identified either with compromised JA biosynthetic capacity (*fad3-2, fad7, fad8, dad1, opr3, dde1 or dde2*) or with defects specifi-cally in JA perception or signal transduction (*jar1, coi1 and jin1*). The mutant plants were severely compromised in their defence against insect attack and succumbed to infection by pathogenic soil fungi. Huffaker et al. (2006) show that in jas-monate-deficient *fad3, 7,8* triple-mutant plants, the *PROPEP1* and *PDF1.2* expressing are down-regulated. *A thaliana* plants overexpressing PROPEP1 show altered root development and enhanced resistance to the root pathogen *Pythium irregulare* (Huffaker et al., 2006). Walia et al. (2007) show that 44 genes are up-regulated during JA treatment and salinity stress given to Hordeum vulgare. These include photosynthesis and stress response-related genes such as the small subunit of ribulose-1,5-bisphosphate carboxylase, arginine decarboxylase 2 (ADC2), carbonic anhydrase (CA), glutathionone S-transferase, fasci-clin-like arabinogalactan (FLA10), jacalin lectin protein and a water stress-induced tonoplast intrinsic protein. JA-responsive *Arabidopsis* genes related to salinity stress tolerance include a salt-tolerance zinc-finger protein (STZ/ZAT10), a sodium/potassium/calcium exchanger (At5g17860), an outward-recti-fying potassium channel (At4g18160) and a mechano-sensitive ion channel protein (At5g19520) (Walia et al., 2007). Like eth-ylene, JA signal pathways are also requisite for defence against UV-B damage. In *jar1* (insensitive to JA) mutant plants, the UV-B-induced up-regulation of *PDF1·2* and *PR-1* gene expres-sion level was considerably reduced compared with wild-type plants, indicating a role of JA in the up-regulation of these genes in response to UV-B exposure (Mackerness, 1999). Plasma membrane intrinsic protein 1 gene, which plays an essential role in the defence of plants against water stress, is strongly expressed in response to JA and to other stresses such as mannitol, NaCl and wounding in *Populus alba* × *P. tremula* var. glandulosa.

Nitric oxide

NO functioning as a signal in hormonal responses, is a gaseous-free radical, diatomic molecule and plays important roles in diverse physiological and pathological processes in plants cellular and developmental mechanisms. NO synthesis in plants takes place by arginine- and nitrite-dependent pathways. Nitric oxide synthase (NOS) catalyses the generation of NO from the input of O_2 and NADPH. NO has dual roles in plant cell physiology. Endogenous NO is stringently controlled in low-level homeostasis through the production and elimination of NO under normal physiological condition. Low concentrations of NO can not only interrupt ROS-triggered chain reaction, but can also improve plant resistance by inducing gene expression (Beligni and Lamattina, 1999, 2000). When the cell alters itself or is exposed to environmental stress or climate change, the rapid increase of NO elicits an oxidative burst resulting in cell death (He et al., 2012). Parani et al. (2004) studied the gene expression analysis in *Arabidopsis* using microarray in response to 0.1 and 1.0 mM sodium nitroprusside (SNP), a donor of NO. They found 342 up-regulated and 80 down-regulated genes in response to NO treatments (Parani et al., 2004). They observed the up-regulation of several genes encoding disease-resistant proteins, WRKY proteins, transcription factors, zinc-finger proteins, glutathione S-transferases, ABC transporters, kinases and biosynthetic genes of ethylene, JA, lignin and alkaloids. The transcript level of several typical pathogens-induced genes (e.g. NBS-LRRs, NDR1) and genes coding for disease-resistant proteins was induced by the NO donor SNP (Parani et al., 2004). The transcript level of several plant defence responses modulating transcription factors, such as WRKYs, EREBPs (ethylene-responsive element-binding proteins), several zinc-finger proteins and dehydration-responsive element-binding proteins (DREB1 and DREB2), were also induced by SNP (Polverari et al., 2003). Zheng et al. (2010) show that the pretreatment of NO effectively contributed to a better balance between carbon and nitrogen metabolism by increasing the total soluble protein and by enhancing the activities of endopeptidase and carboxypeptidase in plants under salt stress (Zheng et al., 2010).

Salicylic acid

The small phenolic compound of SA is a phytohormone, emerging as a paradigm of an important regulatory role in multiple physiological processes, including plant immune response and growth regulators. SA has an intensive role in signal-mediating local and systemic plant defence gene expression and regulation responses against pathogens (An and Mou, 2011; Rivas-San Vicente and Plasencia, 2011). SA signalling is mediated by both NPR1-dependent mechanisms and NPR1-independent mechanisms. SA signalling pathway mediates through the ankyrin repeat containing a BTB/POZ domain protein NPR1 (non-expresser of PR1), which was identified in mutant screens of *Arabidopsis* (Cao et al., 1994). NPR1 is retained in the cytoplasm as an oligomer in the absence of SA or pathogen challenge. Upon SA or pathogen induction, NPR1 monomer is released to enter the nucleus where it activates and regulates defence-related gene expression (Mou et al., 2003). SA application and pathogen challenge experiments augment NPR1 gene expression. Overexpression of *Arabidopsis* NPR1 or its homologues confer broad resistance against diverse pathogens in multiple plant species (Cao et al., 1998; Chern et al., 2001; Lin et al., 2004; Malnoy et al., 2007; Parkhi et al., 2010). MKP1 and PTP1 regulate plant growth homeostasis, acting as repressors of the stress-induced MAPK pathway involving MPK3 and MPK6, which leads to SA biosynthesis and expression of PR genes (Bartels et al., 2009). The *mkp1* and *mkp1 ptp1* mutants have growth defects, increased levels of endogenous SA and constitutive defence responses including *PR* gene expression and resistance to the bacterial pathogen *P. syringae*. SA also mediates the defence-signalling pathway through NPR1-independent mechanism. For instance, *Arabidopsis* protein MYB30 positively regulates the pathogen-induced hypersensitive response in an SA-dependent, NPR1-independent manner (Raffaele et al., 2006). Another example for NPR1-independent, SA-dependent gene expression regulation is the constitutive defence mutants, *cpr5, cpr6* and *hrl1* (Clarke et al., 2000; Devadas et al., 2002).

There are several reviews, which focus on the role of SA and gene expression during plant defence mechanism in adverse environmental condition. However, SA has also been identified as an essential regulatory signal-mediating plant response to abiotic stresses such as drought, chilling, heavy metal tolerance, heat and osmotic stress (Borsani et al., 2001; Kang and Saltveit, 2002; Chini et al., 2004; Freeman et al., 2005; Larkindale et al., 2005). SA also acts as a negative regulator of seed germination, presumably due to an SA-induced oxidative stress (Xie et al., 2007). However, some recent reports suggested that SA recovers the seed germination under abiotic stress conditions. For example, under salt stress (100–150 mM NaCl), only 50% of *Arabidopsis* seeds germinate, but in the presence of SA (0.05–0.5 mM), seed germination increases to 80%. Exogenous application of SA also partially reverses the inhibitory effect of oxidative (0.5 mM paraquat) and heat stress (50°C for 3 h) on seed germination (Alonso-Ramirez et al., 2009). SA is also involved in the regulation of the alternative oxidase (AOX) pathway in plants by inducing and regulating its gene expression (Kapulnik et al., 1992). AOX is an enzyme, which controls the adenosine triphosphate (ATP) synthesis to maintain growth rate homeostasis. SA treatment induces *AOX1* gene expression levels that increases from two- to six-fold after 4 h of induction (Norman et al., 2004). SA regulates the many senescence-associated genes (SAGs) transcript. Transcripts of several SAGs, such as *SAG12*, are considerably reduced or undetectable in SA-deficient *Arabidopsis* plants (Morris et al., 2000). Moreover, SA activates the expression of the *Arabidopsis* senescence-related genes α*VPE*, γ*VPE*, *WRKY6*, *WRKY53* and SEN1 that encode two vacuolar-processing enzymes, two transcription factors and a protease, respectively (Kinoshita et al., 1999; Robatzek and Somssich, 2001; Miao et al., 2004; Schenk et al., 2005).

13.3 Stress-regulated promoters and gene expression

The eukaryotic promoter architecture: The controller

Eukaryotic DNA-dependent RNA polymerase II (RNA Pol II) is responsible for the transcription of the genetic information encoded in the DNA sequence protein-coding genes (Smale and Kadonaga, 2003). The correct spatial and temporal transcription of genes needs to be tightly controlled as it is the first step in differential gene expression (Maston et al., 2006), which is a prerequisite for the execution of biological processes such as cell growth, morphology, the development of multi-cellular organisms, the response to environmental conditions, disease and differentiation in all eukaryotic organisms. Transcription regulation of eukaryotic protein-coding genes (class II) is an orchestrated process that requires the concerted functions of multiple proteins or transcription factors (Martinez, 2002; Maston et al., 2006). The region of the gene upstream of the coding and 5′ UTR regions is called the promoter and is also known as the upstream region or the regulatory region of the gene (Figure 13.2). The eukaryotic promoter structure is responsible to a large extent for the regulation of transcription.

The promoter is composed of a core promoter and proximal promoter elements. The distal elements may contact the

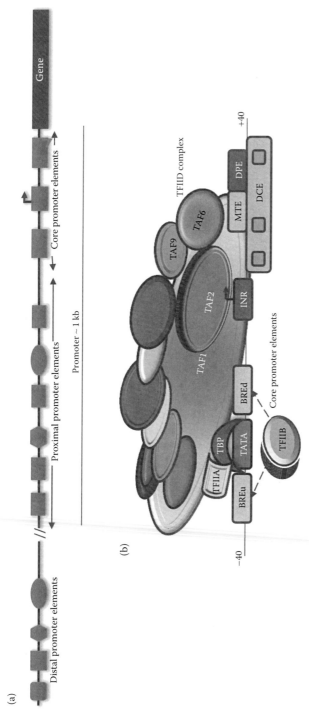

FIGURE 13.2 **(See colour insert.)** Eukaryotic CPEs and interacting basal transcription factors: (a) Schematic of a typical gene regulatory region. The promoter, which is composed of a core promoter and proximal promoter elements, typically spans <1 kb pairs. Distal (upstream) regulatory elements, which can include enhancers, silencers, insulators and locus control regions, can be located up to 1 Mb pairs from the promoter. These distal elements may contact the core promoter or proximal promoter through a mechanism that involves looping out the intervening DNA. (b) The core promoter (−40 to +40 bp) diagram shows the location of known CPEs. TFIID components are known to directly interact with CPEs. The described protein–CPE interactions are indicated by an arrow. The BREu and BREd elements are recognised by TFIIB, the TATA-box is recognised by TBP (and TBP2), the Inr is recognised by TAF1 and TAF2, the DPE is recognised by TAF6 and TAF9 and the DCE is recognised by TAF1. The MTE is known to interact with TFIID in general.

core promoter or proximal promoter through a mechanism that involves looping out the intervening DNA. The core promoter elements (CPEs) are the region at the beginning of a gene that serves as the docking site for the basic transcriptional machinery and pre-initiation complex (PIC) assembly, and defines the position of the transcription start site (TSS) as well as the direction of transcription (Smale and Kadonaga, 2003; Maston et al., 2006; Sandelin et al., 2007). The core promoter includes DNA elements that can extend up to 40 bp upstream and/or downstream of the transcription start site (TSS) (Kadonaga, 2002). The first described CPE was the TATA box, the binding site for the TBP subunit of TFIID. In addition to the TATA box, core promoters can be composed of numerous other elements, including initiator element (INR), downstream promoter element (DPE), downstream core element (DCE), TFIIB-recognition element (BRE) and motif ten element (MTE) (Thomas and Chiang, 2006) (Figure 13.2).

The proximal promoter is defined as the region immediately upstream (up to a few hundred base pairs) from the core promoter (Figure 13.2), and typically contains multiple binding sites for activators (Juven-Gershon and Kadonaga, 2010). The *cis*-elements to which general transcription factors bind are located in this region. These regulatory motifs can promote or suppress the binding of the core promoter components involved in the basal transcription complex. Sequence-specific DNA-binding transcription factors are capable of activating or repressing transcription. They cause genes to be selectively expressed in response to a certain environmental condition in a particular cell. Other transcription factors and co-regulators are specific in their activity and typically bind to promoters of genes that are regulated in response to some stimulus such as a pathogen or an environmental condition.

Distal (upstream) regulatory elements can include enhancers, silencers, insulators, and locus control regions. It is located up to thousands of base pairs away from the TSS, downstream of the gene, and also in the introns of the gene (Juven-Gershon and Kadonaga, 2010). Conceivably, the transcription factors that bind to these elements also bind to the basal transcription complex by bending the DNA into a loop and thus, acting as the transcription. These sequences can suppress or enhance transcription significantly and the strand on which they occur does not act on their influence. They are also responsible for tissue-specific expression as are elements in the proximal promoter, which often consist of repeats of the same elements as in the proximal promoter. Enhancers can act as the transcription of more than one gene, both by changing the chromatin structure and by interacting with the PIC.

Promoters: Precise tools for controlling gene expression in environmental stress

Gene expression changes during environmental stresses are complex phenomena. Promoters are a very powerful tool in genetic engineering and biotechnological applications because the expression of genes operably linked to them can be regulated to function at certain stages of development or a particular cell/tissue or an external/internal stimulus under defined conditions. Promoters regulate and fine tune the expression levels and patterns of transgenic genes. The transgene plays a significant role in determining the phenotypes of transgenic plants. Promoters can be categorised into different categories such as constitutive promoters (active constantly in most or all tissues/parts), tissue-specific promoters (controlling gene expression in a tissue-dependent manner), development-stage-specific promoters (controlling gene expression at certain stages of development), inducible promoters (regulated by the application of an external chemical or physical signal), synthetic promoters (cis-regulatory sequence of DNA that can be used to specifically control gene activity in any cell or tissue type of interest) and bidirectional promoters (two genes that are transcribed in opposite directions) (Peremarti et al., 2010).

In the perspective of environmental stress, necessities to re-engineer the promoters that are induced in response to a specific condition or change in the environment are predominantly valuable. These types of promoters are triggered by one or more stimuli such as hormones (e.g. GA, ABA, JA, SA and auxin), chemicals, environmental conditions (dehydration, heat, water, salt, wounding etc.) and biotic stress (microbes, insects and nematodes). Such promoters not only reduce the genetic load to the plant but are also safe to environments. Table 13.1 provides different types of inducible promoters that have been used to enhance the gene expression of plant transgenes during varied stresses.

Differences in the expression patterns of genes during stress are a result of the diverse architecture of the promoters (Srivastava et al., 2014). Induction of the promoters is regulated by transcription factors, activators and suppressors that bind to cis-regulatory elements present in the promoter regions (Priest et al., 2009). Many such kinds of cis-regulatory elements have been studied in response to the environmental signal. Even with the differences in promoter architecture, functional dissection of the promoter regions has started to reveal the common cis-acting elements and has provided us with the tools to identify distinct DNA-binding proteins required to modulate transcription. A number of web-based and bioinformatic tools have been developed to identify the potential plant cis-elements

Table 13.1 List of some stress-inducible promoters used in different stress conditions

Promoter	Origin	Stress	Reference
Aopr1	*Asparagus officinalis*	Wound and pathogen	Warner et al. (1993)
Apx1	*Arabidopsis*	Heat, oxidative stress	Storozhenko et al. (1998)
Bjchi1	*Brassica juncea*	Wounding, JA, NaCl and PEG treatment	Wu et al. (2009)
Cor6.6	*Arabidopsis*	Low temperature, exogenous ABA and dehydration	Wang et al. (1995)
Eas4	*Nicotiana tabacum*	Elicitor and pathogen	Yin et al. (1997)
Gols1	*Arabidopsis*	Temperature sensistive/ heat inducible	Panikulangara et al. (2004)
Gstf8	*Arabidopsis*	Pathogen	Perl-Treves et al. (2004)
Gn1	*Nicotiana plumbaginifolia*	Elicitor and pathogen	Kooshki et al. (2003)
Gpx1	*Citrus sinensis*	Salt stress	Avsian-Kretchmer et al. (2004)
Gst6	*Arabidopsis*	Oxidative stress and elicitor	Chen and Singh (1999)
Hahb1	*Sunflower*	Chilling, freezing, drought and salinity stresses	Cabello and Chan (2012)
Hahb4	*Sunflower*	Water stress, ABA or NaCl induced	Dezar et al. (2005)
Hp1	*Oryza sativa*	ABA, salt and drought stress	Rai et al. (2009)
Hsp17.6 G1	*Helianthus annuus*	Heat	Rojas et al. (2002)
Hsp17.6.2	*Soybean*	Heat	Lee and Schoffl (1996)
Hsp90-1	*Arabidopsis*	Heat, arsenite	Haralampidis et al. (2002)
Hva1	*Barley*	Drought, cold, heat and salinity	Straub et al. (1994)
Kin1	*Arabidopsis*	Low temperature, exogenous ABA and dehydration	Wang et al. (1995)
Lip9	*Oryza sativa*	High salinity, dehydration and cold stresses	Rabbani et al. (2003); Nakashima et al. (2007)
Nac6	*Oryza sativa*	High salinity, dehydration and cold stresses	Nakashima et al. (2007)

continued

Table 13.1   (continued) List of some stress-inducible promoters used in different stress conditions

Promoter	Origin	Stress	Reference
Osaba2	*Oryza sativa*	ABA, salt and drought stress	Rai et al. (2009)
Osnced3	*Oryza sativa*	Drought and high salinity	Bang et al. (2013)
Pdf1.2	*Arabidopsis*	Pathogen	Manners et al. (1998)
Pdh45	*Pea*	High salinity	Tajrishi et al. (2011)
PR-1a	*Nicotiana tabacum*	Pathogen	Buchel et al. (1996)
Pin2	*Potato*	Wound inducible, JA and ABA	Xu et al. (1993)
Rab16a	*Oryza sativa*	ABA, salt and drought stress	Rai et al. (2009)
Rab17	*Maize*	Water stress or ABA treatment	Vilardell et al. (1991)
Rab21	*Oryza sativa*	Drought	Yi et al. (2010)
Rd22	*Arabidopsis*	ABA and drought stress	Abe et al. (1997)
Rd29a	*Arabidopsis*	ABA, cold, salt and drought stress	Yamaguchi-Shinozaki and Shinozaki (1994)
Rd29b	*Arabidopsis*	ABA, cold, salt and drought stress	Yamaguchi-Shinozaki and Shinozaki (1994)
Sag12	*Arabidopsis*	Elicitor and senescence-specific promoter	Gan and Amasino (1995)
Sag39	*Oryza sativa*	Elicitor and senescence-specific promoter	Liu et al. (2010)
Win3.12	*Nicotiana tabacum*	Wound inducible	Hollick and Gordon (1995)
Wsi18	*Oryza sativa*	Drought	Yi et al. (2010)
Wun1	*Potato*	Wound inducible	Siebertz et al. (1989)

in the regulatory sequences of co-expressed genes (Higo et al., 1999; Lescot et al., 2002). Table 13.2 listed the *cis*-regulatory elements (obtained from PLACE and PlantCARE websites) that are identified in response to the environmental signal or stresses (Higo et al., 1999; Lescot et al., 2002).

13.4   Coordination between transcriptional complex and stress-related promoters

Gene expression is regulated at multiple levels in the cellular and physiological response to environmental stress. Synchronisation of *cis*-regulatory elements and CPEs is very

Table 13.2 *Cis*-regulatory motif present in different stress-regulated plants promoter

ID	Hormone/ stress	Origin species	Sequence	Reference
–141NTG13	Xenobiotic stress	*Nicotiana tabacum*	GCTTTTGATG ACTTCAAACAC	Fromm et al, (1991)
14BPATERD1	Water stress	*Arabidopsis thaliana*	CACTAAATTGTC AC	Simpson et al. (2003)
20NTNTNOS	Auxin; wounding and methyl jasmonate	*Nicotiana tabacum*	TGAGCTAAG CACATACGTCA	Kim et al. (1994)
ABADESI2	ABA	*Oryza sativa*; *Triticum aestivum*	GGACGCGTGGC	Lam and Chua (1991)
ABRE2HVA1	ABA	*Hordeum vulgare*	CCTACGTGG CGG	Straub et al. (1994)
ABREATCONS ENSUS	ABA	*Arabidopsis thaliana*	YACGTGGC	Choi et al. (2000)
ABREATRD22	ABA, dehydration	*Arabidopsis thaliana*	RYACGTGGYR	Iwasaki et al. (1995)
ABREAZMRAB 28	ABA	*Zea mays*	GCCACGTGGG	Busk and Pages (1997)
ABREDISTBBN NAPA	ABA	Brassica napus	GCCACTTGTC	Ezcurra et al. (1999)
ABRELATERD1	ABA Erd	*Arabidopsis thaliana*	ACGTG	Simpson et al. (2003)
ABREOSRGA1	ABA	*Oryza sativa*	CCACGTGG	Seo et al. (1995)
ABRERATCAL	ABE, calcium	*Arabidopsis thaliana*	MACGYGB	Kaplan et al. (2006)
ABRETAEM	ABA	*Triticum aestivum*	GGACACGTGGC	Guiltinan et al. (1990)
ABREZMRAB28	ABR; freezing tolerance	*Zea mays*; *Arabidopsis thaliana Oryza sativa*; *Populus* spp	CCACGTGG	Suzuki et al. (2005)
ACEATCHS	Ace; Uv-A; and Uv-B	*Arabidopsis thaliana*	GACACGTAGA	Hartmann et al. (1998)
ACGTABREMO TIFA2OSEM	ABA; ABRE; motif A and DRE	*Oryza sativa*; *Arabidopsis thaliana*	ACGTGKC	Hattori et al. (2002)

continued

Table 13.2   (continued) *Cis*-regulatory motif present in different stress-regulated plants promoter

ID	Hormone/ stress	Origin species	Sequence	Reference
ACGTROOT1	ABE; cold tolerance	*Nicotiana tabacum*; Glycine max	GCCACGTGGC	Salinas et al. (1992)
AGCBOXNPGLB	Ethylene	*Nicotiana plumbaginifolia*; *Arabidopsis thaliana*; *Nicotiana sylvestris*; *Oryza sativa*	AGCCGCC	Hart et al. (1993)
AS1LIKECSHPRA	Cytokinin	*Cucumis sativus*	AAATGACGAAA ATGC	Jin et al. (1998)
ASF1MOTIFCA MV	Auxin; SA; xenobiotic stress; disease resistance	CaMV; Cauliflower mosaic virus; plant; *Nicotiana tabacum*; *Arabidopsis thaliana*	TGACG	Despres et al. (2003)
ASF1NTPARA	Auxin	*Nicotiana tabacum*	TTACGCAAGC AATGACAT	Sakai et al. (1998)
B2GMAUX28	Auxin	Glycine max	CTTGTCGTCA	Nagao et al. (1993)
CAREOSREP1	GA	*Oryza sativa*	CAACTC	Sutoh and Yamauchi (2003)
CATATGGMSA UR	Auxin	Glycine max	CATATG	
CBFHV	Low temperature	*Hordeum vulgare*	RYCGAC	Xue (2002)
CCAATBOX1	HSE (heat-shock element); CCAAT box	Glycine max	CCAAT	Rieping and Schöffl (1992)
CCTCGTGTCTC GMGH3	Auxin	Glycine max	CCTCGTGTCTC	Ulmasov et al. (1995)
COREOS	Oxidative stress; antioxidant	*Oryza sativa*	AAKAATWYRTA WATAAAAMTT TTATWTA	Tsukamoto et al. (2005)
CPBCSPOR	Cytokinin	*Cucumis sativus*	TATTAG	Fusada et al. (2005)

Table 13.2 (continued) *Cis*-regulatory motif present in different stress-regulated plants promoter

ID	Hormone/stress	Origin species	Sequence	Reference
CURECORECR	Copper; oxygen; and hypoxic	*Chlamydomonas reinhardtii*	GTAC	Quinn et al. (2000)
CYTOSITECSH PRA	Cytokinin	*Cucumis sativus*	AAGATTG ATTGAG	Jin et al. (1998)
DRE1COREZM RAB17	ABA; drought response	*Zea mays*	ACCGAGA	Busk et al. (1997)
DRECRTCORE AT	Drought	*Oryza sativa*; *Zea mays*; *Helianthus annuus* (sunflower)	RCCGAC	Dubouzet et al. (2003)
DREDR1ATRD2 9AB	Drought; water stress; oxidative stress; low temperature; high salt; stresscold; and dehydration	*Arabidopsis thaliana*; *Populus* spp	TACCGACAT	Kasuga et al. (1999)
ELRECOREPCR P1	Elicitor; SA	*Petroselinum crispum* (parsley); *Nicotiana tabacum*	TTGACC	Rushton et al. (1996)
ELRENTCHN50	ELRE; elicitor and chitinase	*Nicotiana tabacum*	GGTCANNNAG TC	Fukuda (1997)
EMBP1TAEM	ABA	*Triticum aestivum*; *Arabidopsis thaliana*	CACGTGGC	Guiltinan et al. (1990)
EREGCC	Ethylene; pathogen	*Nicotiana tabacum*; *Arabidopsis thaliana*	TAAGAGCCGCC	Koyama et al. (2003)
ERELEE4	Ethylene; senescence	*Lycopersicon esculentum*; *Dianthus caryophillus* *Lycopersicon chilense*	AWTTCAAA	Itzhaki et al. (1994)

continued

Table 13.2 (continued) *Cis*-regulatory motif present in different stress-regulated plants promoter

ID	Hormone/ stress	Origin species	Sequence	Reference
GARE1OSREP1	GA	*Oryza sativa*	TAACAGA	Sutoh and Yamauchi (2003)
GAREAT	GA	*Arabidopsis thaliana*	TAACAAR	Ogawa et al. (2003)
GGTCCCATGM SAUR	Auxin	Glycine max	GGTCCCAT	
GREGIONNTPR B1B	Ethylene	*Nicotiana tabacum*	TGGCGGCTCTTA TCTCACGTGA TG	Meller et al. (1993)
HBOXCONSEN SUSPVCHS	Elicitor; stress and wounding	Bean (*Phaseolus vulgaris*); *Nicotiana tabacum*	CCTACCNNNNN NNCT	Loake et al. (1992)
HSELIKENTGL N2	Heat shock	*Nicotiana tabacum*	AGGAATTCCT	Ohme-Takagi and Shinshi (1990)
JASE1ATOPR1	Senescence; JA	*Arabidopsis thaliana*	CGTCAATGAA	He and Gan (2001)
JERECRSTR	Jasmonic acid	Periwinkle (*Catharanthus roseus*)	CTCTTAGACCGC CTTCTTTG AAAG	Menke et al. (1999)
LECPLEACS2	Ethylene	*Lycopersicon esculentum*	TAAAATAT	Matarasso et al. (2005)
LS5ATPR1	Auxin; SA and root	*Arabidopsis thaliana*; *Nicotiana tabacum*	TCTACGTCAC	Despres et al. (2000)
LTRE1HVBLT49	Low temperature	*Hordeum vulgare*	CCGAAA	Dunn et al. (1998)
LTREATLTI78	Low temperature, cold	*Arabidopsis thaliana*; *Hordeum vulgare*	ACCGACA	Nordin et al. (1993)
LTRECOREATC OR15	Low temperature; cold; drought; and ABA	*Arabidopsis thaliana*; *Brassica napus*	CCGAC	Baker et al. (1994)
MEJARELELOX	JA	*Lycopersicon esculentum*	GATACANNAAT NTGATG	Beaudoin and Rothstein (1997)

Table 13.2 (continued) *Cis*-regulatory motif present in different stress-regulated plants promoter

ID	Hormone/ stress	Origin species	Sequence	Reference
MYB2CONSEN SUSAT	MYB; Rd22bp1; ABA; leaf; seed and stress	*Arabidopsis thaliana*	YAACKG	Abe et al. (2003)
MYBATRD22	Dehydration; water stress	*Arabidopsis thaliana*	CTAACCA	Abe et al. (1997)
MYCATERD1	Water stress; Erd	*Arabidopsis thaliana*	CATGTG	Simpson et al. (2003)
MYCATRD22	Dehydration; water stress	*Arabidopsis thaliana*	CACATG	Abe et al. (1997)
MYCCONSENS USAT	ABA; cold	*Arabidopsis thaliana*	CANNTG	Abe et al. (2003)
SARECAMV	SA; SARE; and As-1	CaMV; Cauliflower mosaic virus	CTGACGTAAG GGATGACGCAC	Qin et al. (1994)
T/GBOXATPIN2	Wounding	*Lycopersicon esculentum*; *Arabidopsis thaliana*	AACGTG	Boter et al. (2004)
TATCCAOSAMY	GA; sugar starvation	*Oryza sativa*	TATCCA	Lu et al. (2002)
TCA1MOTIF	SA; stress and TCA-1	*Hordeum vulgare*; *Nicotiana tabacum*	TCATCTTCTT	Goldsbrough et al. (1993)
TGA1ANTPR1A	SA; xenobiotic stress	*Nicotiana tabacum*	CGTCATCGAGAT GACG	Strompen et al. (1998)
WARBNEXTA	Wounding	*Brassica napus*	GTACGTGTTATA AAACGTGT	Elliott and Shirsat (1998)
WBOXATNPR1	Disease resistance; SA and WRKY	*Arabidopsis thaliana*	TTGAC	Yu et al. (2001)
WBOXNTCHN48	W box; WRKY and elicitor	*Nicotiana tabacum*	CTGACY	Yamamoto et al. (2004)
WBOXNTERF3	W box; ERF3 and wounding	*Nicotiana tabacum*	TGACY	Nishiuchi et al. (2004)

continued

Table 13.2   (continued) *Cis*-regulatory motif present in different stress-regulated plants promoter

ID	Hormone/ stress	Origin species	Sequence	Reference
WINPSTPIIIK	Wound; wounding	*Solanum tuberosum* (potato)	AAGCGTAAGT	Palm et al. (1990)
WRECSAA01	Wound; AAO and wounding	*Cucumis sativas*	AAWGTATCSA	Palm et al. (1990)
WRKY71OS	WRKY; GA; MYB; W box; TGAC and PR proteins	*Oryza sativa*; *Petroselinum crispum* (parsley)	TGAC	Zhang et al. (2004)

essential to regulate the transcription of growth- and stress-related genes. Understanding how transcription factors and pre-initiation complex bind to these elements is not only necessary but also the coordination between them in stress gene expression is an indispensable factor. General cofactors are frequently associated in gene stimulation to expedite the communication between gene-specific transcription factors and components of the general transcription machinery (Figure 13.3). These general cofactors include TAFs found in TFIID and mediator. Other than these general cofactors, TBP-containing and TBP-non-containing complexes also influence the gene expression during development and stress condition (Baumann et al., 2010; Juven-Gershon and Kadonaga, 2010).

TBP-associated factors   The TFIID multi-subunit protein complex is thought to be the main component of the pre-initiation complex responsible for promoter recognition and its binding is considered as being the key rate-limiting step in controlling the core promoter activity. The TFIID complex is composed of TBP and 15 TAFs (Albright and Tjian, 2000; Lago et al., 2004). TAFs reveal significant structural and functional conservation between yeast mammals and plants (Lago et al., 2004). Certain TAFs are involved in core promoter recognition. Thus, the same core promoter recognition factor can interact differently with distinct core promoters potentially providing a mechanism of how core promoter architecture can contribute to differential gene activity (Smale and Kadonaga, 2003). Whole-genome analysis of temperature-sensitive mutants of TAFs in budding

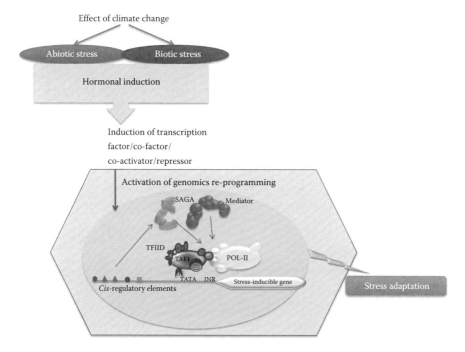

FIGURE 13.3 (**See colour insert.**) A model describing the coordination of different types of transcriptional regulatory complex in plant gene expression in the context of climate change. Various transcription factor/co-factor/co-activator/repressor was induced during stress condition and controls the stress-responsive gene expression. *Cis*-regulatory elements are also involved in stress-responsive transcription shown in different shapes at the proximal promoter.

yeast has revealed that TAFs regulate about 70% of whole-gene expression (Lee et al., 2000). Recently, plant TAFs are becoming the subject of intensive study (Lago et al., 2004, 2005; Bertrand et al., 2005; Furumoto et al., 2005; Benhamed et al., 2006; Gao et al., 2006; Tamada et al., 2007; Kubo et al., 2011; Mougiou et al., 2012), and their molecular and physiological functions in response to stress, development and differentiation are being progressively revealed. Modifications in the levels of *Arabidopsis*-specific TAFs are implicated in plant development affecting the organisation of shoot apical meristems, leaf development, formation of floral organs and leaves, fertility, pollen tube growth and light responses (Gurley et al., 2006). Mutations in TAF1a regulate light-responsive genes by controlling acetylation of histones H3 and H4 at the target promoters (Bertrand et al., 2005; Benhamed et al., 2006). *Arabidopsis* TAF6 plays an important role in the pollen tube function (Lago et al., 2005). *Arabidopsis* TAF10 mutants, with reduced expression, are more sensitive to salt stress, whereas overexpression of TAF10 increased seed germination rate upon osmotic stress (Gao et al., 2006), suggesting that TAF10 is involved in osmotic stress responses. A mutation in *Arabidopsis* TAF12B results in failure to induce a subset of ethylene-regulated genes in etiolated seedlings (Robles et al., 2007). *Arabidopsis* TAF12B protein also regulates a set of genes involved in late signalling processes governing a range of cytokinin responses, including cell proliferation and differentiation (Kubo et al., 2011).

The mediator

Another way by which activators and co-activators may recruit RNA Pol II to a promoter is by interacting with a multi-protein complex, the mediator (Blazek et al., 2005; Kidd et al., 2011; Mathur et al., 2011). The mediator was first discovered in yeast after the finding that purified activators and components of the general transcription machinery were not sufficient for regulated *in vitro* transcription. The activator-dependent transcription also required the addition of crude yeast extracts, and the unknown component mediating the missing function was named the mediator (Kelleher et al., 1990; Flanagan et al., 1991). This complex is required for regulated transcription of nearly all RNA Pol II-dependent genes in *Saccharomyces cerevisiae* and functions as a bridge between regulatory proteins and the basal RNA Pol II transcription machinery, as a regulator of the phosphorylation status of the C-terminal domain (CTD) and possibly as a modulator of the chromatin structure. The mediator complex in plants was found to comprise 27 subunits, 21 of which were conserved between plants and other eukaryotes (Backstrom et al., 2007). In *Arabidopsis*, mediator subunits have essential roles, including plant growth and development. For example, STRUWWELPETER (SWP)/ MED14 controls the cell number during primordia initiation and also regulates the duration of cell proliferation in aerial organs (Autran et al., 2002). The mediator subunits MED12–MED13 have been shown to regulate developmental timing during embryo patterning (Gillmor et al., 2010). Phytochrome and flowering time1 (PFT1)/MED25 is known to play a key role in light signalling and flowering time (Inigo et al., 2012; Klose et al., 2012). MED25 also plays the essential role in JA-mediated pathogen defence (Kidd et al., 2009), drought and salt stress (Elfving et al., 2011) and JA and ABA signalling (Chen et al., 2012).

TBP-containing and TBP-free-containing complexes implicated in transcription

A number of distinct multi-protein complexes containing TBP-associated factors, but lacking TBP, were found in cells from different organisms. The complexes such as TBP-free TAF-containing complex (TFTC), TFTC-related PCAF/GCN5 complexes, Spt-Ada-Gcn5 acetyltransferase (SAGA), SAGA-like complex (SLIK), Spt3-TAF9-GCN5L acetylase (STAGA) and polycomb complex 1 (PRC1) possess multiple and essential functions in the general transcription regulation of RNA POL II genes (Thomas and Chiang, 2006; Baumann et al., 2010). These complexes are implicated in gene activation mainly due to their histone acetyltransferase (HAT) activity and co-activator function of TAFs, whereas PRC1 is implicated in gene silencing (Garrick et al., 2008). Although much information about these complex components in yeast, human and metazoan species is available, there is little information in the case of plants. Yeast and human SAGA complexes are encompassed by 20 sub-units with a total mass of approximately 1.8 MDa (Rodriguez-Navarro, 2009). SAGA complex is involved in 10% RNA Pol II-dependent transcriptional regulation of yeast genes (Bhaumik, 2011). Components of the SAGA complex are also studied in the plant in light, abiotic and biotic stress conditions. SAGA complex regulating gene expression responds to environmental stress conditions (Baker and Grant, 2007). Many reports suggest that SAGA complex is directly or indirectly involved in many developments and stress-induced signalling pathways, for instance, UV induced (Pankotai et al., 2005), high osmotic stress induced (Zapater et al., 2007) and arsenite stress condition (Nagy et al., 2009). Huisinga and Pugh (2004) also reported that the SAGA complex of *S. cerevisiae* is involved in the up-regulation of genes in response to environmental stresses including carbon starvation (Huisinga and Pugh, 2004). *Arabidopsisada2b-1* mutants display more hypersensitivity to salt stress (Hark et al., 2009; Kaldis et al., 2011) and ABA stress than wild-type plants (Hark et al., 2009). Although *sgf29a* mutant shows salt tolerance, the gene expression level of some stress-related genes was reduced in *sgf29a* mutant such as RD29b (responsive to desiccation 29b), RAB18 (responsive to aba 18) and COR78 (cold-regulated 78) (Kaldis et al., 2011). *Arabidopsis* GCN5 and ADA2b proteins show a role in cold acclimation and mutations in these proteins display the reduction of cold-inducible COR gene expression (Stockinger et al., 2001; Vlachonasios et al., 2003; Mao et al., 2006; Hark et al., 2009). Thus, the characteristics of the SAGA complex in regulating stress genes are conserved in the plant like yeasts or humans (Huisinga and Pugh, 2004).

13.5 Future directions and conclusions

The variations in gene expression are likely to play a critical role in both acclimation and adaptation to a changing environment; thus, they influence plant growth and development. Understanding how climate changes will impact plant gene expression and its downstream pathway is of paramount importance for crop productivity and yield. Furthermore, gene expression helps us to understand a few aspects of hormone-regulated downstream events, for instance, transcriptional reprogramming and regulation of the signalling molecule. Molecular and genomic analyses have facilitated gene expression and enabled plant genetic engineering using several functional or regulatory transcription factors or other regulatory genes to activate specific or broad pathways related to stress tolerance in plants. Recent years have marked a period of significant progress in defining both *cis*-acting elements within the regulatory regions of genes. The use of different stress-inducible promoters appeared to improve the plant tolerance by minimising disturbance to development during environmental stress. Significant cross talk and inter-connections are involved between these above factors in stress signalling. Nevertheless, systematic methodologies with molecular and genomic investigates will facilitate the resolution of such complex networks and lead to the finding of other stress factors or mechanisms during the changing environmental scenario. Recent developments have identified several signalling pathway factors related to the stress gene expression response in plants. But, our understanding of the whole-plant climate change gene expression and stress response mechanism is inadequate. Therefore, the challenge to understand the interaction between gene expression and stress response during the changing environmental condition and simultaneously the increase in crop productivity is very essential. So, recognising the chain of molecular events from the changing environmental condition insight by numerous signalling molecules to their physiological function is essential for defining the molecular network of stress-signalling pathway better. Identification of various stress-regulated synthetic and modified promoters as well as *cis*-regulatory elements is also necessary in the context of climate change. To further develop method to gain a better understanding of different signalling components, their functions and how these signals communicate with stress-regulated promoters, imminent experimental studies and assessment must be conducted. Once the mechanism of gene expression is understood better in the context of climate change, new prospects for agricultural biotechnology may become evident for developing stress-tolerant plants.

References

ABE, H., URAO, T., ITO, T., SEKI, M., SHINOZAKI, K. and YAMAGUCHI-SHINOZAKI, K. 2003. *Arabidopsis* AtMYC2 (bHLH) and AtMYB2 (MYB) function as transcriptional activators in abscisic acid signaling. *Plant Cell* 15, 63–78.

ABE, H., YAMAGUCHI-SHINOZAKI, K., URAO, T., IWASAKI, T., HOSOKAWA, D. and SHINOZAKI, K. 1997. Role of *Arabidopsis* MYC and MYB homologs in drought- and abscisic acid-regulated gene expression. *Plant Cell* 9, 1859–1868.

ABELES, F.B., MORGAN, P.W. and SALTVEIT, J.M. 1992. *Ethylene in Plant Biology*. San Diego: Academic Press.

ACHARD, P., CHENG, H., DE GRAUWE, L., DECAT, J., SCHOUTTETEN, H., MORITZ, T., VAN DER STRAETEN, D., PENG, J. and HARBERD, N.P. 2006. Integration of plant responses to environmentally activated phytohormonal signals. *Science* 311, 91–94.

ALBRIGHT, S.R. and TJIAN, R. 2000. TAFs revisited: More data reveal new twists and confirm old ideas. *Gene* 242, 1–13.

ALONSO-RAMIREZ, A., RODRIGUEZ, D., REYES, D., JIMENEZ, J.A., NICOLAS, G., LOPEZ-CLIMENT, M., GOMEZ-CADENAS, A. and NICOLAS, C. 2009. Evidence for a role of gibberellins in salicylic acid-modulated early plant responses to abiotic stress in *Arabidopsis* seeds. *Plant Physiol* 150, 1335–1344.

AN, C. and MOU, Z. 2011. Salicylic acid and its function in plant immunity. *J Integr Plant Biol* 53, 412–428.

ARGUESO, C.T., FERREIRA, F.J. and KIEBER, J.J. 2009. Environmental perception avenues: The interaction of cytokinin and environmental response pathways. *Plant Cell Environ* 32, 1147–1160.

AUTRAN, D., JONAK, C., BELCRAM, K., BEEMSTER, G.T., KRONENBERGER, J., GRANDJEAN, O., INZE, D. and TRAAS, J. 2002. Cell numbers and leaf development in *Arabidopsis*: A functional analysis of the STRUWWELPETER gene. *EMBO J* 21, 6036–6049.

AVSIAN-KRETCHMER, O., GUETA-DAHAN, Y., LEV-YADUN, S., GOLLOP, R. and BEN-HAYYIM, G. 2004. The salt-stress signal transduction pathway that activates the gpx1 promoter is mediated by intracellular H_2O_2, different from the pathway induced by extracellular H_2O_2. *Plant Physiol* 135, 1685–1696.

BACKSTROM, S., ELFVING, N., NILSSON, R., WINGSLE, G. and BJORKLUND, S. 2007. Purification of a plant mediator from *Arabidopsis thaliana* identifies PFT1 as the Med25 sub-unit. *Mol Cell* 26, 717–729.

BAENA-GONZALEZ, E. and SHEEN, J. 2008. Convergent energy and stress signaling. *Trends Plant Sci* 13, 474–482.

BAJGUZ, A. and HAYAT, S. 2009. Effects of brassinosteroids on the plant responses to environmental stresses. *Plant Physiol Biochem* 47, 1–8.

BAKER, S.P. and GRANT, P.A. 2007. The SAGA continues: Expanding the cellular role of a transcriptional co-activator complex. *Oncogene* 26, 5329–5340.

BAKER, S.S., WILHELM, K.S. and THOMASHOW, M.F. 1994. The 5′-region of *Arabidopsis thaliana* cor15a has *cis*-acting elements that confer cold-, drought- and ABA-regulated gene expression. *Plant Mol Biol* 24, 701–713.

BALBI, V. and DEVOTO, A. 2008. Jasmonate signalling network in *Arabidopsis thaliana*: Crucial regulatory nodes and new physiological scenarios. *New Phytol* 177, 301–318.

BANG, S.W., PARK, S.H., JEONG, J.S., KIM, Y.S., JUNG, H., HA, S.H. and KIM, J.K. 2013. Characterization of the stress-inducible OsNCED3 promoter in different transgenic rice organs and over three homozygous generations. *Planta* 237, 211–224.

BARTELS, S. ANDERSON, J.C., GONZALEZ BESTEIRO, M.A., CARRERI, A., HIRT, H., BUCHALA, A., METRAUX, J.P., PECK, S.C. and ULM, R. 2009. MAP kinase phosphatase1 and protein tyrosine phosphatase1 are repressors of salicylic acid synthesis and SNC1-mediated responses in *Arabidopsis*. *Plant Cell* 21, 2884–2897.

BAUMANN, M., PONTILLER, J. and ERNST, W. 2010. Structure and basal transcription complex of RNA polymerase II core promoters in the mammalian genome: An overview. *Mol Biotechnol* 45, 241–247.

BEAUDOIN, N. and ROTHSTEIN, S.J. 1997. Developmental regulation of two tomato lipoxygenase promoters in transgenic tobacco and tomato. *Plant Mol Biol* 33, 835–846.

BECANA, M., MATAMOROS, M.A., UDVARDI, M. and DALTON, D.A. 2010. Recent insights into antioxidant defenses of legume root nodules. *New Phytol* 188, 960–976.

BECK, J. 2013. Predicting climate change effects on agriculture from ecological niche modeling: Who profits, who loses? *Clim Change* 116, 177–189.

BELIGNI, M.A.V.N. and LAMATTINA, L. 1999. Nitric oxide counteracts cytotoxic processes mediated by reactive oxygen species in plant tissues. *Planta* 208, 337–344.

BELIGNI, M.V. and LAMATTINA, L. 2000. Nitric oxide stimulates seed germination and de-etiolation, and inhibits hypocotyl elongation, three light-inducible responses in plants. *Planta* 210, 215–221.

BENHAMED, M., BERTRAND, C., SERVET, C. and ZHOU, D.X. 2006. *Arabidopsis* GCN5, HD1 and TAF1/HAF2 interact to regulate histone acetylation required for light-responsive gene expression. *Plant Cell* 18, 2893–2903.

BERTRAND, C., BENHAMED, M., LI, Y.F., AYADI, M., LEMONNIER, G., RENOU, J.P., DELARUE, M. and ZHOU, D.X. 2005. *Arabidopsis* HAF2 gene encoding TATA-binding protein (TBP)-associated factor TAF1, is required to integrate light signals to regulate gene expression and growth. *J Biol Chem* 280, 1465–1473.

BHAUMIK, S.R. 2011. Distinct regulatory mechanisms of eukaryotic transcriptional activation by SAGA and TFIID. *Biochim Biophys Acta* 1809, 97–108.

BLAZEK, E., MITTLER, G. and MEISTERERNST, M. 2005. The mediator of RNA polymerase II. *Chromosoma* 113, 399–408.

BORSANI, O., VALPUESTA, V. and BOTELLA, M.A. 2001. Evidence for a role of salicylic acid in the oxidative damage generated by NaCl and osmotic stress in *Arabidopsis* seedlings. *Plant Physiol* 126, 1024–1030.

BOTER, M., RUIZ-RIVERO, O., ABDEEN, A. and PRAT, S. 2004. Conserved MYC transcription factors play a key role in jasmonate signaling both in tomato and *Arabidopsis*. *Genes Dev* 18, 1577–1591.

BRAY, E.A., BAILEY-SERRES, J. and WERETILNYK, E. 2000. Responses to abiotic stress. *In Biochemistry and Molecular Biology of Plants*. G.W. Buchanan BB, Jones RL, eds., Rockville: American Society of Plant Physiologists, pp. 1158–1203.

BRUCE, W.B., EDMEADES, G.O. and BARKER, T.C. 2002. Molecular and physiological approaches to maize improvement for drought tolerance. *J Exp Bot* 53, 13–25.

BUCHEL, A.S., MOLENKAMP, R., BOL, J.F. and LINTHORST, H.J. 1996. The PR-1a promoter contains a number of elements that bind GT-1-like nuclear factors with different affinity. *Plant Mol Biol* 30, 493–504.

BUSK, P.K., JENSEN, A.B. and PAGES, M. 1997. Regulatory elements *in vivo* in the promoter of the abscisic acid responsive gene rab17 from maize. *Plant J* 11, 1285–1295.

BUSK, P.K. and PAGES, M. 1997. Protein binding to the abscisic acid-responsive element is independent of VIVIPAROUS1 *in vivo*. *Plant Cell* 9, 2261–2270.

CABELLO, J.V. and CHAN, R.L. 2012. The homologous homeodomain-leucine zipper transcription factors HaHB1 and AtHB13 confer tolerance to drought and salinity stresses via the induction of proteins that stabilize membranes. *Plant Biotechnol J* 10, 815–825.

CAO, H., LI, X. and DONG, X. 1998. Generation of broad-spectrum disease resistance by overexpression of an essential regulatory gene in systemic acquired resistance. *Proc Natl Acad Sci USA* 95, 6531–6536.

CAO, H., BOWLING, S.A., GORDON, A.S. and DONG, X. 1994. Characterization of an *Arabidopsis* mutant that is nonresponsive to inducers of systemic acquired resistance. *Plant Cell* 6, 1583–1592.

CHANG, C.-C., CHEN, C.-C. and McCARL, B. 2012. Evaluating the economic impacts of crop yield change and sea level rise induced by climate change on Taiwan's agricultural sector. *Agric Econ* 43, 205–214.

CHANG, C., KWOK, S.F., BLEECKER, A.B. and MEYEROWITZ, E.M. 1993. *Arabidopsis* ethylene-response gene ETR1: Similarity of product to two-component regulators. *Science* 262, 539–544.

CHAVES, M.M., FLEXAS, J. and PINHEIRO, C. 2009. Photosynthesis under drought and salt stress: Regulation mechanisms from whole plant to cell. *Ann Bot* 103, 551–560.

CHEN, R., JIANG, H., LI, L., ZHAI, Q., QI, L., ZHOU, W., LIU, X. et al. 2012. The *Arabidopsis* mediator subunit MED25 differentially regulates jasmonate and abscisic acid signaling through interacting with the MYC2 and ABI5 transcription factors. *Plant Cell* 24, 2898–2916.

CHEN, W. and SINGH, K.B. 1999. The auxin, hydrogen peroxide and salicylic acid induced expression of the *Arabidopsis* GST6 promoter is mediated in part by an ocs element. *Plant J* 19, 667–677.

CHERN, M.S., FITZGERALD, H.A., YADAV, R.C., CANLAS, P.E., DONG, X. and RONALD, P.C. 2001. Evidence for a disease-resistance pathway in rice similar to the NPR1-mediated signaling pathway in *Arabidopsis*. *Plant J* 27, 101–113.

CHEW, Y.H. and HALLIDAY, K.J. 2011. A stress-free walk from *Arabidopsis* to crops. *Curr Opin Biotechnol* 22, 281–286.

CHINI, A., GRANT, J.J., SEKI, M., SHINOZAKI, K. and LOAKE, G.J. 2004. Drought tolerance established by enhanced expression of the CC-NBS-LRR gene, ADR1, requires salicylic acid, EDS1 and ABI1. *Plant J* 38, 810–822.

CHOI, H., HONG, J., HA, J., KANG, J. and KIM, S.Y. 2000. ABFs, a family of ABA-responsive element binding factors. *J Biol Chem* 275, 1723–1730.

CHOI, J., HUH, S.U., KOJIMA, M., SAKAKIBARA, H., PAEK, K.H. and HWANG, I. 2010. The cytokinin-activated transcription factor ARR2 promotes plant immunity via TGA3/NPR1-dependent salicylic acid signaling in *Arabidopsis*. *Dev Cell* 19, 284–295.

CHOUDHARY, S.P., KANWAR, M., BHARDWAJ, R., YU, J.Q. and TRAN, L.S. 2012. Chromium stress mitigation by polyamine- brassinosteroid application involves phytohormonal and physiological strategies in *Raphanus sativus* L. *PLoS One* 7, e33210.

CLARKE, J.D., VOLKO, S.M., LEDFORD, H., AUSUBEL, F.M. and DONG, X. 2000. Roles of salicylic acid, jasmonic acid, and ethylene in cpr-induced resistance in *Arabidopsis*. *Plant Cell* 12, 2175–2190.

CLOUSE, S.D. and SASSE, J.M. 1998. Brassinosteroids: Essential regulators of plant growth and development. *Annu Rev Plant Physiol Plant Mol Biol* 49, 427–451.

COMPANT, S., VAN DER HEIJDEN, M.G. and SESSITSCH, A. 2010. Climate change effects on beneficial plant–microorganism interactions. *FEMS Microbiol Ecol* 73, 197–214.

CRAMER, G.R., URANO, K., DELROT, S., PEZZOTTI, M. and SHINOZAKI, K. 2011. Effects of abiotic stress on plants: A systems biology perspective. *BMC Plant Biol* 11, 163.

DESCH NES, O. and GREENSTONE, M. 2007. The economic impacts of climate change: Evidence from agricultural output and random fluctuations in weather. *Am Econ Rev* 97, 354–385.

DESPRES, C., DELONG, C., GLAZE, S., LIU, E. and FOBERT, P.R. 2000. The *Arabidopsis* NPR1/NIM1 protein enhances the DNA binding activity of a subgroup of the TGA family of bZIP transcription factors. *Plant Cell* 12, 279–290.

DESPRES, C., CHUBAK, C., ROCHON, A., CLARK, R., BETHUNE, T., DESVEAUX, D. and FOBERT, P.R. 2003. The *Arabidopsis* NPR1 disease resistance protein is a novel cofactor that confers redox regulation of DNA binding activity to the basic domain/leucine zipper transcription factor TGA1. *Plant Cell* 15, 2181–2191.

DEVADAS, S.K., ENYEDI, A. and RAINA, R. 2002. The *Arabidopsis* hrl1 mutation reveals novel overlapping roles for salicylic acid, jasmonic acid and ethylene signalling in cell death and defence against pathogens. *Plant J* 30, 467–480.

DEZAR, C.A., GAGO, G.M., GONZALEZ, D.H. and CHAN, R.L. 2005. Hahb-4, a sunflower homeobox-leucine zipper gene, is a developmental regulator and confers drought tolerance to *Arabidopsis thaliana* plants. *Transgenic Res* 14, 429–440.

DHAUBHADEL, S., BROWNING, K.S., GALLIE, D.R. and KRISHNA, P. 2002. Brassinosteroid functions to protect the translational machinery and heat-shock protein synthesis following thermal stress. *Plant J* 29, 681–691.

DINAKAR, C., DJILIANOV, D. and BARTELS, D. 2012. Photosynthesis in desiccation tolerant plants: Energy metabolism and antioxidative stress defense. *Plant Sci* 182, 29–41.

DIVI, U.K., RAHMAN, T. and KRISHNA, P. 2010. Brassinosteroid-mediated stress tolerance in *Arabidopsis* shows interactions with abscisic acid, ethylene and salicylic acid pathways. *BMC Plant Biol* 10, 151.

DOLFERUS, R., JI, X. and RICHARDS, R.A. 2011. Abiotic stress and control of grain number in cereals. *Plant Sci* 181, 331–341.

DUBOUZET, J.G., SAKUMA, Y., ITO, Y., KASUGA, M., DUBOUZET, E.G., MIURA, S., SEKI, M., SHINOZAKI, K. and YAMAGUCHI-SHINOZAKI, K. 2003. OsDREB genes in rice, *Oryza sativa* L., encode transcription activators that function in drought-, high-salt- and cold-responsive gene expression. *Plant J* 33, 751–763.

DUNN, M.A., WHITE, A.J., VURAL, S. and HUGHES, M.A. 1998. Identification of promoter elements in a low-temperature-responsive gene (blt4.9) from barley (*Hordeum vulgare* L.). *Plant Mol Biol* 38, 551–564.

ELFVING, N., DAVOINE, C., BENLLOCH, R., BLOMBERG, J., BRANNSTROM, K., MULLER, D., NILSSON, A. et al. 2011. The *Arabidopsis thaliana* Med25 mediator subunit integrates environmental cues to control plant development. *Proc Natl Acad Sci USA* 108, 8245–8250.

ELLIOTT, K.A. and SHIRSAT, A.H. 1998. Promoter regions of the extA extensin gene from *Brassica napus* control activation in response to wounding and tensile stress. *Plant Mol Biol* 37, 675–687.

ELSGAARD, L., BORGESEN, C.D., OLESEN, J.E., SIEBERT, S., EWERT, F., PELTONEN-SAINIO, P., ROTTER, R.P. and SKJELVAG, A.O. 2012. Shifts in comparative advantages for maize, oat and wheat cropping under climate change in Europe. *Food Addit Contam Part A: Chem Anal Control Expo Risk Assess* 29(10), 1514–1526.

ETHERIDGE, N., CHEN, Y.F. and SCHALLER, G.E. 2005. Dissecting the ethylene pathway of *Arabidopsis*. *Brief Funct Genomic Proteomic* 3, 372–381.

EZCURRA, I., ELLERSTROM, M., WYCLIFFE, P., STALBERG, K. and RASK, L. 1999. Interaction between composite elements in the napA promoter: Both the B-box ABA-responsive complex and the RY/G complex are necessary for seed-specific expression. *Plant Mol Biol* 40, 699–709.

FLANAGAN, P.M., KELLEHER, R.J., 3RD, SAYRE, M.H., TSCHOCHNER, H. and KORNBERG, R.D. 1991. A mediator required for activation of RNA polymerase II transcription *in vitro*. *Nature* 350, 436–438.

FLEXAS, J., BOTA, J., LORETO, F., CORNIC, G. and SHARKEY, T.D. 2004. Diffusive and metabolic limitations to photosynthesis under drought and salinity in C(3) plants. *Plant Biol (Stuttg)* 6, 269–279.

FREEMAN, J.L., GARCIA, D., KIM, D., HOPF, A. and SALT, D.E. 2005. Constitutively elevated salicylic acid signals glutathione-mediated nickel tolerance in Thlaspi nickel hyperaccumulators. *Plant Physiol* 137, 1082–1091.

FROMM, H., KATAGIRI, F. and CHUA, N.H. 1991. The tobacco transcription activator TGA1a binds to a sequence in the 5′ upstream region of a gene encoding a TGA1a-related protein. *Mol Gen Genet* 229, 181–188.

FUKUDA, Y. 1997. Interaction of tobacco nuclear protein with an elicitor-responsive element in the promoter of a basic class I chitinase gene. *Plant Mol Biol* 34, 81–87.

FURUMOTO, T., TAMADA, Y., IZUMIDA, A., NAKATANI, H., HATA, S. and IZUI, K. 2005. Abundant expression in vascular tissue of plant TAF10, an orthologous gene for TATA box-binding protein-associated factor 10, in *Flaveria trinervia* and abnormal morphology of *Arabidopsis thaliana* transformants on its overexpression. *Plant Cell Physiol* 46, 108–117.

FUSADA, N., MASUDA, T., KURODA, H., SHIMADA, H., OHTA, H. and TAKAMIYA, K. 2005. Identification of a novel *cis*-element exhibiting cytokinin-dependent protein binding *in vitro* in the 5′-region of NADPH-protochlorophyllide oxidoreductase gene in cucumber. *Plant Mol Biol* 59, 631–645.

GAN, S. and AMASINO, R.M. 1995. Inhibition of leaf senescence by autoregulated production of cytokinin. *Science* 270, 1986–1988.

GAO, X., REN, F. and LU, Y.T. 2006. The *Arabidopsis* mutant stg1 identifies a function for TBP-associated factor 10 in plant osmotic stress adaptation. *Plant Cell Physiol* 47, 1285–1294.

GAO, X.H., XIAO, S.L., YAO, Q.F., WANG, Y.J. and FU, X.D. 2011. An updated GA signaling relief of repression regulatory model. *Mol Plant* 4, 601–606.

GARRETT, K.A., DENDY, S.P., FRANK, E.E., ROUSE, M.N. and TRAVERS, S.E. 2006. Climate change effects on plant disease: Genomes to ecosystems. *Annu Rev Phytopathol* 44, 489–509.

GARRICK, D., DE GOBBI, M., SAMARA, V., RUGLESS, M., HOLLAND, M., AYYUB, H., LOWER, K. et al. 2008. The role of the polycomb complex in silencing alpha-globin gene expression in nonerythroid cells. *Blood* 112, 3889–3899.

GECHEV, T.S., VAN BREUSEGEM, F., STONE, J.M., DENEV, I. and LALOI, C. 2006. Reactive oxygen species as signals that modulate plant stress responses and programmed cell death. *Bioessays* 28, 1091–1101.

GILLMOR, C.S., PARK, M.Y., SMITH, M.R., PEPITONE, R., KERSTETTER, R.A. and POETHIG, R.S. 2010. The MED12–MED13 module of mediator regulates the timing of embryo patterning in *Arabidopsis*. *Development* 137, 113–122.

GODA, H., SAWA, S., ASAMI, T., FUJIOKA, S., SHIMADA, Y. and YOSHIDA, S. 2004. Comprehensive comparison of auxin-regulated and brassinosteroid-regulated genes in *Arabidopsis*. *Plant Physiol* 134, 1555–1573.

GOLDSBROUGH, A.P., ALBRECHT, H. and STRATFORD, R. 1993. Salicylic acid-inducible binding of a tobacco nuclear protein to a 10 bp sequence which is highly conserved among stress-inducible genes. *Plant J* 3, 563–571.

GRENE, R. 2002. Oxidative stress and acclimation mechanisms in plants. *Arabidopsis Book* 1, e0036.

GRUN, S., LINDERMAYR, C., SELL, S. and DURNER, J. 2006. Nitric oxide and gene regulation in plants. *J Exp Bot* 57, 507–516.

GUILTINAN, M.J., MARCOTTE, W.R., JR. and QUATRANO, R.S. 1990. A plant leucine zipper protein that recognizes an abscisic acid response element. *Science* 250, 267–271.

GUO, J., YANG, X., WESTON, D.J. and CHEN, J.G. 2011. Abscisic acid receptors: Past, present and future. *J Integr Plant Biol* 53, 469–479.

GURLEY, W.B., O'GRADY, K., CZARNECKA-VERNER, E. and LAWIT, S.J. 2006. General transcription factors and the core promoter: Ancient roots. *In* Regulation of transcription in plants. Grasser KD (ed). *Annu Plant Rev* 29:1–21, Blackwell, Oxford.

HA, S., VANKOVA, R., YAMAGUCHI-SHINOZAKI, K., SHINOZAKI, K. and TRAN, L.S. 2012. Cytokinins:

Metabolism and function in plant adaptation to environmental stresses. *Trends Plant Sci* 17, 172–179.

HANNA, E.G., BELL, E., KING, D. and WOODRUFF, R. 2011. Climate change and Australian agriculture: A review of the threats facing rural communities and the health policy landscape. *Asia Pac J Public Health* 23, 105S–118S.

HAO, X.Y., HAN, X., JU, H. and LIN, E.D. 2010. Impact of climatic change on soybean production: A review. *Ying Yong Sheng Tai Xue Bao* 21, 2697–2706.

HARALAMPIDIS, K., MILIONI, D., RIGAS, S. and HATZOPOULOS, P. 2002. Combinatorial interaction of *cis* elements specifies the expression of the *Arabidopsis* AtHsp90-1 gene. *Plant Physiol* 129, 1138–1149.

HARK, A.T., VLACHONASIOS, K.E., PAVANGADKAR, K.A., RAO, S., GORDON, H., ADAMAKIS, I.D., KALDIS, A., THOMASHOW, M.F. and TRIEZENBERG, S.J. 2009. Two *Arabidopsis* orthologs of the transcriptional coactivator ADA2 have distinct biological functions. *Biochim Biophys Acta* 1789, 117–124.

HART, C.M., NAGY, F. and MEINS, F., JR. 1993. A 61 bp enhancer element of the tobacco beta-1,3-glucanase B gene interacts with one or more regulated nuclear proteins. *Plant Mol Biol* 21, 121–131.

HARTMANN, U., VALENTINE, W.J., CHRISTIE, J.M., HAYS, J., JENKINS, G.I. and WEISSHAAR, B. 1998. Identification of UV/blue light-response elements in the *Arabidopsis thaliana* chalcone synthase promoter using a homologous protoplast transient expression system. *Plant Mol Biol* 36, 741–754.

HATTORI, T., TOTSUKA, M., HOBO, T., KAGAYA, Y. and YAMAMOTO-TOYODA, A. 2002. Experimentally determined sequence requirement of ACGT-containing abscisic acid response element. *Plant Cell Physiol* 43, 136–140.

HE, H., ZHAN, J., HE, L. and GU, M. 2012. Nitric oxide signaling in aluminum stress in plants. *Protoplasma* 249, 483–492.

HE, X.J., MU, R.L., CAO, W.H., ZHANG, Z.G., ZHANG, J.S. and CHEN, S.Y. 2005. AtNAC2, a transcription factor downstream of ethylene and auxin signaling pathways, is involved in salt stress response and lateral root development. *Plant J* 44, 903–916.

HE, Y. and GAN, S. 2001. Identical promoter elements are involved in regulation of the OPR1 gene by senescence and jasmonic acid in *Arabidopsis*. *Plant Mol Biol* 47, 595–605.

HIGO, K., UGAWA, Y., IWAMOTO, M. and KORENAGA, T. 1999. Plant *cis*-acting regulatory DNA elements (PLACE) database: 1999. *Nucleic Acids Res* 27, 297–300.

HOFFMAN, T., SCHMIDT, J.S., ZHENG, X. and BENT, A.F. 1999. Isolation of ethylene-insensitive soybean mutants that are altered in pathogen susceptibility and gene-for-gene disease resistance. *Plant Physiol* 119, 935–950.

HOLLICK, J.B. and GORDON, M.P. 1995. Transgenic analysis of a hybrid poplar wound-inducible promoter reveals developmental patterns of expression similar to that of storage protein genes. *Plant Physiol* 109, 73–85.

HUA, J., SAKAI, H., NOURIZADEH, S., CHEN, Q.G., BLEECKER, A.B., ECKER, J.R. and MEYEROWITZ, E.M. 1998. EIN4 and ERS2 are members of the putative ethylene receptor gene family in *Arabidopsis*. *Plant Cell* 10, 1321–1332.

HUFFAKER, A., PEARCE, G. and RYAN, C.A. 2006. An endogenous peptide signal in *Arabidopsis* activates components of the innate immune response. *Proc Natl Acad Sci USA* 103, 10098–10103.

HUISINGA, K.L. and PUGH, B.F. 2004. A genome-wide housekeeping role for TFIID and a highly regulated stress-related role for SAGA in *Saccharomyces cerevisiae*. *Mol Cell* 13, 573–585.

INIGO, S., ALVAREZ, M.J., STRASSER, B., CALIFANO, A. and CERDAN, P.D. 2012. PFT1, the MED25 subunit of the plant mediator complex, promotes flowering through CONSTANS dependent and independent mechanisms in *Arabidopsis*. *Plant J* 69, 601–612.

ITZHAKI, H., MAXSON, J.M. and WOODSON, W.R. 1994. An ethylene-responsive enhancer element is involved in the senescence-related expression of the carnation glutathione-S-transferase (GST1) gene. *Proc Natl Acad Sci USA* 91, 8925–8929.

IWASAKI, T., YAMAGUCHI-SHINOZAKI, K. and SHINOZAKI, K. 1995. Identification of a *cis*-regulatory region of a gene in *Arabidopsis thaliana* whose induction by dehydration is mediated by abscisic acid and requires protein synthesis. *Mol Gen Genet* 247, 391–398.

JEON, J., KIM, N.Y., KIM, S., KANG, N.Y., NOVAK, O., KU, S.J., CHO, C. et al. 2010. A subset of cytokinin two-component signaling system plays a role in cold temperature stress response in *Arabidopsis*. *J Biol Chem* 285, 23371–23386.

JIN, G., DAVEY, M.C., ERTL, J.R., CHEN, R., YU, Z.T., DANIEL, S.G., BECKER, W.M. and CHEN, C.M. 1998. Interaction of DNA-binding proteins with the 5'-flanking region of a cytokinin-responsive cucumber hydroxypyruvate reductase gene. *Plant Mol Biol* 38, 713–723.

JOHNSON, P.R. and ECKER, J.R. 1998. The ethylene gas signal transduction pathway: A molecular perspective. *Annu Rev Genet* 32, 227–254.

JUVEN-GERSHON, T. and KADONAGA, J.T. 2010. Regulation of gene expression via the core promoter and the basal transcriptional machinery. *Dev Biol* 339, 225–229.

KADONAGA, J.T. 2002. The DPE, a core promoter element for transcription by RNA polymerase II. *Exp Mol Med* 34, 259–264.

KALDIS, A., TSEMENTZI, D., TANRIVERDI, O. and VLACHONASIOS, K.E. 2011. *Arabidopsis thaliana* transcriptional co-activators ADA2b and SGF29a are implicated in salt stress responses. *Planta* 233, 749–762.

KANG, H.M. and SALTVEIT, M.E. 2002. Chilling tolerance of maize, cucumber and rice seedling leaves and roots are differentially affected by salicylic acid. *Physiol Plant* 115, 571–576.

KAPLAN, B., DAVYDOV, O., KNIGHT, H., GALON, Y., KNIGHT, M.R., FLUHR, R. and FROMM, H. 2006. Rapid transcriptome changes induced by cytosolic Ca^{2+} transients reveal ABRE-related sequences as Ca^{2+}-responsive *cis* elements in *Arabidopsis*. *Plant Cell* 18, 2733–2748.

KAPULNIK, Y., YALPANI, N. and RASKIN, I. 1992. Salicylic acid induces cyanide-resistant respiration in tobacco cell-suspension cultures. *Plant Physiol* 100, 1921–1926.

KASUGA, M., LIU, Q., MIURA, S., YAMAGUCHI-SHINOZAKI, K. and SHINOZAKI, K. 1999. Improving plant drought, salt and freezing tolerance by gene transfer of a single stress-inducible transcription factor. *Nat Biotechnol* 17, 287–291.

KELLEHER, R.J., 3RD, FLANAGAN, P.M. and KORNBERG, R.D. 1990. A novel mediator between activator proteins and the RNA polymerase II transcription apparatus. *Cell* 61, 1209–1215.

KIBA, T. and MIZUNO, T. 2003. Hormonal regulation of plant metabolism: Cytokinin action and signal transduction. *Tanpakushitsu Kakusan Koso* 48, 2029–2036.

KIDD, B.N., CAHILL, D.M., MANNERS, J.M., SCHENK, P.M. and KAZAN, K. 2011. Diverse roles of the mediator complex in plants. *Semin Cell Dev Biol* 22, 741–748.

KIDD, B.N., EDGAR, C.I., KUMAR, K.K., AITKEN, E.A., SCHENK, P.M., MANNERS, J.M. and KAZAN, K. 2009. The mediator complex subunit PFT1 is a key regulator of jasmonate-dependent defense in *Arabidopsis*. *Plant Cell* 21, 2237–2252.

KIEBER, J.J., ROTHENBERG, M., ROMAN, G., FELDMANN, K.A. and ECKER, J.R. 1993. CTR1, a negative regulator of the ethylene response pathway in *Arabidopsis*, encodes a member of the raf family of protein kinases. *Cell* 72, 427–441.

KIM, J.C., LEE, S.H., CHEONG, Y.H., YOO, C.M., LEE, S.I., CHUN, H.J., YUN, D.J. et al. 2001. A novel cold-inducible zinc finger protein from soybean, SCOF-1, enhances cold tolerance in transgenic plants. *Plant J* 25, 247–259.

KIM, Y., BUCKLEY, K., COSTA, M.A. and AN, G. 1994. A 20 nucleotide upstream element is essential for the nopaline synthase (nos) promoter activity. *Plant Mol Biol* 24, 105–117.

KINOSHITA, T., YAMADA, K., HIRAIWA, N., KONDO, M., NISHIMURA, M. and HARA-NISHIMURA, I. 1999. Vacuolar processing enzyme is up-regulated in the lytic vacuoles of vegetative tissues during senescence and under various stressed conditions. *Plant J* 19, 43–53.

KLOSE, C., BUCHE, C., FERNANDEZ, A.P., SCHAFER, E., ZWICK, E. and KRETSCH, T. 2012. The mediator complex subunit PFT1 interferes with COP1 and HY5 in the regulation of *Arabidopsis* light signaling. *Plant Physiol* 160, 289–307.

KOH, S., LEE, S.C., KIM, M.K., KOH, J.H., LEE, S., AN, G., CHOE, S. and KIM, S.R. 2007. T-DNA tagged knockout mutation of rice OsGSK1, an orthologue of *Arabidopsis* BIN2, with enhanced tolerance to various abiotic stresses. *Plant Mol Biol* 65, 453–466.

KOOSHKI, M., MENTEWAB, A. and STEWART JR, C.N. 2003. Pathogen inducible reporting in transgenic tobacco using a GFP construct. *Plant Science* 165, 213–219.

KOYAMA, T., OKADA, T., KITAJIMA, S., OHME-TAKAGI, M., SHINSHI, H. and SATO, F. 2003. Isolation of tobacco ubiquitin-conjugating enzyme cDNA in a yeast two-hybrid system with tobacco ERF3 as bait and its characterization of specific interaction. *J Exp Bot* 54, 1175–1181.

KRASENSKY, J. and JONAK, C. 2012. Drought, salt, and temperature stress-induced metabolic rearrangements and regulatory networks. *J Exp Bot* 63, 1593–1608.

KUBO, M., FURUTA, K., DEMURA, T., FUKUDA, H., LIU, Y.G., SHIBATA, D. and KAKIMOTO, T. 2011. The CKH1/EER4 gene encoding a TAF12-like protein negatively regulates cytokinin sensitivity in *Arabidopsis thaliana*. *Plant Cell Physiol* 52, 629–637.

LAGO, C., CLERICI, E., DRENI, L., HORLOW, C., CAPORALI, E., COLOMBO, L. and KATER, M.M. 2005. The *Arabidopsis* TFIID factor AtTAF6 controls pollen tube growth. *Dev Biol* 285, 91–100.

LAGO, C., CLERICI, E., MIZZI, L., COLOMBO, L. and KATER, M.M. 2004. TBP-associated factors in *Arabidopsis*. *Gene* 342, 231–241.

LAM, E. and CHUA, N.H. 1991. Tetramer of a 21-base pair synthetic element confers seed expression and transcriptional enhancement in response to water stress and abscisic acid. *J Biol Chem* 266, 17131–17135.

LARKINDALE, J., HALL, J.D., KNIGHT, M.R. and VIERLING, E. 2005. Heat stress phenotypes of *Arabidopsis* mutants implicate multiple signaling pathways in the acquisition of thermotolerance. *Plant Physiol* 138, 882–897.

LEE, J.H. and SCHOFFL, F. 1996. An Hsp70 antisense gene affects the expression of HSP70/HSC70, the regulation of HSF, and the acquisition of thermotolerance in transgenic *Arabidopsis thaliana*. *Mol Gen Genet* 252, 11–19.

LEE, T.I., CAUSTON, H.C., HOLSTEGE, F.C., SHEN, W.C., HANNETT, N., JENNINGS, E.G., WINSTON, F., GREEN, M.R. and YOUNG, R.A. 2000. Redundant roles for the TFIID and SAGA complexes in global transcription. *Nature* 405, 701–704.

LESCOT, M., DEHAIS, P., THIJS, G., MARCHAL, K., MOREAU, Y., VAN DE PEER, Y., ROUZE, P. and ROMBAUTS, S. 2002. PlantCARE, a database of plant *cis*-acting regulatory elements and a portal to tools for *in silico* analysis of promoter sequences. *Nucleic Acids Res* 30, 325–327.

LIU, L., ZHOU, Y., SZCZERBA, M.W., LI, X. and LIN, Y. 2010. Identification and application of a rice senescence-associated promoter. *Plant Physiol* 153, 1239–1249.

LIN, W.C., LU, C.F., WU, J.W., CHENG, M.L., LIN, Y.M., YANG, N.S., BLACK, L., GREEN, S.K., WANG, J.F. and CHENG, C.P. 2004. Transgenic tomato plants expressing the *Arabidopsis* NPR1 gene display enhanced resistance to a spectrum of fungal and bacterial diseases. *Transgenic Res* 13, 567–581.

LOAKE, G.J., FAKTOR, O., LAMB, C.J. and DIXON, R.A. 1992. Combination of H-box [CCTACC(N)7CT] and G-box (CACGTG) *cis* elements is necessary for feed-forward stimulation of a chalcone synthase promoter by the phenylpropanoid-pathway intermediate *p*-coumaric acid. *Proc Natl Acad Sci USA* 89, 9230–9234.

LU, C.A., HO, T.H., HO, S.L. and YU, S.M. 2002. Three novel MYB proteins with one DNA binding repeat mediate sugar and hormone regulation of alpha-amylase gene expression. *Plant Cell* 14, 1963–1980.

MACKERNESS, A.H.S., SURPLUS, S.L., BLAKE, P., JOHN, C.F., BUCHANAN-WOLLASTON, V., JORDAN, B.R. and THOMAS, B. 1999. Ultraviolet-B-induced stress and changes in gene expression in *Arabidopsis thaliana*: Role of signalling pathways controlled by jasmonic acid, ethylene and reactive oxygen species. *Plant Cell Environ* 22, 1413–1423.

MADAN, P., JAGADISH, S.V., CRAUFURD, P.Q., FITZGERALD, M., LAFARGE, T. and WHEELER, T.R. 2012. Effect of elevated CO_2 and high temperature on seed-set and grain quality of rice. *J Exp Bot* 63, 3843–3852.

MALNOY, M., JIN, Q., BOREJSZA-WYSOCKA, E.E., HE, S.Y. and ALDWINCKLE, H.S. 2007. Overexpression of the apple MpNPR1 gene confers increased disease resistance in *Malus* x *domestica*. *Mol Plant Microbe Interact* 20, 1568–1580.

MANNERS, J.M., PENNINCKX, I.A.M.A., VERMAERE, K., KAZAN, K., BROWN, R.L., MORGAN, A., MACLEAN, D.J., CURTIS, M.D., CAMMUE, B.P.A. and BROEKAERT, W.F. 1998. The promoter of the plant defensin gene PDF1.2 from *Arabidopsis* is systemically activated by fungal pathogens and responds to methyl jasmonate but not to salicylic acid. *Plant Mol Biol* 38, 1071–1080.

MAO, Y., PAVANGADKAR, K.A., THOMASHOW, M.F. and TRIEZENBERG, S.J. 2006. Physical and functional interactions of *Arabidopsis* ADA2 transcriptional coactivator proteins with the acetyltransferase GCN5 and with the cold-induced transcription factor CBF1. *Biochim Biophys Acta* 1759, 69–79.

MARTINEZ, E. 2002. Multi-protein complexes in eukaryotic gene transcription. *Plant Mol Biol* 50, 925–947.

MASTON, G.A., EVANS, S.K. and GREEN, M.R. 2006. Transcriptional regulatory elements in the human genome. *Annu Rev Genom Hum Genet* 7, 29–59.

MATARASSO, N., SCHUSTER, S. and AVNI, A. 2005. A novel plant cysteine protease has a dual function as a regulator of 1-aminocyclopropane-1-carboxylic acid synthase gene expression. *Plant Cell* 17, 1205–1216.

MATHUR, S., VYAS, S., KAPOOR, S. and TYAGI, A.K. 2011. The mediator complex in plants: Structure, phylogeny, and expression profiling of representative genes in a dicot (*Arabidopsis*) and a monocot (rice) during reproduction and abiotic stress. *Plant Physiol* 157, 1609–1627.

MELLER, Y., SESSA, G., EYAL, Y. and FLUHR, R. 1993. DNA–protein interactions on a *cis*-DNA element essential for ethylene regulation. *Plant Mol Biol* 23, 453–463.

MENKE, F.L., CHAMPION, A., KIJNE, J.W. and MEMELINK, J. 1999. A novel jasmonate- and elicitor-responsive element in the periwinkle secondary metabolite biosynthetic gene Str interacts with a jasmonate- and elicitor-inducible AP2-domain transcription factor, ORCA2. *EMBO J* 18, 4455–4463.

MIAO, Y., LAUN, T., ZIMMERMANN, P. and ZENTGRAF, U. 2004. Targets of the WRKY53 transcription factor and its role during leaf senescence in *Arabidopsis*. *Plant Mol Biol* 55, 853–867.

MIRADE ORDUNA, R. 2010. Climate change associated effects on grape and wine quality and production. *Food Res Int* 43, 1844–1855.

MORETTI, C.L., MATTOS, L.M., CALBO, A.G. and SARGENT, S.A. 2010. Climate changes and potential impacts on postharvest quality of fruit and vegetable crops: A review. *Food Res Int* 43, 1824–1832.

MORILLON, R., CATTEROU, M., SANGWAN, R.S., SANGWAN, B.S. and LASSALLES, J.P. 2001. Brassinolide may control aquaporin activities in *Arabidopsis thaliana*. *Planta* 212, 199–204.

MORRIS, K., MACKERNESS, S.A., PAGE, T., JOHN, C.F., MURPHY, A.M., CARR, J.P. and BUCHANAN-WOLLASTON, V. 2000. Salicylic acid has a role in regulating gene expression during leaf senescence. *Plant J* 23, 677–685.

MOU, Z., FAN, W. and DONG, X. 2003. Inducers of plant systemic acquired resistance regulate NPR1 function through redox changes. *Cell* 113, 935–944.

MOUGIOU, N., POULIOS, S., KALDIS, A. and VLACHONASIOS, K. 2012. *Arabidopsis thaliana* TBP-associated factor 5 is essential for plant growth and development. *Mol Breed* 30, 355–366.

MUBAYA, C.P., NJUKI, J., MUTSVANGWA, E.P., MUGABE, F.T. and NANJA, D. 2012. Climate variability and change or multiple stressors? Farmer perceptions regarding threats to livelihoods in Zimbabwe and Zambia. *J Environ Manage* 102, 9–17.

MUSSIG, C., LISSO, J., COLL-GARCIA, D. and ALTMANN, T. 2006. Molecular analysis of brassinosteroid action. *Plant Biol (Stuttg)* 8, 291–296.

NAGAO, R.T., GOEKJIAN, V.H., HONG, J.C. and KEY, J.L. 1993. Identification of protein-binding DNA sequences in an auxin-regulated gene of soybean. *Plant Mol Biol* 21, 1147–1162.

NAGY, Z., RISS, A., ROMIER, C., LE GUEZENNEC, X., DONGRE, A.R., ORPINELL, M., HAN, J., STUNNENBERG, H. and TORA, L. 2009. The human SPT20-containing SAGA complex plays a direct role in the regulation of endoplasmic reticulum stress-induced genes. *Mol Cell Biol* 29, 1649–1660.

NAKASHIMA, K., ITO, Y. and YAMAGUCHI-SHINOZAKI, K. 2009. Transcriptional regulatory networks in response to abiotic stresses in *Arabidopsis* and grasses. *Plant Physiol* 149, 88–95.

NAKASHIMA, K., TRAN, L.-S.P., VAN NGUYEN, D., FUJITA, M., MARUYAMA, K., TODAKA, D., ITO, Y., HAYASHI, N., SHINOZAKI, K. and YAMAGUCHI-SHINOZAKI, K. 2007. Functional analysis of a NAC-type transcription factor OsNAC6 involved in abiotic and biotic stress-responsive gene expression in rice. *Plant J* 51, 617–630.

NEMHAUSER, J.L., HONG, F. and CHORY, J. 2006. Different plant hormones regulate similar processes through largely non-overlapping transcriptional responses. *Cell* 126, 467–475.

NISHIUCHI, T., SHINSHI, H. and SUZUKI, K. 2004. Rapid and transient activation of transcription of the ERF3 gene by wounding in tobacco leaves: Possible involvement of NtWRKYs and autorepression. *J Biol Chem* 279, 55355–55361.

NISHIYAMA, R., LE, D.T., WATANABE, Y., MATSUI, A., TANAKA, M., SEKI, M., YAMAGUCHI-SHINOZAKI, K., SHINOZAKI, K. and TRAN, L.S. 2012. Transcriptome analyses of a salt-tolerant cytokinin-deficient mutant reveal differential regulation of salt stress response by cytokinin deficiency. *PLoS One* 7, e32124.

NISHIYAMA, R., WATANABE, Y., FUJITA, Y., LE, D.T., KOJIMA, M., WERNER, T., VANKOVA, R. et al. 2011. Analysis of cytokinin mutants and regulation of cytokinin metabolic genes reveals important regulatory roles of cytokinins in drought, salt and abscisic acid responses, and abscisic acid biosynthesis. *Plant Cell* 23, 2169–2183.

NORDIN, K., VAHALA, T. and PALVA, E.T. 1993. Differential expression of two related, low-temperature-induced genes in *Arabidopsis thaliana* (L.) Heynh. *Plant Mol Biol* 21, 641–653.

NORMAN, C., HOWELL, K.A., MILLAR, A.H., WHELAN, J.M. and DAY, D.A. 2004. Salicylic acid is an uncoupler and inhibitor of mitochondrial electron transport. *Plant Physiol* 134, 492–501.

OGAWA, M., HANADA, A., YAMAUCHI, Y., KUWAHARA, A., KAMIYA, Y. and YAMAGUCHI, S. 2003. Gibberellin biosynthesis and response during *Arabidopsis* seed germination. *Plant Cell* 15, 1591–1604.

OHME-TAKAGI, M. and SHINSHI, H. 1990. Structure and expression of a tobacco beta-1,3-glucanase gene. *Plant Mol Biol* 15, 941–946.

OKUSHIMA, Y., OVERVOORDE, P.J., ARIMA, K., ALONSO, J.M., CHAN, A., CHANG, C., ECKER, J.R. et al. 2005. Functional genomic analysis of the auxin response factor gene family members in *Arabidopsis thaliana*: Unique and overlapping functions of ARF7 and ARF19. *Plant Cell* 17, 444–463.

OLESEN, J.E., TRNKA, M., KERSEBAUM, K.C., SKJELVAG, A.O., SEGUIN, B., PELTONEN-SAINIO, P., ROSSI, F., KOZYRA, J. and MICALE, F. 2011. Impacts and adaptation of European crop production systems to climate change. *Eur J Agron* 34, 96–112.

OLESEN, J.R.E. and BINDI, M. 2002. Consequences of climate change for European agricultural productivity, land use and policy. *Eur J Agron* 16, 239–262.

OVERVOORDE, P.J., OKUSHIMA, Y., ALONSO, J.M., CHAN, A., CHANG, C., ECKER, J.R., HUGHES, B. et al. 2005. Functional genomic analysis of the auxin/indole-3-acetic acid gene family members in *Arabidopsis thaliana*. *Plant Cell* 17, 3282–3300.

PALM, C.J., COSTA, M.A., AN, G. and RYAN, C.A. 1990. Wound-inducible nuclear protein binds DNA fragments that regulate a proteinase inhibitor II gene from potato. *Proc Natl Acad Sci USA* 87, 603–607.

PANIKULANGARA, T.J., EGGERS-SCHUMACHER, G., WUNDERLICH, M., STRANSKY, H. and SCHOFFL, F. 2004. Galactinol synthase 1. A novel heat shock factor target gene responsible for heat-induced synthesis of raffinose family oligosaccharides in *Arabidopsis*. *Plant Physiol* 136, 3148–3158.

PANKOTAI, T., KOMONYI, O., BODAI, L., UJFALUDI, Z., MURATOGLU, S., CIURCIU, A., TORA, L., SZABAD, J. and BOROS, I. 2005. The homologous *Drosophila* transcriptional adaptors ADA2a and ADA2b are both required for normal development but have different functions. *Mol Cell Biol* 25, 8215–8227.

PARANI, M., RUDRABHATLA, S., MYERS, R., WEIRICH, H., SMITH, B., LEAMAN, D.W. and GOLDMAN, S.L. 2004. Microarray analysis of nitric oxide responsive transcripts in *Arabidopsis*. *Plant Biotechnol J* 2, 359–366.

PARK, C.M. 2007. Auxin homeostasis in plant stress adaptation response. *Plant Signal Behav* 2, 306–307.

PARK, J.E., PARK, J.Y., KIM, Y.S., STASWICK, P.E., JEON, J., YUN, J., KIM, S.Y., KIM, J., LEE, Y.H. and PARK, C.M. 2007. GH3-mediated auxin homeostasis links growth regulation with stress adaptation response in *Arabidopsis*. *J Biol Chem* 282, 10036–10046.

PARKHI, V., KUMAR, V., CAMPBELL, L.M., BELL, A.A., SHAH, J. and RATHORE, K.S. 2010. Resistance against various fungal pathogens and reniform nematode in transgenic cotton plants expressing *Arabidopsis* NPR1. *Transgenic Res* 19, 959–975.

PENG, H.P., LIN, T.Y., WANG, N.N. and SHIH, M.C. 2005. Differential expression of genes encoding 1-aminocyclopropane-1-carboxylate synthase in *Arabidopsis* during hypoxia. *Plant Mol Biol* 58, 15–25.

PENNINCKX, I.A., THOMMA, B.P., BUCHALA, A., METRAUX, J.P. and BROEKAERT, W.F. 1998. Concomitant activation of jasmonate and ethylene response pathways is required for induction of a plant defensin gene in *Arabidopsis*. *Plant Cell* 10, 2103–2113.

PEREMARTI, A., TWYMAN, R.M., GOMEZ-GALERA, S., NAQVI, S., FARRE, G., SABALZA, M., MIRALPEIX, B. et al. 2010. Promoter diversity in multigene transformation. *Plant Mol Biol* 73, 363–378.

PERILLI, S., MOUBAYIDIN, L. and SABATINI, S. 2010. The molecular basis of cytokinin function. *Curr Opin Plant Biol* 13, 21–26.

PERL-TREVES, R., FOLEY, R.C., CHEN, W. and SINGH, K.B. 2004. Early induction of the *Arabidopsis* GSTF8 promoter by specific strains of the fungal pathogen *Rhizoctonia solani*. *Mol Plant Microbe Interact* 17, 70–80.

POLVERARI, A., MOLESINI, B., PEZZOTTI, M., BUONAURIO, R., MARTE, M. and DELLEDONNE, M. 2003. Nitric oxide-mediated transcriptional changes in *Arabidopsis thaliana*. *Mol Plant Microbe Interact* 16, 1094–1105.

POPKO, J., HANSCH, R., MENDEL, R.R., POLLE, A. and TEICHMANN, T. 2010. The role of abscisic acid and auxin in the response of poplar to abiotic stress. *Plant Biol (Stuttg)* 12, 242–258.

PRIEST, H.D., FILICHKIN, S.A. and MOCKLER, T.C. 2009. *Cis*-regulatory elements in plant cell signaling. *Curr Opin Plant Biol* 12, 643–649.

QIAO, W. and FAN, L.M. 2008. Nitric oxide signaling in plant responses to abiotic stresses. *J Integr Plant Biol* 50, 1238–1246.

QIN, F., SHINOZAKI, K. and YAMAGUCHI-SHINOZAKI, K. 2011. Achievements and challenges in understanding plant abiotic stress responses and tolerance. *Plant Cell Physiol* 52, 1569–1582.

QIN, X.F., HOLUIGUE, L., HORVATH, D.M. and CHUA, N.H. 1994. Immediate early transcription activation by salicylic acid via the cauliflower mosaic virus as-1 element. *Plant Cell* 6, 863–874.

QUINN, J.M., BARRACO, P., ERIKSSON, M. and MERCHANT, S. 2000. Coordinate copper- and oxygen-responsive Cyc6 and Cpx1 expression in *Chlamydomonas* is mediated by the same element. *J Biol Chem* 275, 6080–6089.

RABBANI, M.A., MARUYAMA, K., ABE, H., KHAN, M.A., KATSURA, K., ITO, Y., YOSHIWARA, K., SEKI, M., SHINOZAKI, K. and YAMAGUCHI-SHINOZAKI, K. 2003. Monitoring expression profiles of rice genes under cold, drought, and high-salinity stresses and abscisic acid application using cDNA microarray and RNA gel-blot analyses. *Plant Physiol* 133, 1755–1767.

RAFFAELE, S., RIVAS, S. and ROBY, D. 2006. An essential role for salicylic acid in AtMYB30-mediated control of the hyper-sensitive cell death program in *Arabidopsis*. *FEBS Lett* 580, 3498–3504.

RAHMAN, A. 2012. Auxin: A regulator of cold stress response. *Physiol Plant* 147, 28–35.

RAI, M., HE, C. and WU, R. 2009. Comparative functional analysis of three abiotic stress-inducible promoters in transgenic rice. *Transgenic Res* 18, 787–799.

RIEPING, M. and SCHOFFL, F. 1992. Synergistic effect of upstream sequences, CCAAT box elements, and HSE sequences for enhanced expression of chimaeric heat shock genes in transgenic tobacco. *Mol Gen Genet* 231, 226–232.

RIVAS-SAN VICENTE, M. and PLASENCIA, J. 2011. Salicylic acid beyond defence: Its role in plant growth and development. *J Exp Bot* 62, 3321–3338.

ROBATZEK, S. and SOMSSICH, I.E. 2001. A new member of the *Arabidopsis* WRKY transcription factor family, AtWRKY6, is associated with both senescence- and defence-related processes. *Plant J* 28, 123–133.

ROBLES, L.M., WAMPOLE, J.S., CHRISTIANS, M.J. and LARSEN, P.B. 2007. *Arabidopsis* enhanced ethylene response 4 encodes an EIN3-interacting TFIID transcription factor required for proper ethylene response, including ERF1 induction. *J Exp Bot* 58, 2627–2639.

RODRIGUEZ-NAVARRO, S. 2009. Insights into SAGA function during gene expression. *EMBO Rep* 10, 843–850.

ROJAS, A., ALMOGUERA, C., CARRANCO, R., SCHARF, K.D. and JORDANO, J. 2002. Selective activation of the developmentally regulated Ha hsp17.6 G1 promoter by heat stress transcription factors. *Plant Physiol* 129, 1207–1215.

RUSHTON, P.J., TORRES, J.T., PARNISKE, M., WERNERT, P., HAHLBROCK, K. and SOMSSICH, I.E. 1996. Interaction of elicitor-induced DNA-binding proteins with elicitor response elements in the promoters of parsley PR1 genes. *EMBO J* 15, 5690–5700.

RYAN, M.G. 1991. Effects of climate change on plant respiration. *Ecol Appl* 1, 157–167.

SAHER, S., FERNANDEZ-GARCIA, N., PIQUERAS, A., HELLIN, E. and OLMOS, E. 2005. Reducing properties, energy efficiency and carbohydrate metabolism in hyperhydric and normal carnation shoots cultured *in vitro*: A hypoxia stress? *Plant Physiol Biochem* 43, 573–582.

SAKAI, T., TAKAHASHI, Y. and NAGATA, T. 1998. The identification of DNA binding factors specific for as-1-like sequences in auxin-responsive regions of parA, parB and parC. *Plant Cell Physiol* 39, 731–739.

SALINAS, J., OEDA, K. and CHUA, N.H. 1992. Two G-box-related sequences confer different expression patterns in transgenic tobacco. *Plant Cell* 4, 1485–1493.

SANDELIN, A., CARNINCI, P., LENHARD, B., PONJAVIC, J., HAYASHIZAKI, Y. and HUME, D.A. 2007. Mammalian RNA polymerase II core promoters: Insights from genome-wide studies. *Nat Rev Genet* 8, 424–436.

SANTNER, A. and ESTELLE, M. 2009. Recent advances and emerging trends in plant hormone signalling. *Nature* 459, 1071–1078.

SANTOS, A.P., SERRA, T., FIGUEIREDO, D.D., BARROS, P., LOURENCO, T., CHANDER, S., OLIVEIRA, M.M. and SAIBO, N.J. 2011. Transcription regulation of abiotic stress responses in rice: A combined action of transcription factors and epigenetic mechanisms. *OMICS* 15, 839–857.

SCANDALIOS, J.G. 1997. Oxidative stress and defense mechanisms in plants: Introduction. *Free Radic Biol Med* 23, 471–472.

SCHENK, P.M., KAZAN, K., RUSU, A.G., MANNERS, J.M. and MACLEAN, D.J. 2005. The SEN1 gene of *Arabidopsis* is regulated by signals that link plant defence responses and senescence. *Plant Physiol Biochem* 43, 997–1005.

SCHLENKER, W. and ROBERTS, M.J. 2009. Nonlinear temperature effects indicate severe damages to U.S. crop yields under climate change. *Proc Natl Acad Sci USA* 106, 15594–15598.

SEKI, M., KAMEI, A., YAMAGUCHI-SHINOZAKI, K. and SHINOZAKI, K. 2003. Molecular responses to drought, salinity and frost: Common and different paths for plant protection. *Curr Opin Biotechnol* 14, 194–199.

SEO, H.S., KIM, H.Y., JEONG, J.Y., LEE, S.Y., CHO, M.J. and BAHK, J.D. 1995. Molecular cloning and characterization of RGA1 encoding a G protein alpha subunit from rice (*Oryza sativa* L. IR-36). *Plant Mol Biol* 27, 1119–1131.

SHIBASAKI, K., UEMURA, M., TSURUMI, S. and RAHMAN, A. 2009. Auxin response in *Arabidopsis* under cold stress: Underlying molecular mechanisms. *Plant Cell* 21, 3823–3838.

SHINOZAKI, K., YAMAGUCHI-SHINOZAKI, K. and SEKI, M. 2003. Regulatory network of gene expression in the drought and cold stress responses. *Curr Opin Plant Biol* 6, 410–417.

SIEBERTZ, B., LOGEMANN, J., WILLMITZER, L. and SCHELL, J. 1989. *Cis*-analysis of the wound-inducible promoter wun1 in transgenic tobacco plants and histochemical localization of its expression. *Plant Cell* 1, 961–968.

SIMPSON, S.D., NAKASHIMA, K., NARUSAKA, Y., SEKI, M., SHINOZAKI, K. and YAMAGUCHI-SHINOZAKI, K. 2003. Two different novel *cis*-acting elements of erd1, a clpA homologous *Arabidopsis* gene function in induction by dehydration stress and dark-induced senescence. *Plant J* 33, 259–270.

SINGH, E., TIWARI, S. and AGRAWAL, M. 2009. Effects of elevated ozone on photosynthesis and stomatal conductance of two soybean varieties: A case study to assess impacts of one component of predicted global climate change. *Plant Biol (Stuttg)* 11 (Suppl 1), 101–108.

SINGH, P., NEDUMARAN, S., NTARE, B.R., BOOTE, K.J., SINGH, N.P., SRINIVAS, K. and BANTILAN, M.C.S. 2014. Potential benefits of drought and heat tolerance in groundnut for adaptation to climate change in India and West Africa. *Mitigation Adapt Strateg Global Change* 19(5), 509–529.

SMALE, S.T. and KADONAGA, J.T. 2003. The RNA polymerase II core promoter. *Annu Rev Biochem* 72, 449–479.

SRIVASTAVA, R., RAI, K.M., SRIVASTAVA, M., KUMAR, V., PANDEY, B., SINGH, S.P., BAG, S.K. et al., 2014. Distinct role of core promoter architecture in regulation of light-mediated responses in plant genes. *Molecular Plant* 7(4), 626–641.

STOCKINGER, E.J., MAO, Y., REGIER, M.K., TRIEZENBERG, S.J. and THOMASHOW, M.F. 2001. Transcriptional adaptor and histone acetyltransferase proteins in *Arabidopsis* and their interactions with CBF1, a transcriptional activator involved in cold-regulated gene expression. *Nucleic Acids Res* 29, 1524–1533.

STOROZHENKO, S., DE PAUW, P., VAN MONTAGU, M., INZE, D. and KUSHNIR, S. 1998. The heat-shock element is a functional component of the *Arabidopsis* APX1 gene promoter. *Plant Physiol* 118, 1005–1014.

STRAUB, P.F., SHEN, Q. and HO, T.D. 1994. Structure and promoter analysis of an ABA- and stress-regulated barley gene, HVA1. *Plant Mol Biol* 26, 617–630.

STROMPEN, G., GRUNER, R. and PFITZNER, U.M. 1998. An as-1-like motif controls the level of expression of the gene for the pathogenesis-related protein 1a from tobacco. *Plant Mol Biol* 37, 871–883.

SUN, J.Q., JIANG, H.L. and LI, C.Y. 2011. Systemic/jasmonate-mediated systemic defense signaling in tomato. *Mol Plant* 4, 607–615.

SUTOH, K. and YAMAUCHI, D. 2003. Two *cis*-acting elements necessary and sufficient for gibberellin-upregulated proteinase expression in rice seeds. *Plant J* 34, 635–645.

SUZUKI, M., KETTERLING, M.G. and MCCARTY, D.R. 2005. Quantitative statistical analysis of *cis*-regulatory sequences in ABA/VP1- and CBF/DREB1-regulated genes of *Arabidopsis*. *Plant Physiol* 139, 437–447.

TAJRISHI, M.M., VAID, N., TUTEJA, R. and TUTEJA, N. 2011. Overexpression of a pea DNA helicase 45 in bacteria confers salinity stress tolerance. *Plant Signal Behav* 6, 1271–1275.

TAMADA, Y., NAKAMORI, K., NAKATANI, H., MATSUDA, K., HATA, S., FURUMOTO, T. and IZUI, K. 2007. Temporary expression of the TAF10 gene and its requirement for normal development of *Arabidopsis thaliana*. *Plant Cell Physiol* 48, 134–146.

THOMAS, M.C. and Chiang, C.M. 2006. The general transcription machinery and general cofactors. *Crit Rev Biochem Mol Biol* 41, 105–178.

TIRYAKI, İ. 2007. The role of auxin-signaling gene axr1 in salt stress and jasmonic acid inducible gene expression in *Arabidopsis thaliana*. *J Cell Mol Biol* 6, 189–195.

TSUKAMOTO, S., MORITA, S., HIRANO, E., YOKOI, H., MASUMURA, T. and TANAKA, K. 2005. A novel *cis*-element that is responsive to oxidative stress regulates three antioxidant defense genes in rice. *Plant Physiol* 137, 317–327.

TULLUS, A., KUPPER, P., SELLIN, A., PARTS, L., SOBER, J., TULLUS, T., LOHMUS, K., SOBER, A. and TULLUS, H. 2012. Climate change at northern latitudes: Rising atmospheric humidity decreases transpiration, N-uptake and growth rate of hybrid aspen. *PLoS One* 7, e42648.

ULMASOV, T., LIU, Z.B., HAGEN, G. and GUILFOYLE, T.J. 1995. Composite structure of auxin response elements. *Plant Cell* 7, 1611–1623.

VILARDELL, J., MUNDY, J., STILLING, B., LEROUX, B., PLA, M., FREYSSINET, G. and PAGES, M. 1991. Regulation of the maize rab17 gene promoter in transgenic heterologous systems. *Plant Mol Biol* 17, 985–993.

VLACHONASIOS, K.E., THOMASHOW, M.F. and TRIEZENBERG, S.J. 2003. Disruption mutations of ADA2b and GCN5 transcriptional adaptor genes dramatically affect *Arabidopsis* growth, development, and gene expression. *Plant Cell* 15, 626–638.

WALBOT, V. 2011. How plants cope with temperature stress. *BMC Biol* 9, 79.

WALIA, H., WILSON, C., CONDAMINE, P., LIU, X., ISMAIL, A.M. and CLOSE, T.J. 2007. Large-scale expression profiling and physiological characterization of jasmonic acid-mediated adaptation of barley to salinity stress. *Plant Cell Environ* 30, 410–421.

WALLEY, J.W. and DEHESH, K. 2010. Molecular mechanisms regulating rapid stress signaling networks in *Arabidopsis*. *J Integr Plant Biol* 52, 354–359.

WALTER, A., SILK, W.K. and SCHURR, U. 2009. Environmental effects on spatial and temporal patterns of leaf and root growth. *Annu Rev Plant Biol* 60, 279–304.

WANG, H., DATLA, R., GEORGES, F., LOEWEN, M. and CUTLER, A.J. 1995. Promoters from kin1 and cor6.6, two homologous *Arabidopsis thaliana* genes: Transcriptional regulation and gene expression induced by low temperature, ABA, osmoticum and dehydration. *Plant Mol Biol* 28, 605–617.

WANG, W., VINOCUR, B. and ALTMAN, A. 2003. Plant responses to drought, salinity and extreme temperatures: Towards genetic engineering for stress tolerance. *Planta* 218, 1–14.

WARNER, S.A., SCOTT, R. and DRAPER, J. 1993. Isolation of an *Asparagus* intracellular PR gene (AoPR1) wound-responsive promoter by the inverse polymerase chain reaction and its characterization in transgenic tobacco. *Plant J* 3, 191–201.

WASILEWSKA, A., VLAD, F., SIRICHANDRA, C., REDKO, Y., JAMMES, F., VALON, C., FREI DIT FREY, N. and LEUNG, J. 2008. An update on abscisic acid signaling in plants and more. *Mol Plant* 1, 198–217.

WASTERNACK, C. and KOMBRINK, E. 2010. Jasmonates: Structural requirements for lipid-derived signals active in plant stress responses and development. *ACS Chem Biol* 5, 63–77.

WATERER, D., BENNING, N., WU, G., LUO, X., LIU, X., GUSTA, M., McHUGHEN, A. and GUSTA, L. 2010. Evaluation of abiotic stress tolerance of genetically modified potatoes (*Solanum tuberosum* cv. Desiree). *Mol Breed* 25, 527–540.

WOODWARD, A.W. and BARTEL, B. 2005. Auxin: Regulation, action, and interaction. *Ann Bot* 95, 707–735.

WU, L., ZHANG, Z., ZHANG, H., WANG, X.C. and HUANG, R. 2008. Transcriptional modulation of ethylene response factor protein JERF3 in the oxidative stress response enhances tolerance of tobacco seedlings to salt, drought, and freezing. *Plant Physiol* 148, 1953–1963.

WU, X.F., WANG, C.L., XIE, E.B., GAO, Y., FAN, Y.L., LIU, P.Q. and ZHAO, K.J. 2009. Molecular cloning and characterization of the promoter for the multiple stress-inducible gene BjCHI1 from *Brassica juncea*. *Planta* 229, 1231–1242.

XIA, X.J., WANG, Y.J., ZHOU, Y.H., TAO, Y., MAO, W.H., SHI, K., ASAMI, T., CHEN, Z. and YU, J.Q. 2009. Reactive oxygen species are involved in brassinosteroid-induced stress tolerance in cucumber. *Plant Physiol* 150, 801–814.

XIE, Z., ZHANG, Z.L., HANZLIK, S., COOK, E. and SHEN, Q.J. 2007. Salicylic acid inhibits gibberellin-induced alpha-amylase expression and seed germination via a pathway involving an abscisic-acid-inducible WRKY gene. *Plant Mol Biol* 64, 293–303.

XU, D., MCELROY, D., THORNBURG, R.W. and WU, R. 1993. Systemic induction of a potato pin2 promoter by wounding, methyl jasmonate, and abscisic acid in transgenic rice plants. *Plant Mol Biol* 22, 573–588.

XUE, G.P. 2002. Characterisation of the DNA-binding profile of barley HvCBF1 using an enzymatic method for rapid, quantitative and high-throughput analysis of the DNA-binding activity. *Nucleic Acids Res* 30, e77.

YAMAGUCHI-SHINOZAKI, K. and SHINOZAKI, K. 1994. A novel *cis*-acting element in an *Arabidopsis* gene is involved in responsiveness to drought, low-temperature, or high-salt stress. *Plant Cell* 6, 251–264.

YAMAMOTO, S., NAKANO, T., SUZUKI, K. and SHINSHI, H. 2004. Elicitor-induced activation of transcription via W box-related *cis*-acting elements from a basic chitinase gene by WRKY transcription factors in tobacco. *Biochim Biophys Acta* 1679, 279–287.

YI, N., KIM, Y.S., JEONG, M.H., OH, S.J., JEONG, J.S., PARK, S.H., JUNG, H., CHOI, Y.D. and KIM, J.K. 2010. Functional analysis of six drought-inducible promoters in transgenic rice plants throughout all stages of plant growth. *Planta* 232, 743–754.

YIN, S., MEI, L., NEWMAN, J., BACK, K. and CHAPPELL, J. 1997. Regulation of sesquiterpene cyclase gene expression. Characterization of an elicitor- and pathogen-inducible promoter. *Plant Physiol* 115, 437–451.

YU, D., CHEN, C. and CHEN, Z. 2001. Evidence for an important role of WRKY DNA binding proteins in the regulation of NPR1 gene expression. *Plant Cell* 13, 1527–1540.

YUAN, S. and LIN, H.H. 2008. Role of salicylic acid in plant abiotic stress. *Z Naturforsch C* 63, 313–320.

ZAPATER, M., SOHRMANN, M., PETER, M., POSAS, F. and DE NADAL, E. 2007. Selective requirement for SAGA in Hog1-mediated gene expression depending on the severity of the external osmostress conditions. *Mol Cell Biol* 27, 3900–3910.

ZHANG, H., MAO, X. and JING, R. 2011. SnRK2 acts within an intricate network that links sucrose metabolic and stress signaling in wheat. *Plant Signal Behav* 6, 652–654.

ZHANG, T. and HUANG, Y. 2012. Impacts of climate change and inter-annual variability on cereal crops in China from 1980 to 2008. *J Sci Food Agric* 92, 1643–1652.

ZHANG, Z.L., XIE, Z., ZOU, X., CASARETTO, J., HO, T.H. and SHEN, Q.J. 2004. A rice WRKY gene encodes a transcriptional repressor of the gibberellin signaling pathway in aleurone cells. *Plant Physiol* 134, 1500–1513.

ZHENG, C., JIANG, D., DAI, T., JING, Q. and CAO, W. 2010. Effects nitroprusside, a nitric oxide donor, on carbon and nitrogen metabolism and the activity of the antioxidation system in wheat seedlings under salt stress. *Acta Ecol Sinica* 30, 1174–1183.

ZHU, J.K. 2001. Cell signaling under salt, water and cold stresses. *Curr Opin Plant Biol* 4, 401–406.

ZHU, Z. and GUO, H. 2008. Genetic basis of ethylene perception and signal transduction in *Arabidopsis*. *J Integr Plant Biol* 50, 808–815.

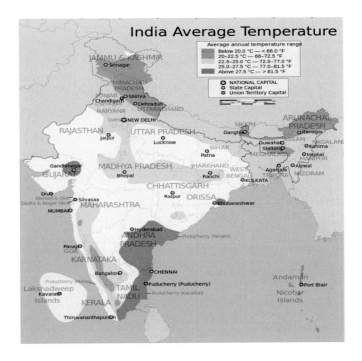

FIGURE 2.5 Temperature pattern of India.

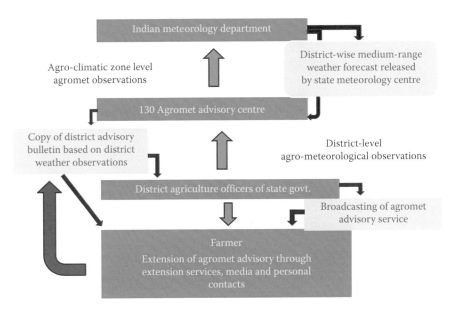

FIGURE 2.8 District-level agromet advisory service.

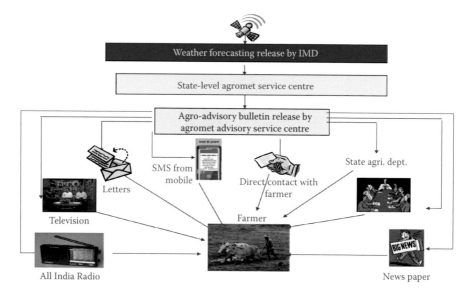

FIGURE 2.9 Effective communication mediums for communication of weather forecasting between agromet service centre and beneficiaries.

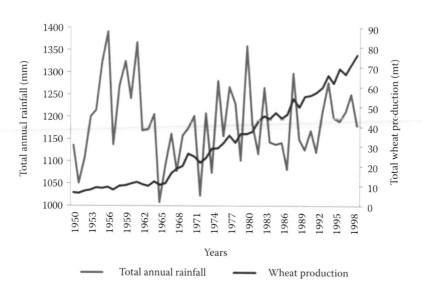

FIGURE 4.1 Trend in total annual rainfall (mm) and production (million tonnes) of wheat over years.

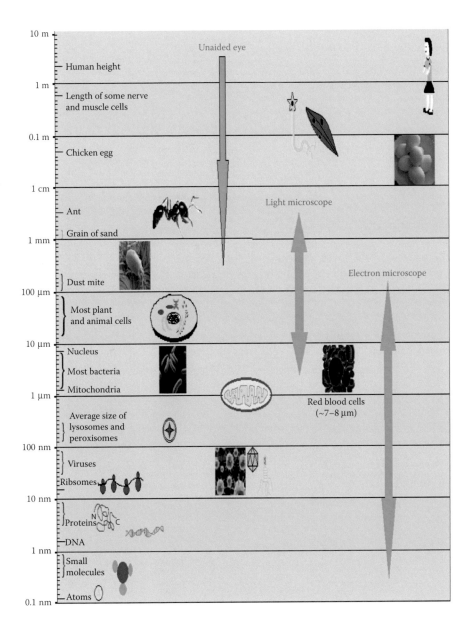

FIGURE 5.1 Size variations in biomolecules and living organisms.

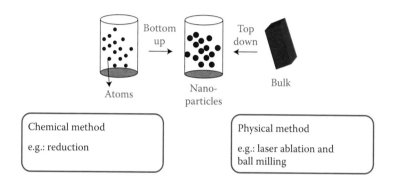

FIGURE 5.2 Classical approaches for nanoparticle synthesis.

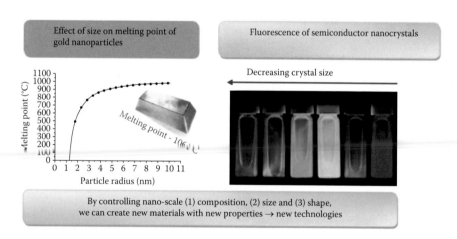

FIGURE 5.3 The effect of particle size on chemical and physical properties of nanoparticles.

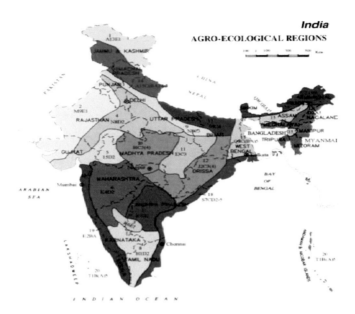

FIGURE 11.2 Map of India showing three ecological zones (Mountains, Hills and Terai) in the region.

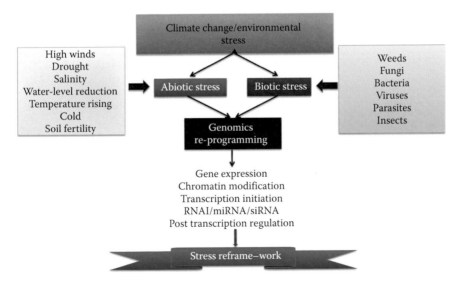

FIGURE 13.1 A model for abiotic and biotic stress signalling leads to genomics reprogramming during climate change or environmental stress.

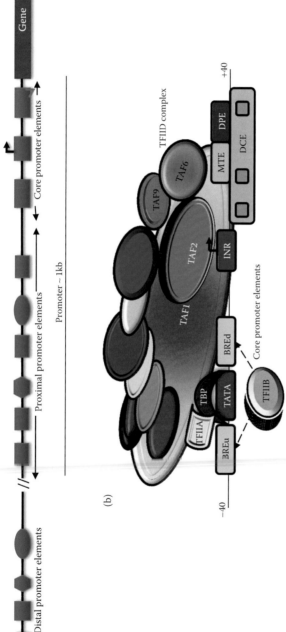

FIGURE 13.2 Eukaryotic CPEs and interacting basal transcription factors: (a) Schematic of a typical gene regulatory region. The promoter, which is composed of a core promoter and proximal promoter elements, typically spans <1 kb pairs. Distal (upstream) regulatory elements, which can include enhancers, silencers, insulators and locus control regions, can be located up to 1 Mb pairs from the promoter. These distal elements may contact the core promoter or proximal promoter through a mechanism that involves looping out the intervening DNA. (b) The core promoter (−40 to +40 bp) diagram shows the location of known CPEs. TFIID components are known to directly interact with CPEs. The described protein–CPE interactions are indicated by an arrow. The BREu and BREd elements are recognised by TFIIB, the TATA-box is recognised by TBP (and TBP2), the Inr is recognised by TAF1 and TAF2, the DPE is recognised by TAF6 and TAF9 and the DCE is recognised by TAF1. The MTE is known to interact with TFIID in general.

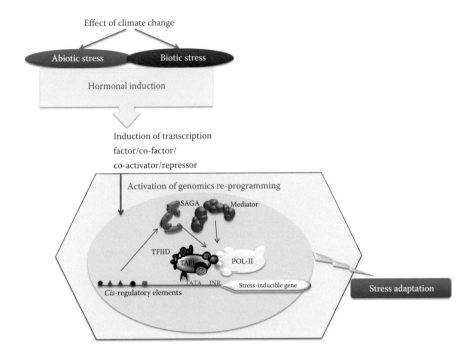

FIGURE 13.3 A model describing the coordination of different types of transcriptional regulatory complex in plant gene expression in the context of climate change. Various transcription factor/co-factor/co-activator/repressor was induced during stress condition and controls the stress-responsive gene expression. *Cis*-regulatory elements are also involved in stress-responsive transcription shown in different shapes at the proximal promoter.

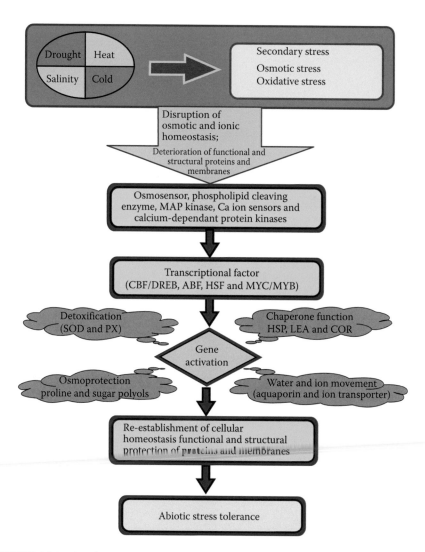

FIGURE 14.2 A schematic representation of plant response to abiotic stress.

CHAPTER FOURTEEN

Acclimation and adaptation of plants to different environmental abiotic stresses

Amit Kumar, Rakesh Singh Sengar and Shivendra Vikram Sahi

Contents

Abstract

Plant stress usually reflects some sudden change in environmental condition. However, in stress-tolerant plant species, exposure to a particular stress leads to acclimation to that specific stress in a time-dependent manner. Plant stress and plant acclimation are intimately linked with each other. The stress-induced modulation of homeostasis can be considered as the signal for the plant to initiated processes required for the establishment of a new homeostasis associated with the acclimated state. Plants exhibit stress resistance or stress tolerance because of their genetic capacity to adjust or to acclimate to the stress and establish a new homeostatic state over time. Furthermore, the acclimation process in stress-resistant species is usually reversible upon removal of the external stress. Stress induces many biochemical, molecular and physiological changes and responses that influence various cellular and whole-plant processes that affect crop yield and quality.

14.1 Introduction

The establishment of homeostasis associated with the new acclimated state is not the result of a single physiological process but rather the result of many physiological processes that the plant integrates over time, that is, integrates over the acclimation period (Figure 14.1). Plants usually integrate

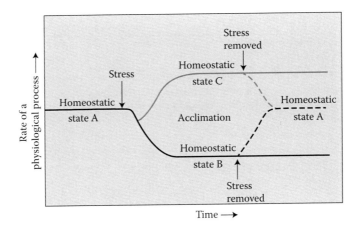

FIGURE 14.1 Schematic relationship between stress and acclimation. (Adapted from Hopkins, W. G., Hüner, N. P. A. 2009. *Introduction to Plant Physiology*, 4th ed., John Wiley Sons, Inc., Hoboken, USA.)

these physiological processes over a short-term as well as a long-term basis. Short-term acclimations involve responses occurring within minutes of environmental change and typically involve pre-existing components within a biochemical pathway; these responses are easily reversible, for example, increases in enzyme activity (i.e. Calvin cycle) in response to increases in temperature. Long-term acclimation, on the other hand, may begin within minutes, but is pronounced within days or weeks following an environmental change. These responses typically involve altered patterns of gene expression, reallocation of resources between the component processes of photosynthesis, and morphological change. The responses are not immediately reversible and often lead to the development of a visually different phenotype. Long-term responses represent acclimation if they improve performance in the altered environment. Acclimation usually involves the differential expression of specific sets of genes associated with exposure to a particular stress. The remarkable capacity to *regulate gene expression* in response to environmental change in a time-nested manner is the basis of plant plasticity. A good example is plants growing in shade develop larger leaves and an enhanced photosynthetic apparatus for improved light capture.

The short-term processes involved in acclimation can be initiated within seconds or minutes but long-term processes are less transient and, thus, usually exhibit a longer lifetime.

14.2 Types of plant stress

Agriculture and climate changes are interrelated processes. One may result in the other. Climate change induces many bio-chemical, molecular and physiological changes and responses that influence various cellular and whole-plant processes that affect crop yield and quality. Abiotic stresses, such as drought, salinity, extreme temperatures, chemical toxicity and oxidative stress, are serious threats to agriculture and the natural status of the environment. Abiotic stress is the primary cause of crop loss worldwide, reducing average yields for most major crop plants by more than 50% (Bray et al. 2000). Drought, salinity, extreme temperatures and oxidative stress are often interconnected, and may induce similar cellular damage. For example, drought and/or salinisation are manifested primarily as osmotic stress, result-ing in the disruption of homeostasis and ion distribution in the cell (Zhu 2001). Oxidative stress, which frequently accompanies high temperature, salinity or drought stress, may cause denatur-ation of functional and structural proteins (Smirnoff 1998).

The complex plant response to abiotic stress, which involves many genes and biochemical–molecular mechanisms, is sche-matically represented in Figure 14.2. The ongoing elucidation of the molecular control mechanisms of abiotic stress tolerance, which may result in the use of molecular tools for engineer-ing more tolerant plants, is based on the expression of specific stress-related genes. These can be divided into three sets of cat-egories as follows:

> *First categories*: These categories are involved in signalling cascades and in transcriptional control, such as MyC, MAP kinase and SOS kinase (Zhu 2001), phospholipases (Frank et al. 2000) and transcriptional factors such as HSF, and the CBF/DREB and ABF/ABAE families (Choi et al. 2000).

> *Secondary categories*: These categories are directly involved in the protection of membranes and proteins, such as heat-shock proteins (HSPs) and chaperones, and late embryogenesis abundant (LEA) proteins (Thomashow 1999; Bray et al. 2000).

> *Third categories*: These categories are involved in water and ion uptake and transport such as aquaporins and ion transporters (Blumwald 2000).

To maintain growth and productivity, plants must adapt to stress conditions and exercise-specific tolerance mechanisms. Plant modification for enhanced tolerance is mostly based on

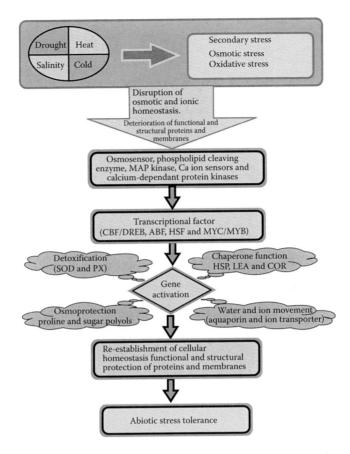

FIGURE 14.2 **(See colour insert.)** A schematic representation of plant response to abiotic stress.

the manipulation of genes that protect and maintain the function and structure of cellular components, the transfer of one or several genes involved in either signalling and regulatory pathways, or encode enzymes present in pathways leading to the synthesis of functional and structural protectants, such as osmolytes and antioxidants, or encode stress-tolerance-conferring proteins.

Plant stress can be divided into two primary categories. Abiotic stress is a physical (e.g. light, temperature) or chemical insult that the environment may impose on a plant. Biotic stress is a biological insult (e.g. insects, disease) to which a plant may be exposed during its lifetime (Figure 14.3). Some plants may be injured by a stress, which means that they exhibit one or more metabolic dysfunctions. If the stress is moderate and short term, the injury may be temporary and the plant may recover when the stress is removed. If the stress is severe enough, it may

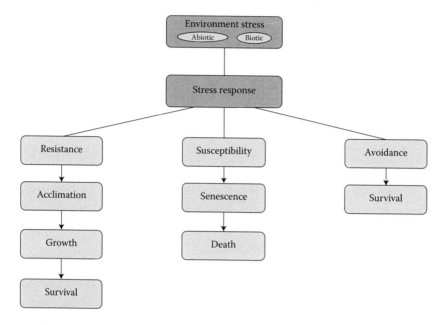

FIGURE 14.3   The effect of environmental stress on plant survival. (Adapted from Hopkins, W. G., Hüner, N. P. A. 2009. *Introduction to Plant Physiology*, 4th ed., John Wiley Sons, Inc., Hoboken, USA.)

prevent flowering, seed formation and induce senescence that leads to plant death. Such plants are considered to be susceptible. Some plants escape the stress altogether, such as ephemeral, or short-lived, desert plants.

Ephemeral plants germinate, grow and flower very quickly following seasonal rains. Thus, they complete their life cycle during a period of adequate moisture and form dormant seeds before the onset of the dry season. In a similar manner, many arctic annuals rapidly complete their life cycle during the short arctic summer and survive over winter in the form of seeds. Because ephemeral plants never really experience the stress of drought or low temperature, these plants survive the environmental stress by stress avoidance. Avoidance mechanisms reduce the impact of a stress, even though the stress is present in the environment. Many plants have the capacity to tolerate a particular stress and, hence, are considered to be stress resistant. Stress resistance requires that the organism exhibit the capacity to adjust or to acclimate to the stress.

Short-term acclimation

The *short-term processes* involved in acclimation can be initiated within seconds or minutes upon exposure to a stress, but may be transient in nature. That means that although these processes can be detected very soon after the onset of a stress, their activities also disappear rather rapidly. As a consequence, the lifetime of these processes is rather short. An example of a short-term response is a change in light, water availability or temperature that are part of the acclimation process.

14.3 Effect of temperature on plant and its environment

Mesophytic plants (terrestrial plants adapted to temperate environments that are neither excessively wet nor dry) have a relatively narrow temperature range of about 10°C for optimal growth and development. Outside of this range, varying amounts of damage occur, depending on the magnitude and duration of the temperature fluctuation. In this section, we will discuss three types of temperature stress: high temperatures, low temperatures above freezing and temperatures below freezing. Most actively growing tissues of higher plants are tillable to survive extended exposure to temperatures above 45°C or even short exposure to temperatures of 55°C or above. However, non-growing cells or dehydrated tissues (e.g. seeds and pollen) remain viable at much higher temperatures. Pollen grains of some species can survive 70°C and some dry seeds can tolerate temperatures as high as 120°C.

Plants and related organisms may be broadly classified according to their ability to withstand temperature. Those that grow optimally at lower temperatures (between 0°C and 10°C) are called psychrophiles. The psychrophiles include primarily algae, fungi and bacteria. Higher plants generally fall into the category of mesophiles, whose optimum temperature lies roughly between 10°C and 30°C. Thermophiles will grow unhindered at temperatures between 30°C and 65°C (Oosterhuis 2002; Zrobek-Sokolnik 2012), although there are reports of cyanobacteria growing at temperature as high as 85°C, these temperature ranges apply t hydrated, actively growing organisms. Dehydrated organisms and organs, such as resurrection plants (*Selaginella lepidophylla*) and dry seeds with moisture contents as low as 5%, are able to withstand a much broader range of temperatures for extended periods of time.

Temperature effects on membrane and its enzyme

Plant membranes consist of a lipid bilayer interspersed with proteins and sterols, and any abiotic factor that alters membrane properties can disrupt cellular processes. The physical properties of the lipids greatly influence the activities of the integral membrane proteins, including H^+-pumping ATPases, carriers and channel-forming proteins that regulate the transport of ions and other solutes. High temperatures cause an increase in the fluidity of membrane lipids and a decrease in the strength of hydrogen bonds and electrostatic interactions between polar groups of proteins within the aqueous phase of the membrane. High temperatures thus modify membrane composition and structure, and can cause leakage of ions. High temperatures can also lead to a loss of the three-dimensional structure required for correct function of enzymes or structural cellular components, thereby leading to loss of proper enzyme structure and activity.

Temperature effects on photosynthesis

Photosynthesis and respiration are both inhibited by temperature stress. Typically, photosynthetic rates are inhibited by high temperatures to a greater extent than respiratory rates. Although chloroplast enzymes such as rubisco, rubiscoactivase, NADP-G3P dehydrogenase and PEP carboxylase become unstable at high temperatures, the temperatures at which these enzymes began to denature and lose activity are distinctly higher than the temperatures at which photosynthetic rates begin to decline. This would indicate that the early stages of heat injury to photosynthesis are more directly related to changes in membrane properties and to uncoupling of the energy transfer mechanisms in chloroplasts.

This imbalance between photosynthesis and respiration is one of the main reasons for the deleterious effects of high temperatures. On an individual plant, leaves growing in the shade have a lower-temperature compensation point than leaves that are exposed to the sun (and heat). Reduced photosynthate production may also result from stress-induced stomatal closure, reduction in leaf canopy area and regulation of assimilate partitioning.

Acclimatisation of plant at high temperatures

High-temperature acclimation involves a considerable reorganisation of the thylakoid membrane, including adaptive changes of lipid composition. They contribute to the optimum physical state of the membrane (microviscosity, permeability, etc.). During heat acclimation, the threshold temperature at which fluidity still maintains the native membrane structure and function, rises (Raison et al. 1982). The more saturated lipid species decrease, noticeably the thylakoid membrane mobility at elevated temperatures, thus keeps a well-arranged lateral movement of electrons carried between the photosystems. The achieved thermotolerance of the majority of thermosensible light reactions is probably the result of both the lipid-induced thermostable conformations of the membrane-connected thyla-koid protein subunits and the adjustment of membrane lipid flu-idity. *In vivo* interaction of xanthophyll-cycle pigments with the membrane lipid matrix is supported by a series of experimental facts (reviewed by Sarry et al. 1994). It has been reported that zeaxanthin synthesis modulates the fluidity (Gruszecki and Strzalka 1991) and the lipid peroxidation status (Sarry et al. 1994) of thylakoid membranes.

High temperature negatively affects both metabolic (Mahan and Mauget 2005) and reproductive (Snider et al. 2010) efficien-cies. Heat stress effects are notable at various levels, including plasma membrane and biochemical pathways operative in the cytosol or cytoplasmic organelles (Sung et al. 2003). Initial effects of heat stress, however, are on plasma lemma, which shows more fluidity of lipid bilayer under heat stress. This leads to the induction of Ca^{2+} influx and cytoskeletal reorganisa-tion, resulting in the upregulation of mitogen-activated protein kinases and calcium-dependent protein kinase. Signalling of these cascades at the nuclear level leads to the production of antioxidants and compatible osmolytes for cell water balance and osmotic adjustment. Production of ROS in the organelles (e.g. chloroplast and mitochondria) is of great significance for signalling as well as production of antioxidants. The antioxi-dant defence mechanism is a part of heat-stress adaptation, and its strength is correlated with acquisition of thermotolerance (Figure 14.4). Accordingly, in a set of wheat genotypes the capacity to acquire thermotolerance was correlated with activi-ties of catalase (CAT) and SOD (Superoxide dis mutase), higher ascorbic acid content and less oxidative damage. One of the most closely studied mechanisms of thermotolerance is the induction of HSPs. Each major HSP family has a unique mechanism of action with chaperonic activity. The protective effects of HSPs can be attributed to the network of the chaperone machinery, in which many chaperones act in concert. An increasing num-ber of studies suggest that the HSPs/chaperones interact with other stress-response mechanisms. The HSPs/chaperones can play a role in stress signal transduction and gene activation as well as in regulating cellular redox state. They also interact with other stress-response mechanisms such as production of osmolytes and antioxidants. HSPs are generally classified into five evolutionarily conserved groups: HSP100, HSP90, HSP70, HSP60 and small HSPs. Most, but not all, HSPs are molecu-lar chaperones, which bind and stabilise proteins at intermedi-ate stages of folding, assembly, degradation and translocation across membranes. Membrane lipid saturation is considered

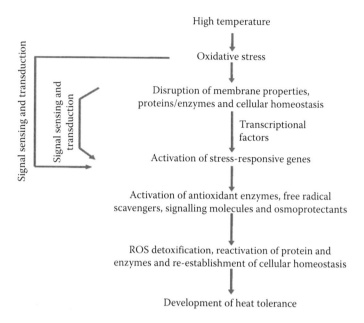

FIGURE 14.4 A schematic illustration of heat-induced signal transduction mechanism and development of heat tolerance in plants.

as an important element in high thermotolerance. In a mutant wheat line with increased heat resistance, heat treatment increased relative quantities of linolenic acid (among galacto-lipids) and *trans*-3-hexaldecanoic acid (among phospholipids), when compared with the wild type. Currently, it is unknown whether a higher or a lower degree of membrane lipid satu-ration is beneficial for high-temperature tolerance. The contri-bution of lipid and protein components to membrane function under heat stress needs further investigation. Localisation of low molecular weight (LMW)-HSPs with chloroplastic mem-branes upon heat stress suggests that they play a role in protect-ing photosynthetic electron transport. An important component of thermotolerance is changes in gene expression. Heat stress is known to swiftly alter patterns of gene expression (Yang et al. 2006), inducing expression of the HSP complements and inhib-iting expression of many other genes.

Acclimatisation of the plant at low temperatures

The study of cold-tolerant, herbaceous plants such as wheat, barley, spinach and the model plant species has enhanced our understanding of the metabolic and molecular events before, during and after acclimation. This has assisted greatly in the search for metabolic and genetic factors involved in cold tolerance. One of the immediate responses of cold-tolerant plants to low temperature is an increase in the proportion of unsaturated fatty acid bound to lipids associated with the plasma membrane, mitochondrial membranes as well as thylakoid membranes. Various biophysical measurements indicate that this ensures the membrane can remain in a more fluid and less gel-like state at lower temperature which enhances membrane stability and function at these low temperatures. A change in the membrane from the fluid state to a more solid state is marked by an abrupt change in the membrane activity. The temperature at which this transition occurs is known as the transition temperature. This means that at temperatures above the transition temperature, the membrane remains fluid but becomes more solid or gel like at temperatures below the transition temperature. This allows higher activity of membrane process at lower temperature.

Cold acclimation of herbaceous plants induces change in gene expression. During acclimation, there are changes in m-RNA transcription and increases in protein synthesised. A major class of cold-induced genes encode homologs of late embryogenesis active proteins that are synthesised late in embryogenesis and during dehydration stress. These polypeptides fall into a number of families based on amino acid sequence similarities. However, these proteins encoded by cold-regulated genes share common physical properties.

The promoter regions contain cold-regulated genes that are activated in response to low temperature and dehydration stress. Analyses of these promoter regions identify a DNA regulatory element called dehydration responsive element (DRE). The DRE has a conserved core C-repeat sequence of CCGAC that imparts responsiveness to low temperature and dehydration. Specific proteins that bind to the DRE are called C-repeat binding factors (CBFs). Thus, CBFs are transcriptional activators that are involved in regulating the expression of cold-regulated genes. It is concluded that cold acclimation is regulated by a family of CBF transcriptional factors.

Q_{10} is the temperature dependent for plant respiration   Enzymes and enzyme reactions are sensitive to temperature. Enzyme reactions typically are considered to have a Q_{10} of about 2, which means that the rate of the reaction doubles for each 10°C rise in temperature. The rate of reaction increases with temperature until an optimum is reached, beyond which the rate usually declines sharply; the decline in enzyme activity is normally caused by thermal denaturation as a result of protein unfolding. It is usually assumed that Q_{10} is independent of the temperature range over which it is measured. However, this is not true for plant respiration. In fact, the Q_{10} increases linearly upon short-term increases in temperatures from 40°C to 0°C. This dynamic temperature response of Q_{10} measures the temperature-dependent change in the respiration rate; an increase in ambient temperature will cause a greater change in rates of respiration in plants native to cold, Arctic climates than in plants native to hotter climates. Furthermore, other abiotic factors such as irradiance and water deficit can also influence the Q_{10} for plant respiration. Short temperature exposures to low temperatures reduces the flux of carbon through glycolysis and the TCA cycle because low temperatures will reduce the activity of the various enzymes involved in these pathways. In addition, low temperatures will decrease the fluidity of the inner membrane, which decreases the rate of respiratory electron transport. As a consequence, short-term exposure to low temperatures will reduce the rates of CO_2 evolution and O_2 consumption. However, at moderate to high temperatures, it is not enzyme activity that limits the rate of reaction but rather the availability of substrates such as ADP and ATP, at high temperature, mitochondrial membranes may become leaky to protons and, therefore, reduce the capacity to synthesise ATP by chemiosmosis.

14.4 Plants adjust osmotically in water stress condition

Osmotic adjustment is the capacity of plant cells to accumulate solutes and use them to lower Ψw (these sign means water potential) during periods of osmotic stress. The adjustment involves a net increase in solute content per cell that is independent of the volume changes that result from loss of water. The decrease in ΨS (=osmotic potential) is typically limited to about 0.2–0.8 MPa, except in plants adapted to extremely dry conditions.

Water stress in many plants is a decrease in osmotic potential resulting from an accumulation of solute. This process is known as osmotic adjustment. While some increase in solute concentration is expected as a result of dehydration and decreasing cell volume, osmotic adjustment refers specifically to a net increase in solute concentration due to metabolic processes triggered by stress. An osmotic adjustment generates a more negative leaf water potential, thereby helping to maintain water movement into the leaf and, consequently, leaf turgor. Osmotic adjustment may also play an important role in partially recoveries by helping to maintain leaf turgor. Osmotic adjustments also enable plants to keep their stomata open and continue taking up CO_2 for photosynthesis under conditions of moderate water stress. One amino acid that appears to be particularly sensitive to stress is proline. A large number of plants synthesise proline from glutamine in the leaves. The role of proline is demonstrated by experiments with tomato cells in culture. Cells subjected to water stress by exposure to hyperosmotic concentration of GEG responded with an initial loss of turgor and rapid accumulation of proline.

14.5 State transition mechanism for energy distribution

State transition is one of the best mechanisms for short-term regulation of energy distribution and is based on reversible phosphorylation of light-harvesting complex II (LHCII) protein. The phosphorylation of protein is a ubiquitous mechanism for regulating many aspects of gene regulation and response to environmental stimuli in all eukaryote organisms. The phosphorylation of proteins is catalysed by a class of enzymes known as protein kinases. Chloroplasts contain a thylakoid membrane-bound protein kinase capable of phosphorylating LHCH. The activity of this kinase is sensitive to the redox state of the thylakoid membrane and is activated when excess energy drives PSII, resulting in a build-up of reduced plastoquinone. Plants that are exposed to conditions that result in the preferential excitation of PSII are considered to be in state second.

The resulting phosphorylation of LHCII increases the negative charge of the protein, causing LHCII to dissociate from PSII. The same negative charge also loosens the appression of the thylakoid membranes in the grana stacks, freeing a certain portion of LHCII to migrate into the PSI-rich stroma thylakoids. This shifts the balance of energy away from the PSII complexes, which remain behind in the appressed region, in favour of PSI. The preferential excitation of PSI is referred to as state I. Recently, it has been shown that the PSI-H subunit of *Arabidopsis thaliana* is required for reversible transitions between state I and state II.

14.6 Long-term acclimation

Long-term processes are less transient and thus usually exhibit a longer lifetime. However, the lifetimes of these processes overlap in time such that the short-term processes usually constitute the initial responses to a stress while the long-term processes are usually detected later in the acclimation process. Long-term response to changes in light, water availability and temperature that are part of the acclimation process results in phenotypic alteration.

Photoacclimation The process whereby adjustments are made to the structure and function of the photosynthetic apparatus in response to changes in growth irradiance is called photoacclimation. One consequence of photoacclimation is a change in pigment composition which results in an altered visible phenotype. It is important to know that photoacclimation requires growth and development. For example, photoautotroph grown under high light typically exhibit a decrease in total chlorophyll per leaf area compared with the same plants grown at low irradiance. Thus, the leaves of high-light plants are usually a pale green or yellow green compared with a dark green phenotype of the same species grown at low light. Functionally, high-light plants exhibits a photosynthetic light response curve for CO_2 assimilation that is distinct from plants grown under low light, when measured as net photosynthesis. Typically, plants grown under high light have a higher photosynthetic capacity, that is, a higher light-saturated rate of photosynthesis than low-light plants. In contrast, high-light plants may have a lower initial slope, compared with the same plants grown at low light.

Gene expression is regulated by light Many green algae may exhibit an even more dramatic change in phenotype in response to growth at either high or low light than terrestrial plants. Experiments utilising single-cell green algal species such as *Dunaliella tertiolecta* and *Chlorella vulgaris* indicate that the light-dependent change in the content of LHCII (light-harvesting complex II) is modulated in response to the redox state of the plastoquinone pool. Photosynthetic electron transport can be inhibited specifically at the *Cyt b_6f* complex with a compound called DBMIB (2,5-dibromo-6-isopropyl-3-methyl-1,4-benzyquinone). In the presence of this compound, there is a net accumulation of PQH_2 because, although PSII is able to convert PQ to PQH_2, PSI cannot oxidise this pool because of the chemical block at the condition, the PQ pool remains largely reduced and transcription of the nuclear *Lbcb* genes coding for the major LHCII polypeptides is repressed. This results in an inhibition of the biosynthesis of LHCII polypeptides, which decreases the LHCII polypeptide content. As a consequence, this results in yellow phenotype typical of high-light grown algal cells. Alternatively, photosynthetic electron transport can also be inhibited specifically at PSII in the presence of DCMU (3-(3,4-diclorophenyl)-1,1-dimethylurea. Under these conditions, PSII is unable to reduce PQ to PQH_2 pool oxidised by PSI. This produces a green phenotype which mimics low-light grown cells.

The maintenance of cellular energy balance is called photostasis, which is dependent chloroplast–mitochondrial interactions. For example, the mitochondrial *Moel* protein is thought to regulate the transcription of mitochondrial genes involved in the maintenance of the mitochondrial respiratory electron transport, however, under high light, the *moc1* mutant, which lacks this mitochondrial protein, is unable to up-regulate rates of respiration to match the production of fixed carbon by photosynthesis. The block in mitochondrial electron transport slows the rate of respiratory carbon metabolism which, in turn, causes a feedback inhibition in the rate of photosynthetic electron transport. This also results in the reduction of the PQ pool in the chloroplast. This is an excellent example of the link between chloroplast and mitochondrial metabolism and its importance in the regulation of the gene expression.

Acclimation to drought effects at different plant parts

One of the long-term effects of water deficit is a reduction in vegetative growth. Shoot growth, and especially the growth of leaves, is generally more sensitive than root growth. In a study in which water was withheld from maize plants, for example, there was a significant reduction of leaf expansion when tissue water potentials reached −0.45 MPa and growth was completely inhibited at −1.00 MPa. At the same time, normal root growth was maintained until the water potential of the root tissues reached −0.85 MPa and was not completely inhibited until the water potential dropped to −1.4 MPa. Reduced leaf expansion is beneficial to a plant under conditions of water stress because it leads to a smaller leaf area and reduced transpiration.

Roots are generally less sensitive than shoots to water stress. Apparently, osmotic adjustment in roots is sufficient to maintain water uptake and growth down to much lower water potentials than is possible in leaves. Relative root growth may actually be enhanced by low water potential, such that the shoot–root ratio will change in favour of the proportion of roots. An increase in the root–shoot ratio as the water supply becomes depleted is clearly advantageous, as it improves the capacity of the root system to extract more water by exploring larger volumes of soil. A changing root–shoot ratio is accompanied by a change in source–sink relationships resulting in a larger proportion of photosynthates partitioned to the roots.

Role of lipids in drought stress conditions

Along with proteins, lipids are the most abundant component of membranes and they play a role in the resistance of plant cells to environmental stresses (Kuiper 1980; Suss and Yordanov 1986). Strong water deficit leads to a disturbance of the association between membrane lipids and proteins as well as to a decrease in the enzyme activity and transport capacity of the bilayer (Caldwell and Whitman 1987). Poulson et al. (2002) established that for *Arabidopsis*, polyunsaturated trienoic fatty acids may be an important determinant of responses of photosynthesis and stomatal conductance to environmental stresses such as vapour pressure deficit. When *Vigna unguiculata* plants were submitted to drought the enzymatic degradation of galacto- and phospholipids increased. The stimulation of lipolytic activities was greater in the drought-sensitive than in drought-tolerant cultivars (cvs) (Sahsah et al. 1998).

Drought stress provoked considerable changes in lipid metabolism in rape (*Brassica napus*) plants (Benhassaine-Kesri et al. 2002). The decline in leaf polar lipid was mainly due to a decrease in MGDG (monogalactosyldiacylglycerol) content. Determination of molecular species in phosphatidylcholine and MGDG indicated that the prokaryotic molecular species of MGDG (C18/C16) decreased after DS while eukaryotic molecular species (C18/C18) remain stable. It was suggested that the prokaryotic pathway leading to MGDG synthesis was strongly affected by DS while the eukaryotic pathway was not. Strong WD results in a profound overall drop in MGDG, the major leaf glycolipid. In drought-sensitive seedlings of *Lotus corniculatus* the ratio of MGDG/DGDG declined threefold, while the relative part of MGDG was 12 fold lower.

Mechanisms of acclimation to water deficit and stress tolerance

Many plant systems can survive dehydration, but to a different extent. According to Hoekstra et al. (2001) on the basis of the critical water level, two types of tolerance are distinguished: drought tolerance can be considered as the tolerance of moderate dehydration, down to moisture content, below which there is no bulk cytoplasmic water present—about 0.3 g H_2O g^{-1} DW. Desiccation tolerance refers to the tolerance of further dehydration, when the hydration shell of the molecules is gradually lost. Desiccation tolerance also includes the ability of cells to rehydrate successfully.

Major alterations in patterns of gene expression are known to occur at the early stages of stresses. Some of these changes are thought to provide a long-term protection against stress damage. According to Bohnert and Shen (1999), a nearly universal reaction under stress conditions, including WD, is the accumulation of 'compatible solutes', many of which are osmolytes (i.e. metabolites whose high cellular concentration increases the osmotic potential significantly) considered to lead to osmotic adjustment. These observations indicate that 'compatible solutes' may have other functions as well, namely in the protection of enzyme and membrane structure and in scavenging of radical oxygen species.

Desiccation induces a zeaxanthin + anteraxanthin-mediated photo-protective mechanism in desiccation-intolerant *Frullania dilatata* (Deltoro et al. 1998). They propose that when CO_2 fixation and, therefore, ATP consumption are decreased at low relative water content (RWC), the functioning electron flow gives rise to an acidification of the thylakoid lumen that induces zeaxanthin and anteratxanthin synthesis. It has been proposed that the photo-protective process results in the diversion of energy away from the reaction centre (Ruban and Horton 1995; Medrano et al. 2002). There are, however, experimental data which do not support the statement that the xanthophyll cycle plays a major or specific role in the direct energy dissipation of absorbed light energy (Schindler and Lichtenthaler 1994). According to Tambussi et al. (2002), the non-photochemical fluorescence quenching (qN), as well as the content of zeaxanthin and anteraxanthin after moderate water stress, increased significantly. However, at severe water stress a further rise in these xanthophylls was not associated with any increase in qN. In addition, the β-carotene content rose significantly during severe WD, suggesting an increase in antioxidant defence. One tentative scheme of photosynthetic control under drought is proposed by Medrano et al. (2002, Figure 14.5).

Acclimatisation of plants exposed to cold temperatures

The ability to tolerate freezing temperatures under natural conditions varies greatly among tissues. Seeds and other partially dehydrated tissues, as well as fungal spores, can be kept indefinitely at temperatures near absolute zero (0 K or $-273°C$), indicating that these very low temperatures are not intrinsically harmful. Hydrated, vegetative cells can also retain viability at freezing temperatures, provided that ice crystal formation can be restricted to the intercellular spaces and cellular dehydration is not too extreme.

Temperate plants have the capacity for *cold acclimation*—a process whereby exposure to low but nonlethal temperatures (typically above freezing) increases the capacity for low-temperature

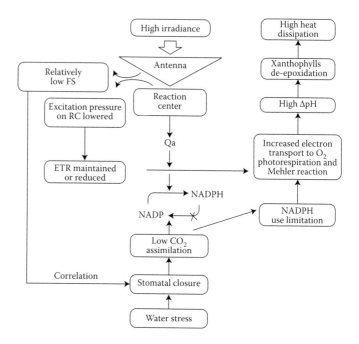

FIGURE 14.5 The relationship between Fs and stomatal conductance provides a method for remote sensing stress. (Adapted from Medrano, H., Escalona, J. M., Boto, J., Gulias, J., Flexas, J. 2002. Regulation of photosynthesis of C3 plants in response to progressive drought: Stomatal conductance as a reference parameter. *Ann. Bot.* 89, 895–905.)

survival. Cold acclimation in nature is induced in the early autumn by exposure to short days and non-freezing, chilling temperatures, which combine to stop growth. A diffusible factor that promotes acclimation, most likely ABA, moves from leaves via the phloem to overwintering stems. ABA accumulates during cold acclimation and is necessary for this process.

Cold acclimation and the development of maximum freezing tolerance in overwintering herbaceous plants such as winter wheat, winger rye, spinach and *Arabidopsis thaliana* require active growth and development at low temperature. As a result, leaves of these plant species developed at low temperatures are anatomically, morphologically, physiologically and biochemically distinct from the same plants developed at warm temperatures. For example, these herbaceous species grown at low temperatures exhibit a short, compact growth habit, thicker leaves due to an increase in leaf mesophyll cell size and an increase in the number of palisade cell layers, an increase in the number of palisade cell layers, and an increase in cell cytoplasm associated with a decrease in leaf water content.

Plants survive freezing temperatures by limiting ice formation

During rapid freezing, the protoplast, including the vacuole, may super-cool, that is, the cellular water remains liquid because of its solute content even at temperatures several degrees below its theoretical freezing point. Super-cooling is common to many species of the hardwood forests. Cells can super-cool to only about −40°C, the temperature at which ice forms spontaneously. Spontaneous ice formation sets the low-temperature limit at which many alpine and sub-Arctic species that undergo deep super-cooling can survive. It may also explain why the altitude of the timberline in mountain ranges is at or near the −40°C minimum isotherm. Several specialised plant proteins, termed antifreeze proteins, limit the growth of ice crystals through a mechanism independent of lowering of the freezing point of water. Synthesis of these antifreeze proteins is induced by cold temperatures. The proteins bind to the surfaces of ice crystals to prevent or slow further crystal growth.

Cold-resistant plants tend to have membranes with more unsaturated fatty acids

As temperatures drop, membranes may go through a phase transition from a flexible liquid-crystalline structure to a solid gel structure. The phase transition temperature varies with species (tropical species: 10–12°C; apples: 3–10°C) and the actual lipid composition of the membranes. Chill-resistant plants tend to have membranes with more unsaturated fatty acids. Chill-sensitive plants, on the other hand, have a high percentage of saturated fatty acid chains, and membranes with this composition tend to solidify into a semi-crystalline state at a temperature well above 0°C. Prolonged exposure to extreme temperatures may result in an altered composition of membrane lipids, a form of acclimation. Certain transmembrane enzymes can alter lipid saturation, by introducing one or more double bonds into fatty acids. This modification lowers the temperature at which the membrane lipids begin a gradual phase change from fluid to semi-crystalline form and allows membranes to remain fluid at lower temperatures, thus protecting the plant against damage from chilling.

Cold-acclimated plants secrete antifreeze proteins

Plant can tolerate freezing because of their ability to control the freezing event itself. As long as the freezing of water is confined to the apoplast, that is the cell wall and the extracellular space, the plant will survive. Alternatively, if freezing occurs intracellularly, the plant will die. Cold acclimation in many plants is associated with the secretion of antifreeze proteins from the cytoplasm into the apoplast. Antifreeze proteins (AFPs) have been reported in ferns, gymnosperms, as well as mono- and dicotyledonous angiosperms. AFPs inhibit ice crystal growth by binding to the surface of a growing ice crystal via hydrogen bonding between specific hydrophilic amino acids present in the AFPs and water within the crystal lattice of ice. The presence of AFPs in cold-tolerant plants is not constitutive but requires exposure to low temperature and they accumulate in virtually all plant tissue, including seeds, stems, leaves, flowers and roots.

The CBFs are the most extensively studied among the stress-related transcription factors because of their critical role in the regulation of low-temperature stress response in *Arabidopsis* and other plant species (reviewed by Thomashow et al. 2001). CBF transcriptional activators, namely Cbf1, Cbf2 and Cbf3 (DREB1a, DREB1b, DREB1c), bind to CRT/DRE elements found in the regulatory regions of many cold inducible genes and induce their transcription activating the plant response to low temperature. The *Arabidopsis* Cbf genes are organised in a tandem arrangement localised on chromosome 4 and their amino acid sequences share a common AP2/EREBPDNA-binding domain (Medina et al. 1999). The expression of the Cbf-like transcripts is transiently up-regulated by cold after 15 min of low-temperature exposure (Medina et al. 1999). A sudden cold stress, transferring the plants directly from 20°C to 4°C, leads to a fast accumulation of Cbf transcripts with a maximum after 3 h of stress. Then a drop of the mRNA steady-state level can be detected, and after 9–21 h of cold stress only a very low amount of Cbf mRNAs can be found. The same expression profile can be recorded after a gradual temperature decrease, suggesting that cold shock is not required to induce Cbf expression; rather, an absolute temperature is being sensed. The threshold temperature, promoting transcript accumulation, is approximately 14°C (Zarka et al. 2003).

Acclimation of the photosynthetic apparatus due to high temperatures

Plants that can acclimate to high temperatures are called thermotolerant plants. Photosynthetic capacity measured as the maximum light saturated rate of CO_2 assimilation is sensitive to temperature. In most cases the productivity of a plant is directly related to the rate of photosynthesis. Photosynthesis is, like all other physiological processes, temperature dependent. For a specific plant there exists an optimal temperature at which the net rate of carbon dioxide fixation is maximal. It was found that among all cell functions, the photosynthetic activity of chloroplasts is one of the most heat sensitive

(Berry and Bjorkman 1980). The damage due to heat stress includes a wide range of structural and functional changes. Their effect on growth and survival depends on the intensity and duration of heat stress. A long period at a moderately high temperature may be as injurious as a brief exposure to an extreme temperature.

It is considered that the primary site of damage is associated with components of the photosynthetic system located in the thylakoid membranes, most probably photosystem II (PSII) (Berry and Bjorkman 1980). The PSII complex is a pigment–protein complex that utilises light energy to drive the transport of electrons and the oxidation of water to oxygen. It is believed that increasing temperature first leads to a blockage of PSII reaction centres and then to a dissociation of the antenna pigment protein complexes from the central core of the PSII (Sundby et al. 1986). Separation of the light-harvesting complex II (LHCII) from the core centre induces destacking of the grana and temperature-induced migration of the reaction centre (PSIIb) or LHCII (state transition) to the non-appressed region, which would have consequences for the energy distribution between PSI and PSII. It has been found that moderately high temperatures stimulate PSI activity *in vivo* and *in vitro* (Sayed et al. 1989). This stimulation appears to be associated with an increased capacity for cyclic electron flow around PSI, which could be an adaptive process, producing ATP under conditions when PSII activity is severely diminished. This ATP synthesis could be important for the survival of plants and necessary for repair of stress-damaged processes, as suggested by Janssen et al. (1992). It is also well known that increasing the temperature at which plants are grown causes an upward shift of the optimal temperature of photosynthesis in numerous species and renders the photosynthetic apparatus more tolerant to heat stress (Berry and Bjorkman 1980). This phenomenon is termed acclimation. Acclimation to a new growth temperature is not instant but requires a certain time period. Under stress, organisms undergo first of all destabilisation followed by normalisation and stability enhancement when limits of tolerance are not exceeded and adaptive capacity is not overtaxed (Larcher 1987).

14.7 Oxygen may protect during acclimation to various stresses

Although the oxygen evolving complex associated with PSII results in the light-dependent evolution of oxygen, O_2 can also act as an alternative electron acceptor; photosynthetic electron transport may also consume oxygen. Even under normal condition, up to 5–10% of the photosynthetic electrons that are generated by PSI may react with molecular oxygen rather than with $NADP^+$. This has important functional consequence for active chloroplasts. The photo-reduction of oxygen by PSI is called the Mehler reaction and results in the production of another toxic, reactive oxygen species known as a super-oxide radical. To counteract the accumulation of this radical, photosynthetic organisms have evolved mechanism to protect themselves from excess light and the potential ravages of O_2. An effective system for the removal of super oxide is the ubiquitous enzyme superoxide dismutase. SOD is found in several cellular compartments including the chloroplast. It is able to scavenge and inactivate superoxide radicals by forming hydrogen peroxide and molecular oxygen.

At the molecular level, the negative effect of high-temperature stress on leaves may be partly a consequence of the oxidative damage to important molecules as a result of the imbalance between production of activated O_2 and antioxidant defences (Foyer et al. 1994). This hypothesis is very plausible because chloroplasts are a major source of activated O_2 in plants (Asada et al. 1998), and because antioxidants, which may play a critical role in preventing oxidative damage, are greatly affected by environmental stresses (Bowler et al. 1994). In chloroplasts, the superoxide radical (O_2•–) is produced by photo-reduction of O_2 at PSI and PSII, and singlet O_2 is formed by energy transfer to O_2 from triplet excited-state chlorophyll (Asada and Takahashi 1987). H_2O_2 can originate, in turn, from the spontaneous or enzyme-catalysed dis-mutation of O_2•–. Fortunately, in optimal conditions leaves are rich in antioxidant enzymes and metabolites and can cope with activated O_2, thus minimising oxidative damage. An increase of the active O_2 forms in plant tissue has been found at high-temperature stress (Foyer et al. 1997; Dat et al. 1998). High temperatures can also influence the antioxidant enzymes: superoxide dismutase (EC 1.15.1.1), the first enzyme in the detoxifying process, converts O_2•– radicals to H_2O_2. In chloroplasts, H_2O_2 is reduced by ascorbate peroxidase (EC 1.11.1.11) using ascorbate as an electron donor. Oxidised ascorbate is then reduced by reactions that are catalysed by monodehydroascorbate reductase (EC 1.8.5.1) and glutathione reductase (EC 1.6.4.2) in a series of reactions known as the Halliwell–Asada pathway (Bowler et al. 1992).

14.8 Plant adaptation to environmental stress

Plants have various mechanisms that allow them to survive and often prosper in the complex environments in which they live. Adaptation to the environment is characterised by genetic changes in the entire population that have been fixed by natural selection over many generations. In contrast, individual plants can also respond to changes in the environment, by directly altering their physiology or morphology to allow them to better survive the new environment. These responses require no new genetic modifications, and if the response of an individual improves with repeated exposure to the new environmental condition, then the response is one of acclimation.

Adaptation to heat stress

Living organisms can be classified into three groups, subject to the preferred temperature of growth. These are (a) *psychrophiles*, which grow optimally at low temperature ranges between 0°C and 10°C; (b) *mesophiles*, which favour moderate temperature and grow well between 10°C and 30°C; and (c) *thermophiles*, which grow well between 30°C and 65°C or even higher. There is a great variation among the plant species in terms of their response and tolerance to high temperature. Plant adaptation to heat stress includes avoidance and tolerance mechanisms which employ a number of strategies (Figure 14.6).

Under high-temperature conditions, plants exhibit various mechanisms for surviving which include long-term evolutionary phenological and morphological adaptations and short-term avoidance or acclimation mechanisms such as changing leaf orientation, transpirational cooling or alteration of membrane lipid compositions. Closure of stomata and reduced water loss, increased stomatal and trichomatous densities, and larger xylem vessels are common heat-induced features in plant (Srivastava et al. 2012). Plants growing in a hot climate avoid heat stress by reducing the absorption of solar radiation. This ability is supported by the presence of small hairs (tomentose) that form a thick coat on the surface of the leaf as well as cuticles, protective waxy covering. In such plants, leaf blades often turn away from light and orient themselves parallel to sun rays (paraheliotropism). Solar radiation may also be reduced by rolling leaf blades. Plants with small leaves are also more likely to avoid heat stress: they evacuate heat to ambient more quickly due to smaller resistance of the air boundary layer in comparison with large leaves. Plants rely on the same anatomical and physiological adaptive mechanisms that are deployed in a water deficit to

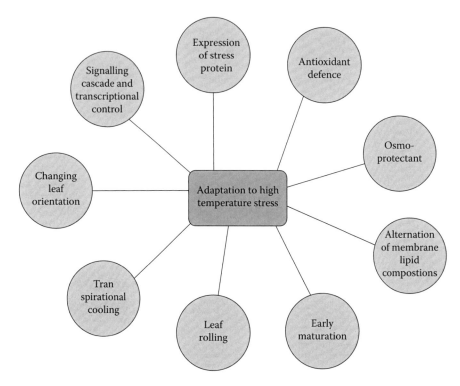

FIGURE 14.6   Different adaptation mechanisms of plants to high temperature.

limit transpiration. In well-hydrated plants, intensive transpiration prevents leaves from heat stress, and leaf temperature may be 6°C or even 10–15°C lower than ambient temperature. Many species have evolved life histories which permit them to avoid the hottest period of the year. This can be achieved by leaf abscission, leaving heat-resistant buds, or in desert annuals, by completing the entire reproductive cycle during the cooler months (Fitter and Hay 2002). Such morphological and phenological adaptations are commonly associated with biochemical adaptations favouring net photosynthesis at HT (in particular C4 and CAM photosynthetic pathways), although C3 plants are also common in desert floras (Fitter and Hay 2002).

Tolerance mechanisms   Heat tolerance is generally defined as the ability of the plant to grow and produce economic yield under high temperature. This is a highly specific trait, and closely related species, even different organs and tissues of the same plant, may vary significantly in this respect. Plants have evolved various mechanisms for thriving under higher prevailing temperatures. They include short-term avoidance/acclimation mechanism or long-term evolutionary adaptations. Some major tolerance mechanisms, including ion transporters, LEA proteins, osmoprotectants, antioxidant defence and factors involved in signalling cascades and transcriptional control are essentially significant to counteract the stress effects (Rodríguez et al. 2005). In case of sudden heat stress, short-term response, that is, leaf orientation, transpirational cooling and changes in membrane lipid composition are more important for survival (Rodríguez et al. 2005). Smaller yield losses due to early maturation in summer show possible involvement of an escape mechanism in heat stress tolerance (Adams et al. 2001). Different tissues in plants show variations in terms of developmental complexity, exposure and responses towards the prevailing or applied stress types (Queitsch et al. 2000). The stress responsive mechanism is established by an initial stress signal that may be in the form of an ionic and osmotic effect or changes in the membrane fluidity. This helps to reestablish homeostasis and to protect and repair damaged proteins and membranes (Vinocur and Altman 2005).

References

ADAMS, S. R., COCKSHULL, K. E., CAVE, C. R. J. 2001. Effect of temperature on the growth and development of tomato fruits. *Ann. Bot.* 88, 869–877.

ASADA, K., TAKAHASHI, M. 1987. Production and scavenging of active oxygen in photosynthesis. In: *Photoinhibition*. Eds. D. J. Kyle, C. B. Osmond and C. J. Arntzen, Elsevier, Amsterdam, pp. 227–287.

ASADA, K., ENDO, T., MANO, J., MIYAKE, C. 1998. Molecular mechanisms of relaxation and protection from light stress. In: *Stress Responses of Photosynthetic Organisms*. Eds. K. Satoh and N. Murata, Elsevier Science, BV, Amsterdam, pp. 37–52.

BERRY, J., BJORKMAN, O. 1980. Photosynthetic response and adaptation to temperature in higher plants. *Ann. Rev. Plant Physiol.* 31, 491–453.

BENHASSAINE-KESRI, G., AID, F., DEMANDRE, C., KADER, J-C., MAZLIAK, P. 2002. Drought stress affects chloroplast lipid metabolism in rape (*Brassica napus*) leaves. *Physiol. Plant.* 115, 221–227.

BLUMWALD, E. 2000. Sodium transport and salt tolerance in plants. *Curr. Opin. Cell Biol.* 12, 431–434.

BOHNERT, H. J., SHEN, B. 1999. Transformation and compatible solutes. *Scientia Hort.* 78, 237–260.

BOWLER, C., VAN MONTAGU, M., INZE, D. 1992. Superoxide dismutase and stress tolerance. *Ann. Rev. Plant Physiol. Plant Mol. Biol.* 43, 83–116.

BOWLER, C., VAN CAMP, W., VAN MONTAGU, M., INZE, D. 1994. Superoxide dismutase in plants. *Crit. Rev. Plant Sci.* 13, 199–218.

BRAY, E. A., BAILEY-SERRES, J., WERETILNYK, E. 2000. Responses to abiotic stresses. In: *Biochemistry and Molecular Biology of Plants*. Eds. W. Gruissem, B. Buchannan R. Jones, American Society of Plant Physiologists, Rockville, MD, pp. 1158–1249.

CALDWELL, C. R., WHITMAN, C. E. 1987. Temperature-induced protein conformational changes in barley root plasma membrane-enriched microsomes. I. Effect of temperature on membrane protein and lipid mobility. *Plant Physiol.* 84, 918–923.

CHOI, H. I., HONG, J. H., HA, J., KANG, J. Y., KIM, S. Y. 2000. ABFs, a family of ABA-responsive element binding factors. *J. Biol. Chem.* 275, 1723–1730.

DELTORO, V. I., CALATAYUD, A., GIMENO, C., ABADIA, A., BARRENO, E. 1998. Changes in chlorophyll a fluorescence, photosynthetic CO_2 assimilation and xanthophyll cycle interconversions during dehydration in desiccation-tolerant and intolerant liverworts. *Planta* 207, 224–228.

DAT, J., FOYER, C., SCOTT, I. 1998. Changes in salicylic acid and antioxidants during induced thermotolerance in mustard seedlings. *Plant Physiol.* 118, 1455–1461.

FITTER, A. H., HAY, R. K. M. 2002. *Environmental Physiology of Plants*, 3rd ed., Academic Press, London, UK.

FRANK, W., MUNNIK, T., KERKMANN, K., SALAMINI, F., BARTELS, D. 2000. Water deficit triggers phospholipase D activity in the resurrection plant carterostigma plantagineur. *Plant Cell.* 12, 111–124.

FOYER, C., DESCOURVIERES, P., KUNERT, K. 1994. Protection against oxygen radicals: Important defence mechanisms studied in transgenic plants. *Plant Cell Environ.* 17, 507–523.

FOYER, C. H., LOPES-DEGADO, H., DAT, J. F., SCOTT, I. M. 1997. Hydrogen peroxide- and glutathione-associated mechanisms of acclimatory stress tolerance and signalling. *Plant Physiol.* 100, 241–254.

GRUSZECKI, W., STRZALKA, K. 1991. Does the xanthophyll cycle take part in the regulation of fluidity of the thylakoid membrane? *Biochim. Biophys. Acta* 1060, 310–314.

HOPKINS, W. G., HÜNER, N. P. A. 2009. *Introduction to Plant Physiology*, 4th ed., John Wiley Sons, Inc., Hoboken, USA.

HOEKSTRA, F., GOLOVINA, E., BUITINK, J. 2001. Mechanisms of plant desiccation tolerance. *Trends Plant Sci.* 8(9), 431–438.

JANSSEN, L. H. J., WAMS, H. W., VAN HASSELT, P. R. 1992. Temperature dependence of chlorophyll fluorescence induction and photosynthesis in tomato as affected by temperature and light conditions during growth. *J. Plant Physiol.* 139, 549–554.

KUIPER, P. J. C. 1980. Lipid metabolism as a factor in environmental adaptation. In: *Biogenesis and Function of Plant Lipids*. Eds. Maliak, P. et al. Elsevier, Amsterdam, pp. 169–176.

LARCHER, W. 1987. Stress bei Pflanzen. *Naturwissenschaften* 74, 158–167.

MAHAN, J. R., MAUGET, S.A. 2005. Antioxidant metabolism in cotton seedlings exposed to temperature stress in the field. *Crop Sci.* 45, 2337–2345.

MEDINA, J, BARGUES, M., TEROL, J., PEREZ-ALONSO, M., SALINAS, J. 1999. The *Arabidopsis* CBF gene family is composed of three genes encoding AP2 domain-containing proteins whose expression is regulated by low temperature but not by abscisic acid or dehydration. *Plant Physiol.* 119, 463–469.

MEDRANO, H., ESCALONA, J. M., BOTO, J., GULIAS, J., FLEXAS, J. 2002. Regulation of photosynthesis of C3 plants in response to progressive drought: Stomatal conductance as a reference parameter. *Ann. Bot.* 89, 895–905.

OOSTERHUIS, D. M. 2002. Day or night high temperatures: A major cause of yield variability. *Cotton Grower* 46, 8–9.

POULSON, M. E., EDWARDS, G. E., BROWSE, J. 2002. Photosynthesis is limited at high leaf to air vapor deficit in a mutant of *Arabidopsis thaliana* that lacks trienoic fatty acids. *Photosynth. Res.* 72, 55–63.

QUEITSCH, C., HONG, S. W., VIERLING, E., LINDQUIST, S. 2000. Hsp101 plays a crucial role in thermotolerance in *Arabidopsis*. *Plant Cell* 12, 479–492.

RAISON, J. K., ROBERTS, J. K. M., BERRY, J. A. 1982. Correlations between the thermal stability of chloroplast thylakoid membranes and the composition and fluidity of their polar lipids upon acclimation of higher plant *Nerium oleander* to growth temperature. *Biochim. Biophys. Acta* 688, 218–228.

RODRÍGUEZ, M., CANALES, E., BORRÁS-HIDALGO, O. 2005. Molecular aspects of abiotic stress in plants. *Biotechnol. Appl.* 22, 1–10.

RUBAN, A. V., HORTON, P. 1995. Regulation on non-photochemical quenching of chlorophyll fluorescence in plants. *Aust. J. Plant Physiol.* 22, 221–230.

SARRY, J.-E., MONTILLET, J.-L., SAUVAIRE, Y., HAVAUX, M. 1994. The protective function of the xanthophylls cycle in photosynthesis. *FEBS Lett.* 353, 147–150.

SAHSAH, Y., CAMPOS, P., GAREIL, M., ZUILY-FODIL, PHAM-THI, A. T. 1998. Enzymatic degradation of polar lipids in *Vigna unguiculata* leaves and influence of drought stress. *Physiol. Plant.* 104, 577–586.

SAYED, O. H., EARNSHAW, M. J., EMES, M. J. 1989. Photosynthetic response of different varieties of wheat to high temperature. II. Effect of heat stress on photosynthetic electron transport. *J. Exp. Bot.* 40, 633–638.

SCHINDLER, C., LICHTENTHALER, H. 1994. Is there a correlation between light-induced zeaxanthin accumulation and quenching of variable chlorophyll a fluorescence? *Plant Physiol. Biochem.* 32, 813–823.

SMIRNOFF, N. 1998. Plant resistance to environmental stress. *Curr. Opin. Biotech.* 9, 214–219.

SRIVASTAVA, S., PATHAK, A. D., GUPTA, P. S., SHRIVASTAVA, A. K., SRIVASTAVA, A. K. 2012. Hydrogen peroxide-scavenging enzymes impart tolerance to high temperature induced oxidative stress in sugarcane. *J. Environ. Biol.* 33, 657–661.

SNIDER, J. L., OOSTERHUIS, D. M., KAWAKAMI, E. M. 2010. Genotypic differences in thermotolerance are dependent upon prestress capacity for antioxidant protection of the photosynthetic apparatus in *Gossypium hirsutum. Physiol. Plantarum* 138, 268–77.

SUNG, D.-Y., KAPLAN, F., LEE, K.-J., GUY, C. L. 2003. Acquired tolerance to temperature extremes. *Trends Plant Sci.* 8, 179–187.

SUNDBY, C., MELIS, A., MAENPAA, P., ANDERSSON, B. 1986. Temperature-dependant changes in the antenna size of photosystem II. *Biochim. Biophys. Acta* 851, 475–483.

SUSS, K.-H., YORDANOV, I. 1986. Biosynthetic cause of *in vivo* acquired thermotolerance of photosynthetic light reactions and metabolic responses of chloroplasts to heat stress. *Plant Physiol.* 81, 192–199.

TAMBUSSI, E. A., CASADESUS, J., MUNNE-BOSH, S. M., ARAUS, J. L. 2002. Photoprotection in water-stressed plants of durum wheat (*Tr. Turgidum* var. *durum*): Changes in chlorophyll fluorescence spectral signature and photosynthetic pigments. *Functional Plant Biol.*, 29, 35–44.

THOMASHOW, M. F. 1999. Plant cold acclimation: Freezing tolerance genes and regulatory mechanisms. *Annu. Rev. Plant Biol.* 50, 571–599.

THOMASHOW, M. F., GILMOUR, S. J., STOCKINGER, E. J., JAGLO-OTTOSEN, K. R., ZARKA, D. G. 2001. Role of the *Arabidopsis* CBF transcriptional activators in cold acclimation. *Physiol. Plant* 112, 171–175.

VINOCUR, B., ALTMAN, A. 2005. Recent advances in engineering plant tolerance to abiotic stress: Achievements and limitations. *Curr. Opin. Biotechnol.* 16, 123–132.

YANG, K. A., LIM, C. J., HONG, J. K., PARK, C. Y., CHEONG, Y. H., CHUNG, W. S., LEE, K. O,, LEE, S Y., CHO, M. J., LIM, C. O. 2006. Identification of cell wall genes modified by a permissive high temperature in Chinese cabbage. *Plant Sci.* 171, 175–182.

ZARKA, D. G., VOGEL, J. T., COOK, D., THOMASHOW, M. F. 2003. Cold induction of Arabidopsis Cbf genes involves multiple ice (inducer of Cbf expression) promoter elements and a cold-regulatory circuit that is desensitized by low temperature. *Plant Physiol.* 133, 910–918.

ZHU, J. K. 2001. Cell signaling under salt, water and cold stresses. *Curr. Opin. Plant Biol.* 4, 401–406.

ZROBEK-SOKOLNIK, A. 2012. Temperature stress and responses of plants. In: *Environmental Adaptations and Stress Tolerance of Plants in the Era of Climate Change.* Eds. P. Ahmad, M. N. V. Prasad, Springer, New York, NY, USA, pp. 113–134.

CHAPTER FIFTEEN

Global warming impact on crop productivity

S.C. Santra, Anusaya Mallick and A.C. Samal

Contents

Abstract

Global warming refers to climate change that causes an increase in the average temperature of the lower atmosphere. Global warming has emerged as the most serious environmental threat to the Earth in the twenty-first century. The global population is now discussing the environmental problems and their remedies. This environmental problem on a global scale can originate from any part of the world yet affect the whole Earth. The global warming problem directly or indirectly affects the ecosystem, agriculture and human beings. It is estimated that temperature has been increasing at an average rate of 0.3% per decade or 5°C in 170 years due to industrialisation, urbanisation, burning of fossil fuel, deforestation and so on. The concentration of carbon dioxide (CO_2) in the atmosphere is expected to double by the end of the twenty-first century. Crop prediction models have been made to study the impact of climate change on agricultural production and food security. In India, a number of experiments have been done to understand the nature and magnitude of change in different crop yield due to climate change.

15.1 Introduction

An increase in the concentration of greenhouse gases (GHGs) is the major cause of global warming. The global atmospheric temperature has increased by about 0.5°C since 1975 (Hansen et al., 1999). This 'burst' of warmth has taken the global temperature to its highest level in the past millennium and has recorded the 1990s as the warmest decade and 1998 as the warmest year in the Northern Hemisphere (Mann et al., 1999). There are many competing natural and anthropogenic climate factors responsible for global warming, but increased GHGs are estimated to be the largest forcing, especially during the past few decades (Intergovernmental Panel on Climate Change (IPCC) 1996). In recent decades, the supporting evidences provided by observing heat storage in the ocean, which is positive and of the magnitude of the energy imbalance estimated from climate forcing (Hansen et al., 1997). During the last century in India, there is a trend in increasing surface temperature (Hingane et al., 1985; Srivastava et al., 1992; Rupa Kumar et al., 1994; De and Mukhopadhyay, 1998; Pant et al., 1999; Singh and Sontakke, 2002; Singh et al., 2001). The IPCC has considered a range of scenarios for future GHGs, which is further expanded in its special report on emission scenarios (Nakicenovic et al., 2000). The climate models suggest that these scenarios yield a steep by continuous increase in global temperature throughout the twenty-first century with warming of several degree Celsius by 2100 AD, if climate sensitivity is 2–4°C for doubled CO_2. Climate forcing by CO_2 is the largest forcing. Forcing by CH_4 (0.7 W . m²) is half as large as that of CO_2 and the total forcing by non-CO_2 GHGs (1.4 W · m²) equals that of CO_2.

However, CO_2 is essential for plant growth. The rise of CO_2 will promote plant growth through intensified photosynthesis. Some reports indicate that a rise in the levels of CO_2 would actually benefit plants, rather than harm them. The growth rates of C3 plants increase in response to elevated concentrations of CO_2. Thus, global warming might increase plant growth, because of higher temperatures and higher levels of atmospheric CO_2. High atmospheric temperatures caused by elevated concentrations of CO_2 will induce heat injury and physiological disorders in some crops, which will decrease the income of farmers and agricultural countries. Photosynthesis is one of the most sensitive physiological processes to high-temperature stress. Reproductive development is more sensitive than vegetative development to high temperatures, and heat sensitivity differs among crops. Thus, global warming can have opposite effects on plant growth. From a long-term viewpoint, however, high atmospheric temperatures will drive the main sites of crop production. Water shortages caused by global warming will be the greatest problem for crop production. Plants fundamentally rely on adequate fresh water, and agricultural water accounts for 70% of water use worldwide. As higher temperatures increase evaporation from water sources and decrease precipitation, arid regions will become further desolated. Overall, the entire crop production will be affected by global warming, resulting in worldwide food shortages.

15.2 Predictions of climate change models

To estimate the impact of long-term global climate changes on agriculture, we should understand the direction and magnitude of climate changes. Climate change projections rely on large, complex computer models, known as global circulation models (GCMs) that have been successful in depicting the gross features of the observed large-scale climatological features. However, there is great uncertainty associated with these projections on a regional scale, since GCMs are yet to realistically reproduce the observed features at a regional scale, particularly over the monsoon region. The most commonly used GCMs are GISS (Goddard Institute for Space Studies, National Aeronautics and Space Administration), GFDL (Geophysical Fluid Dynamics Laboratory, National Oceanic and Atmospheric administration) and UKMO (United Kingdom Meteorological Office). There are inherent limitations of GCMs in predicting climate change. The most significant limitations include

1. Poor spatial resolution
2. Inadequate coupling of atmospheric and oceanic processes
3. Poor simulation of cloud processes
4. Inadequate representation of the biosphere and its feedback

There are now many different models being used to simulate climate change, because several modelling groups constantly revise the GCMs. In general, GCMs can at best be used to suggest the likely direction and rate of change, because they still have significant inherent limitations in simulating current climate.

The rate of increase of global mean surface temperature is predicted to be around 0.3°C before the end of the century. Regional climate changes are different from the global mean. It is predicted that surface air will warm faster over land than over oceans and that the warming is expected to be 50–100% greater than the global mean in high northern latitudes in winter. There is increased precipitation in the order of 5–10% in middle and high latitude continents (35–55°N) in winter. The global mean sea level is expected to rise about 6 cm per decade on average over the next century mainly due to the thermal expansion of the oceans and the melting of some land ice. A sea level rise of about 65 cm is predicted by the end of the next century.

Although GCM predictions are not ideal for agricultural impact analysis, they serve as a suitable benchmark for our global economic analysis directed at evaluating general directions and elative magnitudes of change, In particular, the GCM predictions suggest broad geographical zones across which climate change may affect agriculture. Increased precipitation and warming in the high northern latitudes could enhance agricultural production potential in the northern regions of the erstwhile Soviet Union, Canada and northern Europe. Drying in the interior of continents in the northern middle latitudes combined with warming could lead to negative crop and livestock effects in the United States and Western Europe and the most agriculturally productive regions of Canada. Other northern middle latitude regions, including South-East Asia could suffer from coastal inundation.

There are exceptions to the broadly generalised climate patterns sketched by the IPCC. While China falls within the category of northern middle latitude countries, climate models suggest crop production potential could increase. Regions of agricultural importance in the southern middle latitudes include Argentina and Australia. Projections show a wetter and, therefore, agriculturally more productive climate for the major agricultural regions in Australia (Walker et al., 1989). Much less is known about the possible agricultural effects of climate change in the tropical latitudes encompassing regions of Africa, Latin America and Asia. In general, temperature changes are expected to be smaller in equatorial regions than in higher latitudes, but there is very little agreement on changes in precipitation and soil moisture.

15.3 Global warming impacts on plant growth and agriculture

There has been much talk recently of the rise in CO_2 levels observed over the last few decades and its potential impact on global climate and ecosystems. The buildup of CO_2 and other GHGs can trap heat in the atmosphere, increasing the average temperature of the Earth. An important effect is that of CO_2 in chemical reactions that occur in nature, because it is an essential component in many of these reactions. One of the main biological interactions involving CO_2 is the process of photosynthesis, by which CO_2 from the atmosphere is converted into glucose by plants. Most plants growing in enhanced CO_2 exhibit increased rates of net photosynthesis. The higher photosynthesis rates are then manifested in higher leaf area, dry matter production and yield for many crops (Kimball, 1983). In several cases, high CO_2 has contributed to upward shifts in temperature optima for photosynthesis (Jurik et al., 1984) and to enhanced growth with higher temperatures (Idso et al., 1987); other studies, however, have not shown such benefits (Baker et al., 1989). CO_2 enrichment also tends to close plant stomata, and by doing so, reduces transpiration per unit leaf area while still enhancing photosynthesis. The stomatal conductance of 18 agricultural species has been observed to decrease markedly (by 36%, on average) in an atmosphere enriched by doubled CO_2 (Morison and Gifford, 1984). However, crop transpiration per ground area may not be reduced commensurately, because decreases in individual leaf conductance tend to be offset by increases in crop leaf area (Allen et al., 1998). In any case, higher CO_2 often improves water use efficiency, defined as the ratio between crop biomass accumulation or yield and the amount of water used in evapo-transpiration. Growth rates of C_3 plants have shown to be higher at elevated concentrations of CO_2. Some plants, such as soya beans, have demonstrated faster growth in an environment with high levels of CO_2 and high temperatures.

Agriculture plays a key role in the overall economic and social well-being of India. For any particular crop, the effect of increased temperature will depend on the crop's optimal temperature for growth and reproduction (USGCRP, 2009). The components of the natural ecosystem are very sensitive to changes in weather and climate, particularly to extreme weather events, decreased soil moisture, temperature change and increased CO_2 in the atmosphere. They will also affect the ground water replenishment patterns and evapo-transpiration rates (Allen et al., 1994). So, vegetation as well as agriculture is likely to be affected from such changes in weather and atmosphere. The impact on agriculture could be of two major types. First, by altering production adversely in the main food-producing areas, climate change could enhance food scarcities. The location of main food-producing regions could change. Second, there could be a profound impact on physiological mechanisms regulating plant and animal productivity. The greatest impact is likely to come from changes in the precipitation pattern. Attri and Rathore (2003) suggested the adaptation strategies for sustainable production of wheat and ensuring food security.

15.4   A crop yield response to climate change: Global perspective

Global estimates of climate impacts on agriculture have been fairly rough to date due to lack of consistent methodology and uncertainty about the physiological effects of CO_2. How climate change might affect agriculture was studied by Liverman (1987) and Warrick (1988). Kane et al. (1989) broadly predicted improvements in agricultural production at high latitudes and reductions in Northern Hemisphere mid-continental agricultural regions. The IPCC (1990) concluded that while future food production should be maintained, negative impacts were likely in some regions, particularly where present-day vulnerability is high. An international project was created by the US Environmental Protection Agency, 'Implications of Climate Change for International Agriculture: Global Food Trade and Vulnerable Regions', to estimate the potential effects of greenhouse gas-induced climate change on global food trade, focusing on the distribution and quantity of production of the major food crops for a consistent set of climate change scenarios and CO_2 physiological effects. Other goals of the project were to determine how currently vulnerable, food-deficit regions may be affected by global climate change, to identify the future locations of those regions and the magnitudes of their food deficits and to study the effectiveness of adaptive responses, including the use of genetic resources to global climate change.

As a part of the US Environmental Protection Agency project, crop scientists estimated that the yield changes at over 100 sites in over 20 countries under common climate change scenarios using compatible crop growth models. The crop models were those developed by the International Benchmark Sites Network for Agrotechnology Transfer (IBSNAT, 1990). Preliminary national production changes for wheat based on the IBSNAT crop model have been chronicled (Rosenzweig et al., 1991). These results show that the climate change scenarios without the physiological effects of CO_2 cause a decrease in the estimated national production, while the physiological effects of CO_2 mitigate the negative effects. The UKMO climate change scenario (mean global warming of 5.2°C) generally causes the largest production declines, while the GFDL and GISS (4.0°C and 4.2°C mean global warming, respectively) production changes are more moderate. When embedded in a global agricultural food trade model, the basic linked system, (Fischer et al., 1994), the production change estimates based on the IBSNAT crop model results will allow for projection of potential impacts on food prices, shifts in comparative advantage and altered patterns of global trade flows for a suite of global climate change, population growth and policy scenarios.

15.5 Direct effect

There have been a number of studies in India to understand the nature and magnitude of crop yield at selected sites under elevated atmospheric CO_2 and associated climatic change (Abrol et al., 1991; Sinha and Swaminathan, 1991; Aggarwal and Sinha, 1993; Aggarwal and Kalra, 1994; Gangadhar Rao and Sinha, 1994; Mathauda and Mavi, 1994; Gangadhar Rao et al., 1995; Mohandass et al., 1995; Lal et al., 1998, 1999; Francis 1999; Saseendran et al., 2000, Rathore et al., 2001; Aggarwal and Mall, 2002; Mall and Aggarwal, 2002; Aggarwal, 2003; Attri and Rathore, 2003, Mall et al., 2004, 2006).

An increase in the CO_2 level may result in an increase in food production. It is predicted that a twofold increase in CO_2 will lead to a 10–15% increase in dry matter production provided all other factors remain constant. As C_3 plants respond much more to an increase in the CO_2 level than do C_4 plants, crops in Central and Northern Europe and similar latitudes are expected to perform better than crops in tropical areas, where maize, sorghum, sugarcane and millets are staples and grown in abundance. Generally, it is assumed that increased atmospheric CO_2 could enhance growth rates of certain types of crop plants and that change in temperature and precipitation would affect livestock crops, pest and soils. Plant growth is generally limited by carbon sink rather than carbon source. Temperature-sensitive plant developmental processes may constrain many agricultural yields. In order to draw an idea about crop response under enhanced CO_2 concentration, other factors such as temperature and precipitation must be taken into consideration as their combined effect may be somewhat different. The effect of elevated CO_2 on rice cultivars are studied by Uprety et al. (2000, 2002, 2003). The yields of some crops with elevated CO_2 concentration are shown in Tables 15.1 through 15.3 (Cline, 1999; Aggarwal, 2000; Uprety et al., 2000).

Increases in CO_2 and other GHGs result in climate changes such as temperature increase, more erratic pattern of rainfall, weather change and so on. The uneven monsoon rainfall in India leads to large-scale droughts and floods, having a major effect on Indian crop production (Parthasarathy and Pant, 1985; Parthasarathy et al., 1992; Selvaraju, 2003; Kumar et al., 2004) and on the economy of the country (Gadgil et al., 1999a; Kumar and Parikh, 2001). This certainly has an impact on agriculture and this impact manifests itself in a number of ways, namely, changes in the length of growing season. Scientists suggest that a 1°C increase in average temperature would tend to advance the thermal limit of cereal cropping in the mid-latitude Northern Hemisphere and would bring more land under cultivation than

Table 15.1 Effect of elevated (620 ppmv) CO_2 concentration on yield and yield attributes in rice (*Oryza sativa* var. Pusa 834)

Character	Ambient (310 ppmv)	Elevated (620 ppmv)	% Increase over ambient	Critical difference at 0.05P
Grain yield (g)	28.28	43.22	52.83	8.50
100 grain weight (g)	18.80	21.90	16.49	0.92
Number of panicles	12.00	14.00	16.63	0.76
Grain number/ panicle	139.40	230.40	65.28	23.20

Source: Uprety, D. C. et al., 2000. *Indian J. Plant Physiol.*, 5: 105–107.

Table 15.2   Grain yields, components of yield and harvest index of rice grown at different CO_2 levels and temperature regimes

CO_2 concentration (ppmv)	Mean air temperature (°C)	Grain yield mg/ha	Number of panicles/ plant	Number of filled grains/ panicle	Harvest index (%)
330	24.2	7.9	5.1	34.5	47
	25.1	8.0	5.6	47.5	43
	31.0	4.3	5.4	19.0	26
	34.1	3.2	5.2	15.2	24
660	24.2	8.4	5.0	37.7	46
	25.1	10.1	6.2	54.2	47
	31.0	6.4	6.0	24.6	29
	34.1	3.4	5.6	12.9	18

Source: Cline, R. W. 1999. *Global Warming and Agriculture: Impact Estimation by Country.* Center for Global Development and the Peterson Institute for International Economics, Washington, DC.

Table 15.3   Seed yield, components of yield and harvest index of soya bean grown at two CO_2 concentrations and three temperatures

CO_2 concentration	Day/night temperature (°C)	Grain yield (g/plant)	Seed number/ plant	Seed mass (mg/seed)	Harvest index (%)
330	26/9	9.0	44.7	202	53
	31/24	10.1	52.1	195	51
	36/29	10.1	58.9	172	45
660	26/19	13.1	58.8	223	49
	31/24	12.5	63.2	198	45
	36/29	11.6	70.1	165	44

Source: Aggarwal, P. K. 2000. *Climate Change and Agriculture: Information Needs and Research Priorities.* Indian Agricultural Research Institute, New Delhi.

what it is today under cool climates (Mann et al., 1999). Mixed yield results were obtained from different models and it is quite difficult to summarise them and come to a conclusion. In cool temperature and cold regions, the yield is expected to rise, provided the soil moisture is optimum. But global warming causes soil moisture depletion and this has a serious effect on crop yields. A study predicts that the maize yield will reduce by 20%, even if the crop is irrigated due to severe depletion of soil moisture and an increase in temperature (ICAR, 2005) (Table 15.4). In another study, it is revealed that an increase in rainfall may be beneficial to yields of crops. It is assumed that with an increase

Table 15.4   Estimated impact of heat wave on wheat yield during 2004

State	Yield loss (%)	Production loss (million tonnes)
Uttaranchal	8.60	0.066
Punjab	8.32	1.287
Haryana	7.62	0.704
Uttar Pradesh	6.75	1.720
Himachal Pradesh	5.79	0.033
Bihar	4.73	0.230
Rajasthan	3.87	0.213
Madhya Pradesh	1.11	0.084
Maharashtra	0.00	0.000
West Bengal	0.00	0.000
India	46.79	4.387

Source: ICAR, 2005. Annual Report, 2005, New Delhi.

in the temperature by 1°C, the precipitation will increase by at least 100 mm and this increased precipitation will enhance 10% yield of rice, wheat and maize. On the other hand, yield of cereals and other agricultural produce will come down heavily, particularly in warmer regions, as revealed in studies. Some studies even predict severe drought in places in low latitudes. Whether the effect of enhanced levels of GHGs will be beneficial or not, as far as yield is concerned, is not clear even today.

15.6   Indirect effect

GHGs such as CO_2, CH_4 and N_2O are directly related with agriculture. The global warming potential (GWP) of CH_4 is 20 times and that of N_2O is 300 times more than that of CO_2. These GHGs are nearly transparent to the visible and near-infrared wavelengths of sunlight, but they absorb and re-emit downward a large fraction of the longer infrared radiation emitted by the Earth. As a result of this heat trapping, the atmosphere radiates large amounts of long-wavelength energy downward to the Earth's surface and long-wavelength radiant energy received on the Earth is increased.

It is certain that agriculture is a source of N_2O. Nitrous oxide, which is present in the atmosphere at about 310 ppbv, is slowly increasing at a rate of about 25% annually. But despite its low concentration and slow rise, N_2O is becoming an important greenhouse gas because of its longer lifetime (150 years) and greater GWP. Atmospheric N_2O absorbs thermal radiation. An increase of 0.2–0.3% in N_2O concentration in the atmosphere contributes about 5% to global warming.

Emissions of GHGs, particularly the non-CO_2 gases such as methane, nitrous oxide, carbon monoxide and nitrogen oxides, from the agriculture sector are significant in India. The primary sources are the large agricultural areas under paddy cultivation and large cattle population in India. The overall budget of atmospheric CH_4 emission amounts to 500 Tg/year of which 60 Tg/year from paddy fields worldwide. Methane emissions from the agricultural sector for 1990 in Tg is as follows:

Livestock—0.3 Tg/year (largest contribution is from non-dairy followed by buffaloes); paddy cultivation—4.07 ± 1.25 Tg/year; animal manure—0.9 Tg/year; field burning of agricultural residues—0.116 Tg/year.

1. The Indian domestic livestock population increased from 456 million in 1987 to 467 million in 1992 and is expected to increase to 625 million in 2020.

2. The paddy cultivation area of 42.32 mha in India is the largest in Asia. The global emission of methane from paddy cultivation is 60 Tg.

3. Field burning of agricultural residues also releases $CO = 2531$ Gg and $N_2O = 3$ Gg, and agricultural soils also release $N_2O = 0.24$ Tg/year (1 Tg = 10^{12} g or 1 million tonnes).

15.7 Pests and diseases susceptibility

The impacts of elevated CO_2 should be considered among others in the context of

1. Changes in air temperature, particularly nocturnal temperature due to increase in CO_2 and other trace gases and changes in moisture availability and their effect on vegetative versus reproductive growth.

2. Need for more farm resources (e.g. fertilisers).

3. Survival and distribution of pest populations, thus developing a new equilibrium between crops and pests (Krupa, 2003).

Indirectly, there may be considerable effects on land use due to snow melt, spatial and temporal rainfall variability, availability of irrigation, frequency and intensity of inter- and

intra-seasonal droughts and floods, soil organic matter trans-
formations, soil erosion, change in pest profiles, decline in
arable areas due to submergence of coastal lands and availability of energy. All these can have a tremendous impact on
agricultural production and, hence, food security of any region
(Aggarwal, 2003). The rising temperatures and CO_2 and uncertainties in rainfall associated with global warming may or
may not have serious direct and indirect consequences on crop
production. It is, therefore, important to have an assessment of
these consequences of global warming on different crops, especially on cereals contributing to food security (Gadgil, 1995;
Gadgil et al., 1999a,b). Mechanistic crop growth models are
now routinely used for assessing the impacts of climate change.
There are several crop models now available for the same crop
that can be employed for impact assessment of climate change
(Mall and Aggarwal, 2002). Crop models, in general, integrate
current knowledge from various disciplines, including agrometeorology, soil physics, soil chemistry, crop physiology, plant
breeding and agronomy, into a set of mathematical equations to
predict growth, development and yield of a crop (Aggarwal and
Kalra, 1994; Hoogenboom, 2000).

As there is an intimate relationship between occurrence
and distribution of pests and diseases and the temperature, any
change in temperature will have a significant effect on pest and
disease development and their interaction with agricultural
crops. A number of effects of global warming on insects, pests
and disease-causing organisms have been identified. These are
increases in the number and rate of development in a season,
expansion of area of distribution, earlier establishment in their
population in a favourable season, and more intense attacks,
particularly by midnight and exotic species. Again, with the
shifting agricultural production to new areas and changing of
the agro-climatic regions, the emergence of new insect species
is very possible which will render an additional threat to agricultural production.

15.8 Climate change and water availability

Climate change is likely to intensify, accelerate or enhance the
global hydrological cycle (IPCC, 2008). Water is the primary
medium through which climate change influences the Earth's
ecosystem and maintains the livelihood and well-being of societies. Temperature change due to global warming influences
the availability of water. Higher temperatures and changes in

extreme weather conditions are projected to affect availability and distribution of rainfall, snow melting, river flows and ground water. Water stress is already high, particularly in many developing countries. Water resources management affects almost all aspects of society and the economy, in particular health, food production and food security, domestic water supply and sanitation, health, energy, industry, ecosystem function and environmental sustainability (IPCC, 2007).

Changes in rainfall due to global climate change may effect the surface moisture availability, which becomes important for crop stand establishment in the rain-fed areas. Modelling techniques are not reliable enough to predict precipitation changes and one can expect some increased drought in some regions and increased rainfall in others. Small changes in precipitation can have magnified run-off effects. Doubling of CO_2 predicts a 8–15% increase in global precipitation with a 30% increase in the water-holding capacity, but the atmosphere precipitation will not keep up with potential evaporation. The problem will be more acute at higher altitudes. Run-off is also likely to decrease due to more arid conditions and the increased frequency of droughts.

15.9 Climate change and soil fertility

Soil fertility is very essential for sustaining productivity in the world and depends on a complex network of soil structure, water, oxygen and nutrient availability. The soil's organic matter enables it to support plant life and soil microbes. Soil microbes facilitate the decomposition of organic matter from litter fall and CO_2 is a natural by-product of this process. Rising atmospheric temperatures and/or CO_2 levels are likely to increase photosynthesis and plant productivity. However, the effects of warming on soil's molecular composition have been poorly studied. It is therefore unclear to what extent the carbon-containing components of soil matter will accumulate or degrade and thus how much carbon will be sequestered by the soil and how much will be released into the atmosphere as CO_2.

The doubling of CO_2 increases plant biomass production, soil water use efficiency by the plants and C/N ratios of plants. The change in the C/N ratios of plant residues returned to the soil have impact on soil microbial processes and affects the production of trace gases NO_x and N_2O. Higher soil temperature stimulates microbial respiration and decomposition of organic matter. Plants may take in more nutrients. Brackish-water inundation on coastal areas gives rise to potential acid sulphate soil layers.

Experts are unsure how global warming will affect agriculture. CO_2 increases the efficiency of photosynthesis, only if the temperature increases by less than a few degrees; for example, 2°C in wheat and soya bean yields 10–15%; maize and rice 8%, at 4°C yields decrease. CO_2 increases water use efficiency (antitranspiration effect) as temperatures rise. The increase in precipitation results in a rise in the sea level and leads to flooding and loss of cropping area. The drainage problems and seawater intrusion into fresh water may become too arid and are decrease the production. Other arid areas may get more rainfall and start to produce. The growing season is likely to extend where northern regions may benefit, but there is little benefit for the tropics.

15.10   Climate change impact on Indian agriculture

Global climatic change has its own paradigm shift in its research methodologies and developmental aspects. India being an agri-economy-based country, it is quite obvious that we depend heavily upon the agricultural sector and its productivity. On the other hand, agriculture and environment are mutually interlinked with each other in its entire operation and execution towards the economy regeneration of any country. The impact of climate change on Indian agriculture is being studied to a limited extent. For the Indian subcontinent, it is predicted that the mean atmospheric temperature will increase by 1–40°C (Sinha and Swaminathan, 1991). Although the solar radiation received at the surface will vary geographically, on an average, it is expected to decrease by about 1% (Hume and Cattle, 1990). Rice and wheat are the two most important cereals that fill the Indian breadbasket. While rice and wheat constitute the major cropping system of the Indo-Gangetic plains of northern India, the southern peninsula comprising Godavari and Kauvery delta exclusively depend on rice. For India as a whole, rice may become even more important in the national food security system, since rice can give higher yields than wheat under a wider range of growing conditions. In the subsequent paragraphs, discussion will be centred on the impact of climate change on rice production.

Rice is cultivated in diverse ecologies that differ from each other in water availability and depth of standing water during growth. It is grown over a wide geographic range from 45°N to 40°S to elevations of more than 2500 m but with average daily temperature in the range of 20–30°C. The impact of climate change on rice production is of paramount importance in planning strategies to meet the increasing demands for rice. In recent years, controlled environment studies have enriched our knowledge on the effects of increased temperature and CO_2 level (Cure and Acock, 1986). This is mainly through the stimulation of photosynthesis in the plant and improvement of water use efficiency. The impact of temperature is more complex, with the yield being reduced in both sides of the spectrum. Most present varieties are highly sensitive to daytime temperatures with yield decreasing linearly with increases in daytime temperatures above 33°C (Satake and Yoshida, 1978). Jagadish et al. (2007) showed that high temperatures affect the pattern of flowering and the number of spikelets that reach anthesis in rice. A reduction of the rice-growing areas is a possibility if spikelet sterility would increase under predicted higher temperatures and if water reserves could not meet the increased evapo-transpiration. However, Bachelete and Kropff (1995) predicted a significant increase in the irrigated ecosystem in eastern India based on a modelling study on the impact of climatic change on agro-climatic zones in Asia.

Agriculture is likely to respond initially to climate change through a series of automatic mechanisms. Some of these mechanisms are biological, and others are routine adjustments by farmers and markets. Climate change will impact agriculture by causing damage and gain at scales ranging from individual plants or animals to global trade networks. At the plant or field scale, climate change is likely to interact with rising CO_2 concentrations and other environmental changes to affect crop and animal physiology. Climate change involving alterations in temperature, precipitation and sea level as well as increased incidence of ultraviolet B radiation (280–320 nm) are distinct possibilities in the not too distant future. Impacts and adaptation (agronomic and economic) are likely to extend to the farm and surrounding regional scales. As the Indian economy depends to a great extent on agriculture, the assessment of climate change impacts on agriculture has acquired special significance. In developing countries such as India, climate change could represent an additional stress on ecological and socio-economic systems that are already facing tremendous pressures due to rapid urbanisation, industrialisation and economic development. With its huge and growing population, a 7500-km-long densely populated (DOD, 2002) and low-lying coastline, and an economy that is closely tied to its natural resource base, India is considerably vulnerable to the impacts of climate change.

Table 15.5 Indian food production and projection scenario

Product	Projection (million tonnes)		
	2000	2010	2020
Rice	85.4	103.6	122.1
Wheat	71.0	85.8	102.8
Coarse grains	29.9	34.9	40.9
Total cereals	184.7	224.3	265.8
Pulses	16.1	21.4	27.8
Food grains	200.8	245.7	293.6
Fruits	41.1	56.3	77.0
Vegetables	84.5	112.7	149.7
Milk	75.3	103.7	142.7
Meat and eggs	3.7	5.4	7.8
Marine products	507	8.2	11.8

Source: Aggarwal, P. K. et al., 2004. *Environ. Sci. Policy,* 7(6): 487–498.

India's vulnerability to climate change is widely recognised. It lies in 4n area characterised by seasonal weather patterns, and experiences land degradation, rapid economic development and an increasing population. Since over 70% of India's population relies on agriculture for their livelihood and sustenance, most of the research focus has been on how India's agriculture might be affected by climate change. However, the concentration of research on climate change impacts fails to recognise that there are other complex global processes at work, such as globalisation, which may influence or exacerbate the situation. In the last decade there have been significant changes in India's economy as it moves towards liberalisation. Table 15.5 shows the Indian food production and projection scenarios up to 2020 (Aggarwal et al., 2002).

Agriculture and allied activities constitute the single largest component of India's economy, contributing nearly 27% of the total gross domestic product in the year 1999–2000. Agriculture exports account for 13–18% of the total annual exports of the country. However, given that 62% of the cropped area is still dependent on rainfall, Indian agriculture continues to be fundamentally dependent on weather. A few studies on the impact on agriculture have been reported for India in the IPCC Third Assessment Report. While there is report of a decrease in rice yields by 3–15% under a scenario of 1.5°C rise in temperature and a 2 mm/day increase in precipitation exist, some other reports reflect that yields of soya bean in India would vary between -22% and 18% under different climate scenarios considering $\pm 2^\circ C$ and $\pm 4^\circ C$ change in temperature, ± 20 and $\pm 40\%$ change in precipitation.

Some estimations say that there is a decrease in rice yields at the rate of 0.71 tonne/ha with an increase in minimum temperature from $18^\circ C$ to $19^\circ C$ and a decrease of 0.41 ton/ha with a temperature increase from $22^\circ C$ to $23^\circ C$. Whereas other estimations suggest that a $2^\circ C$ increase in mean air temperature could decrease rice yield by about 0.75 ton/ha in the high yield areas and by about 0.06 ton/ha in the low yield coastal regions. Further, a $0.5^\circ C$ increase in winter temperature would reduce wheat crop duration by 7 days and reduce yield by 0.45 ton/ha. An increase in winter temperature of $0.5^\circ C$ would thereby translate into a 10% reduction in wheat production in the high yield states of Punjab, Haryana and Uttar Pradesh. Hence, a potential rise in temperature will have disastrous consequences on wheat production in India. The study showed that with rice increasing mean daily temperature decreases the period from transplantation to maturity. Such reduction in duration is often accompanied by decreasing crop fields. There are, however, genotypic differences per day yield potential. A breeder can consciously select strains with high per day productivity. Increased levels of CO_2 increase the photosynthetic rate and, hence, dry matter production, but an increase in temperature reduces crop duration and thereby an increase in the yields. In Pantnagar district the irrigated yield was stimulated under doubled CO_2 and increased temperature. This study concluded that the impact on rice production would be positive in the absence of nutrient and water limitations. Another crop simulation study estimated that under elevated CO_2 condition, the wheat yields could decrease by $28-68\%$ without considering the CO_2 fertilisation effects. Researchers suggest that in North India, a $1^\circ C$ rise in mean temperature would have no significant effect on wheat yields, while a $2^\circ C$ increase would reduce yields in most places. Recent studies have examined the adaptation options while estimating the agricultural impacts. The study showed that even with adaptation by farmers of their cropping patterns and inputs, in response to climate change, the losses would remain significant. The loss in farm-level net revenue is estimated to range between 9% and 25% for a temperature rise of $2-3.5^\circ C$.

Estimations are there that India's climate could become warmer under conditions of increased atmospheric CO_2. The average temperature range is predicted to be $2.33-4.78^\circ C$ with a doubling in CO_2 concentrations. It is also likely that there will be an increase in the frequency of heavy rainfall events in South and

South-East Asia. Some researchers presented a climate change scenario for the Indian subcontinent, taking projected emissions of GHGs and sulphate aerosols into account. It predicts an increase in the annual mean maximum and minimum surface air temperatures of 0.7°C and 1.0°C over land in the 2040s with respect to the 1980s. Since the warming over land is projected to be lower in magnitude than that over the adjoining ocean, the land–sea thermal contrast that drives the monsoon mechanism could possibly decline. However, there continues to be considerable uncertainty about the impacts of aerosols on the monsoon.

15.11   Vulnerability assessment in Indian agriculture

These changes will arguably alter India's vulnerability, creating a different set of winners and losers in the climate change game. The multifaceted approach of some research attempts to capture the differing sources of vulnerability by incorporating the concept of 'double exposure', which refers 'to the fact that certain regions, sectors, ecosystems and social groups will be confronted by the impacts of climate change and by the consequences of globalisation'. Another key component of this study is the recognition that climate change and globalisation are dynamic processes. Accordingly, one may assume that the impacts of the global process will change over time as well, altering the location, type and severity of vulnerability.

To get into the realm of this aspect, we need to look into the following components:

- Recent environment changes its impact on Indian agriculture
- Technological advancement
- Ecological impact

The adaptation options are possible at various levels, that is, farmers, economic agents and macro-level policy issues. The potential and costs of adaptation will be possibly through historic analysis of technology penetration. For example, the relative adoption speed of various measures such as adaptation measure, adjustment time (years), variety adoption, 3–14 dams and irrigation, 50–100 variety development, 8–15 tillage systems, 10–12 opening new lands, 3–10 irrigation equipment, 20–25 fertiliser adoption and so on are of significant importance. There were observations of adoption and technological responses in post-independent Indian agriculture, which estimated a response time of 5–15 years for items such as productive life of farm assets, crop rotation cycles and recovery from major disasters. Within broad categories of responses, some of which could be beneficial regardless of how or whether climate changes include

- Improved training and general education of populations dependent on agriculture
- Identification of the present vulnerabilities of agricultural systems
- Agricultural research to develop new crop varieties
- Food programmes and other social security programmes to provide insurance against supply changes
- Transportation, distribution and market integration to provide the infrastructure to supply food during crop shortfalls
- Removal of subsidies, which can, by limiting changes in prices, mask the climate change signal in the market

This analysis consists of four parts:

- A macro-level vulnerability analysis, based on GIS
- A domestic policy analysis
- A micro-level analysis (including village-level case studies in four different agricultural regions)
- Integrated analyses that will synthesise the preceding work and offer policy recommendations for facilitating adaptation in the agricultural sector

This will illustrate the results from the macro-level analysis, which involves the mapping of vulnerability profiles. These profiles are based on indicators that directly or indirectly represent sources of vulnerability. The indicators are combined and weighted to create composite indexes that illustrate how vulnerability varies spatially. A base vulnerability index for India has already been developed by some researchers, using indicators that reflect social, biophysical and technological vulnerability. The base vulnerability layer is overlaid with either a climate sensitivity index/layer, based on climate norms from 1961 to 1990, or a trade sensitivity index/layer. From these maps, districts that have both high climate and economic globalisation vulnerabilities (i.e. are doubly exposed) are identified. Villages for the case studies are selected from these highly vulnerable districts. The macro-level analysis should also include climate scenario data, thus capturing the dynamic aspect of vulnerability.

The GCM predictions are only about the average changes in future climate. It is unable to show changes in the frequency and intensity of hurricanes, frequency of floods or the intensity of monsoons. It is still uncertain how climate variability will vary as a consequence of GHGs. The other types of errors could be due to the weakness of the GCMs in representing physical processes in the atmosphere relating to clouds. Yet, another cause of inaccuracy is the fact that the equilibrium runs assuming instantaneous doubling of CO_2 and subsequent equilibrium state which might not occur in reality. Hence, the predictions based on such assumptions could be quite uncertain.

15.12 Mitigation of climate change

The atmospheric levels of GHGs such as CO_2, CH_4, N_2O, O_3 and CFCs are increasing with rapid industrialisation, intensive agricultural and related activities. The levels of CO_2, the most important GHG, are expected to increase by 3.5°C in the next 50 years. The more buildup of these gas levels would increase the temperature of the Earth and effect climatic changes. If there is an increase of 0.5°C in the mean temperature in Punjab, Haryana and Uttar Pradesh, it would reduce the productivity of wheat crop by 10%. Changes in rainfall patterns will enforce land use changes and alter bio-diversity and whole ecology. If the current rate of fossil fuel burning continues, which is the main source of atmospheric CO_2, we must look for alternative sources of renewable energy.

The overall levels of production can be maintained through a combination of shifts in agricultural zones and adjustments in technology and management. Every effort should be made to arrest deforestation and promote upgrading of degraded land through agro-forestry and other appropriate forms of land use. This will help to increase carbon fixation on the Earth. Establishment of crop-weather watch groups, special groups consisting of meteorologists, agricultural research and extension workers, developmental administrators and mass media representatives, should be established in every agro-ecological region of the country to continuously monitor the weather situation and analyse its implications for crop growth and pest incidence. Such crop-weather watch groups can benefit at the national level from the World Climate Programme of WMO and early warning system of FAO. This will help to equip farmers and fishermen with location-specific information to increase their preparedness for floods, tropical cyclones and drought. Kumar and Parikh (1998a, b) have shown that even with the adaptation by farmers of their cropping patterns and inputs, in response to climate change, the losses would remain significant. The loss in farm-level net revenue is estimated to range between 9% and 25% for a temperature rise of 2–3.5°C.

With the knowledge gathered to date, it is possible to consolidate the gains and adapt to changes in order to tackle the negative effects of climate change on rice production and productivity for assuring the food security of the Indian population. Some of the possible options include

Varietal adaptation: Considerable variation exists between the rice varieties in tolerance to high temperatures. If the sensitivity of spikelet sterility to temperature is increased by 2°C for the new varieties, it can offset the detrimental effect completely. Two possible adaptations are likely to occur: one could be the use of varieties more tolerant to temperature in the low latitude region and the other is the use of late maturing varieties to take advantage of the longer growing season in high latitude areas (Matthews et al., 1995).

Adjustment of planting date: Adjusting planting dates could be another strategy, which is likely to be adopted in the future by rice farmers. At high latitudes, a rise of temperature would lengthen the period in which rice can be grown. In northern China (Shenyang), yield increases of 43% are expected by advancing the planting date of rice by 30 days. At Madurai, India, significant yield decrease was predicted if current planting dates were used under the GISS scenario due to high spikelet sterility. It could be possible to offset these yield reductions if planting was delayed by 1 month.

Environment-friendly cultivation practice: It would be useful to standardise methods of reducing the contribution of agriculture to GHG accumulation in the atmosphere either through efficient water management to control CH_4 release or proper utilisation of fertiliser-N that could contribute to the emission of N_2O.

15.13 Genetic resource centres for adaptation to climate changes

Gene pools occur in nature for adaptation to drought, floods and sea level changes. Unfortunately, gene erosion through a variety of causes is leading to the loss of valuable genetic material. Every effort should be made to establish specialised genetic resource centres for collecting and conserving species and genotypes with the desirable genes. These centres will identify and maintain candidate genes for use in recombinant DNA experiment.

Tools of molecular biology and recombinant DNA techniques make transfer of genes across sexual barriers possible. For example, the 'elongation' gene from rice, which enables it to grow to the needed height for remaining above flood water levels, can transfer to other crops. To achieve such goals, it will be necessary to establish a Genetic Enhance Centre consisting of an expert in molecular biology and genetic engineering. This will be of immense help to the breeding of crop varieties possessing greater tolerance to drought, floods and seawater intrusions.

The large area under wastelands in our country could be gainfully employed for sequestration of GHGs by taking up large-scale plantation. The emissions of CH_4 and N_2O from paddy fields could be brought down by intermittent flooding (aerobic condition) of fields (with no yield reduction) instead of continued submergence. The methane production from rice fields in our country may not be as high as projected by some other countries, as our paddy fields generally have lower levels of carbon and are predominantly rain fed. The nitrate toxicity in ground water increased due to more use of N fertilisers. The environment-friendly integrated nutrient management, integrated pest management and certain other ameliorative/management practices should be adopted as a long-term strategy for keeping our natural resources free of toxins. Replacement of crop residue burning with *in situ* (incorporation) or *ex situ* management is called upon to minimise production of GHGs and climate change. In twenty-first-century agriculture, at least four technological developments or trends have already begun and are likely to intensify in the future: agricultural biotechnology, conservation tillage, organic farming and precision farming. These technologies have the potential or proven role in increasing soil carbon sequestration and thereby soil carbon storage to mitigate global warming and climate change.

15.14 Conclusion

Anthropogenic GHG emissions and climate change have a number of implications for agricultural productivity, but the aggregate impact of these is not yet known and indeed many such impacts and their interactions have not yet been reliably quantified, especially at the global scale. An increase in mean temperature can be confidently expected, but the impacts on productivity may depend more on the magnitude and timing of extreme temperatures. A mean sea level rise can also be confidently expected, which could eventually result in the loss of agricultural land through permanent inundation. The impacts of temporary flooding through storm surges may be large although less predictable.

Freshwater availability is critical, but predictability of precipitation is highly uncertain and there is an added problem of lack of clarity on the relevant metric for drought—some studies including IPCC consider metrics based on local precipitation and temperature such as the Palmer drought severity index, but this does not include all relevant factors. Agricultural impacts in some regions may arise from climate changes in other regions, owing to the dependency on rivers fed by precipitation, snowmelt and glaciers some distance away. Drought may also be offset to some extent by an increased efficiency of water use by plants under higher CO_2 concentrations, although the impact of this is again uncertain especially at large scales. The climate models used here project an increase in annual mean soil moisture availability and run-off in many regions, but nevertheless, across most agricultural areas there is a projected increase in the time spent under drought as defined in terms of soil moisture.

Moreover, the sign of crop yield projections is uncertain as this depends critically on the strength of CO_2 fertilisation and also O_3 damage. Few studies have assessed the response of crop yields to CO_2 fertilisation and O_3 pollution under actual growing conditions, and consequently model projections are poorly constrained. Indirect effects of climate change through pests and diseases have been studied locally, but a global assessment is not yet available. Overall, it does not appear to be possible at the present time to provide a robust assessment of the impacts of anthropogenic climate change on global-scale agricultural productivity.

References

ABROL, Y. P., BAGGA, A. K., CHAKRAVORTY, N. V. K. and WATTAL, P. K. 1991. Impact of rise in temperature on the productivity of wheat in India. In: *Impact of Global Climate*

Changes in Photosynthesis and Plant Productivity. (Y.P. Abrol et al., eds). Oxford & IBH Publishers, New Delhi, pp. 787–789.

AGGARWAL, P. K. 2000. *Climate Change and Agriculture: Information Needs and Research Priorities.* Indian Agricultural Research Institute, New Delhi.

AGGARWAL, P. K. 2003. Impact of climate change on Indian agriculture. *J. Plant Biol.,* 30(2): 189–198.

AGGARWAL, P. K. and KALRA, N. 1994. *Simulating the Effect of Climatic Factors, Genotype and Management on Productivity of Wheat in India.* Indian Agricultural Research Institute Publication, New Delhi, India, p. 156.

AGGARWAL, P. K. and MALL, R. K. 2002. Climate change and rice yields in diverse agro-environments of India. II. Effect of uncertainties in scenarios and crop models on impact assessment. *Clim. Change,* 52(3): 331–343.

AGGARWAL, P. K. and SINHA, S. K. 1993. Effect of probable increase in carbon dioxide and temperature on productivity of wheat in India. *J. Agric. Meteorol.,* 48(5): 811–814.

AGGARWAL, P. K., JOSHI, P. K., INGRAM, J. S. I. and Gupta, R. K. 2004. Adapting food systems of the Indo-Gangetic plains to global environmental change: Key information needs to improve policy formulation. *Environ. Sci. Policy,* 7(6): 487–498.

ALLEN, R. G., SMITH, M., PEREIRA, L. S. and PERRIER, A. 1994. An update for the calculation of reference evapotranspiration. *ICID Bull.,* 43(2): 35–92.

ALLEN, R. G., PEREIRA, L. S., RAES, D. and SMITH, M. 1998. Crop evapotranspiration. Guidelines for computing crop water requirements. In: *FAO Irrigation and Drainage.* FAO, Rome, pp. 300.

ATTRI, S. D. and RATHORE, L. S. 2003. Simulation of impact of projected climate change on wheat in India. *Int. J. Climatol.,* 23: 693–705.

BACHELETE, D. and KROPFF, M. J. 1995. The impact of climatic change on agroclimatic zones in Asia. In: *Modeling the Impact of Climate Change on Rice Production in Asia.* R. B. Matthews et al., eds). CAB International, U.K.

BAKER, J. T., ALLEN, JR., L. H., BOOTE, K. J., JONES, P. and JONES, J. W. 1989. Response of soybean to air temperature and carbon dioxide concentration. *Crop Sci.,* 29: 98–105.

CLINE, R. W. 1999. *Global Warming and Agriculture: Impact Estimation by Country.* Center for Global Development and the Peterson Institute for International Economics, Washington, DC.

CURE, J. D. and ACOCK, B. 1986. Crop responses to carbon dioxide doubling: A literature survey. *Agric. For. Meteorol.,* 38: 127–145.

DE, U. S. and MUKHOPADHYAY, R. K. 1998. Severe heat wave over the Indian subcontinent in 1998, in perspective of global climate. *Curr. Sci.,* 75(12): 1308–1311.

DOD. 2002. Annual Report 2001/02, Department of Ocean Development, New Delhi.

FISCHER, G., FROHBERG, K., PARRY, M. L. and ROSENZWEIG, C. 1994. Climate change and world food supply, demand and trade. Who benefits, who loses? *Global Environ. Change,* 4(1): 7–23.

FRANCIS, M. 1999. Simulating the impact of increase in temperature and CO_2 on growth and yield of rice, M.Sc. Thesis (unpublished), Indian Agricultural Research Institute, New Delhi.

GADGIL, S. 1995. Climate change and agriculture—An Indian perspective. Curr. Sci., 69(8): 649–659.

GADGIL, S., ABROL, Y. P. and RAO, SESHAGIRI, P. R. 1999a.
On growth and fluctuation of Indian food grain production. Curr.
Sci., 76(4): 548–556.

GADGIL, S., RAO, SESHAGIRI, P. R. and SRIDHAR, P. R.
1999b. Modeling impact of climate variability on rainfed
groundnut. Curr. Sci., 76(4): 557–569.

GANGADHAR RAO, D. and SINHA, S. K. 1994. Impact of climate
change on simulated wheat production in India. In: Implications
of Climate Change for International Agriculture: Crop Modeling
Study. (C. Rosenzweig and I. Iglesias, eds). USEPA 230-B-94-
003. USEPA, Washington, DC. pp. 1–17.

GANGADHAR RAO, D., KATYAL, J. C., SINHA, S. K. and
SRINIVAS, K. 1995. Impacts of climate change on sorghum
productivity in India: Simulation study, American Society of
Agronomy, 677 S. Segoe Rd., Madison, WI 53711, USA, Climate
Change and Agriculture: Analysis of Potential International
Impacts. ASA Special Publ. No. 59. pp. 325–337.

HANSEN, J., SATO, M. and RUEDY, R. 1997. Radioactive forcing
and climate response. J. Geophys. Res., 102: 6831–6864.

HANSEN, J., REDDY, R., GLASCOE, J. and SATO, M. 1999.
GISS analysis of surface temperature change. J. Geophys. Res.,
104: 30997–31022.

HINGANE, L. S., RUPA KUMAR, K. and RAMANA MURTHY,
BH. V. 1985. Long-term trends of surface air temperature in
India. J. Climatol., 5: 521–528.

HOOGENBOOM, G. 2000. Contribution of agrometeorology to the
simulation of crop production and its applications. Agric. Forest
Meteorol., 103: 137–157.

HUME, C. J. and CATTLE, H. 1990. The greenhouse effect: Meteorological
mechanisms and models. Outlook Agric., 19: 17–23.

IBSNAT [International benchmark sites network for agrotechnol-
ogy transfer]. 1990. IN: Proceedings of IBSNAT Symposium:
Decision Support System for Agrotechnology Transfer,
University of Hawaii, Honolulu.

ICAR, 2005. Annual Report, 2005, New Delhi.

IDSO, S. B., KIMBALL, B. A., ANDERSON, M. G. and
MAUNEY, J. R. 1987. Effects of Atmospheric CO_2 enrichment
on plant growth: The interactive role of air temperature. Agric.
Ecosys. Environ., 20: 1–10.

IPCC [Intergovernmental Panel on Climate Change]. 1990.
Climate Change, The IPCC Scientific Assessment. WHO and
UNEP, Cambridge University Press, Cambridge, UK.

IPCC [Intergovernmental Panel on Climate Change]. 1996. Revised
1996 IPCC Guidelines for National Greenhouse Gas Inventories.

IPCC [Intergovernmental Panel on Climate Change]. 2007. Impacts,
adaptation and vulnerability. Contribution of Working Group II

to the Fourth Assessment Report of the Intergovernmental Panel on Climate Change. In: *Climate Change 2007*. (M. L. Parry, O. F. Canziani, J. P. Palutikof, P. J. van der Linden and C. E. Hanson, eds). Cambridge, United Kingdom, Cambridge University Press, and New York, NY, p. 976.

IPCC [Intergovernmental Panel on Climate Change]. 2008. Technical Paper VI. In: *Climate Change and Water*. (B. C. Bates, Z. W. Kundzewicz, S. Wu and J. P. Palutikof, eds). IPCC Secretariat, Geneva, pp. 210.

JAGADISH, S. V. K., CRAUFURD P. Q. and WHEELER, T. R. 2007. High temperature stress and spikelet fertility in rice (*Oryza sativa* L.). *J. Exp. Bot.*, 58(7): 1627–1635.

JURIK, T. W., WEBER, J. A. and GATES, D. M. 1984. Short term effects of CO_2 on gas exchange of leaves of bigtooth aspen (*Populus grandidentata*) in field. *Plant Physiol.*, 75: 1022–1026.

KANE, S., REILLY, J. and BUCKLIN, R. 1989. *Implications of the Greenhouse Effect for World Agricultural Commodity Markets*. US Department of Agriculture. Washington, DC.

KIMBALL, B. A. 1983. Carbon dioxide and agricultural yield: An assemblage and analysis of 430 prior observations. *Agron. J.*, 75: 779–788.

KRUPA, S. 2003. Atmosphere and agriculture in the new millennium. *Environ. Pollut.*, 126: 293–300.

KUMAR, K. S. and PARIKH, J. 1998a. Climate change impacts on Indian agriculture: The Ricardian approach. In: *Measuring the Impacts of Climate Change on Indian Agriculture*. (Dinar et al., eds). World Bank Technical Paper No. 402. World Bank, Washington, DC.

KUMAR, K. S. and PARIKH, J. (1998b). Climate change impacts on Indian agriculture: Results from a crop modeling approach. In: *Measuring the Impacts of Climate Change on Indian Agriculture*. (Dinar et al., eds). World Bank Technical Paper No. 402. World Bank, Washington, DC.

KUMAR, K. S. K. and PARIKH, J. 2001. Indian agriculture and climate sensitivity. *Global Environ. Change*, 11: 147–154.

KUMAR, K. K., KUMAR, K. R., ASHRIT, R. G., DESHPANDE, N. R. and HANSEN, J. W. 2001. Climate impacts on Indian agriculture. *Int. J. Climatol.*, 24(11): 1375–1393.

LAL, M., SINGH, K. K., SRINIVASAN, G., RATHORE, L. S. and SASEENDRAN, A. S. 1998. Vulnerability of rice and wheat yields in NW-India to future change in climate. *Agric. For. Meteorol.*, 89: 101–114.

LAL, M., SINGH, K. K., SRINIVASAN, G., RATHORE, L. S., NAIDU, D. and TRIPATHI, C. N. 1999. Growth and yield response of soybean in Madhya Pradesh, India to climate variability and change. *Agric. For. Meteorol.*, 93: 53–70.

LIVERMAN, D. 1987. Forecasting the impact of climate on food systems: model testing and model linkage. *Clim. Change*, 11, 267–285.

MALL, R. K. and AGGARWAL, P. K. 2002. Climate change and rice yields in diverse agro-environments of India. I. Evaluation of impact assessment models. *Clim. Change*, 52(3): 315–331.

MALL, R. K., LAL, M., BHATIA, V. S., RATHORE, L. S. and SINGH, R. 2004. Mitigating climate change impact on soybean productivity in India: A simulation study. *Agric. For. Meteorol.*, 121(1–2): 113–125.

MALL, R. K., SINGH, R., GUPTA, A., SRINIVASAN, G. and RATHORE, L. S. 2006. Impact of climate change on Indian agriculture: A review. *Clim. Change*, 78: 445–478.

MANN, M. E., BRADLEY, R. S. and HUGHES, M. K. 1999. Northern hemisphere temperature during the past millennium: Interface, uncertainties and limitations. *Geophys. Res. Lett.*, 26: 759.

MATHAUDA, S. S. and MAVI, H. S. 1994. Impact of climate change in rice production in Punjab, India. In: *Climate Change and Rice Symposium*, IRRI, Manila, Philippines.

MATTHEWS, R. B., KROPFF, M. J., BACHELET, D. and VAN LAAR, H. H. 1995. Modeling the impact of climate change on rice production in Asia. Executive summary. CABI in association with IRRI, 13.

MOHANDASS, S., KAREEM, A. A., RANGANATHAN, T. B. and JEYARAMAN, S. 1995. Rice production in India under current and future climates. In: *Modeling the Impact of Climate Change on Rice Production in Asia.* (R. B. Matthews, M. J. Kropff, D. Bachelet and H. H. Laar van, eds). CAB International, U.K. pp. 165–181.

MORISON, J. I. L. and GIFFORD, R. M. 1984. Plant growth and water use with limited water supply in high CO_2 concentrations. I. Leaf area, water use and transpiration. *Aust. J. Plant Physiol.*, 11: 361–374.

NAKICENOVIC, N., DAVIDSON, O., DAVIS, G., GRUBLER, A, KRAM, T., LA ROVERE, E. L., METZ, B., MORITA, T., PEPPER, W. and PITCHER, H. 2000. *Special Report on Emission Scenario.* Cambridge University Press, Cambridge, UK.

PANT, G. B., RUPA KUMAR and K. BORGAONKAR, H. P. 1999. Climate and its long-term variability over the western Himalaya during the past two centuries. In: *The Himalayan Environment.* (S. K. Dash and J. Bahadur, eds). New Age International (P) Limited, Publishers, New Delhi, pp. 172–184.

PARTHASARATHY, B. and PANT, G. B. 1985. Seasonal relationship between Indian summer monsoon rainfall and southern oscillation. *J. Clim.*, 5: 369–378.

PARTHASARATHY, B., RUPA KUMAR, K. and MUNOT A. A. 1992. Forecast of rainy season food grain production based on monsoon rainfall. *Indian J. Agric. Sci.*, 62: 1–8.

RATHORE, L. S., SINGH, K. K. SASEENDRAN, S. A. and BAXLA, A. K. 2001. Modelling the impact of climate change on rice production in India. *Mausam*, 52(1): 263–274.

ROSENZWEIG, C., IGLESIAS, A, BAER, B., BAETHGEN, W., BRKLACICH, M., CHOU, T. Y., CURRY, B. et al. 1991. *Climate Change and International Agriculture Crop Modeling Study.* U.S. Environmental Protection Agency, Washington, DC.

RUPA KUMAR, K., KRISHNA KUMAR, K. and PANT, G. B. 1994. Diurnal asymmetry of surface temperature trends over India. *Geophy. Res. Lett.*, 21: 677–680.

SASEENDRAN, A. S., SINGH, K. K., RATHORE, L. S., SINGH, S. V. and SINHA, S. K. 2000. Effects of climate change on rice production in the tropical humid climate of Kerala, India. *Clim. Change*, 44: 495–514.

SATAKE, T. and YOSHIDA, S. 1978. High temperature induced sterility in *Indica* rices at flowering. *Jpn. J. Crop Sci.*, 47: 6–17.

SELVARAJU, R. 2003. Impact of El Nino- Southern Oscillation on Indian food grain production. *Int. J. Climatol.*, 23: 187–206.

SINGH, N. and SONTAKKE, N. A. 2002. On climatic fluctuations and environmental changes of the Indo-Gangetic plains, India. *Clim. Change*, 52: 287–313.

SINGH, R. S., NARAIN, P. and SHARMA, K. D. 2001. Climate changes in Luni river basin of arid western Rajasthan (India). *Vayu Mandal*, 31(1–4): 103–106.

SINHA, S. K. and SWAMINATHAN, M. S. 1991. Deforestation climate change and sustainable nutrients security. *Clim. Change*, 16: 33–45.

SRIVASTAVA, H. N., DEWAN, B. N., DIKSHIT, S. K., RAO, G. S. P., SINGH, S. S. and RAO, R. 1992. Decadal trends in climate over India. *Mausam*, 43: 7–20.

USGCRP 2009. Global climate change impacts in the United States. In: *United States Global Change Research Program*. (T. R. Karl, J. M. Melillo and T. C. Peterson, eds). Cambridge University Press, New York.

UPRETY, D. C., KUMARI, S., DWIVEDI, N. and MOHAN, R. 2000. Effect of elevated CO_2 on the growth and yield of rice. *Indian J. Plant Physiol.*, 5: 105–107.

UPRETY, D. C., DWIVEDI, N., JAIN, V. and MOHAN, R. 2002. Effect of elevated carbon dioxide concentration on the stomatal parameters of rice cultivars. *Photosynthetica*, 40: 315–319.

UPRETY, D. C., DWIVEDI, N., JAIN, V., MOHAN, R., SAXENA, D. C., JOLLY, M., and PASWAN, G. 2003. Responses of rice cultivars to the elevated CO_2. *Biol. Plantarum*, 46(1), 35–39.

WALKER, B., YOUNG, H., PARSLOW, J., CROCKS, K., FEMING, P., MARGULES, C. and LANDSBERG, J. 1989. *Global Climate Change and Australia*. Division of Wildlife and Ecology, Canberra, Australia.

WARRICK, R. A. 1988. Carbon dioxide, climatic change and agriculture. *Geograph. J.*, 154(2): 221–233.

CHAPTER SIXTEEN

Climate change and sustainability of biodiversity

M.K. Sarma, Sangeeta Baruah and
A.K. Sharma

Contents

Abstract

Changing climatic variables leading to threats for bio-diversity include increasing CO_2 concentration, increasing global temperature, altered precipitation pattern and change in the pattern of extreme weather events such as cyclones, fires or storms. The impact of climate change may be direct or indirect on the living system. The direct impacts are those which are predictable and one can think of getting rid of them. Indirect impacts are those which are very slow to monitor and difficult to predict. They have a long-lasting and permanent change in the features of biodiversity.

16.1 Biodiversity

Biodiversity is the degree of variation of life-forms within a given species, ecosystem, biome or the entire planet. Health of an ecosystem can be measured by biodiversity, which is a function of climate. The term biological diversity was used first by the wildlife scientist and conservationist Raymond F. Dasmann in 1968 in the book *A Different Kind of Country*. The term was widely adopted only after more than a decade, when in the 1980s, it came into common usage in science and environmental policy. The biological definition of biodiversity is 'totality of genes, species and ecosystems of a region'. Biological variation can be identified in four levels:

- Species diversity: Species diversity is the effective number of different species that are represented in a collection of individuals. The effective number of species refers to the number of equally abundant species needed to obtain the same mean proportional species abundance as that observed in the dataset of interest. Species diversity consists of two components, species richness and species evenness.

- *Ecosystem diversity*: Ecosystem diversity refers to the diversity of a place at the level of ecosystems. Ecosystem diversity means a variety of ecosystems present in a biosphere, the variety of species and ecological processes that occur in different physical settings.

- *Genetic diversity*: Genetic diversity refers to the extent of genotypic differences existing in the biological system. It may be intraspecific, interspecific and intrageneric. In other words, genetic diversity means the inter se genetic distance among the individuals in the biological community.

Biodiversity is not evenly distributed; rather, it varies greatly across the globe as well as within regions. The distribution of living entities on the Earth depends on climatic and edaphic variables such as temperature, precipitation, altitude, geography and the presence of other species.

16.2 Climate change

Climate change is the long-lasting and significant change in the statistical distribution pattern of weather over a long period of time that can range from decades to millions of years. The significant factors responsible for climate change include oceanic processes (such as oceanic circulation), variations in radiation received by Earth, plate tectonics and volcanic eruptions and human-induced alterations of the natural world. The most general definition of *climate change* is a change in the statistical properties of the climate system when considered over long periods of time, regardless of cause. Nowadays, in most cases, climate change specifically refers to climate change caused by human activity, as opposed to changes in climate that may have resulted as part of Earth's natural processes. Under this scenario, climate change has become synonymous with anthropological global warming. This climate change is being chiefly expressed in terms of fluctuating weather variables including temperature, precipitation, humidity, cyclone, storms and so on.

Causes of climate change Factors that are responsible for climate change are known as climate forcing or 'forcing mechanisms' including processes such as variations in solar radiation, deviations in the Earth's orbit, mountain building and continental drift, and changes in greenhouse gas concentrations. Forcing mechanisms can be either 'internal' or 'external'. Internal forcing mechanisms are natural processes within the climate system itself. External forcing mechanisms can be either natural (e.g. changes in solar output) or anthropogenic (e.g. increased emissions of greenhouse gasses).

16.3 Impacts of climate change on biodiversity

At present-day, climate change has become a major challenge to the sustainability of biodiversity worldwide. In the atmosphere, gasses such as water vapour, carbon dioxide and methane act like the glass roof of a greenhouse by trapping heat and warming the planet. These gasses are known as greenhouse gasses. As a result of human activities (such as farming activities, land use changes, burning of fossil fuels etc.), the normal level of these gasses is being raised up leading to an increase in the temperature of the Earth's surface and lower atmosphere. The changes in climate have a deleterious impact on biodiversity either directly or indirectly including a shift in the distribution of the biotic community, timing of biological behaviour, that is, phenology, assemblage composition, ecological interactions and community dynamics.

Fossil-age context of biodiversity and climate change

If we look into history, it has been seen that from paleo age itself, Earth is facing the problem of climate change. In those historic ages, Earth was cool, warm, dry, wet and CO_2 levels were both high and low. Constant shifting in vegetation resulted in such climate changes during those days. Here, we can take the example of forest communities dominating most areas in interglacial periods and herbaceous communities dominating most areas during the glacial period. There was evidence that climate change influenced the process of speciation and extinction during that time. For example, the collapse of carboniferous rain forest which occurred 350 million years ago resulted in destruction of amphibian population while encouraging the evolution of reptiles.

Present-day context

There is a growing consensus in the scientific community that climate change is occurring. According to the millennium ecosystem assessment, a comprehensive assessment of the links between ecosystem health and human well-being, climate change is likely to become the dominant direct driver of biodiversity loss by the end of this century. Projected changes in climate, combined with land use change and the spread of exotic and alien species, are likely to limit the capability of some species to migrate and, therefore, will accelerate species loss. It is well accepted that the global average surface temperature is increasing day by day and that the snow and ice ranges in the Northern Hemisphere are decreasing. According to IPCC (2001), the global surface temperature has increased nearly 1° over the past century and it is projected that there would be 1.4–5.8° rise in the next century. Many organisations such as IPCC, UNEP/IES and scientists working worldwide have shown a growing interest in the impact of climate change on biodiversity and are trying to make people aware of their dangerous outcomes. *Climate Change and Biodiversity* (by Lovejoy and Hannah, 2005) is a famous book overviewing the past and potential future effects of climate change on biodiversity. In an effort to draw attention to the mounting threats and opportunities, the Convention of Biological Diversity (CBD) had called on nations of the world to celebrate the 'International Day for Biodiversity' on 22nd May, 2007, under the theme 'Climate change and biodiversity'.

Direct impacts of climate change on biodiversity

A shift in distribution of the biotic community If climatic factors such as temperature and precipitation are changed in such a way that certain species' phenotypic plasticity find it beyond tolerance, the obvious outcome will be its shift in natural distribution. This results in migration of the species or community to other habitats responding to the changing conditions. Vegetation zones may move toward higher latitudes or higher altitudes following the change in average temperature. The current migration rate of a species has been falling day by day as compared to that of the past and, thus, a drastic increase in extinction cases. In addition to altering species' distribution, the rapid pace of the current climate change has decreased the ability of some species to follow the climate to which they are adapted. Storm petrel (*Hydrobates pelagicus*) is the smallest Atlantic seabird, which migrates from its breeding grounds in the North Atlantic to Namibia, South Africa and the Indian Ocean. Migrating birds must build up and maintain large fat reserve to fuel their migration journeys. Climate change affects their fat reserve by altering the conditions along their migration rate. This bird feeds upon zooplankton for its fat reserve. Researchers found that the abundance of zooplankton has deceased as a result of change in sea temperature and chemical composition, which ultimately results in reduction of body mass of the migrating bird. Rates of climate change and species adaptation vary at regional and even local levels and they are of great importance. The maximum rates of spread for some sedentary species including large tree species may be slower than the predicted rates of change in climate conditions. This is likely to lead to localised extinction of these species.

Interaction among species Different species respond differently to a changed condition. Certain species may easily adapt to a new condition while some others may face difficulties in adapting. This difference in sensitivity toward the changing condition will give rise to complex species interaction resulting in further complications.

Invasive species From historic times, new animals and plants are being introduced from one part of the world to the other by humans. Newly introduced species may act invasive, affecting the native inhabitant of the area by eating them, hybridising with them, competing with them or by introducing pathogens or parasites. Global climate change may be advantageous for the establishment of some invasive species as they find the climate suitable for their survival.

Alterations in phenology   Phenology is the study of the timing or seasonality of behaviour. The timing of phonological events such as flowering is greatly dependent on environmental variables such as temperature. Climate change has a drastic impact on the phenology of several life events leading to change in asynchrony between species or change in competition between plants. For example, flowering times of British plants have been changed, leading to annual plants flowering earlier than perennials and insect-pollinated plants flowering earlier than wind-pollinated plants, with potential ecological consequences. Earthwatch research on pollination ecology in the Colorado Rocky Mountains, United States, found that snow melt determines the flowering time for the plant, which is ultimately influenced by climate change. According to this research, lower altitudes are affected differently than higher altitudes. So, it is expected for animals exposed to earlier warm weather to exit hibernation earlier.

Precipitation and evaporation pattern   Climate change has a hazardous impact on the pattern of precipitation and evaporation. An increase has been observed in rainfall variability and dry-season severity. Riverine and valley ecosystems will face heavy flood whereas drought and desertification may be prevalent in the tropical and subtropical zones. Earthwatch scientist Dr. Patricia Wright conducted a study in Madagascar to demonstrate the effect of climate change on the reproductive success of the endangered Milne-Edward's Sifaka lemur (*Propithecus edwardsi*). The specificity of Sifaka lemur's reproductive system is that, older female Sifakas reproduce readily, but their infants survive only when there is adequate rain during lactation. The logic behind this is, Sifaka milk production relies on large quantities of water and nutrients drawn from their leaf food. During drier days, old Sifakas with worn teeth find it difficult to chew enough leaves to produce adequate amount of milk for their infants. As a result of this, mortality rate of the infant is raised. This is a strange fact that a little change in climate can impact infant survival so drastically.

Impact on agricultural ecosystem   Climate change has an extensive impact on agricultural ecosystems. Agricultural crops will be more exposed to the climatic stresses caused by extreme climatic events. Climate change enhances the spreading of pests and diseases as well. Climate change may also cause increased exposure to heat stress, changes in rainfall patterns, greater leaching of nutrients from the soil during intense rains, greater erosion due to stronger winds and more wildfires in drier regions. All these result in lower yield. Moreover, due to climate change, many wild species of food crops would become extinct. For example, one-fourth of all wild potato species are predicted to die out within 50 years, which could make it difficult for future plant breeders to ensure that commercial varieties can cope with climate change.

One of the positive impacts of the increased CO_2 level may be the increased photosynthesis which in turn is expected to contribute to enhanced biomass production. On the other, the increased temperature associated with the increased level of CO_2 would have negative physiological impacts leading to increased photorespiration.

The impact on population structure   The genetic structure of a normal random mating population is explained by the Hardy–Weinberg principle which states that the gene and genotype frequencies in a large random mating population remain constant from generation to generation provided there is no specific disturbing forces viz., mutation, migration, selection and random genetic drift, etc. Climate change can disrupt the equilibrium by inducing migration of gene and genotype from one population to other, induce mutation (radiation and chemical) creating new allele and put differential selection pressure on population.

Indirect impacts of climate change on biodiversity

Although direct impacts are easily observable as well as well predicted, there are certain indirect impacts of climate change on biodiversity that are of equal importance. When climate change directly alters symbiotic fungi associated with plant root system, it will indirectly cause a change in that plant's natural distribution. Another example of indirect impact is—a new grass may spread into a region that results in altering the fire regime and ultimately changes the species composition.

All these indirect impacts are very difficult to predict.

Impact of climate change on specific ecosystems

Impact on forest ecosystems

a. *Tropical montane forests*: As a result of a rise in temperatures, tropical montane forests lose humidity or face

drying out leading to invasion or replacement of montane forest species by lower montane or non-montane
species.

b. *Boreal forests*: Owing to high increases in temperature,
forest fires become frequent in boreal forests which create favourable conditions for pests. Boreal forest will
shift towards Arctic areas destroying the biodiversity of
the forest.

c. *Polar region*: Climate change drastically influences the
reproduction rates of polar bears in the Arctic region.
Bears reuse their dens over a long period of time where
dens get extended to permafrost. These permafrosts are
important for the heat stresses of female bears. When the
permafrost is melted because of an increase in temperature collapsing the dens, female bears face a high thermal
disadvantage that has a negative impact on bear reproduction. The ecological function of the entire region is
affected as a result of the disruption in the bear's (key-
stone species) ecosystem.

Impact on marine ecosystem

a. *The coastal margins*: Certain coastal margins are
backed by areas of intense human use. If the level of sea
increases, it reduces important coastal habitats, including
mud flats and salt marshes.

b. *Warmer oceans*: Increase in sea temperature influences the distribution and survival of certain marine
resources. Coral reefs are facing the most hazardous
impact of climate change. High temperature causes
coral bleaching leading to the loss of the coral reef
structure. Around the world, coral reefs have been
dying as a result of climate change. Staghorn and
Elkhorn were found to be the first coral species to be
registered as 'endangered' which are commonly found
in the Caribbean. Researchers found that in the last two
decades, both species have practically become extinct.
Death of corals disrupts the coral ecosystem, which
leads to alterations in the fauna of associated species as
algae take over the reef.

c. *Increase in acidification*: As a result of climate changes,
atmospheric CO_2 is increasing day by day. This increased
CO_2 is getting dissolved in oceans leading to increase in
acidity. Owing to increases in acidity, organisms such as

corals, shellfish and molluscs are unable to produce cal-
careous parts such as shells.

Impacts of climate on the terrestrial realm

a. Range and abundance shifts

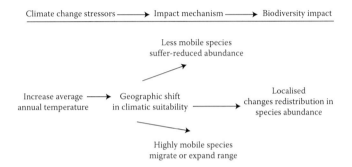

b. Life-cycle changes

c. Evolutionary effect

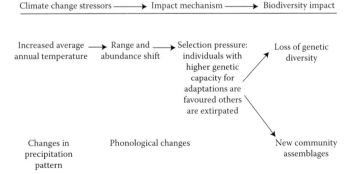

d. Disturbance regimes and ecosystem-level change

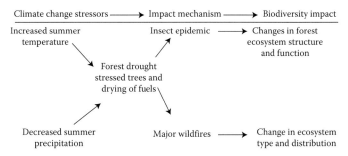

Impact of climate change on the freshwater realm

a. *Hydrological cycle change*: Reduced surface water availability

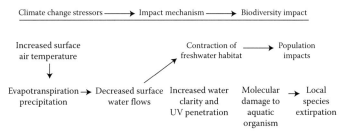

b. *Hydrological cycle change*: Increased surface water availability

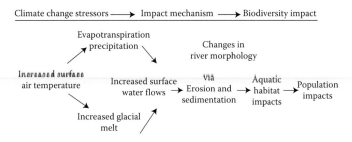

Human activities add fuel to the biodiversity threat

Human-induced pressure on biological diversity is increasing day by day. The burning of coal, oil and natural gas, as well as deforestation and various agricultural and industrial practices are altering the composition of the atmosphere and contributing to climate change. These human activities have led to increased atmospheric concentrations of a number of greenhouse gasses, including carbon dioxide, methane, nitrous oxide, chlorofluorocarbons and ozone in the lower part of the atmosphere. Burning of fossil fuels, releases carbon dioxide gas to the atmosphere. Greenhouse gasses and aerosols affect climate by altering incoming solar radiation and outgoing infrared (thermal) radiation that are part of Earth's energy balance. If the natural abundance of these gasses in the atmosphere is distorted, it causes hazards by warming or cooling the Earth's surface. Since the beginning of the industrial era (about 1750), human activities are greatly influencing climate. This impact is increasing day by day along with the changes in the natural processes such as solar changes and volcanic eruptions.

Management of the biodiversity disaster caused by climate change

Most of the people worldwide are aware of the term global warming. But very few of them know the reasons behind global warming and the measures to control it. Hence, the first and foremost management strategy will be spreading awareness about climate change and its details. We people have to minimise burning of fossil fuel to decrease the emission of CO_2 into the atmosphere. Technologies must be developed to design an alternate energy source. Gerald Meehl (National Centre for Atmospheric Research) has rightly said *Many people don't realize that we are committed right now to a significant amount of global warming and sea level rise...the longer we wait to do something about it, the more change we will have*. People can set an example for others by first reducing the usage of coal, oil, electricity usage around the home and so on. Some of the management strategies are mentioned below:

- Switch to energy-efficient lighting
- Improving the efficiency of home appliances
- Buying energy-efficient appliances when shopping for a new appliance
- Reducing energy needed for heating and cooling
- Practicing fuel-efficient driving
- Buying a fuel-efficient car
- Recycling an air-conditioner coolant
- Driving less
- Painting your home with a light colour if you live in a warm climate, or a dark colour in a cold climate
- Choosing clean energy options
- Buying clean energy certificates and carbon offsets

It is obvious that following all these strategies is not so easy. But at this scenario, even a small step can make a difference and a small change is worthwhile.

Conclusion

Nevertheless, climate change is inevitable. Therefore, we must try to cope with the threats of climate change to maintain the biodiversity for the cause of the human civilisation. The most crucial difficulty in finding out a strategy is that we do not know exactly what is going to happen due to climate change. Whatever predicted is highly speculative and evidences are meagre to formulate an effective strategy to combat the climate change. Therefore, it is time to intensify research on the areas of the effect of changing climate variables on the biological system. This should not be restricted just to species adaptation or performance change, but in a realistic way, the whole physiology of the biological system at the biochemical and molecular level should be taken into consideration to acquire the real understanding about the effect of climate change. This must be taken under a mega global platform for climate change research.

Bibliography

ALTHER, G.R., POST, E., CONVEY, P. et al. (March 2002). Ecological responses to recent climate change. *Nature* 416 (6879): 389–395.

Biodiversity and climate change, Convention on biological diversity, UNEP (www.biodiv.org).

Climate Change: The impact on biodiversity, Earthwatch Educational Resources, (www.earthwatch.org).

Climate change affects biodiversity, Global issues. (www.globalissues.org).

LYNCH, M. AND LANDE, R. 1993. Evolution and extinction in response to environmental change. In: Huey, R.B., Kareiva, P.M., and Kingsolver, J.G., (eds.), *Biotic Interactions and Global Change*. Sunderland, MA: Sinauer Associates. pp. 234–250.

Major impacts: Climate change, Compass resource management (www.compassrm.com).

PARMESAN, C. and YOHE, G. (January 2003). A globally coherent fingerprint of climate change impacts across natural systems. *Nature* 421 (6918): 37–42.

SAHNEY, S., BENTON, M.J. and FALCON-LANG, H.J. 2010. Rain forest collapse triggered Pennsylvanian tetrapod diversification in Euramerica (PDF). *Geology* 38(12): 1079–1082.

CHAPTER SEVENTEEN

Marker-assisted breeding approaches for enhancing stress tolerance in crops in changing climate scenarios

Uma Maheswar Singh, Gunjan Tiwari,
B. Kalyana Babu and Rakesh Srivastava

Contents

Abstract

Climate change threatens not only the yields and livelihoods of traditional farmers, but also the ability of agri-
culturalists worldwide. In addition to the challenge of
abiotic stresses such as temperature extremes (hot and
cold), drought or water stress, the incidence and severity of biotic stresses such as pests, diseases and the invasion of alien weed species are also likely to be greater.
Several strategies for adaptation to climate change have
been proposed to address crop productivity. One strategy
emphasises changing cultural practices. A second strategy involves adopting methods to increase the resilience
of agro-ecosystems to environmental variability.

17.1 Introduction

Climate change is a significant and lasting change in the weather
patterns over periods ranging from decades to millions of years.
It is caused by factors that include oceanic processes, variations in solar radiation received by Earth, plate tectonics and
volcanic eruptions, and human-induced alterations of the natural world. Global climate change is likely to increase the problems of food insecurity, hunger and malnutrition for millions
of people throughout the world. A recent report shows that the
global maize yield reduced by 3.8% between 1980 and 2010
due to climate change (Lobell et al. 2011). The mechanisms
that threaten the crop yield do not affect farmers evenly. The
local landraces mostly grown by subfarmers contribute little
carbon emission into the atmosphere and they tend to be vulnerable to climatic flux (Monterroso et al. 2011). Small farmers
who grow a wide variety of landraces serve as custodians of
crop diversity; their vulnerabilities have implications for the
in situ conservation of diverse crop landraces (Bellon et al.
2011). A study said that agro-biodiversity remains the main

raw material for agro-ecosystems to cope with climate change because it can provide traits for plant breeders and farmers to select resilient, climate-ready crop germplasm (Ortiz 2011). A third strategy involves improving seed varieties to develop new traits, such as drought resistance (Pray et al. 2011). One variant of this third strategy has received considerable attention recently: that is employing marker-assisted breeding (MAB) approaches to develop seeds that are well adapted to climate change. The second variant of this strategy is innovation in plant breeding to develop crop varieties that are more resilient to climate change. The objective of molecular plant breeding is to accumulate favourable alleles that contribute to stress tolerance in a plant genome. In this chapter, we are addressing the role of molecular plant breeding in abiotic constraints to enhance the crop productivity.

Genes that confer stress resistance can be sourced from wild relatives of crops that are held in gene banks or in the live habitats of water deficit or excess, extreme temperature and salinity that have evolved to cope with those conditions. Although some progress has been made through conventional breeding, but due to the complex nature of abiotic stress tolerance undesirable genes are also transferred along with desirable traits that limit the transfer of favourable alleles from diverse genetic resources. MAB is an emerging area that involves transfer of superior genes or alleles, where they were tightly linked to a particular trait into elite genotypes of locally adopted germplasm. However, genetic engineering (GE) involves transferring useful genes or alleles across different species from the animal or plant kingdoms. As a result, biotechnology approaches offer novel strategies for producing suitable crop genotypes that are able to resist drought, high temperature, submergence and salinity stresses. The key strategies where genetic enhancement for stress tolerance has led to crop improvement are outlined in this section.

17.2 Key biotechnological strategies for improving stress tolerance

Molecular breeding approach

1. The development of genomic resources such as molecular markers, including simple sequence repeats (SSRs), single-nucleotide polymorphisms (SNPs) and marker genotyping platforms

2. The development of bi-parental mapping populations by using genetically and phenotypically diverse parental

lines or the selection of a natural population representing diversity for stress tolerance traits

3. The use of linkage mapping or association mapping approaches to identify the quantitative trait loci (QTL) or markers associated with stress tolerance-related parameters, such as leaf water retention, high rates of leaf photosynthesis, stomatal conductance, osmotic adjustment, and faster canopy and root development.

4. The validation of the QTL or markers in a breeding germplasm that have a different genetic background

5. The use of an appropriate MB approach such as MABC, MARS or GWS to develop superior crop genotypes.

Genetic engineering approach

1. The identification of genes encoding signalling proteins, TFs and effector proteins, and novel stress responsive promoters controlling multiple stress tolerance

2. The identification of genes regulating stomatal opening and closure and stress-induced expression to enhance water use efficiency in crops

3. The genetic transformation and development of elite crop genotypes with tolerance to high-temperature stress and other environmental stresses

4. The assessment of promising transgenic lines for multiple stress tolerance under field conditions

5. The deregulation of transgenic lines to enable the release of a superior line or variety.

In this chapter, we are only emphasising different molecular breeding approaches and the role of these approaches for enhancing stress tolerance in crops in a changing climate.

17.3 Section I: Molecular markers and their applications for crop improvement

Before starting the discussion on molecular markers, we should be familiar with genetic markers. Genetic markers represent genetic differences between individual organisms or species; they do not represent the target genes themselves but act as 'signs' or 'flags'. Such markers themselves do not affect the phenotype of the trait of interest because they are located near or 'linked' to genes controlling the trait. The genetic markers occupy specific genomic positions within chromosomes (like genes) called 'loci' (singular 'locus') and are broadly differentiated into three categories (Collard et al. 2005):

1. *Morphological markers:* Morphological markers are usually visually characterised phenotypic characters, such as flower colour, seed shape, growth habits or pigmentation.

2. *Biochemical markers:* Biochemical markers, also known as 'Isozyme' markers, are allelic variants of enzymes, which express differences in enzymes that are detected by electrophoresis and specific staining.

3. *DNA (or molecular) markers:* Reveal sites of variation at the DNA level.

The major disadvantages of morphological and biochemical markers are that they may be limited in number and are influenced by environmental factors or the developmental stage of the plant (Winter and Kahl 1995). However, despite these limitations, morphological and biochemical markers have been extremely useful to plant breeders (Eagles et al. 2001; Weeden et al. 1993). In contrast to the instability of morphological and biochemical markers to different environmental conditions, molecular markers are stable and are more in number.

Molecular markers   DNA markers are the most widely used type of markers due to their abundance in the genome. They arise from different classes of DNA mutations, such as substitution mutations (point mutations), rearrangements (insertions or deletions) or errors in replication of tandemly repeated DNA (Paterson 1996). These markers are selectively neutral because they are usually located in noncoding regions of DNA. Unlike morphological and biochemical markers, DNA markers are basically unlimited in number and are not affected by environmental factors and/or the developmental stage of the plant (Winter and Kahl 1995). Apart from the use of DNA markers in the construction of linkage maps, they have numerous applications in plant breeding, such as assessing the level of genetic diversity within germplasm and cultivar identity (Collard et al. 2005). A list of commonly used molecular markers and their properties are summarised in Table 17.1. On the basis of advancement in molecular markers used, these markers can be categorised into first-, second- and new-generation molecular markers.

First-generation molecular markers   The concept of utilising variations at the DNA level as genetic markers initiated with restriction fragment-length polymorphism (RFLP).

Table 17.1 Comparisons of the most commonly used molecular markers in plants

S. no.	Feature	RFLP	RAPD	AFLP	SSRs	SNPs
1	DNA require (μg)	10	0.02	0.5–1.0	0.05	0.05
2	DNA quality	High	High	Moderate	Moderate	High
3	PCR based	No	Yes	Yes	Yes	Yes
4	Inheritance	Co-dominant	Dominant	Dominant	Co-dominant	Co-dominant
5	No. of polymorphic loci	1–3	1.5–50	20–100	1–3	1
6	Ease of use	Not easy	Easy	Easy	Easy	Easy
7	Amenable to automation	Low	Moderate	Moderate	High	High
8	Reproducibility	High	Low	High	High	High
9	Development cost	Low	Low	Moderate	Moderate	High
10	Cost per analysis	High	Low	Moderate	High	Low

The first official recognition of RFLP came from viruses (Grodzicker et al. 1975), followed by a subsequent demonstration made in the human-globin gene cluster (Jeffreys 1979). Since then, most organisms have been used for the presence of RFLP, and the application of this technology has evolved in various fields. Subsequent to RFLP, several other methods such as variable number of tandem repeats, allele-specific oligonucleotide, allele-specific polymerase chain reaction (PCR), oligonucleotide polymorphism, single-stranded conformational polymorphism (SSCP) and sequence-tagged sites (STS) have been documented. The PCR-based assay has evolved and can detect variations at the DNA level by replacing conventional hybridisation-based assay of detecting DNA-level variations.

Second-generation molecular markers The second-generation molecular markers are microsatellite arrays of tandemly repeated di-, tri-, tetra- and penta-nucleotide DNA sequences, which are dispersed throughout the genomes of all eukaryotic organisms investigated to date. These markers are responsible for various revolutions in the field of molecular breeding. The microsatellites are also called as sequence-tagged microsatellite sites or SSRs. Currently, SSRs are considered as the molecular markers of choice within the genome mapping community and are frequently being adopted by plant researchers as well. SSR contains around 10–50 copies of motifs from 1 to 5 base pairs that may occur in perfect tandem repetition, as imperfect (interrupted) repeats or together with another repeat type. These repeated motifs are flanked by unique or single copy sequences, which give a base clutch for specific amplification via PCR. Primers that are complementary to the unique sequences in those flanking regions can be designed to amplify single copy products. The other marker systems that have been developed during this period include restriction landmark genome scanning, cleaved amplified polymorphic sequence (CAPS), degenerate oligonucleotide primer PCR, SSCP, multiple arbitrary amplicon profiling and sequence characterised amplified region (SCAR). The usage of these marker systems was not realised as new SSRs.

New-generation molecular markers Recent advances in molecular biology have opened the opportunity of utilising various types of molecular tools to identify and use genomic variation improvement of various organisms. Information concerning the basis of these techniques and their applications involve the technology spill over of several genome projects. The last 10 years have witnessed the origin of an array of molecular markers with high-throughput performance coupled with shift from manual mode of detection to complete automation. Inter simple sequence repeats (ISSRs), selective amplification of microsatellite polymorphic loci, SNPs, amplified fragment-length polymorphism (AFLP), selective restriction fragment amplification, allele-specific associated primers, cleavage fragment-length polymorphism, inverse sequence-tagged repeats, directed amplification of mini satellite DNA-PCR, sequence-specific amplified polymorphism, retrotransposon-based insertional polymorphism, inter-retrotransposon amplified polymorphism, retrotransposon-microsatellite amplified polymorphism, methylation-sensitive amplification polymorphism, miniature inverted-repeat transposable element (MITE), three endonuclease AFLP, inter-MITE polymorphisms sequence-related amplified polymorphism, and so on are the markers of recent origin with great potential in understanding the variation at the DNA level.

Genetic differences observed by DNA markers can be visualised by using a vertical and horizontal gel electrophoresis or staining by means of ethidium bromide, silver nitrate, detection with radioactive or colorimetric probes. DNA markers showing differences between individuals of the same or different species (polymorphic markers) are generally more useful than markers that do not discriminate between genotypes (monomorphic markers) of same or different species. Polymorphic markers can also be described as co-dominant or dominant. This description is based on whether markers can differentiate between homozygotes and heterozygotes. Co-dominant markers indicate differences in the size of DNA whereas dominant markers are either present or absent. In the strictest sense, different forms of a DNA marker (e.g. different sized bands on gels) are called marker 'alleles'. Co-dominant markers may have many different alleles whereas a dominant marker has only two alleles. It is beyond the scope of this chapter to discuss the technical method of how DNA markers are generated.

This section provides a brief classification of these molecular markers on the basis of its functions. Molecular markers can be classified into two major groups: (1) based on DNA/DNA hybridisation (e.g. RFLP) and (2) based on PCR amplification of genomic DNA fragments (RAPD, ISSR, SSR, SCAR, AFLP, SNP, CAPS, etc.).

1. *Hybridisation-based molecular markers.* RFLP markers are the most widely used hybridisation-based molecular markers in humans and plant genome analysis. These markers were first used in 1975 to identify DNA sequence polymorphisms for genetic mapping of a temperature-sensitive mutation of adeno-virus serotypes (Grodzicker et al. 1975). It was then utilised for human genome mapping (Botstein et al. 1980), and later adopted for plant genomes (Helentjaris et al. 1986; Weber and Helentjaris 1989). The technique is based on restriction enzymes that reveal a pattern of difference between DNA fragment sizes in individual organisms. Although two individuals of the same species have almost identical genomes, they will always differ at a few nucleotides due to one or more of the following causes: point mutation, insertion/deletion, translocation, inversion and duplication. Some of the differences in DNA sequences at the restriction sites can result in the gain, loss or relocation of a restriction site. Hence, digestion of DNA with restriction enzymes results in fragments whose number and size can vary among individuals, populations and species. The brief steps involved in RFLP molecular marker assay are as follows (Semagn et al. 2006):

    a. Digestion of the DNA with one or more restriction enzyme(s)

    b. Separation of the restriction fragments in agarose gel

c. Transfer of separated fragments from agarose gel to a filter by Southern blotting

d. Detection of individual fragments by nucleic acid hybridisation with a labelled probe(s)

e. Autoradiography

2. *PCR-based markers.* PCR is a molecular biology technique for enzymatically replicating (amplifying) small quantities of DNA without the use of a living organism. It is used to amplify a short (usually up to 10 kb), well-defined part of a DNA strand from a single gene or part of a gene. Since its invention by Kary Mullis in 1983, this technique enabled the development of various types of PCR-based techniques, which honoured him to get the Nobel Prize in 1993. However, the basic PCR procedure was described in 1968 by Kleppe and his co-workers in Khorana's group.

The basic protocol for PCR is simple and is as follows:

a. Double-stranded DNA is denatured at a high temperature (95°C) to form single strands (templates).

b. Short, single strands of DNA (known as primers) bind at specific annealing temperatures (which vary with different conditions) to the single-stranded complementary templates at ends flanking the target sequences.

c. The temperature is raised usually to 72°C for the DNA polymerase enzyme to catalyse the template-directed syntheses of new double-stranded DNA molecules that are identical in sequence to the starting material.

d. The newly synthesized double-stranded DNA target sequences are denatured at high temperature, and the cycle is repeated.

Although the basic protocol of PCR is straightforward, each application requires optimising the various parameters for the species to be studied. During the early days of PCR work, the DNA polymerase would need to be added fresh to the reaction at each temperature cycle, because thermostable (high-temperature tolerant) DNA polymerases were not commercially available. The discovery of *Taq* DNA polymerase, the DNA polymerase present in the bacterium *Thermus aquaticus* in hot springs, was decisive for the immense utility and popularity of PCR-based techniques because of its stability at high temperature, whereas other DNA polymerases became become denatured. Nowadays, the PCR technology is much more advanced with a wide range of thermostable DNA polymerases (such as *Taq*, *Pfu* and *Vent* polymerase) and automation of reactions can be done by a thermocycler, which has found its way into nearly every molecular biology lab in the world. The major advantages of PCR techniques compared to hybridisation-based methods include (Semagn et al. 2006):

1. A small amount of DNA is required.
2. Elimination of radio-isotopes in most techniques which are health hazardous.
3. The ability to amplify DNA sequences from preserved tissues.
4. Accessibility of methodology for small labs in terms of equipment, facilities and cost.
5. No prior sequence knowledge is required for molecular markers like AP-PCR, RAPD, DAF, AFLP and ISSR.
6. High polymorphism that enables generating many genetic markers within a short time.
7. The ability to screen many genes simultaneously either for direct collection of data or as a feasibility study prior to nucleotide sequencing efforts (Wolfe and Liston 1998).

These advantages, however, can vary depending on the specific technique chosen by the researcher. The PCR-based techniques can be categorised into two types, depending on the primers used for amplification:

1. Arbitrary or semi-arbitrary primed PCR techniques that are developed without prior sequence information (e.g. AP-PCR, DAF, RAPD, AFLP, ISSR).
2. Site-targeted PCR techniques that are developed from known DNA sequences (e.g. EST, CAPS, SSR, SCAR, STS).

17.4 Section II: Molecular markers for crop improvement

The rising global population will require increased crop production and productivity with sustainable agriculture. This required increase in crop production needs to occur in the context of mounting water scarcity, decreasing area and environmental degradation of arable land, increasing pollution and possible adverse effects of climate change. Thus, the task of increasing crop yields represents an extraordinary challenge for agricultural scientists. Plant breeding has made remarkable progress in increasing crop yields for over a century by developing suitable cultivars, varieties and hybrids for various crops. Plant breeding along with biotechnological approaches like MAS will play a key role in this coordinated effort for increased food production. Despite optimism about continued yield improvement from conventional breeding, with the advancement of agricultural biotechnology, there is a need to maximise the probability of success (Ortiz 1998). One area of biotechnology, that is, DNA marker technology, derived from research in molecular genetics and genomics, offers great promise for plant breeding. Owing to genetic linkage, the DNA markers can be used to detect the presence of allelic variation in the genes underlying these traits.

Molecular markers are widely accepted as potentially valuable tools for crop improvement in major cereal crops like rice (Mackill et al. 1999), wheat (Koebner and Summers 2003), maize (Tuberosa et al. 2003), barley (Williams 2003), and also in other important crops like tubers (Barone 2004), pulses (Kelly et al. 2003; Svetleva et al. 2003), oilseeds (Snowdon and Friedt 2004), horticultural crop species (Baird 1995; Baird et al. 1997; Mehlenbacher 1994) and pasture species (Jahufer et al. 2002). Some studies suggest that DNA markers will play a vital role in enhancing global food production by improving the efficiency of conventional plant breeding programs (Kasha 1999; Ortiz 1998). Although there has been some concern that the outcomes of DNA marker technology as proposed by initial studies may not be as effective as first thought, many plant breeding institutions have adopted the capacity for marker development and/or marker-assisted selection (MAS) (Eagles et al. 2001). An understanding of the basic concepts and methodology of DNA marker development and MAS, including some of the terminology used by molecular biologists, will enable plant breeders and researchers working in other relevant disciplines to work together towards a common goal—increasing the efficiency of global food production.

Construction of linkage maps

A variation observed in the field does not always give any conclusive indication about variation at genome level. These genotypic variations cannot be visualised by the naked eye. Then comes the need for certain landmarks that can be linked to our trait of interest. Hence, mapping (i.e. determining the exact location of the gene influencing a trait) is necessary for constructing a framework for a particular plant genome with the help of markers. So, the foremost method used was the bi-parental mating method to link a trait with the marker (linkage map). A linkage map determines not only the position of various genes but also their relative distances along the chromosome. The basic principle behind linkage mapping is the close association of trait of interest along with marker that could be screened in the mapping population formed by mating between the parents P1 and P2. The schematic diagram representing recombination event is illustrated in Figure 17.1. The determination of tight association between a gene of interest and the marker could be carried out by knowing the recombination frequency between them that can be traced with the help of recombinant genotypes recovered in the segregating generations. A minimum recombination frequency (<10 cM) depicts tight linkage.

The main steps involved in the linkage map construction are:

- Construction of mapping population and deciding the sample size

- Selection of molecular markers for genotyping the mapping population

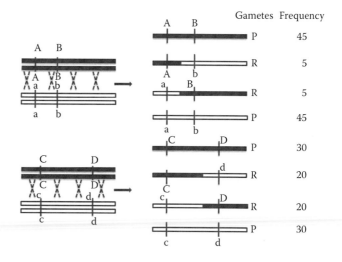

FIGURE 17.1 Diagram indicating cross-over or recombination events between homologous chromosomes. Gametes that are produced after meiosis are either parental (P) or recombinant (R). The smaller the distance between two markers, the smaller the chance of recombination occurring between the two markers. Therefore, recombination between markers C and D should occur more frequently than recombination between markers A and B. This can be observed in a segregating mapping population. By analysing the number of recombinants in a population, it could be determined that markers A and B are closer together compared to C and D.

- Screening of polymorphism between parents and segregants
- Linkage analysis to determine gene order and map distance with the help of different statistical tools

Mapping populations

A population used for gene mapping is commonly called a mapping population. The commonly used mapping populations are obtained from controlled crosses. The selection of a population for genome mapping involves choosing parents and determining a mating scheme. Decisions on the selection of parents, mating designs and the type of markers depend upon the objectives of the experiments. Parents of mapping populations must have sufficient variation for the traits of interest at both the DNA sequence and phenotypic level. The variation at the DNA level is essential to trace the recombination events. The more DNA sequence variation exists, the easier it is to find polymorphic informative markers.

The selection of parents for developing a mapping population is critical to successful map construction. Since a map's economic significance will depend upon marker-trait association, as many qualitatively inherited morphological traits as possible should be included in the genetic stocks chosen as parents for generating the mapping population. Diagrams representing different types of mapping populations are shown in Figure 17.2. Different types of mapping populations that are most frequently used in linkage mapping are as follows:

F_2 population Such populations are produced by selfing or sib mating the individuals in segregating populations generated by crossing the selected parents. The expected ratio of dominant and co-dominant markers from F_2 population is 3:1 and 1:2:1, respectively. These are considered to be the best population for preliminary mapping and require minimum effort and time for development. Such populations are temporary populations because they are highly heterozygous and cannot be propagated indefinitely through seeds, and therefore, are of limited use for fine mapping.

F_2 derived F_3 ($F_{2:3}$) population $F_{2:3}$ populations are obtained by selfing the F_2 individuals for a single generation. Such populations are suitable for specific situations like mapping quantitative traits and recessive genes. The $F_{2:3}$ families can be used for reconstituting the genotype of respective F_2 plants, if needed, by pooling the DNA from plants in the family. Like F_2 population, it is not 'immortal'.

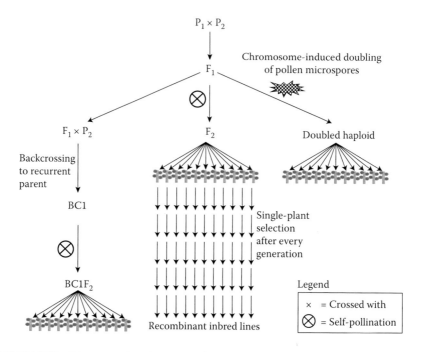

FIGURE 17.2 Diagram of main types of mapping populations for self-pollinating species. (From Collard, B. C. Y., Jahufer, M. Z. Z., Brouwer, J. B., and Pang, E. C. K. 2005. *Euphytica*, 142(1–2), 169–196.)

Backcross population Backcross populations are generated by crossing the F_1 with either of the parents. Usually in genetic analysis, backcross with recessive parent (testcross) is used. With respect to molecular markers, the backcross with dominant parent (B1) would segregate in the ratio 1:0 and 1:1 for dominant and co-dominant markers, respectively. However, backcross with recessive parent (B2) or testcross would segregate in the ratio of 1:1, irrespective of the nature of marker. Like an F_2 population, the backcross populations require less time to be developed, but are not immortal. The specific advantage of backcross populations is that, the populations can be further utilised for marker-assisted backcross breeding.

Double haploids Chromosome doubling of anther culture-derived haploid plants from F_1 generates DHs. In DHs, the expected ratio for the marker is 1:1, irrespective of genetic nature of marker (whether dominant or co-dominant). DHs are permanent mapping populations and, hence, can be replicated and evaluated over locations and years and maintained without any genotypic change. They are useful for mapping both qualitative and quantitative characters. But the main demerit of such population is that since it involves *in vitro* techniques, relatively more technical skills are required in comparison with the development of other mapping populations.

Recombinant inbred lines Recombinant inbred lines (RILs) are produced by continuous selfing or sib mating the progeny of individual members of an F_2 population until complete homozygous is achieved. Once homozygosity is achieved, RILs can be propagated indefinitely without further segregation. The single-seed descent method is best suited for developing RILs. The bulk method and pedigree methods without selection can also be used. RILs also equalise marker types like DHs, so the genetic segregation ratio for both dominant and co-dominant markers would be 1:1. Since RILs are immortal population, they can be replicated over locations and years and therefore are of immense value in mapping QTLs. RILs, being obtained after several cycles of meiosis, are very useful in identifying tightly linked markers. The major disadvantage associated with this population is that it requires many seasons/generations to develop and is relatively difficult in crops with high inbreeding depression.

Near-isogenic lines Near-isogenic lines (NILs) are generated either by repeated selfing or backcrossing the F_1 plants to the recurrent parents. NILs developed through backcrossing are similar to a recurrent parent but for the gene of interest, while NILs developed through selfing are similar in pair, but for the gene of interest (however, they differ a lot with respect to the recurrent parent). The expected segregation ratio of the markers is 1:1, irrespective of the nature of marker. Like DHs and RILs, NILs are also 'immortal mapping population'. These are quite useful in functional genomics. NILs are directly useful only for molecular tagging of the gene concerned, but not for linkage mapping and require many generations for development. Along with these, linkage drag is a potential problem in constructing NILs, which has to be taken care. Generally, a larger population size is needed for high-resolution fine mapping, but for preliminary genetic mapping studies a population size of 50–250 individuals can be used.

Identification of polymorphism The presence of sufficient polymorphism between selected parents is essential before the construction of a linkage map. Therefore, identification of polymorphic DNA markers, which show significant differences between parents, is the second most important step in the construction of a linkage map. DNA polymorphism levels are generally found to be less in inbreeding species as compared to cross-pollinated species; therefore, selection of distantly related parents is useful in inbreeding species. Once polymorphic markers have been identified between parents, they must be used to genotype the entire mapping population. This is known as marker 'genotyping' of the population. Different molecular markers like RFLPs, SSRs, ESTs, CAPs, RAPD, AFLP, ISSR, DArT and SNPs are used for the construction of a linkage map in several plants. Each marker system has several advantages and disadvantages, but generally, highly reproducible, high-throughput, co-dominant and transferable molecular markers are used to increase the utility of genetic maps. The expected segregation ratios for co-dominant and dominant markers are presented in Table 17.2. Remarkable deviations from expected ratios can be analysed using chi-square tests.

Generally, markers segregate in a Mendelian fashion even though distorted segregation ratios can be encountered. Significant deviation from expected segregation ratio in a given marker/population combination is referred to as segregation distortion. There are several reasons for segregation distortion, including: gamete/zygote lethality, meiotic drive/preferential segregation, sampling/selection during population development and differential responses of parental lines to tissue culture in case of DHs. In some polyploidy species such as sugarcane, identifying a polymorphic marker is more complicated. The mapping of diploid relatives of polyploidy species may be of great benefit in developing maps for polyploidy species. However, diploid relatives do not exist for all polyploidy species. Generally, mapping of polyploidy species is based on the use of single-dose restriction fragments (Ripol et al. 1999; Wu et al. 1992).

Table 17.2 Expected segregation ratios for markers in different population types

Population type	Co-dominant markers	Dominant markers
F$_2$	1:2:1 (AA:Aa:aa)	3:1 (B:bb)
Backcross	1:1 (Cc:cc)	1:1 (Dd:dd)
Recombinant inbred or doubled haploid	1:1 (EE:ee)	1:1 (FF:ff)

Linkage analysis of markers

Coding of data for each DNA marker on each individual of a population and conducting linkage analysis using computer programs is the final step in the construction of a linkage map. Missing marker data can also be acknowledged by mapping programs. Computer programs are required to analyse linkages between large numbers of markers, which are not possible manually. Linkage among markers is usually calculated using odds ratios. Markers are assigned to linkage groups using the odd ratios, which refers to the ratio of the probability that two loci are linked with a given recombination value over a probability that the two are not linked. This ratio is more conveniently expressed as the logarithm of the ratio, and is called a logarithm of odds (LOD) value or LOD score (Risch 1992). The LOD values of >3 are typically used to construct linkage maps. A LOD value of 3 between two markers indicates that the linkage is 1000 times more likely (i.e., 1000:1) than no linkage (null hypothesis). The LOD values may be lowered in order to detect a greater level of linkage or to place additional markers within maps constructed at higher LOD values. The commonly used software programs include Mapmaker/EXP (Lander et al. 1987; Lincoln et al. 1992) and MapManager QTX (Manly et al. 2001), which are freely available from the Internet. JoinMap is another commonly used program for constructing the linkage maps (Stam 1993).

A typical output of a linkage map is shown in Figure 17.3. Referring to the road map analogy, linkage groups symbolise roads and markers represent signs or landmarks. A difficulty linked with obtaining an equal number of linkage groups and chromosomes is that the polymorphic markers detected are not necessarily uniformly distributed over the chromosome, but clustered in some regions and absent in others. Along with this, the frequency of recombination is not equal along chromosomes. The accuracy of measuring the genetic distance and determining marker order is directly related to the number of individuals examined in the mapping population. Ideally, the mapping population should consist of a minimum of 50 individuals for constructing the linkage maps (Phillips et al. 2001).

Genetic distance and mapping functions

Generally, frequency of recombination between genetic markers is used to measure the distance along a linkage map. Mapping functions are essential to convert recombination fractions into centimorgans (cM) because recombination frequencies and the frequencies of crossing-over are not related linearly. When the map distance is small (<10 cM), the map distance equals the recombination frequency. However, this relationship does not

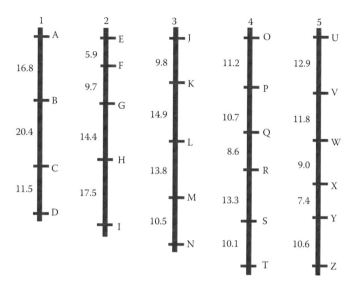

FIGURE 17.3 Hypothetical 'framework' linkage map of five chromosomes (represented by linkage groups) and 26 markers. Ideally, a framework map should consist of evenly spaced markers for subsequent QTL analysis. If possible, the framework map should also consist of anchor markers that are present in several maps, so that they can be used to compare regions between maps. (From Collard, B. C. Y., Jahufer, M. Z. Z., Brouwer, J. B., and Pang, E. C. K. 2005. *Euphytica*, 142(1–2), 169–196.)

apply for the map distance that is greater than 10 cM. There are two commonly used mapping functions:

1. Kosambi mapping function, which assumes that recombination events influence the occurrence of adjacent recombination events, that is, partial interference, and measured as $(\frac{1}{4})\ln[(1 + 2\theta)/(1 - 2\theta)]$.

2. Haldane mapping function, which assumes no interference between crossover events. Haldane's mapping function is based on the Poisson distribution of the number crossing over, so that the genetic distance (m) and observed recombination fraction (θ) containing an odd number of crossover is $m = -\ln(1 - 2\theta)/2$ (Kearsey and Pooni 1996; Paterson 1996; Hartl and Jones 2001).

It should be noted that the distance between genetic markers depends on the genome size of the plant species. It is not related to physical distance of DNA between genetic markers. Moreover, the relationship between genetic and physical distance varies along a chromosome. For example, there are recombinations of 'hot spots' and 'cold spots', which are chromosomal regions where recombination occurs more frequently or less frequently, respectively (Faris et al. 2000; Ma et al. 2001; Yao et al. 2002).

QTL analysis   Quantitative traits in crop plants are controlled and regulated by polygenes, which make their study difficult by Mendelian methods of genetic analysis. In recent years, the availability of polymorphic molecular markers facilitated the genetic analysis of those quantitative attributes by treating polygenes as QTLs, which also segregate in a Mendelian manner. A QTL (a term first coined by Gelderman 1975) is defined as 'a region of the genome that is associated with an effect on a quantitative trait'. Conceptually, a QTL can be a single gene, or it may be a cluster of linked genes that affect the trait. QTL mapping studies have been reported in most crop plants for diverse traits, including yield, quality, disease and insect resistance, abiotic stress tolerance and environmental adaptation (Singh et al. 2012).

Principle of QTL analysis   Identification and mapping of a good number of segregating markers (10–50) per chromosome are not difficult in the populations of most crop plants. However, most of those markers would be in the noncoding regions of the genome and might not affect the trait of interest directly, but a few of these markers might be linked to genomic regions (QTLs) that do influence the trait of interest. Wherever such linkage occurs, the marker locus and also the QTL will co-segregate. Therefore, the basic principle of determining whether a QTL is linked to a marker is to partition the mapping population into different genotype classes based on genotypes at the marker locus, and apply correlative statistics to determine whether the individuals of one genotype differ significantly with the individuals of another genotype with respect to the trait being measured. Circumstances where genes fail to segregate independently are said to display 'linkage disequilibrium (LD)'. QTL analysis, thus, depends on the LD. With natural populations, consistent association between QTL and marker genotype will not frequently exist, except in a very rare condition wherever the marker is completely linked to the QTL. Therefore, QTL analysis is undertaken in segregating mapping populations, such as F_2-derived populations, RILs, near-isogenic lines, DHs and backcross populations

(http://iasri.res.in/ebook/EB_SMAR/e-book_pdf%20files/
Manual%20IV/10-QTL.pdf).

The principle behind the QTL mapping implicates partitioning the mapping population into different genotypic groups based on the genotype data on the mapping population for determining whether significant differences exist between groups with respect to the trait being measured (Tanksley 1993; Young 1996). A major dissimilarity between phenotypic means of the groups (either 2 or 3), depending on the marker system and type of population, indicates that the marker locus being used to partition the mapping population is linked to a QTL controlling the trait. The P value obtained from the differences between mean trait values indicates linkage between marker and QTL and is due to recombination. The diagram representing the linkage between marker and QTL are

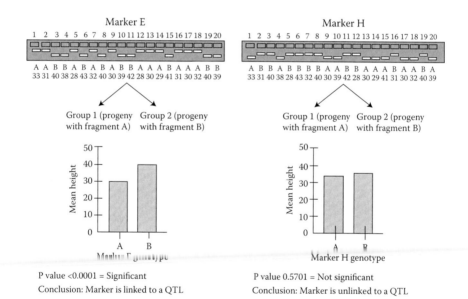

FIGURE 17.4 Principle of QTL mapping. Markers that are linked to a gene or QTL controlling a particular trait (e.g. plant height) will indicate significant differences when the mapping population is partitioned according to the genotype of the marker. Based on the results in this diagram, Marker E is linked to a QTL because there is a significant difference between means. Marker H is unlinked to a QTL because there is no significant difference between means. The closer the marker is to the QTL of interest, the lower the chance for recombination between marker and QTL. (Adapted from Collard, B. C. Y., Jahufer, M. Z. Z., Brouwer, J. B., and Pang, E. C. K. 2005. *Euphytica*, 142(1–2), 169–196.)

illustrated in Figure 17.4. The closer a marker is from a QTL, the lower the prospect of recombination occurring between marker and QTL. Therefore, the QTL and marker will usually be inherited together in the progeny, and the mean of the group with the tightly linked are going to be considerably totally different ($P < 0.05$) to the mean of the group without the marker and will be the reverse in case if there is no significant P value for unlinked QTLs. If the unlinked markers are located far apart or on different chromosomes for the QTL of a particular trait indicates non-significant variation among the genotypes.

Statistical methods to detect QTLs The fundamental objective in QTL mapping studies is to identify QTL, whereas minimising the prevalence of false positives (Type I errors, i.e. declaring an association between a marker and QTL once in reality one does not exist). Tests for QTL/trait association are usually performed by the subsequent approaches:

Single-marker analysis (also 'single-point analysis') This is the simplest method for detecting QTLs related to single markers. Linear regression, analysis of variance and *t*-tests are generally used for this analysis. Among these, the most frequently used technique is linear regression because the coefficient of determination (R2) from the marker explains the phenotypic variation generating from the QTL linked to the marker. This method does not need a complete linkage map and can be accomplished with basic statistical software programs. However, the key disadvantage associated with this process is that the further a QTL is from a marker, the less likely it will be detected. This can be as a result of recombination that could occur between the marker and also the QTL. This causes the magnitude of the effect of a QTL to be underestimated. The utilisation of a large number of segregating DNA markers covering the entire genome (usually at intervals <15 cM) may minimise both problems. The results from single-marker analysis are commonly presented in a table, which indicates the chromosome (if known) or linkage group encompassing the markers, probability values and the percentage of phenotypic variation elucidated by the QTL (R2) (Table 17.3). Sporadically, the allele size of the marker is also described. To execute a single-marker analysis, Q Gene and MapManager QTX most frequently used computer programs (Manly et al. 2001; Nelson 1997).

Table 17.3 Single-marker analysis of markers associated with QTLs using QGene

Marker	Chromosome or linkage group	P value	R^2
E	2	<0.0001	91
F	2	0.0001	58
G	2	0.023	26
H	2	0.5701	2

Source: Nelson, J. 1997. *Molecular Breeding*, 3(3), 239–245.

Simple interval mapping   In order to mitigate the matter related to single marker analysis, these techniques create use of linkage maps and analyses intervals between adjacent pairs of linked markers along chromosomes simultaneously (Lander and Botstein 1989). This approach was considered statistically more powerful compared to single-point analysis because of the use of linked markers for recombination between the markers and the QTL (Lander and Botstein 1989; Liu 1998). Several investigators have used Map Maker/ QTL (Lincoln et al. 1993) and Q Gene (Nelson 1997) to conduct SIM.

Composite interval mapping   Recently, this method has become prevalent for mapping QTLs. This method combines features of interval mapping with linear regression and includes additional genetic markers in the statistical model additionally to an adjacent pair of linked markers for interval mapping (Jansen 1993; Jansen and Stam 1994; Zeng 1993, 1994). The main advantage associated with CIM is that it is more precise and effective at mapping QTLs compared to single-point analysis and interval mapping, especially when linked QTLs are involved. Many researchers have used QTL Cartographer (Basten et al. 1994, 2004), MapManager QTX (Manly et al. 2001) and PLABQTL (Utz and Melchinger 1996) to perform CIM.

Understanding interval mapping results   Interval mapping strategies generate a profile of the sites for a QTL between adjacent linked markers. The result of the test statistic for interval mapping is often conferred employing a LOD score or likelihood ratio statistic (LRS). There is an immediate one-to-one transformation between LOD scores and LRS scores (the conversion can be calculated by: LRS = 4.6 × LOD) (Liu 1998). These LOD or LRS profiles are used to recognise the most possible position for a QTL in relation to the linkage map, which is the position where the highest LOD value is achieved. A typical output from interval mapping is a LOD graph, with markers comprising linkage groups on the x-axis and the test statistic (LOD scores) on the y-axis are illustrated in Figure 17.5. The minimum value of LOD is 2.5, which can be considered as the threshold limit for considering a QTL to be real. The determination of significance thresholds is most commonly accomplished using permutation tests. Briefly, the phenotypic values of the population are 'shuffled' while the marker genotypic values are held constant (i.e. all marker/trait associations are broken) and the QTL analysis is performed to assess the level of false-positive marker/trait associations (Churchill and Doerge 1994; Haley and Anderson 1997; Hackett 2002). This process is then repeated (e.g. 500 or 1000 times) and the significant levels can then be determined based on the level of false-positive marker/trait associations.

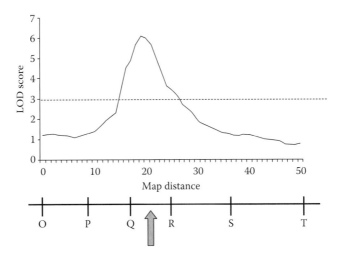

FIGURE 17.5 Hypothetical output showing a LOD profile for chromosome 4. The dotted line represents the significance threshold determined by permutation tests. The output indicates that the most likely position for the QTL is near marker Q (indicated by an arrow). The best flanking markers for this QTL would be Q and R. (Adapted from Collard, B. C. Y., Jahufer, M. Z. Z., Brouwer, J. B., and Pang, E. C. K. 2005. *Euphytica*, 142(1–2), 169–196.)

Reporting and describing QTLs detected from interval mapping The most common approach of reporting QTLs is by demonstrating the most closely linked markers in a table and/or bars (or oval shapes or arrows) on the linkage maps. The chromosomal regions marked by rectangles are typically the regions that exceed the significance threshold. Usually, a pair of markers—the most tightly linked markers on either side of QTLs—are also reported in a table; these markers are referred to as 'flanking' markers. The motive for reporting flanking markers is that selection based on two markers would be more reliable than selection based on a single marker. The rationale for the increased reliability is that there will be a much lower chance of recombination between two markers and QTL compared to the chance between a single marker and QTL. It should also be noted that QTLs can only be identified for traits of interest that segregate between the parents used to construct the mapping population. Therefore, in order to take advantage of the data obtained from a QTL mapping study, several criteria might be used for phenotypic evaluation of a single trait. QTLs that are identified in common regions (based on different criteria for a single trait) are expected to be important QTLs for controlling the trait. Mapping populations can also be constructed based on parents that segregate for multiple traits. This is advantageous because QTLs controlling the different traits can be located on a single map. However, this is not always possible for many parental genotypes used to construct mapping populations, because the parents may only segregate for one trait of interest.

Furthermore, for marker genotyping, and resultant QTL analysis, the same set of lines of the mapping population used for phenotypic evaluation must be available, which may be difficult with completely or semi-destructive bioassays (e.g. screening for resistance to necrotrophic fungal pathogens). In broad terms, an individual QTL may also be described as 'major' or 'minor'. This description relies on the proportion of the phenotypic variation explained by a QTL (based on the R2 value): major QTLs will account for a comparatively large amount (e.g. >10%) and minor QTLs will frequently account for <10%. Sometimes, major QTLs may refer to QTLs that are stable over a wide range of environments, whereas minor QTLs may refer to QTLs that may be environmentally sensitive, especially for QTLs that are associated with disease resistance (Lindhout 2002; Pilet-Nayel et al. 2002). In more strict terms, QTLs may be classified as: (1) suggestive; (2) significant and (3) highly significant. Lander and Kruglyak (1995) suggested this classification in order to 'avoid a flood of false positive claims' and also certify that 'true hints of linkage' were not missed (Lander and Kruglyak 1995). Significant and highly significant QTLs were given significance levels of 5% and 0.1%, respectively, whereas a suggestive QTL is one that would be expected to occur once at random in a QTL mapping study (in other words, there is a warning relating to the dependability of suggestive QTLs). The mapping program MapManager QTX reports QTL mapping results with this classification (Manly et al. 2001).

Confidence intervals for QTLs Although the map position is that the most feasible position of a QTL at which the highest LOD or LRS score is detected, in fact QTLs occur within the confidence intervals. There are numerous ways by which confidence intervals can be calculated. 'One-LOD support interval' is the simplest one, which is determined by finding the region on both sides of a QTL peak that corresponds to a decrease of 1 LOD score (Lander and Botstein 1989; Hackett 2002). 'Bootstrapping', a statistical method for resampling, is another method to determine the confidence interval of QTLs (Visscher et al. 1996), and can be effortlessly applied within some mapping software programs such as MapManager QTX.

Number of markers and marker spacing The number of markers required for a genetic map varies with the number and length of chromosomes within the organism. For the detection of QTLs, a comparatively thin 'framework' (or 'skeletal' or 'scaffold') map with evenly spaced markers is adequate, and preliminary genetic mapping studies generally contain between 100 and 200 markers (Mohan et al. 1997). However, this depends on the genome size of the species; for mapping species with large genomes more markers are desired. Darvasi et al. (1993) reported that the power of recognising a QTL was effectively the same for a marker spacing of 10 cM as for an infinite number of markers, and only slightly decreased for marker spacing of 20 or even 50 cM (Darvasi et al. 1993).

Factors influencing the detection of QTLs The key factors influencing the detection of QTLs segregating in a population are the genetic properties of QTLs that control traits, environmental effects, population size and experimental error.

The genetic properties of QTLs controlling traits comprise the magnitude of the effect of individual QTLs. Only QTLs with sufficiently large phenotypic effects would be detected; QTLs with small effects may fall below the significance threshold of detection. Another genetic property is the distance between linked QTLs. Closely linked QTLs (~20 cM or less) will usually be detected as a single QTL in typical population sizes (<500). Environmental effects may have a profound influence on the expression of quantitative traits. Experiments that are replicated across sites and over time (e.g. different seasons and years) may enable the researcher to investigate environmental influences on QTLs affecting trait(s) of interest. RI or DH populations are ideal for these purposes. The principal experimental design factor is the size of the population used in the mapping study. The larger the population size, the more accurate the mapping study and the more likely it is to allow detection of QTLs with smaller effects. An increase in population size provides gains in statistical power, estimates of gene effects and confidence intervals of the locations of QTLs (Darvasi et al. 1993; Beavis 1998). Error in phenotypic evaluation and mistakes in marker genotyping are the main sources of experimental errors during QTL analysis. Genotyping errors and missing data may affect the order and distance between markers within the linkage maps (Hackett 2002). The phenotyping of the mapping population is of paramount importance for the accuracy of QTL mapping studies. A reliable QTL map can only be produced from reliable phenotypic data. Replicated phenotypic measurements can be used to improve the accuracy of QTL mapping by reducing background 'noise' (Danesh et al. 1994; Haley and Anderson 1997).

Confirmation of QTL   Ideally, due to the factors described above, QTL mapping studies should be independently confirmed or verified. Such confirmation studies referred to as 'replication studies' (by Lander and Kruglyak 1995) involve independent populations constructed from the same parental genotypes or closely related genotypes used in the primary QTL mapping study. Sometimes, larger population sizes may be used. Furthermore, some recent studies have suggested that QTL positions and effects should be evaluated in independent populations, because QTL mapping based on typical population sizes result in a low power of QTL detection and a large bias of QTL effects. Unfortunately, due to constraints such as lack of research funding and time, and possibly a lack of understanding of the need to confirm results, QTL mapping studies are rarely confirmed and validated. Validation is a very important aspect that needs to be taken care by the molecular markers. Some notable exceptions are the confirmation of QTLs associated with root-knot nematode resistance (Li et al. 2001) and bud blight resistance in soybean (Fasoula et al. 2003). QTLs can also be confirmed by using a specific type of population called NILs. NILs are generated by crossing a donor parent (e.g. wild parent possessing a specific trait of interest) to a recurrent parent (e.g. an elite cultivar). The F_1 hybrids are then backcrossed to the recurrent parents to produce first backcross generation (BC_1). The BC_1s are then repeatedly backcrossed to the recurrent parents for a number of generations (at least 6–7 generations). The final BC_7 will contain practically all of the recurrent parent genome except for the small chromosomal region containing a gene or QTL of interest. Homozygous F_2 lines can be obtained by self-pollinating the BC_7 plants. It should be noted that in order to produce NILs containing target genes, the genes have to be selected for during each round of backcrossing. By genotyping NILs with important markers, and comparing mean trait values of particular NIL lines with the recurrent parent, the effects of QTLs could be confirmed.

Short cuts for gene/QTL mapping   The construction of linkage maps and QTL analysis require considerably more time and effort, and may be cost-effective. Therefore, other methods would be of use that can save time and money. The 'short-cut' methods that tag QTLs to discover markers are bulked segregant analysis (BSA) and selective genotyping. The requirement of both the methods is mapping populations. BSA detects markers located in specific chromosomal regions (Michelmore et al. 1991). In the BSA method, two pools or 'bulks' of DNA samples are pooled from 10 to 20 individual plants from a segregating population, but these two bulks should be different for a trait of interest. DNA bulks are made to randomise every loci, except for the region enclosing the gene of interest. Across the two bulk markers are screened. The polymorphic markers identified may represent markers that are linked to a gene or QTL of interest. The identification of linked markers by using BSA is given in Figure 17.6. These polymorphic markers are then used for the genotyping of the entire population, and a localised linkage map may be generated. This enables QTL analysis to be performed and the location of a QTL to be determined (Ford et al. 1999). Generally, BSA is used to tag genes controlling simple traits, but the method may also be used to identify markers linked to major QTLs (Wang and Paterson

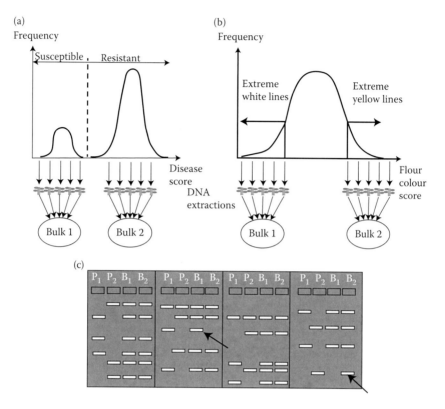

FIGURE 17.6 The preparation of DNA bulks for a simple disease resistance trait (a) and a quantitative quality trait (flower colour) (b). In both cases, two bulks (B1 and B2) are made from individuals displaying extreme phenotypic scores. (c) Polymorphic markers (indicated by arrows) that are identified between bulks may represent markers that are linked to genes or QTLs controlling the traits. Such markers are then used to genotype the entire mapping population and QTL analysis performed. (Adapted from Tanksley et al., 1995. *Trends in Genetics*, 11, 63–68.)

1994). 'High-volume' or 'high-throughput' marker techniques (e.g. RAPD or AFLP) are generally preferred for BSA that can generate multiple markers from a single DNA preparation.

The selective individuals representing the phenotypic extremes or trials of the trait being analysed is known as selective genotyping (Foolad and Jones 1993; Lander and Botstein 1989; Zhang et al. 2003). Individuals with phenotypic extreme are only used for the linkage map construction and QTL analysis. The method of selective genotyping is generally used when phenotyping is costlier or if there are many individuals in the population. The main drawback of this method is that only one trait can be tested at a time because the individuals selected for extreme phenotypic values will usually not represent extreme phenotypic values for other traits and it is not proficient in deter-mining the effects of QTLs (Tanksley 1993). In addition, only interval mapping can be used for the QTL detection because the phenotypic effects would be extremely overvalued in case of single-point analysis (Lander and Botstein 1989).

Conventional breeding and marker-assisted backcrossing

The trait governed by a single gene or by a gene that accounts for a high proportion of the phenotypic variance could be sig-nificantly transferred from donor to recipient line. Since long time, traditional backcross breeding programmes were per-formed for introgressing the qualitative traits on the assumption that the proportion of the recurrent parent genome is recovered at a rate of $1-(1/2)^{t+1}$ for each t generations of backcrossing. Thus, 96.9% recovery of recurrent parent genome is expected after four backcrosses. The deviation from this expectation is entirely due to chance and linkage between the genes from the donor parent being selected for with nearby genes. A good example of the surprising amount of linkage drag that accom-panies backcross breeding programmes was reported by Young and Tanksley in 1989, who genotyped the chromosome carry-ing the tomato mosaic virus (*Tm2*) disease-resistance gene in several tomato cultivars that were developed by introgressing the gene from a wild relative, *Lycopersicon peruvianum* via backcross breeding. They found that even cultivars developed after 20 backcrosses contained introgressed segments as large as 4 cM and one cultivar developed after 11 backcrosses still contained the entire chromosome arm carrying the gene from the donor parent. A minimum of six backcross generations would normally be required to recover 99% of the recurrent parent genome, for the transfer of a single dominant gene. This procedure is too lengthy, particularly in the perspective of the competitive nature of modern hybrid breeding programmes, where the turnover times for new lines and hybrids are fast.

Marker-assisted selection Selecting a phenotype after manipulating genomic regions that are involved in the appear-ance of that phenotype through molecular marker is known as marker-assisted selection. With the advent of an array of molecular tools and techniques, and subsequently reasonably dense molecular genetic maps in various crop plants, marker-assisted selection has become feasible for traits both governed by major genes and QTLs. It may greatly increase the efficiency and effectiveness of selection in modern plant breeding through the precise transfer of genomic regions of interest and by fast recovery of recurrent parent genome compared to conventional breeding methods. In general, the success of a marker-based breeding system depends on four main factors: (i) a definite genetic map with an adequate range of uniformly spaced poly-morphic markers to accurately locate desired QTLs or major factor(s); (ii) close association between the QTL or a significant gene of interest and adjacent markers; (iii) Adequate recombi-nation between the markers and remainder of the genome; and (iv) a capability to analyse a larger range of plants in a time- and cost-effective manner. The success of MAS depends on the association of the markers with various factor of interest. There are three sorts of relationships between the markers and vari-ous genes that could be distinguished:

1. Marker is found inside the gene of interest, which is the most favourable state of affairs for MAS and during this case, it might be ideally mentioned as gene-assisted selection. This might be notably helpful for traits that have laborious or time-consuming phenotypic screening procedures. Whereas this type of relationship is the most preferred one, it is also tough to search out this type of allele-specific markers. For example, SSR markers are designed using the available nucleotide sequence infor-mation for the opaque2 allele that confers high lysine and tryptophan content within the maize kernel. This has offered an efficient means of tracking the opaque2 allele in breeding for nutritionally superior maize genotypes, since the marker is found inside the gene sequence itself and co-segregates with the target gene.

2. The marker is not in linkage equilibrium (LE) with the gene of interest throughout the population, called LD. LD is the tendency of certain combination of alleles to be inherited together. Once markers and genes of interest are physically close to each other, population-wide LD may be found. The selection using these markers might be referred to as LD-MAS.

3. The marker is in LE with the gene of interest through-out the population, that is the most tough and difficult situation for applying MAS. However, in most cases, particularly for the inheritable traits, the target gene(s) inside a QTL has not been characterised at the molecular level. Therefore, the genomic regions to be selected using MAS are usually chromosome segments carrying QTLs in case of polygenic traits. It is desirable either to have two polymorphic DNA markers flanking the target gene (or a QTL), or a marker inside a QTL (if the chromosome

segment is more than 20 cM) to eliminate the chance of genotypes presenting a double recombination between the two flanking markers. Depending on the character of the genomic region (cloned factor, major or minor QTL) involved within the expression of a target trait and the range of selected QTLs or genomic regions that require to be manipulated, many MAS schemes have been proposed that will be discussed later.

In the context of MAS, DNA-based markers may be effectively utilised for two basic purposes: (i) identifying favourable allele(s) (dominant or recessive) across generations and (ii) tracing the most appropriate individual(s) among the segregating progeny, based on allelic composition across a part of or the whole genome. The fundamental advantages of MAS over conventional phenotypic selection are as follows:

- Target alleles that are difficult to score phenotypically and environmentally sensitive and cost and time ineffective is selected with the assistance of markers. Therefore, marker-assisted selection is easier than phenotypic selection or screening procedure. Classical examples of traits that are difficult and laborious to measure are cereal cyst nematode and root lesion nematode resistance in wheat (Eagles et al. 2001; Eastwood et al. 1991; Zwart et al. 2004). Other examples are quality traits that usually need expensive screening procedures.

- DNA markers enable early selection for traits that are expressed in later developmental stages because plants can be screened early at the seedling stage or as seeds. Thus, undesirable plant genotypes are quickly eliminated. This might have tremendous advantages in rice breeding as a result of typical rice production practices that involve sowing pre-germinated seeds and transplanting seedlings into rice paddies, making it easy to transplant solely selected seedlings to the main field.

- Many traits that could not be selected on single plant basis by conventional phenotypic screening strategies because of environmental errors are selected with the help of molecular markers. Co-dominant markers in MAS will discriminate between homo and heterozygous plants for a few traits by individual plant selection that is not possible by conventional phenotypic screening methodology. These advantages of MAS can be exploited by breeders to accelerate the breeding method. This might facilitate certain traits to be 'fast-tracked', leading to faster line development and variety release. Markers may be used to reduce the time for phenotyping that permits selection in off-season nurseries making it a lot of cost and time effective to grow more generations per year (Ribaut and Hoisington 1998).

Another benefit from MAS is that the total number of lines that need to be phenotypically tested can be reduced early in the breeding scheme, which permits more efficient use of glasshouse and/or field space, which is often limited because only important breeding material is maintained. Despite having these potential advantages over conventional breeding, a marker will not necessarily be useful or more effective for every trait and they require a substantial investment in time, money and resources for their development. For many traits, effective phenotypic screening methods already exist and these will often be less expensive for selection in large populations. However, when whole-genome scans are being used, even these traits can be selected for if the genetic control is understood.

Foreground selection and background selection   In marker-assisted selection, molecular markers are increasingly being used to trace the presence of target genes (foreground selection), further so as to speed up the recovery of the recurrent parent genome (background selection) in backcross programmes. Conventional backcrossing in plant breeding is employed to introgress favourable traits from a donor plant into a recurrent parent. During this continual crossing procedure, large segment of donor genome containing some undesirable gene along with target allele additionally is introduced into recurrent parent genome and reconstruction of recurrent parent genome needs a minimum of six backcross generations. So as to minimise this linkage drag, marker assay is advantageous. A marker is employed in terms of marker-assisted backcrossing to either mark out the target gene or to recover the recurrent parent genotype to enhance the efficiency of backcross breeding. Conventional backcross breeding programmes needs extra selfing generations after every backcross generation for the transfer of recessive genes that reduces the effectiveness of most conventional breeding processes. Melchinger has effectively used marker-assisted foreground selection for introgression of disease-resistance genes by presenting a priori approach for calculating the minimum range of individuals and family size needed in recurrent backcrossing. Still, because of lack of allele-specific markers, practical examples of this approach in plant breeding are limited. One successful example is the conversion of traditional maize lines into quality protein maize (QPM) through marker-assisted transfer of a recessive mutant allele, opaque2, and using allele-specific molecular markers. In animal breeding, an array of allele-specific markers has been available facilitating the applications of this approach on a commercial scale to eliminate disease and stress-susceptibility genes. 'Marker-assisted background selection', a term coined by Hospital and Charcosset in 1997, was initially proposed by Young and Tanksley (1989), and experimented by numerous researchers. This strategy has been extensively used in commercial maize breeding programmes, particularly for the selection of lines carrying transgenes conferring herbicide tolerance or insect resistance. Within the background programme, several parameters need to be optimised. Flanking markers for the target allele are essential to get rid of linkage drag. The optimal distance between the target gene and flanking markers govern the selection intensity that will be exerted. The equations given by Hospital and Charcosset (1997) and Frisch and Melchinger (2005) are useful in determining the quantity of BC plants that need to be generated and typed with a special set of flanking markers. Variety of gene/marker associations are reported in crop plants that may probably be used in MAS strategies.

Improving quantitative traits using MAS: Case study

Most of the traits of agronomic importance are complex and controlled by many genes. Improvement of such traits through MAS could be an advanced endeavour, unlike the case of merely inheritable traits. The genetic quality of quantitative trait creates difficulty in their manipulation mainly because of the quantity of genes involved in their expression and interactions among genes (epistasis). Since many genes are involved in the expression of a quantitative trait, these genes, in general, have smaller individual effects on the phenotype, and the effect of the individual genes are not simply identifiable. This needs repetitions of field tests to characterise the exact results of QTLs and to evaluate their stability across environments. Assessment of QTL by environment interaction $(Q \times E)$ continues to be a serious limitation on the efficiency of MAS. Furtrhermore, epistatic interaction among totally different regions of a genome will induce a skew evaluation of QTL effects. Also, if the genomic regions concerned in the interactions are not incorporated in the selection scheme, they will probably bias the selection process. Despite the explosion of QTL mapping experiments in recent years, a number of

constraints have imposed severe limitations on efficient utilisation of QTL mapping information in plant breeding through
MAS. Salient among these constraints are

1. Identification of a limited number of major 'QTLs with
   more phenotypic variance controlling specific traits
2. The notion that QTL identification is required whenever
   additional germplasm is used
3. Inadequacies/experimental deficiencies in QTL analysis
   resulting due to either overestimation or underestimation
   of the number and effects of QTLs
4. Lack of universally valid QTL marker associations applicable over different sets of breeding materials
5. Strong QTL–environment interaction; and difficulty in
   precisely evaluating epistatic effects

Increasing the potency of MAS for quantitative traits need
improved field experimentations/designs, robust mathematical
models and comprehensive statistical methods. As an example, with composite interval mapping (CIM), field data from
different environments are often integrated into a joint analysis to evaluate the $Q \times E$ interactions; thus, enabling identification of stable QTLs across environments. Besides, with
a detailed linkage map, CIM permits an explicit identification of the QTL in the genome and better identification of
linked QTL (in coupling phase) from the identical parental
line (Babu et al. 2004).

Favourable QTL from even a phenotypically inferior parental line (in repulsion phase) also can be effectively identified and
utilised by DNA-based markers. Tanksley and Nelson (1996)
proposed an advanced backcross-QTL (AB-QTL) approach for
enhancing the QTL mapping in tandem with MAS. This analysis
involves crossing between elite germplasm and unadapted gen-
otype/wild relatives with favourable genes/QTLs, followed by
two generations of backcrossing for developing several hundred
sibling lines. These lines, each containing different genomic
segments of the wild relative/unadapted genotype, are then
genotyped using DNA markers. In effect, these lines become
a set of NILs that individually dissect the effects of potential
QTL in the background of the elite parent. At the same time,
the BC lines additionally give comparatively mounted material for an essential step of replicated phenotypic evaluations.
The core of this approach—revealing and accessing the desirable alleles from wild relatives or unadapted genotype—indicates that QTL mapping can go hand-in-hand with MAS rather

than as sequential steps. QTL/marker associations, however, need to be tattered through more intensive research efforts. It would be essential to find out whether extremely tight linkages between marker loci and QTL may lead to highly conserved allele associations. If so, observation of change in marker allele frequency in long-term selection experiments or determination of markers that explain significant portions of the combining ability variance in diallel or factorial crosses might reveal universally applicable markers. Another approach would be to perform QTL analyses in genetically broad-based random mating populations by means of extremely saturated integrated genetic marker maps. Direct QTL-allele-specific markers (such as STS markers derived from cloned QTL alleles) are needed for maximal efficiency of MAS. Even though success in terms of cloning of QTL alleles is very limited, map-based cloning and candidate gene approaches would increasingly facilitate isolation and characterisation of agronomically significant QTLs, owing to the fast advancement being made in genome sequencing of many plants. Cloning of genetic determinants of QTLs is anticipated to bridge the missing link in our understanding of the association among genotype and phenotype (Geiger and Welz 1999).

MAS for drought stress tolerance in maize: CIMMYT investigators have made substantial efforts throughout the past three decades for augmentation of pre- and post-flowering drought tolerance in maize and up to certain limits, energetic significant progress has been achieved for cultivating drought tolerance in CIMMYT maize germplasm through conventional breeding, but still, the approach is slow and long. To accelerate the breeding procedure, molecular markers and QTL information based on precisely managed replicated tests have been used to show the potential to improve the issues related to inconsistent and unpredictable onset of moisture stress or the confounding impact of different stresses such as heat. For this, first of all, a complex trait of drought tolerance was counteracted into simpler components, such as an anthesis-silking interval, that are closely associated with drought tolerance. After that, CIMMYT conducted a series of experiments on QTL analysis and MAS for transfer of drought tolerance to tropical maize, and obtained encouraging results. An associate integrated strategy of QTL-mapping, MAS and functional genomics are currently being employed to additionally provide genomic information and tools to commendably complement conventional selection for rising drought stress tolerance in maize (Bänziger 2000).

Specific significant considerations throughout scheming of a BC/MAS structure

Transfer of specific favourable alleles at a target locus from a donor line to a recipient line with marker assistance is the basic strategy of BC/MAS system. The employment of DNA markers will increase the rapidity of the selection process by separation of progeny at the genotypic level in every generation. In a wide range, the BC scheme targets either at complete or partial line conversion. Complete line conversion involves development of a line that has precisely the same genetic composition as the recipient line, excluding target loci where the presence of homozygous alleles from a donor line is anticipated. The objective of partial line conversion is to develop a line that will have a small proportion of the scattered donor genome over its genome (recipient parent) along with desirable homozygous alleles of the target gene. Numerous factors influence the efficiency and effectiveness of a BC/MAS scheme like the number of target genes, the distance between the flanking markers and also the target gene (2–20 cM), and similarly the variety of genotypes designated in every BC generation. Relying upon the objectives, the experimental design for line conversion through BC/MAS desires has to be designed based on the obtainable resources, the nature of the germplasm (e.g. agronomical quality and variety of lines to be improved) and technical options available at the marker level. As soon as the number of target genes to be transferred are outlined, a future step would be to work out the population size that must be screened at every generation, giving a target-selectable population size of 50–100 genotypes. Henceforth, one ought to confirm the fascinating recombination frequency between the flanking markers and the target. Also the number of genotypes designated at each generation supported the target and also the constraints of the experiment. The number of BC generations required to accomplish the introgression can be foreseen based on simulations. Whereas resources are limited or introgression from a donor line into an outsized range of recipient lines is desired, strategies based on BC/MAS at one target locus exclusively at one advanced BC generation should be considered. Selection in later generations is additionally helpful because the ratio of the standard deviation of the mean of the donor genome contribution increases as the backcrossing proceeds. The backcross procedure can be finished after four, rather than six, backcross generations, even with small population sizes and limited number of marker data points (MDP). Therefore, the marker technology is useful even when the resources in a very breeding programme are limited. MAS has the potential to recover a maximum proportion of recurrent parent genomes up to BC3 generation, which is not attainable via a typical backcross breeding process (the same level of the recurrent parent genome (RPG) as reached in BC7). Nevertheless, for this, large numbers of MDP and more efficient marker systems are required. In the above scheme, the screening of the whole population has to be conducted at least once at the beginning of each BC generation; therefore, it is essential to spot the foremost convenient set of markers for the allele(s) of interest. With the population size running into hundreds or thousands, such screening can be laborious and expensive. However, this can be optimised by using an appropriate combination of DNA markers. With the recent advances in molecular technology, particularly SNPs, a substantial improvement within the capability of expeditiously screen larger populations can be achieved (Babu et al. 2004).

Efficiency of MAS

Computer simulation has provided a robust tool for analysing the planning and potency of MAS programme. Three different selection strategies in a marker-assisted background selection programme, namely two-step, three-step and four-step, which, were compared by computer simulation in terms of faster recovery of an outsized proportion of the RPG. The simulations were based on maize genetic map ($n = 10$) with markers spaced about 20 cM apart and with the assumption that the target locus could be scored directly either through phenotype or a marker completely linked to the target gene. Major conclusions from this simulation experiment are as follows (Babu et al. 2004):

1. A four-stage sampling strategy that includes (a) selecting individuals carrying the target allele; (b) selecting individuals homozygous for recurrent parent genotype at loci flanking the target locus; (c) selecting the individuals homozygous for recurrent parent genotype at the remaining loci on the same chromosome as the target allele; and (d) selecting one individual that is homozygous for the recurrent parent genotype at most loci (across whole genome) among those that remain, is the most efficient procedure in general.

2. With the four-stage sampling strategy and reasonable population size (50–100), one can expect to find BC3 progeny with at least 96% RPG with 90% probability. It would take six generations of traditional backcrossing to reach this stage, besides the risk of a larger probability of linkage drag around the target gene.

3. Increasing the number of markers genotyped at each generation had little effect. Once the threshold of one marker per 20 cM is reached, additional markers (except

perhaps around the target locus) would not be required. The frequency of recombination, and not the quantity of markers, is an additionally significant limiting factor in reducing linkage drag, which proposes that sampling larger populations with fewer markers makes more sense than the reverse.

Recent advances in MAS strategies and genotyping techniques

Single large-scale MAS In the single large-scale MAS strategy, the recipient genotypes which were locally well adapted with good yield characteristics are selected for the MAS to introgress the genes responsible for the target trait to be improved. The donor parent should be selected in such a way that it should show the polymorphism to the targeted gene. Then the donor and recipient parental lines are used for crossing to generate segregating populations. A genomic region of interest for each parental line is identified by combining favourable alleles in the segregating populations (e.g. F_3 families and RILs). Foreground selection is conducted on these large-scale segregated populations for the targeted alleles of genomic regions with the help of molecular markers. It offers a few advantages over other strategies, namely, (1) it is more suitable for gene pyramiding of two or more cloned genes or QTLs, (2) it assures good allelic variability for further line development in different environmental situations and (3) no pressure of selection is required (Babu et al. 2004).

Pedigree MAS The pedigree MAS strategy is particularly suitable to those crops where pedigree of germplasm is available. These selective genotypes must be characterised at the molecular level for their effective utilisation in the breeding programmes. At each segregating generation, along with foreground selection, phenotypic selection is also conducted to identify the desirable genotypes with gene of interest for a trait. Then the molecular markers which were closely linked to the trait of interest can be used to enhance fixation of favourable alleles in the next generations (offspring 1 and offspring 2). This MAS strategy was suggested to be most efficient when conducted on F_2 or F_3 segregating populations (Babu et al. 2004).

Precautions to be followed during MAS In recent years, molecular marker technologies such as MAS are found to be a supplemental technology for traditional breeding strategies, to achieve genetic gains with greater speed and precision. Although MAS is currently used more commonly for simply inherited traits than for polygenic traits, with the development and introduction of reliable PCR-based markers such as SSRs and SNPs, in several crop plants, efficiency of genotyping large populations or breeding materials have been significantly increased. Through crop genomics research, marker-assisted foreground selection is already gaining rapid momentum as allele-specific markers on a number of agronomically important traits (Collard and Mackill 2008). The assurance of MAS for improving polygenic traits in a quick time-frame and in a cost-effective manner is still elusive. There is a wider appreciation that by simply indicating that a complex trait can be dissected into QTLs and mapped to approximate genomic locations using DNA markers would not serve the ultimate goal of plant trait improvement. According to Young, research on quantitative traits need to employ larger population sizes, multiple replications, better scoring methods and environments, appropriate quantitative genetic analysis, various genetic backgrounds and, whenever possible, independent verification through advanced generations or parallel populations. The MAS strategies may put more emphasis on reducing the number of crosses and simple selection steps to maximise their impact in the agriculture (Collard and Mackill 2008).

In the present scenario, 'functional genomics' is making rapid changes for dissecting the function of genes through genome-wide experimental approaches. The tools such as DNA chips, microarrays and expressed sequence tags aid in the quantitative estimation of RNA levels. These RNA expression profiles will help the breeders in selecting the better genotypes with desired gene influencing a trait (Collard and Mackill 2008).

In the future, a greater coordination is required between the workers of molecular breeders and quantitative geneticists to build up and validate hypotheses involving complex gene interactions. Bioinformatics also plays an important role in facilitating these two branches. Thus, by integrating functional genomics, bioinformatics and molecular breeding may create fundamental revolutions in varietal improvement of crops. Solutions to the above-mentioned obstacles of MAS need to be developed in order to achieve a greater impact. The following points need to be considered for high efficiency of marker-assisted selection (Collard and Mackill 2008).

- The workers of conventional breeding and molecular breeders should work collectively to bridge the gap.
- QTL mapping studies need to be implemented very carefully, and QTLs with high LOD with greater phenotypic

variance under multiple environments should be selected for MAS programmes.

- Optimisation of methods used in MAS such as DNA extraction and marker genotyping, especially in terms of cost reduction and efficiency.
- Efficient system for data storage, and more efficient computer simulation programmes may be developed to maximise genetic gain and minimise costs (Kuchel et al. 2005).

17.5 Conclusions

For a long time, plant breeders have played an important role in the crop improvement programmes for generating new improved varieties. It seems clear that the current breeding programmes continue to make progress through commonly used breeding approaches. MAS could greatly assist plant breeders in reaching this goal although, to date, the impact on variety development has been minimal. For an effective use of MAS techniques for varietal improvement, there should be a greater integration and cooperation between the breeders and molecular biotechnologists to bridge the gap that exists presently as a barrier in the development of new improved varieties. The exploitation of the advantages of MAS relative to conventional breeding could have a great impact on crop improvement. The high cost of MAS will continue to be a major obstacle in its adoption for some crop species and plant breeding in developing countries in the near future. Specific MAS strategies may need to be tailored to some crops where the breeding is difficult and it should also be cost-effective. New marker technology can potentially reduce the cost of MAS considerably. If the effectiveness of the new methods are validated and the equipment can be easily obtained, this should allow MAS to become more widely applicable for crop breeding programmes.

References

BAIRD, V. 1995. Progress in Prunus mapping and application of molecular markers to germplasm improvement. *HortScience*, 30(4), 748–749.

BAIRD, V., ABBOTT, A., BALLARD, R., SOSINSKI, B., and RAJAPAKSE, S. 1997. DNA diagnostics in horticulture. In: P. Gresshoff (Ed.), *Current Topics in Plant Molecular Biology:*

Technology Transfer of Plant Biotechnology (pp. 111–130): CRC Press, Boca Raton, FL.

BABU, R., NAIR, S. K., PRASANNA, B. M., and GUPTA, H. S. 2004. Integrating marker-assisted selection in crop breeding– prospects and challenges. *Current Science*, 87(5), 607–619.

BÄNZIGER, M. 2000. *Breeding for drought and nitrogen stress tolerance in maize: From theory to practice*. CIMMYT, Mexico, D. F., 68pp.

BARONE, A. 2004. Molecular marker-assisted selection for potato breeding. *American Journal of Potato Research*, 81(2), 111–117.

BASTEN, C. J., WEIR, B. S., and ZENG, Z.-B. 1994. Zmap-a QTL cartographer. In: J. S. G. C. Smith, B. B. J. Chesnais, W. Fairfull, J. P. Gibson, B. W. Kennedy, and E. B. Burnside (Eds.), *Proceedings of the 5th World Congress on Genetics Applied to Livestock Production: Computing Strategies and Software* (Vol. 22). Guelph, Ontario, Canada: Organizing Committee, 5th World Congress on Genetics Applied to Livestock Production.

BASTEN, C. J., WEIR, B. S., and ZENG, Z.-B. 2004. *QTL Cartographer*, Version 1.17. Department of Statistics, North Carolina State University, Raleigh, NC.

BEAVIS, W. 1998. QTL Analyses: Power, precision and accuracy. In: A. H. Paterson (Ed.), *Molecular Dissection of Complex Traits* (pp. 145–162): CRC Press, Boca Raton, FL.

BELLON, M. R., HODSON, D., and HELLIN, J. 2011. Assessing the vulnerability of traditional maize seed systems in Mexico to climate change. *Proceedings of the National Academy of Sciences of the USA*, 108(33), 13,432–13,437.

BOTSTEIN, D., WHITE, R. L., SKOLNICK, M., and DAVIS, R. W. 1980. Construction of a genetic linkage map in man using restriction fragment length polymorphisms. *American Journal of Human Genetics*, 32, 314–331.

CHURCHILL, G. A., and DOERGE, R. W. 1994. Empirical threshold values for quantitative trait mapping. *Genetics*, 138(3), 963–971.

COLLARD, B. C. Y., JAHUFER, M. Z. Z., BROUWER, J. B., and PANG, E. C. K. 2005. An introduction to markers, quantitative trait loci (QTL) mapping and marker-assisted selection for crop improvement: The basic concepts. *Euphytica*, 142(1–2), 169–196.

COLLARD, B. C., and MACKILL, D. J. 2008. Marker-assisted selection: An approach for precision plant breeding in the twenty-first century. *Philosophical Transactions of the Royal Society B: Biological Sciences*, 363(1491), 557–572.

DANESH, D., AARONS, S., McGILL, G. E., and YOUNG, N. D. 1994. Genetic dissection of oligogenic resistance to bacterial wilt in tomato. *Molecular Plant-Microbe Interaction*, 7(4), 464–471.

DARVASI, A., WEINREB, A., MINKE, V., WELLER, J. I., and SOLLER, M. 1993. Detecting marker-QTL linkage and estimating QTL gene effect and map location using a saturated genetic map. *Genetics*, 134(3), 943–951.

EAGLES, H. A., BARIANA, H. S., OGBONNAYA, F. C., REBETZKE, G. J., HOLLAMBY, G. J., HENRY, R. J., HENSCHKE, P. H., and CARTER, M. 2001. Implementation of markers in Australian wheat breeding. *Australian Journal of Agricultural Research*, 52(12), 1349–1356.

EASTWOOD, R. F., LAGUDAH, E. S., APPELS, R., HANNAH, M., and KOLLMORGEN, J. F. 1991. *Triticum tauschii*: A novel source of resistance to cereal cyst nematode (*Heterodera avenae*). *Australian Journal of Agricultural Research*, 42(1), 69–77.

FARIS, J. D., HAEN, K. M., and GILL, B. S. 2000. Saturation mapping of a gene-rich recombination hot spot region in wheat. *Genetics*, 154(2), 823–835.

FASOULA, V. A., HARRIS, D. K., BAILEY, M. A., PHILLIPS, D. V., and BOERMA, H. R. 2003. Identification, mapping, and confirmation of a Soybean gene for Bud Blight resistance. *Crop Science*, 43(5), 1754–1759.

FOOLAD, M. R., and JONES, R. A. 1993. Mapping salt-tolerance genes in tomato (*Lycopersicon esculentum*) using trait-based marker analysis. *Theoretical and Applied Genetics*, 87(1–2), 184–192.

FORD, R., PANG, E. C. K., and TAYLOR, P. W. J. 1999. Genetics of resistance to ascochyta blight (*Ascochyta lentis*) of lentil and the identification of closely linked RAPD markers. *Theoretical and Applied Genetics*, 98(1), 93–98.

FRISCH, M., and MELCHINGER, A. E. 2005. Selection theory for marker-assisted backcrossing. *Genetics*, 170(2), 909–917.

GEIGER, H. H., and WELZ, H. G. 1999. Principles of marker-assisted selection.

GELDERMAN, H. 1975. Investigations on inheritance of quantitative characters in animals by gene markers I. Methods. *Theoretical and Applied Genetics*, 46(7), 319–330.

GRODZICKER, T., WILLIAMS, J., SHARP, P., and SAMBROOK, J. 1975. Physical mapping of temperature-sensitive mutations of adenoviruses. *Cold Spring Harbor Symposia Quantitative Biology*, 39 Pt 1, 439–446.

HACKETT, C. A. 2002. Statistical methods for QTL mapping in cereals. *Plant Molecular Biology*, 48(5–6), 585–599.

HALEY, C., and ANDERSON, L. 1997. Linkage mapping quantitative trait loci in plants and animals. In: P. Dear (Ed.), *Genome Mapping A Practical Approach* (pp. 49–71). New York: Oxford University Press.

HARTL, D., and JONES, E. 2001. *Genetics: Analysis of Genes and Genomes*. Sudbury, MA: Jones and Bartlett Publisher.

HELENTJARIS, T., SLOCUM, M., WRIGHT, S., SCHAEFER, A., and NIENHUIS, J. 1986. Construction of genetic linkage maps in maize and tomato using restriction fragment length polymorphisms. *Theoretical and Applied Genetics*, 72(6), 761–769.

HOSPITAL, F., and CHARCOSSET, A. 1997. Marker-assisted introgression of quantitative trait loci. *Genetics*, 147, 1469–1485.

JAHUFER, M. Z. Z., COOPER, M., AYRES, J. F., and BRAY, R. A. 2002. Identification of research to improve the efficiency of breeding strategies for white clover in Australia—A review. *Australian Journal of Agricultural Research*, 53, 239–257.

JANSEN, R. C. 1993. Interval mapping of multiple quantitative trait loci. *Genetics*, 135(1), 205–211.

JANSEN, R. C., and STAM, P. 1994. High resolution of quantitative traits into multiple loci via interval mapping. *Genetics*, 136(4), 1447–1455.

JEFFREYS, A. J. 1979. DNA sequence variants in the G gamma-, A gamma-, delta- and beta-globin genes of man. *Cell*, 18(1), 1–10.

KASHA, K. J. 1999. Biotechnology and world food supply. *Genome*, 42(4), 642–645.

KEARSEY, M., and POONI, H. 1996. *The Genetical Analysis of Quantitative Traits*. London: Chapman & Hall.

KELLY, J. D., GEPTS, P., MIKLAS, P. N., and COYNE, D. P. 2003. Tagging and mapping of genes and QTL and molecular marker-assisted selection for traits of economic importance in bean and cowpea. *Field Crops Research*, 82(2–3), 135–154.

KOEBNER, R. M., and SUMMERS, R. W. 2003. 21st century wheat breeding: Plot selection or plate detection? *Trends Biotechnology*, 21(2), 59–63.

KUCHEL, H., YE, G., FOX, R., and JEFFERIES, S. 2005. Genetic and economic analysis of a targeted marker-assisted wheat breeding strategy. *Molecular Breeding*, 16(1), 67–78.

LANDER, E., and KRUGLYAK, L. 1995. Genetic dissection of complex traits: Guidelines for interpreting and reporting linkage results. *Nature Genetics*, 11(3), 241–247.

LANDER, E. S., and BOTSTEIN, D. 1989. Mapping Mendelian factors underlying quantitative traits using RFLP linkage maps. *Genetics*, 121(1), 185–199.

LANDER, E. S., GREEN, P., ABRAHAMSON, J., BARLOW, A., DALY, M. J., LINCOLN, S. E., and NEWBERG, L. A. 1987. MAPMAKER: An interactive computer package for constructing primary genetic linkage maps of experimental and natural populations. *Genomics*, 1(2), 174–181.

LI, Z., JAKKULA, L., HUSSEY, R. S., TAMULONIS, J. P., and BOERMA, H. R. 2001. SSR mapping and confirmation of the QTL from PI96354 conditioning soybean resistance to southern root-knot nematode. *Theoretical and Applied Genetics*, 103(8), 1167–1173.

LINCOLN, S., DALY, M., and ES, L. 1993. *Mapping Genes Controlling Quantitative Traits Using MAPMAKER/QTL Version 1.1: A Tutorial and Reference Manual*. Cambridge, Massachusetts: Whitehead Institute.

LINCOLN, S., DALY, M. J., and LANDER, E. 1992. *Constructing genetic maps with MAPMAKER/EXP 3.0*. Cambridge, Massachusetts: Whitehead Institute.

LINDHOUT, P. 2002. The perspectives of polygenic resistance in breeding for durable disease resistance. *Euphytica*, 124(2), 217–226.

LIU, B. H. 1998. *Statistical Genomics: Linkage, Mapping, and QTL Analysis*. CRC Press, New York.

LOBELL, D. B., BANZIGER, M., MAGOROKOSHO, C., and VIVEK, B. 2011. Nonlinear heat effects on African maize as evidenced by historical yield trials. *Nature Climate Change*, 1(1), 42–45.

MA, X. F., ROSS, K., and GUSTAFSON, J. P. 2001. Physical mapping of restriction fragment length polymorphism (RFLP) markers in homologous groups 1 and 3 chromosomes of wheat by *in situ* hybridization. *Genome*, 44(3), 401–412.

MACKILL, D. J., NGUYEN, H. T., and ZHANG, J. 1999. Use of molecular markers in plant improvement programs for rainfed lowland rice. *Field Crops Research*, 64(1), 177–185.

MANLY, K. F., CUDMORE, R. H., JR., and MEER, J. M. 2001. Map Manager QTX, cross-platform software for genetic mapping. *Mammalian Genome*, 12(12), 930–932.

MEHLENBACHER, S. A. 1994. 998 Classical and molecular approaches to breeding fruit and nut crops for disease resistance. *HortScience*, 29(5), 572.

MICHELMORE, R., PARAN, I., and KESSELI, R. V. 1991. Identification of markers linked to disease-resistance genes by bulked segregant analysis: A rapid method to detect markers in specific genomic regions by using segregating populations. *Proceedings of the National Academy of Sciences*, 88(21), 9828–9832.

MOHAN, M., NAIR, S., BHAGWAT, A., KRISHNA, T. G., YANO, M., BHATIA, C. R., and SASAKI, T. 1997. Genome mapping, molecular markers and marker-assisted selection in crop plants. *Molecular Breeding*, 3(2), 87–103.

MONTERROSO, R., CONDE, C., ROSALES, D., GÓMEZ, J., and GAY, C. 2011. Assessing current and potential rainfed maize suitability under climate change scenarios in México. *Atmósfera*, 24, 53–67.

NELSON, J. 1997. QGENE: Software for marker-based genomic analysis and breeding. *Molecular Breeding*, 3(3), 239–245.

ORTIZ, R. 1998. Critical role of plant biotechnology for the genetic improvement of food crops: Perspectives for the next millennium. *Electronic Journal of Biotechnology*, 1(3), 16–17.

ORTIZ, R. 2011. Agro biodiversity management for climate change. In: J. M. Lenné and D. Wood (Eds.), *Agrobiodiversity Management for Food Security: A Critical Review* (pp. 189–211): CAB International, London, UK.

PATERSON, A. H. 1996. Making genetic maps. In: A. H. Paterson (Ed.), *Genome Mapping in Plants* (pp. 23–39): R. G. Landes Company, San Diego, California; Academic Press, Austin, Texas.

PHILLIPS, R., VASIL, I., and YOUNG, N. 2001. Constructing a plant genetic linkage map with DNA markers. In: R. L. Phillips and I. Vasil (Eds.). *DNA-Based Markers in Plants* (Vol. 6, pp. 31–47): Springer, Netherlands.

PILET-NAYEL, L., MUEHLBAUER, F. J., MCGEE, R. J., KRAFT, J. M., BARANGER, A., and COYNE, C. J. 2002. Quantitative trait loci for partial resistance to Aphanomyces root rot in pea. *Theoretical and Applied Genetics*, 106(1), 28–39.

PRAY, C., NAGARAJAN, L., LI, L., HUANG, J., HU, R., SELVARAJ, K. N., NAPASINTUWONG, O., and BABU, R. C. 2011. Potential impact of biotechnology on adaption of agriculture to climate change: The case of drought tolerant rice breeding in Asia. *Sustainability*, 3(10), 1723–1741.

RIBAUT, J.-M., and HOISINGTON, D. A. 1998. Marker-assisted selection: New tools and strategies. *Trends in Plant Science*, 3(6), 236–239.

RIPOL, M. I., CHURCHILL, G. A., DA SILVA, J. A., and SORRELLS, M. 1999. Statistical aspects of genetic mapping in autopolyploids. *Gene*, 235(1–2), 31–41.

RISCH, N. 1992. Genetic linkage: Interpreting LOD scores. *Science*, 255(5046), 803–804.

SEMAGN, K., BJØRNSTAD, Å., and NDJIONDJOP, M. N. 2006. An overview of molecular marker methods for plants. *African Journal of Biotechnology*, 5(25), 2540–2568.

SNOWDON, R. J., and FRIEDT, W. 2004. Molecular markers in Brassica oilseed breeding: Current status and future possibilities. *Plant Breeding*, 123(1), 1–8.

SINGH, U. M., BABU, B. K., and KUMAR, A. 2012. Quantitative trait loci (QTL) mapping: Concepts and applications. *Biotech Today: A Pulse of Global Science*, 2(1), 37–38.

STAM, P. 1993. Construction of integrated genetic linkage maps by means of a new computer package: Join Map. *The Plant Journal*, 3(5), 739–744.

SVETLEVA, D., VELCHEVA, M., and BHOWMIK, G. 2003. Biotechnology as a useful tool in common bean (*Phaseolus vulgaris* L.) improvement. *Euphytica*, 131(2), 189–200.

TANKSLEY, S. D. 1993. Mapping polygenes. *Annual Review of Genetics*, 27, 205–233.

TANKSLEY, S. D., and NELSON, J. C. 1996. Advanced backcross QTL analysis: A method for the simultaneous discovery and transfer of valuable QTLs from unadapted germplasm into elite breeding lines. *Theoretical and Applied Genetics*, 92(2), 191–203.

TANKSLEY, S. D., GANAL, M., and MARTIN, G. B. 1995. Chromosome landing-a paradigm for map-based cloning in plants with large genomes. *Trends in Genetics*, 11, 63–68.

TUBEROSA, R., SALVI, S., SANGUINETI, M., MACCAFERRI, M., GIULIANI, S., and LANDI, P. 2003. Searching for quantitative trait loci controlling root traits in maize: A critical appraisal. *Plant and Soil*, 255(1), 35–54.

UTZ, H. F., and MELCHINGER, A. E. 1996. PLABQTL: A program for composite interval mapping of QTL. *Journal of Agricultural Genomics*, 2, 1–6.

VISSCHER, P. M., THOMPSON, R., and HALEY, C. S. 1996. Confidence intervals in QTL mapping by bootstrapping. *Genetics*, 143(2), 1013–1020.

WANG, G. L., and PATERSON, A. H. 1994. Assessment of DNA pooling strategies for mapping of QTLs. *Theoretical and Applied Genetics*, 88(3–4), 355–361.

WEBER, D., and HELENTJARIS, T. 1989. Mapping RFLP loci in maize using B-A translocations. *Genetics*, 121(3), 583–590.

WEEDEN, N. F., TIMMERMAN, G. M., and LU, J. 1993. Identifying and mapping genes of economic significance. *Euphytica*, 73(1–2), 191–198.

WILLIAMS, K. J. 2003. The molecular genetics of disease resistance in barley. *Australian Journal of Agricultural Research*, 54(12), 1065–1079.

WINTER, P., and KAHL, G. 1995. Molecular marker technologies for plant improvement. *World Journal of Microbiology and Biotechnology*, 11(4), 438–448.

WOLFE, A. D., and LISTON, A. 1998. Contributions of PCR-based methods to plant systematics and evolutionary biology. In: D. E. Soltis, P. S. Soltis, and J. J. Doyle (Eds.). *Molecular Systematics of Plants II: DNA Sequencing* (pp. 43–86): Kluwer, New York.

WU, K., BURNQUIST, W., SORRELS, M., TEW, T. L., MOORE, P., and TANKSLEY, S. 1992. The detection and estimation of linkage in polyploids using single-dose restriction fragments. *Theoretical and Applied Genetics*, 83, 294–300.

YAO, H., ZHOU, Q., LI, J., SMITH, H., YANDEAU, M., NIKOLAU, B. J., and SCHNABLE, P. S. 2002. Molecular characterization of meiotic recombination across the 140-kb multigenic a1-sh2 interval of maize. *Proceedings of the National Academy of Science of the USA*, 99(9), 6157–6162.

YOUNG, N. D. 1996. QTL mapping and quantitative disease resistance in plants. *Annual Review of Phytopathology*, 34, 479–501.

YOUNG, N. D., and TANKSLEY, S. D. 1989. RFLP analysis of the size of chromosomal segments retained around the Tm-2 locus of tomato during backcross breeding. *Theoretical and Applied Genetics*, 77(3), 353–359.

ZENG, Z. B. 1993. Theoretical basis for separation of multiple linked gene effects in mapping quantitative trait loci. *Proceedings of the National Academy of Science of the USA*, 90(23), 10,972–10,976.

ZENG, Z. B. 1994. Precision mapping of quantitative trait loci. *Genetics*, 136(4), 1457–1468.

ZHANG, L. P., LIN, G. Y., NIÑO-LIU, D., and FOOLAD, M. R. 2003. Mapping QTLs conferring early blight (*Alternaria solani*) resistance in a *Lycopersicon esculentum* X *L. hirsutum* cross by selective genotyping. *Molecular Breeding*, 12(1), 3–19.

ZWART, R. S., THOMPSON, J. P., and GODWIN, I. D. 2004. Genetic analysis of resistance to root-lesion nematode (*Pratylenchus thornei*) in wheat. *Plant Breeding*, 123(3), 209–212.

CHAPTER EIGHTEEN

Global climate change with reference to microorganisms in soil–agriculture ecosystem

Vivek Kumar

Contents

Abstract

Microorganisms found in the soil are vital to many of the ecological processes that sustain life, such as nutrient cycling, decay of plant matter, consumption and production of trace gases, and transformation of metals (Panikov, 1999). Although climate change studies often focus on life at the macroscopic scale, microbial processes can significantly shape the effects that global climate change has on terrestrial ecosystems. According to the International Panel on Climate Change (IPCC) report (2007), warming of the climate system is occurring at unprecedented rates and an increase in anthropogenic greenhouse gas concentrations is responsible for most of this warming. Soil microorganisms contribute significantly to the production and consumption of greenhouse gases, including carbon dioxide (CO_2), methane (CH_4), nitrous oxide (N_2O) and nitric oxide (NO), and human activities such as waste disposal and agriculture have stimulated the production of greenhouse gases by microbes.

18.1 Introduction

From the primary molecules of oxygen formed by oceanic cyanobacteria ~3.5 billion years back to the methanogens luxuriating in the warm, carbon-rich swamps of the Carboniferous period, microbial processes have long been the key drivers of, and responders to, climate change Schopf and Packer (1987). It is widely accepted that microorganisms have played a key part in determining the atmospheric concentration of greenhouse gases, including CO_2, CH_4 and nitrous oxide N_2O (which have the greatest impact on radioactive forcing), throughout much of Earth's history. What is more open to debate is the part that they will play in the coming decades and centuries, the climate feedbacks that will be important, and how humankind might harness microbial processes to manage climate change. The feedback responses of microorganisms to climate change in terms of greenhouse gas flux may either amplify (positive feedback) or reduce (negative feedback) the rate of climate change. With the twenty-first century projected to experience some of the most rapid climatic changes in our planet's history, and with biogenic fluxes of the main anthropogenic greenhouse gases being tied integrally to microorganisms, improving our understanding of microbial processes has never been so important.

In terrestrial ecosystems, the response of plant communities and symbiotic microorganisms, such as mycorrhizal fungi and nitrogen-fixing bacteria, to climate change is well understood; both in terms of physiology and community structure (Bardgett et al., 2009). However, the response of the heterotrophic microbial communities in soils to climate change, including warming and altered precipitation, is less clear. This is a crucial factor, as it determines the nature and extent of terrestrial ecosystem feedback responses. However, understanding the responses of microbial communities to climate change is complicated by the vast and largely unexplored diversity of microbiota found in the terrestrial environment, for which only a few examples of food webs have been fully constructed (Morgan, 2002). Also, different terrestrial ecosystems comprise different microbial communities, and this is further compounded by the effects of land use, other disturbances (such as management practices) and different biogeographical patterns (distribution of microbial communities over space and time).

18.2 Greenhouse gas emissions by microbial control

Understanding the physiology and dynamics of microbial communities is essential if we are to increase our knowledge of the control mechanisms involved in greenhouse gas fluxes (Schimel and Gulledge, 1998; Allison et al. 2010). This topic has received little attention owing to the assumption that microbial community structure has little relevance to large-scale ecosystem models (Schimel, 1995) and to the lack of theoretical background and technologies to measure the vast diversity of microbial communities in natural environments and to determine their link to ecosystem functioning. Nevertheless, recent advances in molecular techniques and their application to the characterisation of the so-called uncultivable microorganisms have started to provide an improved understanding of microbial control of greenhouse gas emissions.

Carbon dioxide gas

In the global carbon cycle, annual emissions of CO_2 from the burning of fossil fuels are dwarfed by the natural fluxes of CO_2, to and from the land, oceans and atmosphere. Current levels of atmospheric CO_2 depend largely on the balance between photosynthesis and respiration. In oceans, photosynthesis is primarily carried out by phytoplankton, whereas autotrophic and heterotrophic respiration return much of the carbon taken up during photosynthesis to the dissolved inorganic carbon pool (Del Giorgio and Duarte, 2002; Arrigo, 2005). For terrestrial ecosystems, the uptake of CO_2 from the atmosphere by net primary production is dominated by higher plants, but microorganisms contribute greatly to net carbon exchange through the processes of decomposition and heterotrophic respiration, as well as indirectly, through their role as plant symbionts or pathogens and by modifying nutrient availability in the soil (van der Heijden et al., 2008).

Approximately 120 billion tonnes of carbon are taken up each year by primary production on land (Hymus and Valentini, 2007) and ~119 billion tonnes of carbon are emitted, half by autotrophic (mainly plant) respiration and half by heterotrophic soil microorganisms (Reay and Grace, 2007). Together, the land and oceans constitute a net sink of ~3 billion tonnes of carbon each year, effectively absorbing about 40% of the current CO_2 emissions from fossil fuel use (Figure 18.1). In addition, 1–2 billion tonnes of carbon are added to the atmosphere each year (Drigo et al., 2007) through changes in land use (predominantly tropical deforestation). Furthermore, because soils store ~2000 billion tonnes of organic carbon, their disturbance by agriculture and other land uses can greatly stimulate the rates of organic matter decomposition and net emissions of CO_2 to the atmosphere (Smith, 2008a,b). For example, deep ploughing or drainage

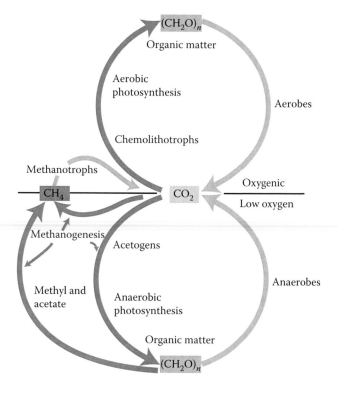

FIGURE 18.1 Impact on global cycle (carbon and nitrogen cycle).

of organic, carbon-rich soils is known to stimulate rates of decomposition and respiration, because it gives microorganisms a greater access to both buried organic carbon and oxygen (Smith, 2008b). Through such cultivation and disturbance, soils are estimated to have already lost 40–90 billion tonnes of carbon since human intervention began (Lal, 1999). Although these responses are mediated by microbial activity, it is generally thought that changes in the structure and diversity of terrestrial microbial communities will have little effect on CO_2 production at the ecosystem level because, unlike CH_4 and N_2O production, CO_2 production results from numerous microbial processes Staddon et al. (2004). However, recent findings have challenged this assumption by providing evidence of a direct link between CO_2 fluxes and changes in the structure and physiology of the microbial community (Carney et al., 2007).

A prime cause of this uncertainty is the inherent complexity and diversity of soil organic matter and the likelihood that the temperature dependence of microbial decomposition of soil carbon compounds of differing chemical composition and substrate quality will vary (Rillig et al., 2002; Davidson and Janssens, 2006). For example, there is evidence that the temperature sensitivity of litter decomposition increases as the quality of organic carbon consumed by microbes declines (Fierer et al., 2005), which is consistent with kinetic theory and indicates a greater temperature sensitivity for decomposition of recalcitrant carbon pools (Knorr et al., 2005). There is also a considerable potential for various environmental constraints, such as physical and chemical protection of organic matter, to decrease substrate availability for microbial attack, thereby dampening microbial responses to warming (Davidson and Janssens, 2006).

Nitrous oxide gas

Similar to CO_2 and CH_4 emissions, global N_2O emissions have a predominantly microbial basis. Natural and anthropogenic sources are dominated by emissions from soils, primarily as a result of microbial nitrification and denitrification (Reay and Grace, 2007). For each tonne of reactive nitrogen (mainly fertiliser) deposited on the Earth's surface, either naturally or deliberately, 10–50 kg are emitted as N_2O (Crutzen, 2007). Several studies have been carried out to distinguish the relative contributions of nitrification and denitrification to net N_2O flux, although little is known about the degree of microbial control of these processes at the ecosystem level (Figure 18.1). Most N_2O produced by nitrification is a result of the activity of autotrophic ammonia (NH_3)-oxidising bacteria, belonging to the class beta-proteobacteria (Teske, 1994). However, recent studies suggest that some archaea also have an important role in nitrification (Leininger, 2006), although their relative contribution to this process is still debated.

By contrast, denitrification is a multistep process in which each step is mediated by a specific group of microorganisms that have the enzymes necessary to catalyse that particular step. The production of N_2O is typically the result of incomplete denitrification. Denitrifying activity is distributed among phylogenetically diverse bacterial populations, although each denitrifying enzyme catalysing a specific step in the process (e.g. nitrate reductase) is highly conserved genetically (Ye et al., 1994). A recent study provided direct evidence of a strong link between denitrifying bacterial communities and the rate of N_2O emission from soils (Salles et al., 2009).

Methane gas

Global emissions of CH_4 are arguably even more directly controlled by microorganisms than emissions of CO_2. Natural emissions (~250 million tonnes of CH_4 per year) are dominated by microbial methanogenesis, a process that is carried out by a group of anaerobic archaea in wetlands, oceans, rumens and termite guts. However, these natural sources are exceeded by emissions from human activities (mainly rice cultivation, landfill, fossil fuel extraction and livestock farming) (~320 million tonnes of CH_4 per year), which aside from some emissions from fossil fuel extraction are also predominately driven by microorganisms Mclain and Ahmann (2008). Methanotrophic bacteria serve as a crucial buffer to the huge amounts of CH_4 produced in some of these environments. The so-called low-affinity methanotrophs (active only at a CH_4 concentration of >40 parts per million; also called type I methanotrophs), which mainly belong to the class Gammaproteobacteria, can often consume a large proportion of the CH_4 produced in soils before it escapes to the atmosphere. For CH_4 already in the atmosphere, methanotrophic bacteria may also act as a net CH_4 sink. The so-called 'high-affinity' methanotrophs (active at a CH_4 concentration of <12 parts per million), which mainly belong to the class Alphaproteobacteria (also known as type II methanotrophs), remove approximately 30 million tonnes of CH_4 from the atmosphere each year (Reay and Grace, 2007).

Reducing CO_2 emissions by microbial communities

Currently, soils contain about 2000 Pg of organic carbon, which is twice the amount of carbon in the atmosphere and thrice the quantity found in vegetation (Smith, 2004). The capacity of different land types (e.g. woodland, pasture and arable land) to store carbon differs, and it has been suggested that land use can be managed to sequester a further 1 Pg of carbon per year in soils (Houghton, 2007); this potential has received considerable scientific attention (Lal, 2008). However, this may not be easily achievable on a global scale owing to the complex biological mechanisms that control the incorporation of organic carbon into soil, as well as the influence of changing abiotic factors, such as moisture, temperature, land use and nitrogen enrichment, which also affect soil carbon pools (Six et al., 2006; Smith et al., 2008). Forest soils are considered to be especially effective at storing carbon, in part because of a high abundance of fungi in the soil relative to bacteria, which favours carbon sequestration (Bailey et al., 2002; De Deyn et al., 2008; Busse, 2009; Castro et al., 2010).

To manage the soil microbial communities to increase carbon sequestration, it will be important to understand their ecology and function. This is a challenge in itself, because of our inability to characterise the species diversity and function of soil microbial communities and our lack of theoretical principles in microbial ecology, such as the definition of a species and the factors driving community formation and structure (Castro et al., 2010). Nevertheless, there is some evidence that bacteria can be categorised on the basis of their carbon mineralisation capacity and can be divided into copiotrophic (characterised by high growth rates on labile carbon and dominant in nutrient-rich environments) and oligotrophic (slow-growing and dominant in nutrient-limited ecosystems) species (Fierer et al., 2007). It has been suggested that the acidobacteria are oligotrophic, whereas the proteobacteria and the actinobacteria form copiotrophic communities.

One may disagree that manipulating land use (e.g. changing from arable land to forestry) and land management practices (e.g. using low-nitrogen-input agriculture) may promote the growth of oligotrophic communities. However, the ecological strategies of other dominant microbial taxa need to be understood. It is true that not all taxa in a phylum will be either copiotrophic or oligotrophic (Allison and Treseder, 2008) and, thus, phyla alone may not be a predictor of carbon loss from the soil (Fierer et al., 2007). It is therefore essential that we use rapidly developing technologies such as high-throughput sequencing to better understand soil microbial diversity. Moreover, emerging technologies such as metagenomics, metatranscriptomics, metaproteomics and stable-isotope probing (SIP) must be used to examine the physiological abilities and roles of individual taxa in a given ecosystem Bardgett et al. (2008). Only then can we begin to predict whether a particular soil is a net carbon emitter or sink based on microbial ecology.

This approach can be further expanded by combining metagenomics with SIP to find out the specific functions of a microbial population in a community. Future work should attempt to use this approach to differentiate between populations that use labile carbon and those that promote carbon sequestration. In agriculture, the often large losses of soil organic carbon owing to cultivation can be reduced by low- and no-tillage practices, which favour soil communities dominated by fungi (Castro et al., 2010); such agroecosystems prevent the increase in microbial decomposition and respiration.

Reducing CH$_4$ emissions by microbial communities

Our understanding of the microbiology of greenhouse gas cycling is more complete for CH$_4$ than for CO$_2$ or N$_2$O, as the pathway is simple and specialised microorganisms are involved. However, many of the above uncertainties also apply to the management of terrestrial CH$_4$ fluxes, because most atmospheric CH$_4$ is produced by microorganisms, it is theoretically feasible to control a substantial proportion of CH$_4$ emissions from terrestrial ecosystems by managing microbial community structure and processes Phillips et al. (2001). The biological oxidation of CH$_4$ by methanotrophs accounts for only ~5% of the global sink of atmospheric CH$_4$ (~30 million tonnes per year) (Hanson and Hanson, 1996) and may therefore seem less important. However, methanotrophs are also responsible for the oxidation of up to 90% of the CH$_4$ produced in soil before it can escape to the atmosphere (Oremland and Culbertson, 1992). It is well established that conversion of arable land or grassland into a forest results in a substantial reduction in CH$_4$ flux (Kolbs, 2009), and it is evident that both the type and abundance of methanotrophs are important for predicting CH$_4$ flux. However, no current climate model considers this finding, so future research must focus on incorporating these data and interactions to improve predictions of CH$_4$ fluxes across various ecosystems. This knowledge can also be applied to the reduction of CH$_4$ emissions by changing land use and management. In rice cultivation, for example, methanotrophs have long played a crucial part in absorbing a proportion of the CH$_4$ produced and, as a result, improved management of flooding frequency and duration could reduce net emissions by increasing oxygen availability in soils (Yagi, 1996; Reay, 2003). There is also a great potential to make effective use of inhibitors of methanogenesis, such as ammonium sulphate fertilisers, in managed systems to promote the growth of sulphate reducers at the expense of methanogens (Neue, 2007). To reduce methane emissions from ruminant livestock, strategies include improving feed quality and directly inhibiting methanogen communities in the rumen using antibiotics, vaccines and alternative electron acceptors (Smith et al., 2008).

Reducing N_2O emissions by microbial communities

A major source of anthropogenic N_2O emission is the use of nitrogen fertilisers in agriculture. As a substantial proportion of applied fertilisers are emitted in the form of N_2O, better-targeted fertiliser applications, which reduce the availability of nitrogen to microorganisms, can substantially decrease N_2O emissions. Schimel and Gulledge (1998) showed potential strategies that include reducing the amount of fertiliser and applying it at an appropriate time (when crop demand for nitrogen is high and leaching-loss rates are low), using slow-release fertilisers, and avoiding nitrogen forms that are likely to produce large emissions or leaching losses (such as nitrate in wet soil). Similarly, improved land drainage and better management practices to limit anaerobic conditions in soils (e.g. land compaction and excessive wetness) could reduce denitrification rates and, thus, N_2O emissions. Finally, for the mitigation of N_2O fluxes from agriculture, the use of nitrification inhibitors in fertilisers to limit nitrate production and subsequent leaching or denitrification losses is now a well-established strategy (Smith et al., 2008). These and similar microorganism-mediated strategies have great potential to reduce greenhouse gas emissions from the land use and agricultural sectors.

Soil-borne pathogens and climate change

As per the IPCC (2007) report, climate change will alter patterns of infectious disease outbreaks in humans and animals. Soil pathogens are no exception: case studies support the claim that climate change is already changing patterns of infectious diseases caused by soil pathogens. For example, over the last 20 years, 67% of the 110 species of harlequin frogs (*Atelopus*) native to tropical regions in Latin America have gone extinct from chytridiomycosisthe, a lethal disease spread by the pathogenic chrytid fungus (*Batrachochytrium dendrobatidis*) (Willey et al., 2009). Research suggests that mid- to high elevations provide ideal temperatures for *B. dendrobatidis*. However, as global warming progresses, *B. dendrobatidis* is able to expand its range due to increasing moisture and warmer temperatures at higher elevations (Muths et al., 2008). This expansion exposes more amphibian communities in previously unaffected or minimally affected areas, specifically at higher elevations, by chytridiomycota. As seen in the case of *Atelopus* harlequin frogs, the spread of soil pathogens due to climatic changes can significantly affect life at the macroscale and ultimately lead to species extinction Briones et al. (2004).

Indirect climate–microbe feedbacks

Climate change can also have marked indirect effects on soil microbial communities and their activity—and hence the potential for microbial feedback to climate change—through its influence on plant growth and vegetation composition. Such plant-mediated indirect effects of climate change on soil microbes operate through a variety of mechanisms, with differing routes of feedback to climate change, but these can be broadly separated into two. The first mechanism concerns the indirect effects of rising atmospheric concentrations of carbon dioxide on soil microbes, through increased plant photosynthesis and transfer of photosynthate carbon to fine roots and mycorrhizal fungi (Johnson et al., 2005; Högberg and Read, 2006; Keel et al., 2006) and heterotrophic microbes (Zak et al., 1993; Bardgett et al., 2005). It is well established that elevated carbon dioxide increases plant photosynthesis and growth, especially under nutrient-rich conditions (Curtis and Wang 1998) and this in turn increases the flux of carbon to roots, their symbionts and heterotrophic microbes through root exudation of easily degradable sugars, organic acids and amino acids (Díaz et al., 1993; Zak et al., 1993). The consequences of an increased carbon flux from roots to soil for microbial communities and carbon exchange are difficult to predict, because they will vary substantially with factors such as plant identity, soil–food–web interactions, soil fertility and a range of other ecosystem properties (Wardle 2002; Bardgett, 2005). But some potential outcomes for soil microbes and carbon exchange include increases in soil carbon loss by respiration and in drainage waters as dissolved organic carbon due to the stimulation of microbial abundance and activity, and enhanced mineralisation of recent and old soil organic carbon (Zak et al., 1993).

Future perspectives and wrapping up

An understanding of soil microbial ecology is central to our ability to assess terrestrial carbon cycle–climate feedbacks. However, the complexity of the soil microbial community and its many roles, coupled with the myriad ways that climate and other global changes can affect soil microbes, hampers our ability to draw firm conclusions on this topic. Despite this uncertainty, we argue that progress can be made in understanding the potential negative and positive contributions of soil microbes to global warming through consideration of both direct and indirect impacts of climate change on microorganisms, and the capacity for such effects to amplify or dampen carbon cycle feedbacks. This is a major challenge, but we believe that progress can be made through the use of long-term multifactor field experiments in relevant biomes, which incorporate consideration of direct and indirect impacts of climate change on soil microbes and their contribution to land–atmosphere carbon exchange, measured at the whole ecosystem scale. Such studies require a collaborative approach to link microbial ecology to the whole ecosystem-scale flux measures and modelling of carbon cycle feedbacks.

There is a consensus among scientists that global climate change is happening and that the increases in global average temperatures since 1900 can be largely attributed to human activities. However, there remains much uncertainty about predictions of future greenhouse gas emissions and the response of these emissions to further changes in the global climate and atmospheric composition. To help tackle this uncertainty, there is a need to better understand terrestrial microbial feedback responses and the potential to manage microbial systems for the mitigation of climate change. There is an urgent need to improve the mechanistic understanding of microbial control of greenhouse gas emissions and the interactions between the different abiotic and biotic components that regulate them. This understanding will help to remove large uncertainties about the prediction of feedback responses of microorganisms to climate change and will enable the knowledge to be incorporated into future models of climate change and terrestrial feedbacks.

It is currently difficult to know whether changes in processes that are associated with climate change are brought about by the effect of climate change on soil microbial communities, by changes in soil abiotic factors or by interactions between the two. Moreover, it is unclear how microorganisms respond to climate change and therefore what their potential is to influence climate feedbacks across ecosystems and along environmental gradients. Another issue that needs to be taken into consideration is that, to date, most studies have focused on one greenhouse gas, whereas evidence suggests that microorganism-mediated fluxes of different greenhouse gases respond differently to climate change. For example, it is assumed that conservation of peatland will enhance carbon sequestration, but this may also increase CH_4 fluxes, so the effect on net greenhouse gas flux is still unclear Woodward et al. (2004).

We have only scratched the surface of the contribution of soil microbes to climate change, and, as highlighted above, there are many uncertainties and challenges. In addition to what is mentioned above, we have identified three major challenges. First, soil microbial communities are extremely diverse, and one of the greatest challenges is understanding how microbial diversity responds to climate change and the functional consequences of this for ecosystem carbon exchange, including the uptake, stabilisation and release of carbon from soil as greenhouse gas. The second major hurdle here is that many microbes are uncultivable, and the function of these non-cultivable microbes is poorly understood because it is difficult to test how they respond to, or modify, their environment (van der Heijden et al., 2008). Third challenge is how environment modifies the microbial functions, is still not understood exactly. However, new molecular and SIP tools are being developed that enable linking of changes in microbial diversity to ecosystem function, by focussing on functional genes that are important for biogeochemical processes and through directly labelling DNA, RNA and phospholipid fatty acids (PLFA) of organisms participating in particular pathways (Zak et al., 1993, 2006; Drigo et al., 2007). These tools have changed the way microbial ecologists explore the ecophysiology of microbial populations in the natural environment, because they enable the study of the metabolic capabilities of uncultivable microorganisms, thus providing insights into the underlying processes regulating carbon flow through different components of the soil microflora.

In the end, as discussed in this chapter, soil microbes and their activities are inextricably linked to the aboveground communities, including plants, herbivores, pathogens and parasites. Understanding the effects of climate transformation on carbon, methane and nitrous oxide dynamics, therefore, requires explicit consideration of the feedbacks that occur between aboveground and belowground communities and their response to climate change.

References

ALLISON, S. D. and TRESEDER, K. K. 2008. Warming and drying suppress microbial activity and carbon cycling in boreal forest soils. *Glob. Chang. Biol.* **14**: 2898–2909.

ALLISON, S. D., WALLENSTEIN, M. D. and BRADFORD, M. A. 2010. Soil carbon response to warming dependent on microbial physiology. *Nat. Geosci.* **3**: 336–340.

ARRIGO, K. 2005. Marine microorganisms and global nutrient cycles. *Nature.* **437**: 349–355.

BAILEY, V. L., SMITH, J. L. and BOLTON, H. 2002. Fungal-to-bacterial ratios in soils investigated for enhanced C sequestration. *Soil Biol. Biochem.* **34**: 997–1007.

BARDGETT, R. D. 2005. *The Biology of Soil. A Community and Ecosystem Approach.* Oxford University Press, Oxford.

BARDGETT, R. D., BOWMAN, W. D., KAUFMANN, R. and SCHMIDT, S. K. 2005. A temporal approach to linking aboveground and belowground ecology. *Trends Ecol. Evol.* **20**: 634–641.

BARDGETT, R. D., DE DEYN, G. B. and OSTLE, N. J. 2009. Plant–soil interactions and the carbon cycle. *J. Ecol.* **97**: 838–839.

BARDGETT, R. D., FREEMAN, C. and OSTLE, N. J. 2008. Microbial contributions to climate change through carbon cycle feedbacks. *ISME J.* **2**: 2805–2814.

BRIONES, M. J. I., POSKITT, J. and OSTLE, N. 2004. Influence of warming and enchytraeid activities on soil CO_2 and CH_4 fluxes. *Soil Biol. Biochem.* **36**: 1851–1859.

BUSSE, M. D. 2009. Soil carbon sequestration and changes in fungal and bacterial biomass following incorporation of forest residues. *Soil Biol. Biochem.* **41**: 220–227.

CARNEY, K. M., HUNGATE, B. A., DRAKE, B. G. and MEGONIGAL, J. P. 2007. Altered soil microbial community at elevated CO_2 leads to loss of soil carbon. *Proc. Natl. Acad. Sci. USA* **104**: 4990–4995.

CASTRO, H. F., CLASSEN, A. T., AUSTIN, E. E., NORBY, R. J. and SCHADT, C. W. 2010. Soil microbial community responses to multiple experimental climate change rivers. *Appl. Environ. Microbiol.* **76**: 999–1007.

CRUTZEN, P. J. 2007. N_2O release from agro-biofuel production negates global warming reduction by replacing fossil fuels. *Atmos. Chem. Phys. Discuss.* **7**: 11191–11205.

CURTIS, P. S. and WANG, X. Z. 1998. A meta-analysis of elevated CO_2 effects on woody plant mass, form, and physiology. *Oecolegia.* **113**: 299–313.

DAVIDSON, E. A. and JANSSENS, I. A. 2006. Temperature sensitivity of soil carbon decomposition and feedbacks to climate change. *Nature.* **440**: 165–173.

DE DEYN, G. B., CORNELISSEN, J. H. C. and BARDGETT, R. D. 2008. Plant functional traits and soil carbon sequestration in contrasting biomes. *Ecol. Lett.* **11**: 516–531.

DEL GIORGIO, P. A. and DUARTE, C. M. 2002. Respiration in the open ocean. *Nature* **420**: 379–384.

DÍAZ, S., GRIME, J. P., HARRIS, J. and MCPHERSON, E. 1993. Evidence of a feedback mechanism limiting plant response to elevated carbon-dioxide. *Nature.* **364**: 616–617.

DRIGO, B., KOWALCHUK, G. A. and VAN VEEN, J. A. 2007. Climate change goes underground: Effects of elevated atmospheric CO_2 on microbial community structure and activities in the rhizosphere. *Biol. Fertil. Soils.* **44**: 667–679.

FIERER, N., CRAINE, J. M., MCLAUCHLAN, K. and
SCHIMEL, J. P. 2005. Litter quality and the temperature sensi-
tivity of decomposition. *Ecology*. **86**: 320–326.
FIERER, N., BRADFORD, M. A. and JACKSON, R. B. 2007.
Toward an ecological classification of soil bacteria. *Ecology* **88**:
1354–1364.
HANSON, R. S. and HANSON, T. E. 1996. Methanotrophic bacte-
ria. *Microbiol. Rev.* **60**: 439–471.
HÖGBERG, P. and READ, D. J. 2006. Towards a more plant physi-
ological perspective on soil ecology. *Trends Ecol. Evol.* **21**:
548–554.
HOUGHTON, R. A. 2007. Balancing the global carbon budget.
Annu. Rev. Earth Planet. Sci. **35**: 313–347.
HYMUS, G. and VALENTINI, R. 2007. In: *Greenhouse Gas
Sinks* (eds Reay, D. S. et al.) pp. 11–30 (CABI Publishing,
Oxfordshire).
IPCC. 2007. Climate Change 2007: Synthesis Report. Available at
http://www.ipcc.ch/publications_and_data/publications_ipcc_
fourth_assessment_report _synthesis_report.htm.
JOHNSON, D., KRESK, M., STOTT, A. W., COLE, L.,
BARDGETT, R. D. and READ, D. J. 2005. Soil invertebrates
disrupt carbon flow through fungal networks. *Science.* **309**: 1047.
KEEL, S. G., SIEGWOLF, R. T. W. and KÖRNER, C. 2006.
Canopy CO_2 enrichment permits tracing the fate of recently
assimilated carbon in a mature deciduous forest. *New Phytol.*
172: 319–329.
KNORR, W., PRENTICE, I. C., HOUSE, J. I. and HOLLAND,
E. A. 2005. Longterm sensitivity of soil carbon turnover to
warming. *Nature.* **433**: 298–301.
KOLBS, S. 2009. The quest for atmospheric methane oxidisers in forest soils. *Environ. Microbiol. Rep.* **1**: 336–346.
LAL, R. 1999. Soil management and restoration for C sequestration
to mitigate the accelerated greenhouse effect. *Prog. Environ. Sci.*
1: 307–326.
LAL, R. 2008. Carbon sequestration. *Phil. Trans. R. Soc. B Biol. Sci.*
363: 815–830.
LEININGER, S. 2006. Archaea predominate among ammonia-oxidizing prokaryotes in soils. *Nature.* **442**: 806–809.
MCLAIN, J. E. T. and AHMANN, D. M. 2008. Increased moisture
and methanogenesis contribute to reduced methane oxidation in
elevated CO_2 soils. *Biol. Fertil. Soils* **44**: 623–631.
MORGAN, J. A. 2002. Looking beneath the surface. *Science* **298**:
1903–1904.
MUTHS, E., PILLIOD, D. S. and LIVO, L. J. 2008. Distribution and environmental limitations of an amphibian pathogen in the Rocky
Mountains, USA. *Biol. Conserv.* **141**: 1484–1492.
NEUE, H. U. 2007. Fluxes of methane from rice fields and potential
for mitigation. *Soil Use Manag.* **13**: 258–267.
OREMLAND, R. S. and CULBERTSON, C. W. 1992. Importance
of methane-oxidizing bacteria in the methane budget as revealed
by the use of a specific inhibitor. *Nature.* **356**: 421–423.

PANIKOV, N. S. 1999. Understanding and prediction of soil microbial community dynamics under global change. *Appl. Soil Ecol.* **11**: 161–176.

PHILLIPS, R. L., WHALEN, S. C. and SCHLESINGER, W. H. 2001. Influence of atmospheric CO_2 enrichment on methane consumption in a temperate forest soil. *Glob. Chang. Biol.* **7**: 557–563.

REAY, D. S. 2003. Sinking methane. *Biologist.* **50**: 15–19.

REAY, D. S. and GRACE, J. 2007. In *Greenhouse Gas Sinks* (eds. Reay, D. S. et al.) pp. 1–10 (CABI Publishing, Oxfordshire).

RILLIG, M. C., HERNANDEZ, G. Y. and NEWTON, P. C. D. 2002. Arbuscular mycorrhizae respond to elevated atmospheric CO_2 after long-term exposure: Evidence from a CO_2 spring in New Zealand supports the resource balance model. *Ecol. Lett.* **3**: 475–478.

SALLES, J. F., POLY, F., SCHMID, B. and LE ROUX, X. 2009. Community niche predicts the functioning of denitrifying bacterial assemblages. *Ecology.* **90**: 3324–3332.

SCHIMEL, J. 1995. In: *Arctic and Alpine Biodiversity: Patterns, Causes and Ecosystem Consequences* (eds Chapin, F. S. III and Körner, C.) pp. 239–254 (Springer, Berlin).

SCHIMEL, J. P. and GULLEDGE, J. 1998. Microbial community structure and global trace gases. *Glob. Chang. Biol.* **4**: 745–758.

SCHOPF, J. W. and PACKER, B. M. 1987. Early Archean (3.3-billion to 3.5-billion-year-old) microfossils from Warrawoona Group, Australia. *Science.* **237**: 70–73.

SIX, J., FREY, S. D., THIET, R. K. and BATTEN, K. M. 2006. Bacterial and fungal contributions to carbon sequestration in agroecosystems. *Soil Sci. Soc. Am. J.* **70**: 555–569.

SMITH, P. 2004. Soils as carbon sinks: The global context. *Soil Use Manag.* **20**: 212–218.

SMITH, P. 2008a. Greenhouse gas mitigation in agriculture. *Phil. Trans. R. Soc. B Biol. Sci.* **363**: 789–813.

SMITH, P. 2008b. Land use change and soil organic carbon dynamics. *Nutr. Cycl. Agroecosyst.* **81**: 169–178.

SMITH, P., FANG, C. M., DAWSON, J. J. C. and MONCRIEFF, J. B. 2008. Impact of global warming on soil organic carbon. *Adv. Agronomy.* **97**: 1–43.

STADDON, P. L., JAKOBSEN, I. and BLUM, H. 2004. Nitrogen input mediates the effect of free-air CO_2 enrichment on mycorrhizal fungal abundance. *Glob. Chang. Biol.* **10**: 1678–1688.

TESKE, A. 1994. Evolutionary relationships among ammonia-oxidizing and nitrite-oxidizing bacteria. *J. Bacteriol.* **176**: 6623–6630.

VAN DER HEIJDEN, M. G. A., BARDGETT, R. D. and VAN STRAALEN, N. M. 2008. The unseen majority: Soil microbes as drivers of plant diversity and productivity in terrestrial ecosystems. *Ecol. Lett.* **11**: 296–310.

WARDLE, D. A. 2002. *Communities and Ecosystems: Linking the Aboveground and Belowground Components.* Princeton University Press, Princeton, New Jersey.

WILLEY, J. M., SHERWOOD, L. M. and WOOLVERTON, C. J. 2009. *Prescott's Principles of Microbiology*. McGraw-Hill, New York, NY.

WOODWARD, F. I., LOMAS, M. R. and KELLY, C. K. 2004. Global climate and the distribution of plant biomes. *Phil. Trans. R. Soc. Lond. B Biol. Sci.* **359**: 1465–1476.

YAGI, K. 1996. Effect of water management on methane emission from a Japanese rice paddy field: Automated methane monitoring. *Global Biogeochem. Cycles* **10**: 255–267.

YE, R. W., AVERILL, B. A. and TIEDJE, J. M. 1994. Denitrification: Production and consumption of nitric-oxide. *Appl. Environ. Microbiol.* **60**: 1053–1058.

ZAK, D. R., BLACKWOOD, C. B. and WALDROP, M. P. 2006. A molecular dawn for biogeochemistry. *Trends Ecol. Evol.* **21**: 288–295.

ZAK, D. R., GRIGAL, D. F. and OHMANN, L. 1993. Elevated atmospheric CO_2 and feedback between carbon and nitrogen cycles. *Plant Soil.* **151**: 105–117.

CHAPTER NINETEEN

Climate change impacts on agricultural productivity in Norway

Asbjørn Torvanger, Michelle Twena and Bård Romstad

Contents

Abstract

Climate change is likely to affect agricultural productivity. In this study, a biophysical statistical model is used to analyse the relationship between yields of potatoes, barley, oats and wheat per decare, and temperature (growing degree days) and precipitation, for the period 1958–2001 at the county level in Norway. If a climate signal can be detected at the county level, this should be of interest for climate policy planners, agricultural authorities and farmers preparing for a warmer climate. We find that in 18% of (the crop and county) cases, there is a positive impact on yield from increased temperature. In the case of crops, the effect is strongest for potatoes. Regionally, the correlations are strongest in Northern Norway, where temperature is likely to be more important as a limiting factor for crop growth than other regions of the country. The effect of increased precipitation is negative in 20% of the cases, which could be due to excess soil moisture or reduced sun radiation associated with more cloud cover. Predictions based on the RegClim scenario for 2040 indicate that potato yields will increase by around 30% in Northern Norway.

19.1 Introduction

Climate change may have significant impacts on society and ecosystems over the next decades. Since a substantial part of expected climate change is likely to be man-made, we are faced with a challenge to decide on emission mitigation policies at international, national and local level [6]. Furthermore, adaptation policies have the potential to lower the overall costs associated with climate change. Given the large number of uncertainties in future emissions, climate system responses and potential impacts, policy design must be based on best available knowledge, and regularly updated when new results become available. For a number of years, impacts research has been hindered by a lack of climate change scenarios with resolution high enough to capture sub-national variations.

Such scenarios are now available from downscaled results of global circulation models (GCMs). In this study, we analyse the effects on agricultural productivity using a regional climate change scenario for Norway for the period

2030–2050—RegClim.* Agriculture is one of the sectors that is most likely to be sensitive to the primary effects of climate change, such as changes in growing season, temperature and precipitation. We seek to establish a statistical relationship between yield per decare for four crops, based on meteorological data from 1958 until 2001, through regression analysis at county level in Norway. In addition, we undertake analyses at the national level. The four crops we investigated were potatoes, wheat (spring and winter), oats and barley. The meteorological data consist of growing degree days (GDD) and annual precipitation. In addition, a time trend was included to account for long-term technology and productivity changes in agriculture. It accounted for, in part, the CO_2 fertilisation response due to the steady rise in the CO_2 concentration level in the atmosphere. Assuming that there were no major changes in agricultural production technologies and practices during this period, we made a prediction of yields per decare for 2040 (as a representative year for the period 2030–2050) based on the RegClim scenario. Through this analysis we tried to detect a climate signal in the annual weather variation and agricultural yield data at a relatively aggregated level (county) in Norway. If such a signal is found, the estimated impacts on agricultural production across regions and four major crops in Norway should be of interest for climate policy planners, agricultural authorities and farmers in preparing for a warmer future.

The main methodological approaches studying impacts on agriculture from climate change are presented in a handbook by the UNEP and IVM [4]. There are two categories of tools, biophysical and economic. Biophysical tools can be divided into experimentation, agro-climatic indices, statistical models, process-based models and spatial or temporal analogues. Economic tools can be divided into economic regression models, microeconomic models and macroeconomic models.

In this study, we have chosen a biophysical statistical model, which links the primary climate change impacts on temperature and precipitation to changes in yield per unit of land. This choice gives priority to the secondary impacts of climate change. A weakness of this approach is its limited ability to predict the effect of future climate change that lies outside the climate variability of the last decades (upon which the estimates of the model parameters are based); another is that there is an implied assumption of mixed technology [4].

---

* See http://regclim.met.no.

Furthermore, the method is founded on correlation analysis and not necessarily on causal mechanisms. There may be dependency between explaining variables (multicollinearity), and relationships between yield, precipitation and temperature may be non-linear. Moreover, the simple model we have chosen is not able to account for effects caused by variability in weather and extreme weather events on yields [7]. Since we are studying a smaller change in climate (as defined by the RegClim scenario), a linear model is probably an acceptable approximation even if the relationships are non-linear. In addition, data availability has put strong restrictions on which variables could be included in the analysis. One example of an important weather variable for plant growth that could not be included was sun radiation, which could be represented through a measure of cloud cover. Through the chosen approach we were able to link changes in climate variables at local level (weather stations) to secondary climate change impacts in terms of changes in agricultural productivity for some crops at county level in Norway. Some major benefits of the approach are simplicity, limited data requirements and the ability to get some control over the significance of various explaining factors. The study is in line with the call of Zilberman et al. [25] to analyse the impact of climate change on agriculture within a disaggregated modelling framework and a focus on empirical research. The results should indicate if county level is a suitable aggregation level to disclose significant effects, or if this is an aggregation level that only produces moderate effects since more distinct local effects are averaged out [25].

An overview and assessment of climate change impacts in Europe, including agriculture, can be found in Parry (2000) [14]. NILF [11] provides a comprehensive survey of climate change impacts for the agricultural sector in Norway. Based on average yields in various climate zones, the climate change impact on agricultural productivity is analysed through a shift in climate zones leading to increased yields for most crops.

An early application of a statistical model is Warrick [24], who simulated wheat yields on the US Great Plains, assuming technology as in 1975 and climate conditions as under the 1936 drought. Leemans and Soloman [8] studied the potential yield changes for spring and winter wheat and other major crops at a global scale under a warmed climate. Using a crop-prediction model with geographic information systems (GIS), they reported that high-latitude regions will be the beneficiaries of climate change, enjoying extended growing seasons and increased productivity. Rötter and Van de Geijn [20] provided a comprehensive review of climate change impacts on livestock and crops yields, including wheat, potatoes, barley and oats. They emphasised the importance of elevated CO_2 concentration and quantified potential yield responses to predicted rises. The authors gave a detailed overview of the findings concerned with crop growth, physiology and phenology. Bootsma et al. [1] used linear regression analysis to examine the relationship between barley yields (among others) and climate variables in Atlantic Canada. They concluded that climate change is unlikely to have a significant impact on barley yields, though a doubling of CO_2 could lead to a 10–15% increase. Nonhebel [12] examined the effects of rising temperature and increases in CO_2 concentration on simulated wheat yields in Europe. She found that higher temperatures caused faster crop growth, leading to a shorter growing period and a decline in yield. CO_2 has the opposite effect, with a doubling of atmospheric concentration leading to a 40% rise in yields. Nonhebel also suggested that in general, changes in the availability of water can have a greater impact on yield than changes in temperature, but summarised that where precipitation patterns remain largely constant, negative effects of higher temperature are offset by positive effects of CO_2 enrichment. Riha et al. [18] and Mearns et al. [9] stress the importance of taking variability in temperature and precipitation into account when making crop yield predictions; both studies demonstrate that increased inter-seasonal variability can reduce yields. Ozkan and Akcaoz [13] analysed the impacts of annual and season variation of 27 climatic variables on the yield of wheat, maize and cotton in the Cukurova region of Turkey based on data from 1975 to 1999. They found that the most significant climatic factors for wheat yields were maximum temperature during planting time and maximum rainfall during flowering time. The wheat model could explain 46% of the variation of yield.

Parry and Carter [16] provide an overview of higher-order impacts of climate change on agriculture following first-order impacts. They report the results of impact and adjustment experiments conducted in five case studies (Iceland, Finland, Japan, Saskatchewan in Canada and northern parts of the former USSR), employing farm simulations and input–output models. They discuss the consequences of biophysical effects for farm income and profitability, food production, regional production costs and the wider economy. They then go on to consider potential managerial, technological and policy responses to these possible outcomes. Mendelsohn et al. [10] use Ricardian analysis to examine the impact of global warming on agriculture in the United States. They report negative climate impacts using a 'farm land' model, but a positive outcome using a 'crop revenue' approach. Their findings highlight the importance of taking adaptation factors into account when evaluating climate effects.

19.2 Description of the model

Statistical model A statistical model relating yield per decare to meteorological data is employed. The relationship between yield per decare, Y, and temperature, T, precipitation, P, and a time trend, τ, is assumed to be linear. Temperature is measured in GDD. The equation is

$$Y_{ijt} = \alpha_{ij} + \beta_{ij} T_{ijt} + \gamma_{ij} P_{ijt} + \theta\tau + \omega_{ijt}$$

where i is the index for crop (potatoes, wheat, oats and barley), j is the county index and t is the time index denoting annual observations from 1958 until 2001. ω_{ijt} is the error term.[*] GDD is defined as the annual sum of degrees accumulated above 5°C threshold. Through an ordinary least squares (OLS) regression we seek to correlate variations from year to year, in yield per decare, to the variability in GDD and precipitation. The estimated parameters are $\hat{\alpha}_{ij}, \hat{\beta}_{ij}, \hat{\gamma}_{ij}$ and $\hat{\theta}_{\tau}$ where the indices are left out for simplicity.

We were unable to take an explicit account of a number of non-climate factors. However, a time trend variable which was included in the regression runs to account for general long-term time trends, which may have been influenced by a number of other factors. Examples of such influences are technological change and innovations (e.g. improvements in agricultural inputs and/or practices, and/or changes in production patterns), increased productivity due to other climate variables, and a fertiliser effect from increased CO_2 concentration in the atmosphere. As an alternative to the time trend we included CO_2 concentrations in some of the regressions (see Annex 19.3 for a closer description of this model variant). Sunlight is another important weather variable for crop yields since it provides

---

[*] We assume that the error variances are constant and that the errors are not autocorrelated. Given that these assumptions are fulfilled, the ordinary least squares estimators are the best linear unbiased estimators. Checking the Durbin Watson statistic for some country cases revealed no indications of autocorrelation problems.

energy for photosynthesis. However, as meteorological stations were unable to provide relevant proxy data (i.e. cloud cover observations) for the complete period of our study, we were not able to include this variable in the analyses.

We carried out regressions at the national level by merging county data into two different variants of the model. In the first model variant, we allowed different constant terms for each county, whereas we assumed that the marginal effect of changes in weather data was the same for all counties. This model variant implies that there are differences in the yield level across counties, but no differences in the marginal yield of changes in the weather (i.e. GDD and precipitation). This is modelled through an additive dummy variable for each county with the exception of Akershus/Oslo, which is taken as the reference county. In the second model variant, different constant terms are retained, but in addition we allow for a shift in the marginal effect (slope) of annual precipitation by adding a multiplicative dummy variable to the precipitation variable for each county. The latter model variant implies that there are systematic differences between counties with respect to the level of yield per decare for a crop, as well as with respect to the marginal effect on yield of changes in precipitation, but with no differences in the marginal effect of changes in GDD. The different treatment of GDD and precipitation is based on regression results at county level, which indicated that there is a larger variance in the marginal effect of precipitation across counties than in temperature (GDD).

Variants of the model

The main model contains GDD, annual precipitation and a time trend as independent variables, and was employed on each crop at county level and at national level. However, a number of model variants were tested on the crop yield and weather data before ending up with this model. The chosen model produced more significant coefficients and a better fit to the data than the alternatives. The model variants included growing season precipitation, carbon dioxide concentration (in different data formats), frost events in the spring (in different data formats), fertiliser use for the latter part of the estimation period, and logarithmic or quadratic weather variables.* See Annex 19.3 for a more detailed account of the model variants that were tested.

---

* Thompson [22] advocates the use of quadratic terms for weather variables. Parry and Carter [16] also find changes in climate to have non-linear effects.

Yield predictions for the RegClim scenario

The equation for predicting yield per decare for crop i in county j under the RegClim climate change scenario, $\hat{Y}_{ijR}$, is

$$\hat{Y}_{ijR} = \hat{\alpha}_{ij} + \hat{\beta}_{ij} \, \hat{T}_{jR} + \hat{\gamma}_{ij} \, \hat{P}_{jR} + \hat{\theta} \, \tau_R$$

where $\hat{T}_{jR}$ is GDD and $\hat{P}_{jR}$ is precipitation in the RegClim scenario in county j, and τ_R is the time trend in 2040 (representing the RegClim period 2030–2050). R is the index for the RegClim scenario.

19.3 Data

The dependent variable is yield per decare for each of the crops potatoes, barley, oats and wheat. The independent variables are the weather data GDD and annual precipitation, in addition to the time trend.

Time periods

For each crop and county, analyses were undertaken for the main period 1958–2001, given that the required data were available. In the absence of sufficiently comprehensive data at county level to enable the incorporation in the model of a variable for technological change, national fertiliser-use figures were examined for clues as to what sort of impact one might expect farming practices to have had on crop yields from the 1950s until today.[*] It appeared that the 44-year period of our study could be split into three 'phases' with respect to fertiliser consumption (in terms of the total value of all varieties sold). The first phase, from 1958–1973, saw a slow, steady increase in the amount of fertiliser bought, the second, from 1974–1988, demonstrated a continuous, sharp rise in sales, while the third phase, 1989–2001, was less clearly defined, but illustrated an overall declining trend. In light of this information, separate regressions were conducted for each of these three time periods. If yields were found to have responded differently during the three phases, this might be detected when we compared each sub-set of the analysis.

Crop data

Annual yield data were supplied by Statistics Norway and collected at county level for each of the four crops in this study

---

[*] Budsjettnemnda for jordbruket, NILF (Norwegian Agricultural Economics Research Institute), 2002.

[21].*,† In Norway, there are 19 counties. However, since yield data for Akershus/Oslo are reported together, there are 18 geographical units in this study. Annual yield was calculated by dividing the total production of each crop per county by the agricultural area employed in the cultivation of that crop (in that county), and was measured in kilograms per decare.

A complete set of crop data for the years 1958–2001 for each county was not available, most notably in northern and western regions. In such cases, one of three approaches was taken: where a single value was missing from a time series, it was interpolated by calculating the average of the recordings directly preceding and following it; where more than one consecutive figure for a crop was unavailable, the missing years were removed from our analysis and the data series was broken up into two shorter time periods; and finally, where there were more than two consecutive breaks in the data, the entire crop for that county was omitted from the analysis.

Weather data   The analysis required data on two climate variables important for crop growth, namely, temperature and precipitation, at county level in Norway. The data were obtained from the Norwegian Meteorological Institute as retrospectively as records permitted, allowing our period of study to extend from 2001 as far back as 1958.

The chosen parameter for temperature was GDD, which is the annual sum of degrees accumulated above 5°C threshold. It was calculated by aggregating the number of degrees that the

---

* Approximately 70% of wheat grown in Norway is sown in the spring and the remainder is planted in the autumn. Annual and regional variations are largely determined by weather conditions, though a general rule, winter wheat production is confined to the counties of South-Eastern Norway (Østfold, Vestfold and Akershus), where the climate is milder and thus more suitable for crops with a high sensitivity to low temperatures.

† In the period 1957–1983, the area data were based on annual sample surveys, except in 1959, 1969 and 1979, when full censuses were carried out. Since 1984, administrative sources have been used, that is, applications for governmental production subsidies, except 1989, when a full census was carried out. In terms of production and yield, up until the mid-1970s, the best judgement by officials in agricultural administration at the municipality level has been used. From the mid-1970s until 1989, the source has been annual sample surveys. Since 1990, cereals production has been based on an administrative source, that is, deliveries reported to the Norwegian Grain/Norwegian Agricultural Authority. Potato production is still based on annual sample surveys.

daily mean temperature fell above 5°C [23].* This is a useful temperature parameter as it gives an indication of the quality of the growing season over a defined period ([23], p. 17).†,‡ Given that the Norwegian climate restricts the growing season for most crops from April to September, it was decided to exclude recorded GDD from months outside this period.§

Annual precipitation, measured in millimeters, is the second weather variable. Precipitation accumulated outside the growing season was included for two reasons. First, it is likely that a significant part of the precipitation falling outside this period would be retained as moisture in the soil, and thereby eventually affecting crop growth when the growing season begins. Second, as a large proportion of precipitation commonly falls in the form of snow during the Norwegian winter, when the onset of spring causes it to melt, a large share of it is likely to serve as a water supply, potentially feeding both soil and crops, before and during the growing season. As temperature increases some of the effect of increased precipitation will disappear due to increased evaporation [15].

As the Norwegian Meteorological Institute (DNMI) collects data from weather stations that are located on the basis of meteorological interest rather than along county boundary lines, it was necessary to make some decisions regarding which stations to use and how to aggregate station data to the county level. This process was made more precise with the use of GIS mapping. A digital land use map of Norway, identifying areas of agricultural activity, was obtained from the Norwegian Institute of Land and Forest Mapping (NIJOS), and geographical coordinates of weather station locations were provided by

---

* To give a simple example, if a month contained just 2 days where the average temperature rose above 5°C, and the average temperature was 7°C on the first day and 9°C on the second, then GDD for that month would be 6°C (i.e. 2°C + 4°C).

† See http://www.smhi.se/hfa_coord/nordklim/report06_2001.pdf.

‡ An alternative temperature parameter is effective growing degree days (EGDD), employed by Bootsma et al. [1]. The authors justify their use of GDD, explaining that GDD 'are designed to represent the growth period for perennial forage crops, while EGDD are specifically designed to be more applicable to the growth period for spring-seeded small grains cereals'. EGDD is defined as the sum of GDD from 10 days after the start of the growing season until the day preceding the average date of the first frost. They find a negative correlation between yield and EGDD, and suggest that this might be due to a higher development rate of crops under warmer temperatures.

§ In Norway, the length of the growing season is defined as the annual sum of days in which the mean temperature exceeds 5°C. The growing season can also be understood as the actual time period (e.g. April–September).

DNMI [3]. With the use of GIS software, these two maps were overlaid, allowing stations in closest proximity to the main area(s) of agricultural activity in each county to be identified and selected. This choice was heavily constrained by the availability of continuous time series data over our period of study (due to some stations being built after 1958, some being taken out of service for some years, and others being closed down), and by the fact that not all weather stations had the facilities to collect both precipitation and temperature data. In some cases, output from more than one station was averaged to produce the data set for a county, for example, where it spanned a broad geographic area and no single weather station was thought to be solely representative. In other cases, data from neighbouring counties were also incorporated, based on the assumption that they contributed relevant information about the weather conditions, which stations situated in the county may not have captured due to their location. Where data were simply unavailable and there were no suitably placed stations in neighbouring counties to provide proxy data, the time period in question was omitted from our analysis for that county.[*] Finally, on three occasions, individual observations were interpolated.[†] In these instances, only 1 month's data were missing from an otherwise complete series.

Analysis at the national level

In order to conduct regression analyses at the national level, it was necessary to produce aggregate weather and crop figures based on the county data used in previous analyses. Production of each crop per county was calculated as a proportion of total national output (for that crop), and then weather data were weighted accordingly. This gave weather data in counties producing a larger share of the national yield (such as in South-Eastern Norway) a higher weight than in those counties where production of that crop was lower. Where data were omitted from analysis at the county level, it was, by necessity, also excluded at the national level.

The RegClim scenario

Projected future values for GDD and annual precipitation were obtained from the RegClim Project—a regional climate scenario for Northern Europe until 2050 [17]. Regional Climate Development Under Global Warming Project (RegClim) uses an 'Atmospheric Regional Climate Model to estimate the regional climate in Northern Europe and adjacent sea areas,

---

[*] That is, Telemark 1990–2001 and Hedmark 1999–2001.
[†] That is, Telemark: precipitation, August 1989; Hedmark: GDD, August 1987 and May 1989.

given the best estimates of climate scenarios from a coupled Atmospheric-Oceanic GCM' (RegClim website, 2002).[*] RegClim predictions consist of a single, average figure for each weather variable for the 20-year period from 2030 to 2050. The RegClim scenario only presents one climate change outcome for Northern Europe, whereas other outcomes can be just as likely given a large number of uncertainties involved in such climate scenario estimates.

Predicting future yields

The crop and county cases where the model was able to explain a sizeable proportion of the annual yield variation through changes in annual precipitation and/or GDD during the growing season, and yielding significant coefficients, were selected for the RegClim projections (see Table 19.1). RegClim data, which forecasts the average percentage change in climate variables between two time periods, 1980–2000 and 2030–2050, were then used as the basis for future predictions. We take 2040 as a representative mid-year for the RegClim period.

Before any calculations could take place, however, it was necessary to adjust both model and RegClim weather data to improve their compatibility. As RegClim figures were only available for individual 50 km^2 grid cells throughout Norway, data were first of all aggregated up to county level. Furthermore, to bring figures in line with model data, predicted weather values were calculated to correspond to regions of agricultural activity, rather than to the county as a whole. Then, using RegClim data, average figures for the relative, forecast percentage change in GDD and annual precipitation between 1980–2000 and 2030–2050 were calculated for almost every county (with the exception of Vestfold). The next step was to find model estimates of the yield for all relevant crops and counties based on average GDD and precipitation for the period 1980–2000. In some cases, our interest extended to all four crops in a particular county, while in others, it was restricted to just one or two. Similarly, in some counties, the model referred to the entire time period of the study; in others it was limited to one or two sub-periods. Next, the average GDD and precipitation for each county was multiplied by the percentage change given by the RegClim scenario. Finally, RegClim GDD and precipitation values were entered into the model to give yield predictions for the selected crops and counties. The effects of changes in GDD and precipitation were calculated separately to measure the independent impact of each variable

---

[*] For further details of the RegClim Project, visit http://regclim.met.no.

Table 19.1 Summary of regression results at county level

County/crop	Observation	R²	Constant		Growing degree days		Precipitation		Time trend	
			Coefficient	t-stat	Coefficient	t-stat	Coefficient	t-stat	Coefficient	t-stat
Østfold										
Barley	43	0.46	**506.602**	4.61	-0.122	-1.71	**-0.098**	-1.83	**3.704**	5.57
Akershus/Oslo										
Potato	44	0.43	712.304	1.05	0.330	0.74	**0.952**	2.53	**18.302**	4.72
Hedmark										
Barley P3	10	0.50	12.602	0.05	0.137	1.06	**0.478**	2.32	2.485	-0.72
Oats P3	10	0.46	-203.027	-0.53	0.288	1.59	**0.632**	2.18	-4.756	-0.97
Potato P3	10	0.41	-184.553	-0.10	**1.774**	1.97	2.121	1.47	-18.381	-0.76
Oppland										
Potato	42	0.24	839.194	1.23	0.530	1.09	**1.237**	2.49	**8.098**	1.95
Buskerud										
Potato P2	15	0.66	1790.157	1.61	-1.046	-1.57	**1.879**	2.53	14.205	0.62
Potato P3	13	0.55	**3452.009**	2.49	0.346	0.38	**-1.775**	-2.73	-1.969	-0.08
Telemark										
Wheat P1	16	0.74	-174.219	-1.46	**0.264**	3.35	0.021	0.28	**7.376**	4.13
Barley P1	16	0.69	-111.873	-0.81	**0.229**	2.52	0.037	0.43	**8.293**	4.01
Oats P1	16	0.87	-110.622	-1.22	**0.186**	3.13	0.081	1.45	**10.548**	7.78
Potato P1	16	0.40	-105.132	-0.12	**1.120**	1.88	0.780	1.40	**27.838**	2.06
Aust-Agder										
Potato P1	16	0.45	-154.192	-0.16	0.936	1.44	**0.647**	1.99	**41.801**	2.74
Potato P3	13	0.35	-967.087	-0.49	**2.394**	1.84	-0.560	-1.31	6.071	0.20

continued

Table 19.1 (continued) Summary of regression results at county level

County/crop	Observation	R^2	Constant Coefficient	t-stat	Growing degree days Coefficient	t-stat	Precipitation Coefficient	t-stat	Time trend Coefficient	t-stat
Vest-Agder										
Barley P1	16	0.49	48.918	0.45	**0.156**	1.98	−0.021	−0.55	**3.491**	2.35
Potato P1	16	0.42	−694.517	−0.62	**1.976**	2.45	0.119	0.30	25.334	1.67
Barley P2	15	0.30	291.789	0.90	0.140	0.63	**−0.318**	−1.99	8.585	1.49
Rogaland										
Wheat1[a]	14	0.82	**348.063**	3.67	0.073	1.10	**−0.137**	−4.75	**6.102**	3.97
Wheat2[b]	21	0.34	**490.947**	2.47	0.102	0.70	**−0.246**	−2.99	1.662	0.64
Oats	44	0.34	**384.630**	3.27	0.134	1.52	**−0.193**	−4.06	**1.580**	2.19
Potato	44	0.29	**1880.336**	2.85	**1.233**	2.51	**−0.804**	−3.01	5.526	1.36
Barley P1	16	0.68	**323.103**	2.07	0.135	1.24	**−0.171**	−3.67	**4.200**	2.07
Barley P2	15	0.70	154.652	0.82	**0.258**	2.29	**−0.285**	−4.10	**10.212**	3.28
Barley P3	13	0.62	**524.713**	3.31	0.132	0.81	**−0.265**	−3.28	0.222	0.07
Hordaland										
Potato	44	0.22	**2378.405**	3.06	0.378	0.58	**−0.343**	−2.08	−7.526	−1.46
Barley P1	16	0.68	156.510	1.25	**0.174**	1.91	−0.072	−3.06	**6.706**	3.82
Oats P1	16	0.57	175.261	1.02	0.139	1.10	−0.069	−2.13	**9.038**	3.72
Barley P2	15	0.53	71.068	0.20	0.192	0.73	**−0.163**	−2.52	**15.328**	2.98
Sogn & Fjordane										
Potato	44	0.23	**1526.346**	2.18	**1.048**	1.84	**−0.236**	−1.81	−7.238	−1.56
Barley P1	16	0.66	−24.730	−0.18	**0.243**	2.49	−0.014	−0.59	**6.709**	3.63

Møre & Romsdal									
Potato	43	0.29	1126.926	1.66	**1.612**	2.66	**−0.563**	−2.47	7.984	1.73
Sør-Trøndelag									
Barley	44	0.41	**163.147**	2.04	**0.144**	2.19	**−0.081**	−2.20	**2.048**	4.13
Oats	44	0.29	**182.735**	1.88	**0.157**	1.96	**−0.099**	−2.19	**1.592**	2.63
Potato	44	0.45	**1394.896**	2.46	**1.605**	3.44	**−0.783**	−2.98	**−10.047**	−2.84
Nord-Trøndelag									
Barley	44	0.38	**173.116**	2.59	**0.125**	2.34	**−0.073**	−2.21	**1.384**	3.20
Potato	44	0.42	**1579.955**	3.35	1.269	3.38	**−0.732**	−3.13	5.212	1.71
Oats P1	16	0.40	**390.134**	2.20	−0.011	−0.08	**−0.200**	−2.44	5.150	1.77
Wheat P3	13	0.47	−107.354	−0.47	0.033	0.22	**0.292**	2.69	4.320	1.21
Nordland									
Barley	37	0.55	93.765	1.10	**0.239**	3.55	**−0.083**	−3.23	0.245	0.31
Oats	26	0.49	101.807	0.77	**0.233**	2.30	**−0.089**	−2.15	1.607	1.09
Potato	44	0.64	578.412	1.45	**2.051**	5.98	**−0.442**	−3.45	**−9.656**	−3.18
Troms									
Potato	44	0.51	157.064	0.35	**2.290**	5.36	0.054	0.21	**−14.297**	−3.67
Finnmark									
Potato	44	0.64	253.329	0.61	**2.678**	7.24	−0.982	−1.37	**−14.616**	−4.20
Potato P1	16	0.66	−560.858	−0.53	**3.005**	4.61	0.041	0.02	3.459	0.17
Potato P2	15	0.74	**1884.200**	2.17	**2.271**	3.45	**−2.474**	−2.07	**−46.213**	−2.89
Potato P3	13	0.44	558.698	0.93	**1.516**	2.08	0.480	0.65	−18.200	−1.45

Note: Data in bold: t-stat >= 1.8.
P1: 1958–1973, P2: 1974–1988, P3: 1989–2001.
a Wheat 1: 1958–1971.
b Wheat 2: 1974–1994.

on agricultural production, and were expressed as a percentage change in estimated average yield in the period 1980–2000.

19.4　Discussion of results

General findings

The regression results show that there is a positive effect of increased GDD (temperature) on yield per decare only for some crops, counties and time periods; confer Table 19.1 (see Annex 19.1 for a detailed account of results). Overall, about 18% of the 236 cases have a significant and positive GDD coefficient. For 3% of the cases, the GDD coefficient is negative and significant. In the case of crops, there are most significant results for potatoes. In terms of regions, the most significant results are found for Northern, Mid-, Western and Southern Norway. Sunlight and high temperatures are more likely to be a limiting factor in northern and western counties than in the south and east. Coefficients for potatoes are between 1.0 and 3.0, with the highest values evident in Northern Norway. This means that an increase of one GDD unit induces a yield increase of 1–3 kg per decare.[*] In addition, there are positive coefficients for barley in seven counties situated in Western and Mid-Norway, and in Nordland. The coefficients are between 0.13 and 0.27. There are also a few significant coefficients for oats ranging from 0.16 to 0.31. These results are consistent with the findings of Leemans and Soloman [8] since high-latitude regions are the primary beneficiaries of a warmer climate. They also reinforce the hypothesis that temperature is a more important limiting factor for crop growth in Northern and Western Norway than in other regions of the country such as Southern and Eastern Norway, where the weather conditions provide higher temperatures during the growth season.

The effect of increased annual precipitation on yield is negative and significant for many counties and crops, in particular, for Western and Mid-Norway, and for Nordland (20% of all cases). On the other hand, 5% of the cases give a positive and significant precipitation coefficient. Another study that finds a negative impact from increased precipitation on agricultural production is Rosenzweig et al. [19], where a dynamic crop model is modified to simulate effects of heavy precipitation and excess soil moisture on corn production in the US Corn Belt. The few positive coefficients are found in Eastern Norway. The coefficients range from −2.5 to 1.9 for potatoes,

---

[*] GDD increases by one unit if the average temperature on a particular day in the growing season increases by 1°C from a minimum base of 5°C.

whereas the coefficients for the cereals range between −0.34 and 0.63 (see Table 19.1 and Annex 19.1 for details). There are two possible explanations for the interesting finding that coefficients have, in some instances, been negative. The first is that precipitation may become so abundant that it leads to excess soil moisture. The second could be a result of the positive correlation between increased precipitation and cloud cover. Thus, increased precipitation means reduced radiation from the sun, leading to reduced photosynthesis, and thereby reduced yield. Both explanations go some way towards explaining the negative correlations between precipitation and yield evident in Western, Mid, and parts of Northern Norway.

The time trend is positive in most significant cases (overall 37% of instances), with the exception of potatoes in Northern Norway (and Sør-Trøndelag), where it is negative (which is equivalent to 4% of the cases). The positive trend can be attributed to long-term productivity gains in agriculture, which can include structural changes (fewer and larger farms), better crop varieties, improved farming techniques and equipment, and more efficient fertiliser use. On the other hand, the negative time trend may reflect structural changes in agriculture that affect productivity negatively; these could be related to government policies.

The national level analyses only provided significant results for potatoes and barley in the model variant allowing for different constant terms (but with the same marginal effect of GDD and precipitation, see Table 19.2). For potatoes the sign of coefficients is the same as in county-level analyses, though the size of coefficients is smaller. Instead, the model provides for different constant yields across counties (i.e. the yield component that is not influenced by GDD, precipitation or time), where the highest significant yield is found in Rogaland (1871 kg), and the lowest in Finnmark (904 kg). For barley, the GDD effect is not significant. Instead the significant constant terms vary between 378 kg in Sogn & Fjordane, and 229 kg in Nordland.* The precipitation coefficient is close to zero, but negative and significant.

Predictions   Using the model to give predictions for the RegClim climate change scenario in 2040, we find that the positive contribution from increased GDD in most of the significant cases (shown in Table 19.1) dominates the negative contribution from increased precipitation. The predictions for potatoes are shown in Table 19.3 (details for all crops are found in Annex 19.2). Only robust

---

* There is no barley yield in Finnmark and there are too few observations in Troms to include in the analysis.

Table 19.2   Regression results at national level

	Potato		Barley	
	Coefficient	t-stat	Coefficient	t-stat
National—GDD	0.864	6.89	0.002	0.11
National—Precipitation	−0.316	−5.19	−0.062	−6.53
National—Time trend	0.304	0.28	2.673	15.13
Constant term				
Akershus/Oslo	1450.814	8.10	308.209	10.64
Østfold	1408.290	−0.55	334.091	2.24
Hedmark	1569.159	1.45	315.359	0.59
Oppland	1634.395	2.24	298.217	−0.82
Buskerud	1330.441	−1.55	301.974	−0.54
Vestfold	1621.466	2.19	347.441	3.38
Telemark	1179.064	−3.23	295.154	−1.04
Aust-Agder	1242.762	−2.53	301.376	−0.56
Vest-Agder	1300.017	−1.86	288.074	−1.67
Rogaland	1870.702	5.24	366.499	4.88
Hordaland	1626.031	1.68	364.073	3.40
Sogn & Fjordane	1717.985	2.65	377.930	4.38
Møre & Romsdal	1702.704	2.80	273.754	−2.53
Sør-Trøndelag	1491.112	0.48	276.644	−2.51
Nord-Trøndelag	1764.810	3.88	268.485	−3.28
Nordland	1238.523	−2.24	229.810	−5.27
Troms	1203.093	−2.33	N/A	N/A
Finnmark	904.066	−4.89	N/A	N/A

Note: Potato: 733 observations ($R^2 = 0.50$). Barley: 660 observations ($R^2 = 0.48$).

predictions are presented, which we calculated to ±20% (at 95% interval levels). In these cases, the predicted yield is higher than in the reference situation, which is based on the model's estimated yield for average GDD and average annual precipitation in the period 1980–2000. However, in many cases the yield increase is small, and in some cases yield is reduced. The largest effect is found in Northern Norway, where the predicted yield increase for potatoes is between 30% and 35%. Other cases where the yield increase is more than 20% is potatoes in Aust-Agder (1989–2001), potatoes in Vest-Agder (1958–1973) and barley in Sogn & Fjordane (1958–1973). In the remaining cases, the change is less than 20% and not considered robust. The relative large prediction intervals reflect that the model can only explain part of the year-to-year variation in yield per

Table 19.3 Yield predictions for potatoes in the RegClim scenario

County	Period	Estimated yield from model	Predicted % change in yield under RegClim scenario		Predicted% change in yield
			GDD (%)	Precipitation (%)	Net effect (%)
Aust-Agder	P3	2830	26		26
Vest-Agder	P1	2375	24		24
Nordland	All	2165	32	−2	30
Troms	All	1987	33		33
Finnmark	All	2285	35		35

decare. If adaptation is taken into account, however, it may well be the case that this figure turns out to be an underestimate, as farmers may choose to dedicate more resources to potato cultivation as climate change improves productivity.

19.5 Further analysis

The estimated (significant) GDD and precipitation coefficients could be used as inputs to estimation of climate change damage functions for the agricultural sector in a cost–benefit economic modelling framework. In terms of expanding the model, important crop yield variables such as sunlight (e.g. using cloud cover as a proxy), fertiliser use and soil quality could be included. Owing to limited data availability, such factors could not be incorporated in this study. Where such data did exist, it was either restricted geographically (e.g. only collected at local sites or at national level) or temporally (only available for limited time periods). Furthermore, the chosen statistical model limited the type of data that could be incorporated. An alternative could be to use a crop model, where a more extensive set of relevant plant growth variables could be introduced. However, this approach, together with limited data availability, would limit the representativeness of the results, and lead to difficulties when trying to aggregate findings to the county level. On the other hand, one could choose an economic model that represents larger regions, but that would limit the model's ability to account for weather variables that are decisive for yield per decare, see, for example, Ref. [5]. The model approach employed in the study could be transferred to other weather-dependent production activities in the primary sectors, for example, other crops, and in forestry. And the same modelling could be used for similar studies in other Scandinavian countries.

19.6 Conclusions

This study shows that climate change is likely to affect agriculture in Norway. The effect on yield per decare varied with geography and crop. There was a positive yield response to temperature increases in most parts of Norway, with the exception of Eastern Norway. Furthermore, there were indications of a North–South gradient, in the sense that the climate change effects grew stronger as we moved from south to north. This finding suggests that growing season temperature was more important as a growth-limiting factor in colder regions (i.e. Northern and Western Norway) than in warmer regions. In terms of crops, the strongest effect was evident for potatoes. Barley yields, and in particular oats and wheat yields, were less responsive to changes in temperature. There was a negative yield response to increased precipitation in many parts of Norway, particularly in the west, and in Trøndelag and Nordland. This negative effect could be caused by excess soil moisture, which can be harmful to plant growth, or be related to reduced incoming sunlight due to the link between increased precipitation and cloud cover. Western Norway has the highest precipitation rate in the country. Therefore, additional precipitation may do crops more harm than good. This negative effect is most pronounced for barley, sometimes apparent for potatoes, but occurs more rarely for oats and wheat. On the other hand, there have been instances where increased precipitation has had a positive effect on productivity, though this has been restricted to potato crops. Indeed, building on the RegClim scenario for 2040, there were robust predictions for increased potato yields in Northern Norway by around 30%, and for some sub-periods in Aust-Agder and Vest-Agder by around 25%. Through adaptation, the negative effects of climate change could be reduced and the positive effects enhanced. Examples of potential adaptive measures include the introduction of new crops and crop variants, earlier sowing, ditching to drain more water from the soil and the utilisation of land that has previously been considered too marginal for agricultural cultivation.

Acknowledgements

We gratefully acknowledge financial support from the Research Council of Norway. We also thank NIJOS (Norwegian Institute of Land Inventory) for providing us with a digital map of Norwegian agricultural resources.

Appendix 19.A: Detailed regression output

County/crop	Observations	R^2	Constant		GS-GDD		Ann-pre		Time trend	
			Coefficient	t-stat	Coefficient	t-stat	Coefficient	t-stat	Coefficient	t-stat
Østfold										
Wheat	43	0.65	**420.535**	3.63	−0.067	−0.90	−0.102	−1.79	**5.958**	8.49
Barley	43	0.46	**506.602**	4.61	−0.122	−1.71	**−0.098**	−1.83	**3.704**	5.57
Oats	43	0.46	**475.532**	3.51	−0.129	−1.48	−0.064	−0.96	**4.626**	5.64
Potato	43	0.06	**2819.000**	3.74	−0.281	−0.58	−0.341	−0.92	6.412	1.40
Wheat P1	16	0.79	110.467	0.68	0.086	0.88	−0.041	−0.56	**12.993**	6.04
Barley P1	16	0.47	346.769	1.76	−0.029	−0.24	−0.090	−1.02	**6.384**	2.45
Oats P1	16	0.64	341.646	1.69	−0.050	−0.41	−0.089	−0.99	**10.023**	3.75
Potato P1	16	0.28	2442.532	1.28	0.003	0.00	−0.603	−0.71	40.414	1.61
Wheat P2	15	0.04	367.937	1.68	−0.010	−0.09	0.011	0.08	1.753	0.37
Barley P2	15	0.15	**495.260**	2.52	−0.113	−1.05	0.106	0.87	−2.716	−0.64
Oats P2	15	0.08	**492.571**	1.84	−0.105	−0.72	0.070	0.42	−1.046	−0.18
Potato P2	15	0.16	**3078.539**	2.16	−0.700	−0.90	0.996	1.13	−29.686	−0.96
Wheat P3	12	0.31	**670.946**	2.18	−0.331	−1.67	−0.084	−0.70	8.061	1.20
Barley P3	12	0.22	**769.197**	2.62	−0.230	−1.22	−0.070	−0.61	−0.498	−0.08
Oats P3	12	0.18	526.531	1.45	−0.263	−1.13	−0.002	−0.02	6.153	0.78
Potato P3	12	0.12	1606.259	1.17	0.238	0.27	−0.498	−0.93	22.946	0.77
Akershus/Oslo										
Wheat	44	0.57	**438.737**	3.22	−0.127	−1.41	−0.065	−0.86	**5.655**	7.28
Barley	44	0.47	**459.357**	4.03	−0.128	−1.71	−0.072	−1.14	**3.775**	5.82
Oats	44	0.42	**411.233**	2.99	−0.132	−1.46	−0.016	−0.21	**4.250**	5.43

continued

County/crop	Observations	R^2	Constant Coefficient	t-stat	GS-GDD Coefficient	t-stat	Ann-pre Coefficient	t-stat	Time trend Coefficient	t-stat
Potato	44	0.43	712.304	1.05	0.330	0.74	**0.952**	2.53	**18.302**	4.72
Wheat P1	16	0.69	208.541	1.25	0.075	0.81	-0.127	-1.27	**8.754**	3.46
Barley P1	16	0.41	283.689	1.39	0.003	0.03	-0.088	-0.72	**6.293**	2.03
Oats P1	16	0.53	266.847	1.38	0.006	0.06	-0.088	-0.76	**7.811**	2.67
Potato P1	16	0.24	801.484	0.62	0.558	0.78	0.423	0.54	34.888	1.77
Wheat P2	15	0.21	364.396	1.67	-0.078	-0.56	-0.025	-0.17	5.687	1.32
Barley P2	15	0.44	**579.153**	3.40	**-0.252**	-2.30	0.205	1.76	-3.767	-1.12
Oats P2	15	0.29	**599.107**	2.44	-0.275	-1.73	0.182	1.08	-2.260	-0.47
Potato P2	15	0.34	987.290	0.72	-0.158	-0.18	**1.844**	1.96	-0.881	-0.03
Wheat P3	13	0.30	**786.092**	2.10	-0.427	-1.74	0.086	0.45	3.322	0.46
Barley P3	13	0.07	**565.182**	2.04	-0.135	-0.75	-0.053	-0.38	0.586	0.11
Oats P3	13	0.13	308.256	0.85	-0.117	-0.50	0.086	0.47	3.953	0.56
Potato P3	13	0.50	79.881	0.06	0.227	0.27	0.778	1.19	43.284	1.74
Hedmark										
Wheat	41	0.74	**307.340**	2.51	-0.062	-0.79	0.023	0.23	**6.836**	9.89
Barley	41	0.63	**262.052**	2.65	-0.013	-0.21	0.001	0.01	**4.280**	7.69
Oats	41	0.53	**244.141**	2.22	0.007	0.09	-0.009	-0.10	**3.861**	6.22
Potato	41	0.37	544.453	0.74	0.775	1.65	1.068	1.77	**14.016**	3.40
Wheat P1	16	0.58	136.315	0.93	0.024	0.28	0.107	0.85	**8.344**	3.88
Barley P1	16	0.34	142.106	0.70	0.051	0.45	0.042	0.24	**6.655**	2.22
Oats P1	16	0.42	100.608	0.52	0.072	0.65	0.055	0.33	**7.471**	2.62
Potato P1	16	0.24	-365.797	-0.25	0.952	1.16	1.966	1.58	32.463	1.53

Wheat P2	15	0.35	**354.476**	1.81	−0.103	−0.76	0.152	0.83	4.691	1.31
Barley P2	15	0.11	**366.487**	2.96	−0.054	−0.63	0.087	0.76	−0.181	−0.08
Oats P2	15	0.09	**376.344**	2.50	−0.072	−0.69	0.059	0.42	0.619	0.22
Potato P2	15	0.13	1264.622	0.98	0.365	0.40	1.451	1.20	−5.222	−0.22
Wheat P3	10	0.17	230.246	0.46	0.078	0.33	0.373	0.98	−2.620	−0.41
Barley P3	10	0.50	12.602	0.05	0.137	1.06	**0.478**	2.32	−2.485	−0.72
Oats P3	10	0.46	−203.027	−0.53	0.288	1.59	**0.632**	2.18	−4.756	−0.97
Potato P3	10	0.41	−184.553	−0.10	**1.774**	1.97	2.121	1.47	−18.381	−0.76
Oppland										
Wheat	42	0.59	**384.311**	3.16	−0.074	−0.86	−0.037	−0.41	**5.443**	7.35
Barley	42	0.39	**266.007**	2.73	−0.008	−0.11	−0.002	−0.04	**2.875**	4.84
Oats	42	0.30	**306.820**	2.27	−0.032	−0.34	−0.035	−0.35	**3.298**	4.01
Potato	42	0.24	839.194	1.23	0.530	1.09	**1.237**	2.49	**8.098**	1.95
Wheat P1	16	0.68	175.427	1.29	0.004	0.04	0.079	0.67	**10.335**	4.72
Barley P1	16	0.35	110.083	0.68	0.069	0.72	0.067	0.47	**6.043**	2.31
Oats P1	16	0.45	211.122	1.23	0.019	0.18	−0.026	−0.18	**7.282**	2.62
Potato P1	16	0.19	266.849	0.21	0.671	0.87	1.772	1.57	18.048	0.86
Wheat P2	15	0.18	**552.990**	2.38	−0.173	−1.05	0.074	0.45	1.172	0.28
Barley P2	15	0.13	**336.610**	2.05	−0.052	−0.44	0.110	0.95	−0.876	−0.30
Oats P2	15	0.19	**463.605**	1.96	−0.156	−0.92	0.148	0.89	−2.076	−0.49
Potato P2	15	0.19	1231.143	0.78	0.263	0.23	1.684	1.52	−9.852	−0.35
Wheat P3	11	0.00	411.889	1.36	0.018	0.09	0.033	0.16	−0.162	−0.02
Barley P3	11	0.09	**512.271**	1.87	−0.012	−0.06	0.055	0.28	−4.986	−0.81
Oats P3	11	0.04	251.114	0.57	0.138	0.47	−0.043	−0.14	−0.645	−0.06
Potato P3	11	0.27	1492.394	1.35	0.946	1.29	0.897	1.15	−15.895	−0.64

County/crop	Observations	R^2	Constant		GS-GDD		Ann-pre		Time
			Coefficient	t-stat	Coefficient	t-stat	Coefficient	t-stat	Coefficient
Buskerud									
Wheat	44	0.53	**467.927**	3.19	−0.149	−1.60	−0.037	−0.46	**5.596**
Barley	44	0.35	**438.658**	3.41	−0.124	−1.52	−0.037	−0.52	**3.426**
Oats	44	0.43	**376.877**	2.59	−0.096	−1.04	−0.030	−0.37	**4.495**
Potato	44	0.13	**2550.186**	3.16	−0.349	−0.68	−0.052	−0.12	**11.219**
Wheat P1	16	0.86	171.141	1.32	0.059	0.90	−0.083	−0.92	**12.200**
Barley P1	16	0.60	**265.168**	1.91	0.037	0.53	−0.115	−1.19	**5.380**
Oats P1	16	0.78	**255.249**	2.09	0.043	0.70	−0.144	−1.69	**7.613**
Potato P1	16	0.19	1034.077	0.86	0.253	0.42	0.813	0.97	30.146
Wheat P2	15	0.30	**645.011**	2.01	−0.297	−1.55	0.185	0.86	0.451
Barley P2	15	0.49	**738.290**	3.29	**−0.359**	−2.67	**0.280**	1.86	−5.669
Oats P2	15	0.34	**747.746**	2.59	**−0.349**	−2.01	0.239	1.24	−5.098
Potato P2	15	0.66	1790.157	1.61	−1.046	−1.57	**1.879**	2.53	14.205
Wheat P3	13	0.23	539.589	1.73	−0.253	−1.25	−0.002	−0.01	6.254
Barley P3	13	0.04	455.528	1.24	−0.123	−0.51	−0.059	−0.34	3.114
Oats P3	13	0.19	136.246	0.34	−0.074	−0.28	−0.034	−0.18	9.935
Potato P3	13	0.55	**3452.009**	2.49	0.346	0.38	**−1.775**	−2.73	−1.969
Vestfold									
Wheat	44	0.54	**385.673**	2.61	−0.056	−0.58	−0.072	−1.05	**5.687**
Barley	44	0.41	**489.148**	3.98	−0.096	−1.21	−0.093	−1.64	**3.565**
Oats	44	0.42	**510.762**	3.29	−0.144	−1.43	−0.065	−0.91	**4.656**
Potato	44	0.26	**2971.869**	3.81	−0.483	−0.96	−0.152	−0.42	**16.152**
Wheat P1	16	0.76	86.570	0.49	0.089	0.92	−0.023	−0.25	**11.842**

Barley P1	16	0.58	**340.251**	1.86	0.023	0.22	−0.135	−1.43	**6.313**
Oats P1	16	0.74	**363.453**	2.11	−0.019	−0.20	−0.139	−1.55	**9.600**
Potato P1	16	0.29	2005.287	1.34	−0.086	−0.10	0.175	0.22	**40.168**
Wheat P2	15	0.08	**494.882**	2.00	−0.098	−0.68	0.074	0.46	−0.116
Barley P2	15	0.31	**609.572**	3.14	**−0.216**	−1.90	0.179	1.43	−4.146
Oats P2	15	0.17	**678.968**	2.22	−0.244	−1.37	0.125	0.63	−2.786
Potato P2	15	0.45	1705.025	1.18	−0.665	−0.79	1.228	1.31	28.847
Wheat P3	13	0.24	**771.762**	1.96	−0.330	−1.49	−0.014	−0.12	3.357
Barley P3	13	0.17	617.106	1.51	−0.216	−0.94	−0.105	−0.83	4.538
Oats P3	13	0.25	584.101	1.23	−0.324	−1.22	−0.038	−0.26	8.186
Potato P3	13	0.14	3466.811	1.74	−0.034	−0.03	−0.550	−0.90	−5.108
Telemark									
Wheat	32	0.42	219.660	1.39	−0.041	−0.38	0.037	0.46	**5.217**
Barley	32	0.28	**298.467**	2.13	−0.057	−0.60	0.017	0.24	**3.351**
Oats	32	0.19	**409.203**	2.73	−0.140	−1.37	0.014	0.18	**2.483**
Potato	32	0.17	**2395.118**	2.93	−0.685	−1.23	0.726	1.76	−3.378
Wheat P1	16	0.74	−174.219	−1.46	**0.264**	3.35	0.021	0.28	**7.376**
Barley P1	16	0.69	−111.873	−0.81	**0.229**	2.52	0.037	0.43	**8.293**
Oats P1	16	0.87	−110.622	−1.22	**0.186**	3.13	0.081	1.45	**10.548**
Potato P1	16	0.40	−105.132	−0.12	**1.120**	1.88	0.780	1.40	**27.838**
Wheat P2	15	0.33	**649.128**	3.02	**−0.282**	−2.01	0.144	1.05	−2.944
Barley P2	15	0.39	**636.209**	3.08	**−0.299**	−2.22	0.170	1.29	−3.246
Oats P2	15	0.44	**741.976**	3.66	**−0.368**	−2.78	0.078	0.60	−1.416
Potato P2	15	0.61	**4065.881**	4.04	**−2.119**	−3.22	**1.449**	2.26	−23.926

continued

County/crop	Observations	R^2	Constant		GS-GDD		Ann-pre		Time
			Coefficient	t-stat	Coefficient	t-stat	Coefficient	t-stat	Coefficient
Aust-Agder									
Wheat	22	0.05	219.660	1.39	−0.213	−0.79	0.027	0.17	−1.854
Barley	44	0.08	**298.467**	2.13	−0.041	−0.50	−0.018	−0.48	**1.144**
Oats	44	0.04	**409.203**	2.73	−0.043	−0.39	−0.011	−0.23	1.077
Potato	44	0.16	**2395.118**	2.93	0.351	0.68	−0.119	−0.52	**−10.486**
Barley P1	16	0.39	60.139	0.41	0.102	1.03	0.028	0.56	**5.668**
Oats P1	16	0.33	18.321	0.11	0.119	1.10	0.033	0.61	**5.302**
Potato P1	16	0.45	−154.192	−0.16	0.936	1.44	**0.647**	1.99	**41.801**
Wheat P2	15	0.08	511.305	0.85	−0.078	−0.22	−0.179	−0.67	7.393
Barley P2	15	0.17	404.742	1.59	−0.123	−0.83	−0.037	−0.33	4.051
Oats P2	15	0.11	**630.031**	1.90	−0.183	−0.95	−0.094	−0.63	−0.554
Potato P2	15	0.19	**3228.416**	2.67	−0.893	−1.28	0.324	0.60	−22.959
Barley P3	13	0.02	367.801	1.00	0.003	0.01	0.001	0.01	−2.116
Oats P3	13	0.03	276.814	0.50	−0.087	−0.24	0.021	0.18	2.686
Potato P3	13	0.35	−967.087	−0.49	**2.394**	1.84	−0.560	−1.31	6.071

County/crop	Observations	R²	Constant Coefficient	t-stat	GS-GDD Coefficient	t-stat	Ann-pre Coefficient	t-stat	Time trend Coefficient	t-stat
Vest-Agder										
Wheat1	12	0.67	123.686	1.11	0.117	1.48	-0.072	-1.71	**6.825**	3.24
Wheat2	19	0.12	104.151	0.24	0.038	0.11	0.059	0.28	7.835	1.21
Barley	44	0.28	**256.312**	1.98	0.053	0.51	-0.091	-1.76	**2.810**	3.58
Oats	44	0.32	**317.180**	2.16	-0.043	-0.36	-0.029	-0.49	**3.805**	4.28
Potato	44	0.06	**1839.660**	3.13	0.467	0.99	-0.317	-1.35	-1.664	-0.47
Wheat P1	12	0.67	123.686	1.11	0.117	1.48	-0.072	-1.71	**6.825**	3.24
Barley P1	16	0.49	48.918	0.45	**0.156**	1.98	-0.021	-0.55	**3.491**	2.35
Oats P1	16	0.79	86.865	0.70	0.135	1.50	-0.048	-1.09	**9.510**	5.62
Potato P1	16	0.42	-694.517	-0.62	**1.976**	2.45	0.119	0.30	25.334	1.67
Wheat P2	15	0.14	47.187	0.09	0.156	0.45	-0.082	-0.33	11.676	1.32
Barley P2	15	0.30	291.789	0.90	0.140	0.63	**-0.318**	-1.99	8.585	1.49
Oats P2	15	0.10	261.498	0.74	0.033	0.14	-0.114	-0.66	6.609	1.06
Potato P2	15	0.01	2126.762	1.62	-0.183	-0.20	-0.066	-0.10	5.326	0.23
Barley P3	13	0.16	436.274	1.32	-0.199	-0.84	-0.011	-0.13	4.146	0.87
Oats P3	13	0.27	671.182	1.62	-0.450	-1.51	0.126	1.19	3.113	0.52
Potato P3	13	0.28	**2761.331**	2.36	0.114	0.14	-0.446	-1.49	-9.730	-0.58
Rogaland										
Wheat1	14	0.82	**348.063**	3.67	0.073	1.10	**-0.137**	-4.75	**6.102**	3.97
Wheat2	21	0.34	**490.947**	2.47	0.102	0.70	**-0.246**	-2.99	1.662	0.64
Barley	44	0.58	**438.401**	5.12	0.096	1.51	**-0.221**	-6.37	**2.602**	4.95
Oats	44	0.34	**384.630**	3.27	0.134	1.52	**-0.193**	-4.06	**1.580**	2.19
Potato	44	0.29	**1880.336**	2.85	**1.233**	2.51	**-0.804**	-3.01	5.526	1.36

continued

County/crop	Observations	R^2	Constant		GS-GDD		Ann-pre		Time trend	
			Coefficient	t-stat	Coefficient	t-stat	Coefficient	t-stat	Coefficient	t-stat
Wheat P1	14	0.82	**348.063**	3.67	0.073	1.10	**-0.137**	-4.75	**6.102**	3.97
Barley P1	16	0.68	**323.103**	2.07	0.135	1.24	**-0.171**	-3.67	**4.200**	2.07
Oats P1	16	0.82	**286.609**	1.98	0.166	1.64	**-0.194**	-4.48	**8.901**	4.74
Potato P1	16	0.41	**4127.854**	2.09	-0.254	-0.18	**-1.478**	-2.50	44.234	1.73
Wheat P2	15	0.52	467.480	1.71	0.134	0.82	**-0.335**	-3.33	5.724	1.27
Barley P2	15	0.70	154.652	0.82	**0.258**	2.29	**-0.285**	-4.10	**10.212**	3.28
Oats P2	15	0.38	236.458	1.06	0.203	1.54	**-0.175**	-2.14	4.612	1.26
Potato P2	15	0.36	1755.915	1.32	1.368	1.72	-0.470	-0.96	-11.066	-0.50
Barley P3	13	0.62	**524.713**	3.31	0.132	0.81	**-0.265**	-3.28	0.222	0.07
Oats P3	13	0.34	314.682	1.00	0.160	0.50	-0.257	-1.61	4.019	0.68
Potato P3	13	0.42	**1734.081**	2.71	0.973	1.48	-0.577	-1.77	9.683	0.81
Hordaland										
Barley	35	0.41	**443.832**	2.84	-0.005	-0.04	**-0.120**	-3.96	**5.051**	4.00
Oats	34	0.15	213.439	1.04	0.121	0.75	-0.060	-1.60	**3.099**	2.08
Potato	44	0.22	**2378.405**	3.06	0.378	0.58	**-0.343**	-2.08	-7.526	-1.46
Barley P1	16	0.68	156.510	1.25	**0.174**	1.91	**-0.072**	-3.06	**6.706**	3.82
Oats P1	16	0.57	175.261	1.02	0.139	1.10	**-0.069**	-2.13	**9.038**	3.72
Potato P1	16	0.46	2512.254	1.31	0.567	0.41	**-0.874**	-2.42	**71.090**	2.63
Barley P2	15	0.53	71.068	0.20	0.192	0.73	**-0.163**	-2.52	**15.328**	2.98
Oats P2	15	0.23	533.612	1.22	0.017	0.05	-0.139	-1.73	-0.354	-0.06
Potato P2	15	0.34	1921.248	1.50	1.065	1.10	-0.381	-1.61	-10.473	-0.56
Potato P3	13	0.24	**2405.011**	2.15	0.927	0.91	-0.131	-0.53	-40.303	-1.62

Sogn & Fjordane

Barley	35	0.46	**288.050**	2.52	−0.008	−0.09	−0.028	−1.26	**4.859**	4.85
Oats	29	0.07	335.359	1.55	−0.035	−0.21	−0.033	−0.80	2.556	1.38
Potato	44	0.23	**1526.346**	2.18	**1.048**	1.84	**−0.236**	−1.81	−7.238	−1.56
Barley P1	16	0.66	−24.730	−0.18	**0.243**	2.49	−0.014	−0.59	**6.709**	3.63
Oats P1	16	0.48	148.321	0.60	0.143	0.80	−0.064	−1.44	**9.695**	2.87
Potato P1	16	0.43	2457.754	1.34	0.474	0.36	**−0.664**	−2.00	**51.424**	2.04
Barley P2	15	0.30	**404.657**	3.22	−0.113	−1.10	−0.030	−0.98	**4.937**	1.97
Oats P2	13	0.15	550.801	1.57	−0.328	−1.17	0.022	0.29	2.068	0.29
Potato P2	15	0.60	**1963.218**	2.72	**1.538**	2.60	**−0.328**	−1.88	**−41.907**	−2.91
Potato P3	13	0.10	920.793	0.59	−0.254	−0.17	0.144	0.58	28.416	0.92

Møre & Romsdal

Barley	43	0.10	**280.494**	2.72	0.024	0.26	−0.059	−1.70	1.207	1.71
Oats	34	0.14	259.926	1.40	0.040	0.24	−0.071	−1.23	**3.316**	2.14
Potato	43	0.29	1126.926	1.66	**1.612**	2.66	**−0.563**	−2.47	7.984	1.73
Barley P1	15	0.28	202.780	0.99	0.102	0.59	−0.075	−1.48	1.757	0.59
Oats P1	15	0.36	175.884	0.87	0.140	0.81	−0.086	−1.71	2.781	0.93
Potato P1	15	0.50	**3657.735**	1.95	−0.233	−0.15	**−1.384**	−2.98	47.282	1.72
Barley P2	15	0.35	169.343	0.99	0.093	0.71	−0.065	−1.18	**5.112**	1.97
Oats P2	15	0.07	34.657	0.08	0.175	0.53	−0.015	−0.11	5.138	0.78
Potato P2	15	0.36	1331.990	1.18	**1.589**	1.83	−0.565	−1.54	−1.527	−0.09
Barley P3	13	0.19	−64.950	−0.30	0.133	0.83	0.058	0.77	2.305	0.57
Potato P3	13	0.47	−730.269	−0.68	1.183	1.48	0.533	1.43	27.294	1.36

County/crop	Observations	R²	Constant		GS-GDD		Ann-pre	
			Coefficient	t-stat	Coefficient	t-stat	Coefficient	t-stat
Sør-Trøndelag								
Wheat1	13	0.1	103.921	0.38	0.149	0.81	−0.019	−0.13
Wheat2	28	0.7	**303.361**	1.89	0.216	1.56	0.006	0.07
Barley	44	0.1	**163.147**	2.04	**0.144**	2.19	**−0.081**	−2.20
Oats	44	0.9	**182.735**	1.88	**0.157**	1.96	**−0.099**	−2.19
Potato	44	0.5	**1394.896**	2.46	**1.605**	3.44	**−0.783**	−2.98
Wheat P1	13	0.1	103.921	0.38	0.149	0.81	−0.019	−0.13
Barley P1	16	0.6	319.918	1.59	0.054	0.35	**−0.168**	−2.03
Oats P1	16	0.5	**444.136**	1.85	0.010	0.05	**−0.258**	−2.61
Potato P1	16	0.9	1761.191	1.22	1.685	1.52	**−1.468**	−2.46
Wheat P2	15	0.4	105.716	0.52	0.231	1.32	−0.018	−0.18
Barley P2	15	0.4	**225.539**	2.46	0.089	1.14	**−0.081**	−1.87
Oats P2	15	0.5	**273.126**	2.19	0.127	1.19	−0.082	−1.38
Potato P2	15	0.8	**1761.212**	2.83	**1.873**	3.51	**−0.656**	−2.22
Wheat P3	13	0.4	356.385	1.31	−0.086	−0.43	0.159	1.11
Barley P3	13	0.8	−116.090	−0.59	**0.269**	1.85	−0.006	−0.06
Oats P3	13	0.4	−260.093	−1.24	0.229	1.48	0.072	0.65
Potato P3	13	0.5	−516.451	−0.59	−0.052	−0.08	**1.030**	2.25
Nord-Trøndelag								
Wheat	44	0.2	199.965	1.77	0.049	0.55	0.023	0.42
Barley	44	0.3	**173.116**	2.59	**0.125**	2.34	**−0.073**	−2.21
Oats	44	0.2	**181.596**	2.02	0.124	1.74	−0.076	−1.69
Potato	44	0.4	**1579.955**	3.35	1.269	3.38	**−0.732**	−3.13
Wheat P1	16	0.0	**452.693**	1.92	−0.103	−0.58	−0.107	−0.98

Barley P1	16	0.27	263.451	1.52	0.057	0.44	−0.119	−1.48
Oats P1	16	0.40	**390.134**	2.20	−0.011	−0.08	**−0.200**	−2.44
Potato P1	16	0.56	1314.999	1.12	1.661	1.89	**−1.060**	−1.94
Wheat P2	15	0.69	**202.729**	2.52	0.047	0.70	−0.067	−1.56
Barley P2	15	0.53	**204.710**	2.67	0.112	1.73	**−0.098**	−2.40
Oats P2	15	0.32	239.383	1.74	0.087	0.75	−0.125	−1.70
Potato P2	15	0.36	**1837.539**	2.13	1.309	1.79	−0.686	−1.49
Wheat P3	13	0.47	−107.354	−0.47	0.033	0.22	**0.292**	2.69
Barley P3	13	0.32	157.078	0.94	0.209	1.93	−0.087	−1.09
Oats P3	13	0.29	89.933	0.47	0.220	1.77	0.004	0.04
Potato P3	13	0.13	**2268.755**	3.74	0.420	1.06	−0.173	−0.60
Nordland								
Barley	37	0.55	93.765	1.10	**0.239**	3.55	**−0.083**	−3.23
Oats	26	0.49	101.807	0.77	**0.233**	2.30	**−0.089**	−2.15
Potato	44	0.64	578.412	1.45	**2.051**	5.98	**−0.442**	−3.45
Barley P1	16	0.59	10.771	0.11	**0.257**	3.15	−0.034	−1.11
Oats P1	16	0.57	−41.369	−0.35	**0.308**	3.11	−0.032	−0.88
Potato P1	16	0.70	39.688	0.06	**2.354**	4.01	−0.314	−1.44
Barley P2	15	0.66	194.787	1.09	**0.207**	1.88	**−0.130**	−2.68
Oats P2	10	0.61	545.561	1.62	0.045	0.19	**−0.222**	−1.91
Potato P2	15	0.66	**1990.438**	1.87	**1.235**	1.89	**−0.817**	−2.83
Potato P3	13	0.64	−183.956	−0.22	**2.453**	3.23	−0.618	−1.64
Troms								
Potato	44	0.51	157.064	0.35	**2.290**	5.36	0.054	0.21
Potato P1	16	0.59	−661.821	−0.80	**3.014**	3.98	0.544	1.10

continued

County/crop	Observations	R^2	Constant Coefficient	t-stat	GS-GDD Coefficient	t-stat	Ann-pre Coefficient	t-stat
Potato P2	15	0.63	528.527	0.68	**1.694**	2.75	-0.652	-1.75
Potato P3	13	0.38	-1681.857	-1.24	**2.194**	1.88	0.875	1.29
Finnmark								
Potato	44	0.64	253.329	0.61	**2.678**	7.24	-0.982	-1.37
Potato P1	16	0.66	-560.858	-0.53	**3.005**	4.61	0.041	0.02
Potato P2	15	0.74	**1884.200**	2.17	**2.271**	3.45	-2.474	-2.07
Potato P3	13	0.44	558.698	0.93	**1.516**	2.08	0.480	0.65

Notes: Data in bold: t-stat >= 1.8.

P1: 1958–1973, P2: 1974–1988, P3: 1989–2001.

GS-GDD: Growing season growing degree days.

Ann-pre: Annual precipitation.

Appendix 19.B: RegClim data and predictions

County	Period	Crop	Average yield (observed)	Estimated yield (model derived)	Average annual GDD		Estimated coefficient	Predicted change in yield	Predicted % change in yield from RegClim
					1980–2000 value[a]	RegClim % change			
Østfold	All	Barley	341	592	1439	0.15			
Akershus/Oslo	All	Potato	2306	1663	1370	0.17			
Hedmark	P3	Barley	395	489	1206	0.18			
	P3	Oats	376	372	1206	0.18			
	P3	Potato	2617	2126	1206	0.18	1.8	394	19%
Oppland	All	Potato	2393	1844	1176	0.24			
Buskerud	P2	Potato	2138	3444	1472	0.20			
	P3	Potato	2504	2221	1472	0.20			
Telemark	P1	Wheat	231	356	1363	0.19	0.3	68	19%
	P1	Barley	269	371	1363	0.19	0.2	59	16%
	P1	Oats	268	314	1363	0.19	0.2	48	15%
	P1	Potato	2137	1591	1363	0.19	1.1	290	18%
Aust-Agder	P1	Potato	2197	823	1515	0.20			
	P3	Potato	1866	2830	1515	0.20	2.4	725	26%
Vest-Agder	P1	Barley	242	449	1467	0.20	0.2	46	10%
	P1	Potato	2057	2375	1467	0.20	2.0	580	24%
	P2	Barley	264	62	1467	0.20			
Rogaland	All	Oats	355	317	1437	0.18			
	All	Potato	2563	2832	1437	0.18	1.2	319	11%
	1958–1971	Wheat	323	350	1437	0.18			
	1974–1994	Wheat	365	358	1437	0.18			

continued

County	Period	Crop	Average yield (observed)	Estimated yield (model derived)	Average Annual GDD		Estimated coefficient	Predicted change in yield	Predicted % change in yield from RegClim
					1980–2000 value[a]	RegClim % change			
	P1	Barley	327	283	1437	0.18	0.3	67	19%
	P2	Barley	367	345	1437	0.18			
	P3	Barley	371	369	1437	0.18			
Hordaland	All	Potato	1956	1822	1243	0.21	0.2	45	12%
	P1	Barley	267	391	1243	0.21			
	P1	Oats	269	199	1243	0.21			
Sogn & Fjordane	All	Pot to	2100	2491	1213	0.30	1.0	381	15%
	P1	Barey	268	441	1213	0.30	0.2	89	20%
Møre & Romsdal	All	Potato	2142	2414	1174	0.24	1.6	454	19%
Sør-Trøndelag	All	Barey	278	421	1143	0.23	0.1	39	9%
	All	Oats	284	439	1143	0.23	0.2	42	10%
	All	Potato	2087	2658	1143	0.23	1.6	431	16%
Nord-Trøndelag	All	Barley	275	431	1212	0.23	0.1	35	8%
	All	Potato	2435	1115	1212	0.23			
	P1	Oats	249	387	1212	0.23			
	P3	Wheat	339	316	1212	0.23			
Nordland	All	Barley	203	393	934	0.36	0.2	80	20%
	All	Oats	206	389	934	0.36	0.2	78	20%
	All	Potato	1602	2165	934	0.36	2.1	689	32%
Troms	All	Potato	1492	1987	724	0.39	2.3	647	33%
Finnmark	All	Potato	1375	2285	695	0.43	2.7	800	35%

Note: Table based on model estimates, changes in GDD and precipitation under the RegClim scenario and yield predictions for 2040.
[a] Average temperature calculation based on available data: Hedmark: 1980–1999; Oppland: 1980–1999; Telemark: 1980–1989.

County	Period	Crop	Average annual precipitation		Estimated coefficient	Predicted change in yield	Predicted % change in yield from RegClim	Predicted % change in yield: net effect
			1980–2000 value[a]	RegClim % change				
Østfold	All	Barley	866	0.05	−0.1	−4	−1%	−1%
Akershus/Oslo	All	Potato	819	0.04	1.0	31	2%	2%
Hedmark	P3	Barley	639	0.05	0.5	15	3%	3%
	P3	Oats	639	0.05	0.6	19	5%	5%
	P3	Potato	639	0.05				19%
Oppland	All	Potato	674	0.04	1.2	33	2%	2%
Buskerud	P2	Potato	789	0.02	1.9	30	1%	1%
	P3	Potato	789	0.02	−1.8	−28	−1%	−1%
Telemark	P1	Wheat	816	0.02				19%
	P1	Barley	816	0.02				16%
	P1	Oats	816	0.02				15%
	P1	Potato	816	0.02				18%
Aust-Agder	P1	Potato	1246	0.05	0.6	40	5%	5%
	P3	Potato	1246	0.05				26%
Vest-Agder	P1	Barley	1260	0.08				10%
	P1	Potato	1260	0.08				24%
	P2	Barley	1260	0.08	−0.3	−32	−52%	−52%
Rogaland	All	Oats	1231	0.18	−0.2	−43	−14%	−14%
	All	Potato	1231	0.18	−0.8	−178	−6%	5%
	1958–1971	Wheat	1231	0.18	−0.1	−30	−9%	−9%
	1974–1994	Wheat	1231	0.18	−0.2	−55	−15%	−15%
	P1	Barley	1231	0.18	−0.2	−38	−13%	−13%

continued

County	Period	Crop	Average annual precipitation		Estimated coefficient	Predicted change in yield	Predicted % change in yield from RegClim	Predicted % change in yield: net effect
			1980–2000 value[a]	RegClim % change				
	P2	Barley	1231	0.18	−0.3	−63	−18%	1%
	P3	Barley	1231	0.18	−0.3	−59	−16%	−16%
Hordaland	All	Potato	2117	0.18	−0.3	−131	−7%	−7%
	P1	Barley	2117	0.18	−0.1	−27	−7%	5%
	P1	Oats	2117	0.18	−0.1	−26	−13%	−13%
Sogn & Fjordane	All	Potato	2023	0.14	−0.2	−67	−3%	13%
	P1	Barley	2023	0.14				20%
Møre & Romsdal	All	Potato	1380	0.11	−0.6	−85	−4%	15%
Sør-Trøndelag	All	Barley	947	0.10	−0.1	−7	−2%	7%
	All	Oats	947	0.10	−0.1	−9	−2%	8%
	All	Potato	947	0.10	−0.8	−71	−3%	14%
Nord-Trøndelag	All	Barley	868	0.08	−0.1	−5	−1%	7%
	All	Potato	868	0.08	−0.7	−51	−5%	−5%
	P1	Oats	868	0.08	−0.2	−14	−4%	−4%
	P3	Wheat	868	0.08	0.3	20	6%	6%
Nordland	All	Barley	1129	0.08	−0.1	−8	−2%	18%
	All	Oats	1129	0.08	−0.1	−8	−2%	18%
	All	Potato	1129	0.08	−0.4	−40	−2%	30%
Troms	All	Potato	1040	0.06				33%
Finnmark	All	Potato	407	0.06				35%

Note: Table based on model estimates, changes in GDD and precipitation under the RegClim scenario and yield predictions for 2040.
[a] Average precipitation calculation based on available data: Hedmark: 1980–1998; Telemark: 1980–1989.

Appendix 19.C: Model variants

Annual or growth season data

With respect to temperature, we initially considered annual GDD as an alternative to growth season GDD (defined as April to September in our study). However, given the Norwegian climate, the difference between these two measures would have been minimal, as there are few days where the temperature rises above 5°C between late autumn and early spring. Conversely, in the case of precipitation, we considered growth season data as an alternative to annual data, but an annual precipitation figure seemed more appropriate than a growing season figure, since a significant proportion of precipitation falling outside the growing season is likely to feed crops during it. This is because a large share of precipitation during winter months is likely to be released as water when the snow melts in spring and early summer, even if some water is lost to runoffs to rivers, and so on.

Inclusion of CO_2 concentration

Data was obtained at the global level (in parts per million) from the Scripps Institution of Oceanography.[*] Different data formats for CO_2 concentration were explored: atmospheric concentration in ppmv; transformation of atmospheric concentration to a normalised series starting at 0 in 1957 and ending at 56 in 2001; logarithmic transformation of atmospheric concentration; and finally, a quadratic term from a second-order polynomial was fitted to atmospheric concentration through regression. These data formats were included either alone as part of the regressions, or in addition to the linear trend. It turned out that the simple time trend behaved as well or better in the regressions than the various CO_2 formats, so we chose to only include the former in the main model. The major reason for this finding is the dominating linear part of CO_2 concentration in the atmosphere.

Frost events

Frost events can be harmful to crops, grains in particular. Wheat is especially sensitive to sub-zero conditions during its vegetative period, when germination and leaf growth take place. Cromey et al. [2] found that a late frost event reduced yields 13–33% for the affected winter wheat crops in the

---

[*] Atmospheric CO_2 concentrations (ppmv) were derived from flask and *in situ* air samples collected at the South Pole. Source: C.D. Keeling, T.P. Whorf and the Carbon Dioxide Research Group, Scripps Institution of Oceanography, University of California, La Jolla, California USA 92093-0444, July 25, 2002; http://cdiac.esd.ornl.gov/trends/co2/sio-spl.htm.

Southland region of New Zealand. With this background, we expected that a weather event, such as a late spring frost episode, would likely have negative consequences for yield. To capture an element of this vulnerability, a dummy variable was introduced in the model, with '1' indicating the occurrence of one or more 'frost events' during that year in a given county and a '0' representing the absence of one. Given the sensitivity of crops to low temperatures during the early phases of their development, a 'frost event' was said to have taken place when the minimum air temperature during one or more days in May was equal to, or fell below $-2°C$ (or $-4°C$ in a second variant of the model). May was selected as a key month as grains are commonly sown in April in Norway.[*][†] In cases where observations from several weather stations had been used to compile weather data for a particular county, the records of all relevant stations were examined for evidence of frost events. Weather stations were initially chosen due to their proximity to areas of agricultural activity in a county; therefore, a frost event occurring at any one of the stations would be likely to have some relevance for at least part of the crop area under cultivation in that county. In terms of our results, we found no evidence to suggest that frost events influence crop yields. This suggests that the model was not well suited to incorporate such a variable.

Fertiliser application to grain production

The limited fertiliser use data that was available at the county level was integrated into the model for the brief period, 1989–1996. Sample surveys provided figures for the application of commercial nitrogen and phosphorus fertilisers to grain and oil seeds in the form of average kilograms per decare for most counties.[‡] Based on the assumption that farmers used both nitrogen and phosphorus optimally, the sum of the two was calculated and included as a third independent variable in addition to the two central climate variables—GDD during the growing season and annual precipitation. The analyses showed that fertiliser use—for the limited period data was available—did not have any significant positive effects on yield.

---

[*] Note that if spring arrives late, sowing can be delayed.

[†] Thirty percent of wheat in Norway is sown in the autumn.

[‡] Resultatkontroll jordbruk, Statistics Norway, 1993, 1995 and 1997. A complete data set for the 8-year period was not available for some Northern and Western counties, that is, Telemark, Hordaland, Sogn & Fjordane, Nord-Trøndelag, Nordland, Troms and Finnmark.

| Quadratic and logarithmic variable | The use of quadratic and logarithmic forms of the independent variables (GDD and precipitation) did not appear to improve the model's capabilities for the four test counties we selected in our analysis (Akershus/Oslo, Rogaland, Sør-Trøndelag and Nordland). Results provided fewer significant coefficients than our main model. |

References

1. BOOTSMA, A. et al. Adaptation of Agricultural Production to Climate Change in Atlantic Canada. Final Report A214, Climate Change Action Fund Project, 2001.
2. CROMEY, M. G., D. S. C. WRIGHT and H. J. BODDINGTON. Effects of frost during grain filling on wheat yield and grain structure, *New Zealand Journal of Crop and Horticulture Science*, 26(4): 279–290, 1998.
3. DNMI. Map of Weather Stations (Stasjonskart). Technical Report, Norwegian Meteorological Institute, 2000.
4. FEENSTRA, J. F. et al. *Handbook on Methods for Climate Change Impact Assessment and Adaptation Strategies.* UNEP/IVM, 1998.
5. GAASLAND, I. Can Warmer Climate Save the Northern Agriculture? Report, Centre for Research in Economics and Business Administration, 2003.
6. IPCC, Intergovernmental Panel on Climate Change. Climate Change 2001: The Scientific Basis—Contribution of Working Group I to the IPCC Third Assessment Report. Cambridge, UK: Cambridge University Press, 2001.
7. KANE, S. et al. An empirical study of the economic effects of climate change on world agriculture, *Climatic Change*, 21: 17–35, 1992.
8. LEEMANS, R. and A. M. SOLOMAN. Modelling the potential change in yield distribution of the earth's crop under a warmer climate, *Climate Research*, 3: 79–96, 1993.
9. MEARNS, L. O. et al. The effect of changes in daily and interannual climatic variability on CERES-wheat: A sensitivity study, *Climatic Change*, 32: 257–292, 1996.
10. MENDELSOHN, R. et al. The impact of global warming on agriculture: A Ricardian analysis, *American Economic Review*, 84(4): 753–771, 1994.
11. NILF. Konsekvenser for jordbruksproduksjonen av økte klimagassutslipp. Rapport C-005-90, Norsk institutt for landbruksøkonomisk forskning, 1990.
12. NONHEBEL, S. Effects of temperature rise and increase in CO_2 concentration on simulated wheat yields in Europe, *Climatic Change*, 34(1): 73–90, 1996.
13. OZKAN, B. and H. AKCAOZ. Impacts of climate factors on yields for selected crops in Southern Turkey, *Mitigation and Adaptation Strategies for Global Change*, 7(4): 367–380, 2003.

14 PARRY, M. L. (ed.), The Europe Acacia Project: Assessment of Potential Effects and Adaptations for Climate Change in Europe, University of East Anglia/European Commission, 2000.
15. PARRY, M. L. *Climate Change and World Agriculture.* London: Earthscan, 1990.
16 PARRY, M. L. and T. R. CARTER. An assessment of the effects of climatic change on agriculture, *Climatic Change*, 15: 95–116, 1989.
17. REGCLIM. Mer variabelt vær om 50 år, mer viten om usikkerheter. Technical Report November, Regional Climate Development under Global Warming Project, 2002.
18. RIHA, S. J. et al. Impact of temperature and precipitation variability on crop model predictions, *Climatic Change*, 32(3): 293–311, 1996.
19. ROSENZWEIG, C. et al. Increased crop damage in the US from excess precipitation under climate change, *Global Environmental Change*, 12(3): 197–202, 2002.
20. RÖTTER, R. and S. C. VAN DE GEIJN. Climate change effects on plant growth, crop yield and livestock, *Climatic Change*, 43(4): 651–681, 1999.
21. SSB. Agricultural Statistics (Jordbruksstatistikk) 1957–2002. NOS, Statistics Norway, 2003.
22. THOMPSON, L. M. Evaluation of weather factors in the production of wheat, *Journal of Soil and Water Conservation*, 17: 149–156, 1962.
23. TVEITO, O. E. et al. Nordic Climate Maps. Report 06/01, Norwegian Meteorological Institute, 2001.
24. WARRICK, R. A. The possible impacts on wheat production of the 1930s drought in the US great plains, *Climatic Change*, 6: 5–26, 1984.
25. ZILBERMAN, D. et al. The economics of climate change in agriculture, *Mitigation and Adaptation Strategies for Global Change*, 9(4): 365–382, 2004.

Index

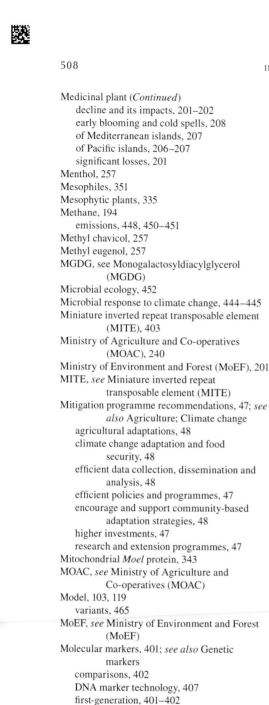

508 INDEX

Printed and bound by CPI Group (UK) Ltd, Croydon, CR0 4YY
22/10/2024

01777647-0006